Numerical Methods for Atmospheric and Oceanic Sciences

Numerical Methods for Atmospheric and Oceanic Sciences deals with various numerical methods that are applied to fluid systems such as the atmosphere and hydrosphere. With a detailed and comprehensive overview of the various numerical methods that are applied to fluid systems in general and the atmospheric and oceanic sciences in particular, this book will be useful for students of atmospheric and oceanic sciences in both senior undergraduate and graduate courses. It provides details of the application of finite difference methods to various problems that involve processes like advection, barotropic, shallow water, baroclinic oscillation, and decay. The concepts of consistency, stability, and convergence are also emphasized. The book provides clear exposition of concepts such as stability, staggered grid, and nonlinear computational instability. The book also provides broad details and applications of advanced numerical methods such as the spectral method, finite element method, and finite volume method.

A. Chandrasekar is Dean (Academics and Continuing Education) and Outstanding Professor at the Department of Earth and Space Sciences, Indian Institute of Space Science and Technology, Thiruvananthapuram. He is a leading expert in atmospheric science in India and has published widely on the above topics throughout his career. In 2010, he authored the book *Basics of Atmospheric Science*.

Numerical Methods for Atmospheric and Oceanic Sciences

A. Chandrasekar

CAMBRIDGE
UNIVERSITY PRESS

CAMBRIDGE
UNIVERSITY PRESS

University Printing House, Cambridge CB2 8BS, United Kingdom

One Liberty Plaza, 20th Floor, New York, NY 10006, USA

477 Williamstown Road, Port Melbourne, VIC 3207, Australia

314 to 321, 3rd Floor, Plot No.3, Splendor Forum, Jasola District Centre, New Delhi 110025, India

103 Penang Road, #05–06/07, Visioncrest Commercial, Singapore 238467

Cambridge University Press is part of the University of Cambridge.

It furthers the University's mission by disseminating knowledge in the pursuit of education, learning and research at the highest international levels of excellence.

www.cambridge.org
Information on this title: www.cambridge.org/9781009100564

First published 2022

Printed and bound at Thomson Press India Ltd

A catalogue record for this publication is available from the British Library

Library of Congress Cataloging-in-Publication Data

Names: Chandrasekar, A., author.
Title: Numerical methods for atmospheric and oceanic sciences / A.
    Chandrasekar, Department of Earth and Space Sciences, Indian Institute
    of Space Science & Technology, Tiruvananthapuram, India.
Description: Cambridge, United Kingdom ; New York, NY : Cambridge
    University Press, 2022. | Includes bibliographical references and index.
Identifiers: LCCN 2021050095 (print) | LCCN 2021050096 (ebook) | ISBN
    9781009100564 (hardback) | ISBN 9781009119238 (ebook)
Subjects: LCSH: Atmospheric physics–Mathematical models. |
    Oceanography–Mathematical models. | BISAC: SCIENCE / Earth Sciences /
    Meteorology & Climatology
Classification: LCC QC880 .C418 2022 (print) | LCC QC880 (ebook) | DDC
    551.5101/1–dc23/eng/20211203
LC record available at https://lccn.loc.gov/2021050095
LC ebook record available at https://lccn.loc.gov/2021050096

ISBN 978-1-009-10056-4 Hardback

... to the memory of Professor T. N. Krishnamurti

Contents

Figures

Foreword

It is with great pleasure and admiration for the author, Professor A. Chandrasekar, that I undertake this task of writing the Foreword to his second book, entitled *Numerical Methods for Atmospheric and Oceanic Sciences* published by Cambridge University Press. After completing his PhD in Applied Mathematics from the Indian Institute of Science, Bangalore, India in 1988, his journey into the vast landscape of atmospheric and oceanic sciences began with a faculty assignment in the Department of Physics and Meteorology, Indian Institute of Technology, Kharagpur, India, where he was a member of the faculty for a little over two decades. Soon, he was called on to direct the newly created Center for Oceans, Rivers, Atmosphere and Land Sciences at the Indian Institute of Technology, Kharagpur, India, a position he held for a year. Over the recent past twelve years he has been on the faculty at the Indian Institute of Space Science and Technology, Tiruvananthapuram, India, where he has held numerous teaching and administrative positions. Despite the demands of administration at various levels at different institutions of higher learning in India, he never lost sight of the fact he was a teacher first and has taught a two-level system of graduate courses – basics of atmospheric and ocean dynamics and numerical methods for solving the atmospheric model equations – on a continuous basis for over three decades. It is these years of experience in classroom teaching that resulted in his first book on basic dynamics and now this second book on the latter topic. For an aspiring graduate student this book represents a one stop shopping option. The style is very engaging, and it is as though he is talking directly to the students. The opening two chapters provide a broad overview of the standard models of interest in atmospheric and ocean sciences. A detailed account of various grid configurations, attendant discretization schemes and the associated error and stability analyses are covered in the next two chapters. Examples of discretization of equations for damped oscillators, linear advection, shallow water barotropic, and baroclinic models along with the use of staggered grids are systematically and thoroughly covered in the next set of seven chapters. Extensive discussion of the ways to handle the boundary conditions along with Lagrangian and semi-Lagrangian methods are contained in the following two chapters. Then there are two chapters dealing with topics of great importance – the spectral methods, finite element, and finite volume methods. The last chapter contains a comprehensive discussion of the ocean models. A short appendix describes an algorithm for solving tridiagonal system that often arise in discretization using implicit methods and there is a good collection of references for further reading. Simultaneous coverage of finite-difference, finite-element, finite volume, and spectral methods and illustrating their use using different classes of models of interest in atmospheric and ocean sciences under one cover is a welcome and unique feature of this book. My own interests are in Dynamic Data Assimilation (DDA) where we assume that we have a model code that can run forward in time. The vast discipline of DDA rests solidly on the foundation laid by the work described in this book. I am confident that aspiring graduate students and those working in the field will greatly benefit from this comprehensive work for years to come. Congratulations on a project well accomplished.

Norman, Oklahoma   S. Lakshmivarahan
August 2021

Preface

The mathematical equations that govern the evolution of the atmosphere and the oceans are essentially a system of coupled nonlinear partial differential equations that do not provide for closed form analytical exact solutions. In situations where closed form analytical solutions are not available, one employs numerical methods for solving the governing mathematical equations that are responsible for the evolution of the atmospheric and oceanic systems. It is clear that advances in numerical methods have contributed greatly to our current understanding of the science of the Earth system in general and the sciences of the atmosphere and oceans, in particular. This book, *Numerical Methods for Atmospheric and Oceanic Sciences*, is written with an objective to provide a detailed and broad overview of the various numerical methods that are applied to fluid systems in general, and in particular to the fluid systems that manifest in the natural environment such as the atmosphere and hydrosphere. Most of the material included in this book has evolved from a single semester course that I taught on "Numerical Weather Prediction" as well as another course that I had taught earlier titled "Numerical Weather Prediction and Modeling." The approach followed in writing this book is to provide adequate theoretical and background discussions that go beyond mere outlining and mentioning the various numerical schemes.

I have dedicated this book to the memory of Professor T. N. Krishnamurti, former Lawton Distinguished Professor, Department of Earth, Ocean and Atmospheric Science, Florida State University, USA, for his pioneering research contributions in the areas of numerical weather prediction, short and long range monsoon prediction, inter-seasonal and inter-annual variability of the tropical atmosphere. Professor Krishnamurti visited my institute twice, once in 2014 and again in 2015. I was truly amazed with Professor Krishnamurti's depth of knowledge, his prolific research output as well as the huge impact of his research on the science of the atmosphere. I am extremely indebted to Professor S. Lakshmivarahan, George Lynn Cross Research Professor Emeritus of the University of Oklahoma, USA, for readily agreeing to my request to write a Foreword for this book. I am certain that a Foreword from someone of his stature and eminence would contribute immensely to the popularity of the book. I greatly value the comments and suggestions from five anonymous peer experts on the book that led to the inclusion of two new chapters and also to minor reorganization of the book. The aforementioned comments also resulted in the inclusion of programming examples using python language in the book, a feature that students, researchers and teachers would find most useful. I am obliged to Professor Geoffrey Vallis of the University of Exeter, UK, for allowing me to modify and use a couple of python codes from his book.

I would be failing in my duty if I did not acknowledge the help and assistance that I received from my colleague Professor A. Salih, and from other scientists such as Dr. R. Krishnan of the Indian Institute of Tropical Meteorology, Pune, Professor A. D. Rao of Indian Institute of Technology, Delhi, Professor G. Bala of the Indian Institute of Science, Bangalore, and Professor B. Chakrapani of Cochin University of Science and Technology in the preparation of this book. While Professor Salih and Professor Chakrapani almost read through the entire book, others took time off from their busy schedule and read through a few chapters of this book. Their suggestions and comments have contributed to the overall improvement of the book. I also received very specific and pertinent comments from my nephew Dr. S. Prahlad working at the National Institute of Aerospace, Hampton, USA, that I found to be very helpful.

I am extremely thankful to my PhD student Mr. Vibin Jose who assisted me greatly during the preparation of this book, especially in the production of the python programs and the figures. I thank my former post-graduate students Mr. Vikram and Ms. Akshaya Nikumbh who also helped with a few python programs.

I have received excellent and unstinted support from Dr. Vaishali Thapliyal, Ms. Qudsiya Ahmed, and Mr. Vikash Tiwari of Cambridge University Press who have been extremely helpful during the period of this book project. I also thank Mr. R. Adarsh of the Thiruvanathapuram branch of Cambridge University Press who patiently heard and cleared several of my queries in the initial stage of the book proposal.

I have received a lot of support and encouragement from Dr. B. N. Suresh, the present chancellor of my institute who was the founding director of my institute when I wrote my first book. I wish to thank Dr. V. K. Dadhwal, former director of my institute who provided all help and assistance and encouraged me to complete the project. I also thank Shri S. Somanath, present director of my institute for his help and support. I have also benefited from the encouragement that I have received from several well-wishers, collaborators, friends, and faculty colleagues from my institute and other institutions on the book project, a list too long to mention and include here. I have been extremely fortunate to be blessed by my parents and Almighty at all stages of my career. Finally, I wish to acknowledge the quiet support and encouragement that I received from my family and thank them for their love, patience, and forbearance.

1

Partial Differential Equations

1.1 Introduction

Most physical as well as engineering systems one encounters in real life can be mathematically modeled using a system of partial differential equations subject to appropriate boundary conditions. These partial differential equations are coupled as well as nonlinear in nature. Owing to their nonlinearity, systems of partial differential equations that represent physical and engineering phenomena do not have closed-form or analytical solutions. Thus, the only alternative available to a scientist or a engineer is to seek a numerical solution for the aforementioned systems of partial differential equations.

There are countless examples of the manifestation of partial differential equations with appropriate boundary conditions in various fields of physics, including magnetism, optics, statistical physics, general relativity, superconductivity, liquid crystals, turbulent flow in plasma and solitons. Furthermore, diverse fields such as fluid mechanics, atmospheric physics, and ocean physics have rich and exhaustive examples of partial differential equations. In this book an effort has been made to familiarize the readers to a general introduction of partial differential equations as well as equations of fluid motion before acquainting them with the various numerical methods. The well-known method of finite differences is introduced and important aspects such as consistency and stability are discussed while applying the above method to standard partial differential equations of the parabolic, hyperbolic, and elliptic types. The method of finite differences is then applied to equations of motion of the atmosphere and oceans. The book also introduces the readers to advanced numerical methods such as semi-Lagrangian methods, spectral method, finite volume, and finite element methods and provides for the application of the above methods to the equations of motion of the atmosphere and oceans.

Towards this end, it is important to introduce partial differential equations (PDE) and the various numerical methods that can be employed to solve PDEs numerically. A PDE is an equation that represents a relationship between an unknown function of two or more independent variables and the partial derivatives of this unknown function with respect to the independent variables. Although the independent variables are either space (x, y, z) or space and time (x, y, z, t) related, the nature of the unknown function depends on the physical/engineering problem being modeled.

The function $f(x)$ is defined as a linear function of x if $f(x)$ can be expressed as $f(x) = mx + b$, where m and b are constants. The order of a PDE is determined by the highest-order derivative that appears in the PDE.

If $u(x, y)$ is a dependent variable, which is a function of two independent variables x and y, then the general second-order PDE can be written as

$$A\frac{\partial^2 u}{\partial x^2} + B\frac{\partial^2 u}{\partial x \partial y} + C\frac{\partial^2 u}{\partial y^2} + F\left(x, y, u, \frac{\partial u}{\partial x}, \frac{\partial u}{\partial y}\right) + G(x, y) = 0, \tag{1.1}$$

where A, B, C are functions of $x, y, u, \frac{\partial u}{\partial x}, \frac{\partial u}{\partial y}$, F may be a nonlinear function, and G may be a function of x and y. In such cases, Equation (1.1) is known as a second-order quasilinear PDE. A quasilinear PDE is a PDE that is linear in the highest derivative. A partial differential equation is called a quasilinear PDE if all the terms with the highest-order derivatives of dependent variables are linear. The coefficients of the highest-order derivative terms in the PDE are functions of only the lower order derivatives of the dependent variables. However, for the quasilinear PDE, the terms in the PDE with lower order derivatives can occur in any manner.

A partial differential equation is called a semilinear PDE if all the terms with the highest-order derivatives of dependent variables are functions of independent variables only. In such cases, the coefficients of the highest-order derivative terms in the PDE are functions of only the independent variables. Equation (1.1) is known as a second-order semilinear PDE if A, B, and C are functions of x and y only.

If the dependent variable and all its partial derivatives appear linearly in any PDE, i.e., there are no terms in the PDE that involve the product of the dependent variables with itself or with its derivatives, then such an equation is called a linear PDE. If F is a linear function, and A, B, and C are functions of only x and y, then Equation (1.1) is called a linear PDE.

If all the terms of a PDE contain the dependent variable or its partial derivatives, then such a PDE is called a homogeneous partial differential equation.

If function F involves the dependent variable u and its derivatives $\frac{\partial u}{\partial x}, \frac{\partial u}{\partial y}$ and also $G = 0$, the Equation (1.1) is called a homogeneous PDE; if $G \neq 0$, then Equation (1.1) is called a nonhomogeneous PDE.

Equation (1.1) can also be written in the following form known as the implicit form

$$f(x, y, u, u_x, u_y, u_{xx}, u_{xy}, u_{yy}) = 0 \tag{1.2}$$

If f is a linear function of u and its derivatives, then the PDE is said to be linear. It is necessary to classify PDE, as different types of PDE arise naturally in very different physical problems; dissimilar types of PDE have different nature of conditions (boundary/initial) to be satisfied and hence, dissimilar types of PDE need to employ different numerical methods for their solution.

It is known that the general solution of ordinary differential equations (ODEs) involve arbitrary constants of integration; in contrast, the general solution of PDEs involves arbitrary functions. Consider, for example, the equation

$$\frac{\partial^2 u}{\partial x \partial y} = 0. \tag{1.3}$$

Integrating Equation (1.3) with respect to y, one gets $\partial u / \partial x = F(x)$, where $F(x)$ is an arbitrary function of x. Integrating the equation once again with respect to x, one gets

$$u(x, y) = f(x) + g(y), \tag{1.4}$$

where $f(x) = \int F(x)dx$ and $g(y)$ are arbitrary functions of x and y respectively. To obtain $f(x)$ and $g(y)$, one needs to have additional information, for example, the initial conditions (if time is one of the independent variables) and/or boundary conditions.

To be specific, suppose that one were to find $u(x, y)$ satisfying Equation (1.4) in the region $x \geq 0, y \geq 0$ and that one is given the following boundary conditions $u = x$, when $y = 0$ and $u = y$, when $x = 0$. Then, the surface $u(x, y)$ must intersect the plane $x = 0$ in the line $u = y$ and the plane $y = 0$ in the line $u = x$. The functions $f(x)$ and $g(y)$ in Equation (1.4) are determined in the following manner. As $u(x, 0) = f(x) + g(0) = x$ and $u(0, y) = f(0) + g(y) = y$, it follows that $u(x, y) = f(x) + g(y) = x - g(0) + y - f(0) = x + y - g(0) - f(0)$. The only way this can satisfy the PDE and the boundary conditions are if $f(0)$ and $g(0)$ are both zero, which implies $u(x, y) = x + y$.

It can be easily verified that the equation satisfies the PDE and the two boundary conditions. The aforementioned example clearly illustrates the importance of the boundary conditions in obtaining the solution of the PDE. For an ordinary differential equation of the second-order, it is known that two conditions are required to obtain an unique solution. It is clear that depending on the nature of the PDE, the sufficient set of boundary conditions that are required for a meaningful solution may vary.

The question that is posed is as follows: what is a sufficient set of boundary conditions for a given PDE? The answer to this question depends on the type of PDE, the latter in turn, depending on the nature of the associated physical problem. Two different types of boundary conditions applied to the same PDE, will invariably lead to two different types of solution. Hence, methods of solution of PDEs will depend on the nature and type of the boundary conditions.

One expects that a given PDE subject to suitable boundary conditions will possess an unique solution. Any physical or engineering problem defined by Equation (1.1) in a given two-dimensional domain is said to be "well-posed" if

1. there exists at least one solution (existence)

2. there exists atmost one solution (uniqueness)

3. the solution is stable.

Three types of PDEs arise when one classifies PDE and these are (i) parabolic type, (ii) elliptic type, and (iii) hyperbolic type. Examples of the parabolic type of PDEs are the diffusion equation whereas examples of elliptic and hyperbolic PDEs are the Laplace equation and wave equation, respectively.

1.2   Diffusion Equation

The most common form of diffusion equation is as follows:

$$\frac{\partial c}{\partial t} = D\nabla^2 c, \tag{1.5}$$

where c is the concentration, which is in general a function of space and time, D is the diffusion coefficient, and ∇^2 is the Laplacian operator, which in Cartesian coordinates is

$$\nabla^2 \equiv \frac{\partial^2}{\partial x^2} + \frac{\partial^2}{\partial y^2} + \frac{\partial^2}{\partial z^2}.$$

By definition, a flux $\bar{J}$ is a movement of particles (or other quantities) through a unit measure (point, length, area) per unit time. From Ficks' law of diffusion, it follows that flux $\bar{J}$ is related to concentration c through the following equation

$$\bar{J} = -D\nabla c, \tag{1.6}$$

where ∇ is the gradient operator. The negative sign signifies that the flux is always in the direction opposite to the gradient operator. The direction of the gradient operator, also known as the "ascendant," is in the maximum rate of change of 'c' and is always directed from low values to high values of c. Hence Equation (1.6) clearly shows that the direction of flux is always directed from high values to low values of c.

For a normal diffusion process, particles cannot be created or destroyed. This implies that the flux of particles into one region must be the sum of the particle flux flowing out of the surrounding regions. The aforementioned statement can be easily expressed mathematically by the continuity equation given by

$$\frac{\partial c}{\partial t} + \nabla \cdot \bar{J} = 0. \tag{1.7}$$

Using Equation (1.7) in Equation (1.6), one gets

$$\frac{\partial c}{\partial t} - \nabla \cdot (D\nabla c) = 0. \tag{1.8}$$

If the diffusion coefficient D is a constant, then Equation (1.8) becomes the diffusion equation (1.5). The diffusion equation can be applied to solving problems in mass diffusion, momentum diffusion, and heat diffusion. It is clear that under different situations, the diffusion equation assumes different forms. For example, in the case of heat diffusion, c will be the temperature T whereas D will become the coefficient of thermal diffusivity α. Equation (1.5) for the case of heat diffusion is also known as the heat conduction equation, whose one-dimensional form is given by

$$\frac{\partial T}{\partial t} = \alpha \frac{\partial^2 T}{\partial x^2}, \tag{1.9}$$

where T is the temperature of a heated rod, α is the coefficient of thermal conductivity, x is the distance along the rod, and t is the time. In Figure 1.1, the heated rod extends from $x = 0$ to $x = a$ with $T(x,t)$ being the temperature of the rod at location x, the distance from the end $x = 0$ and time t.

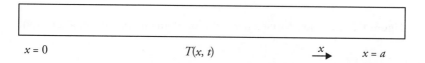

Figure 1.1 Temperature distribution of a heated rod of length a.

It is extremely helpful to picturize the solution of a PDE. In the case of Equation (1.9), the solution can be expressed as a surface, $z = T(x,t)$ in a three-dimensional space (x,t,z), as shown in Fig 1.2. The domain of the solution, Ω, is the region $0 \leq t < \infty$ and

$0 \leq x \leq a$. The temperature distribution at some time $t_o > 0$ is the curve $z = T(x,t_o)$, where the plane $t = t_o$ intersects the solution curve. The curve $z = T(x,0)$ is the initial temperature distribution that is assumed to be given. Equation (1.9) states that at any point (i.e., at any point x,t) in the solution surface, the slope of the surface in the t-direction is related locally to the rate of change of the slope in the x-direction. It is abundantly clear that in order to obtain a unique solution, there is a need to prescribe the nature of the solution (the behaviour of the surface) at the edges of the solution domain: at $t = 0$ and at $x = 0$, and $x = a$. It makes sense to expect, on the basis of physical reasoning, that in order to predict the future evolution of the temperature, one needs to have knowledge of the initial state, i.e., the initial temperature distribution in the rod, $T(x,0)$. In a similar manner, it makes sense to expect that the temperature values at the ends of the rod at any particular time would affect the temperature distribution in the rod.

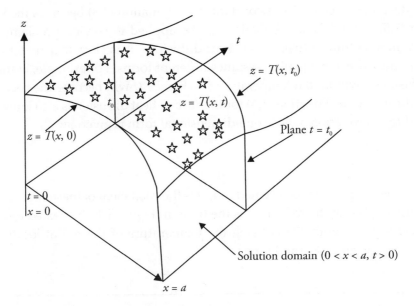

Figure 1.2   Solution surface and solution domain, Ω, for Equation (1.9).

1.3   First-order Equations

One of the most important first-order PDE is the one-dimensional advection equation,

$$\frac{\partial \rho}{\partial t} + u \frac{\partial \rho}{\partial x} = 0 \tag{1.10}$$

where ρ is the air density, u is a constant velocity, t is the time, and x is space coordinate. Equation (1.10) expresses the statement that the rate of change of air density of an air parcel with respect to time is zero following the motion, i.e., the air density of an air parcel is constant following the motion. Alternatively, Equation (1.10) states that the total or substantive derivative of air density is zero, following the motion. A fluid flow is said to be incompressible if the fractional change in density of an air parcel, associated with a change in pressure, following the motion is very small. In effect, incompressible fluid flow is one for which the rate of change of air density of an air parcel with respect to time is zero, following the motion. Hence, Equation (1.10) is valid for an incompressible fluid flow.

Consider a one-dimensional flow of an incompressible fluid. The continuity equation that expresses the principle of conservation of mass for a one-dimensional flow of density $\rho(x,t)$ for an incompressible fluid is expressed as

$$\frac{d\rho}{dt} = 0, \tag{1.11}$$

where $d\rho/dt$ signifies the rate of change of density $\rho(x,t)$ following the motion as expressed in the Lagrangian description of fluid motion. Its equivalent expression in the Eulerian description of motion is given by (1.10), where u is the nonzero constant velocity component in the x direction. The aforementioned equation called the advection equation can also be easily derived from the following consideration

Consider a one-dimensional flow of an incompressible fluid. Assuming that the fluid density $\rho(x,t)$ changes only due to convective/advective processes, one can write the following

$$\rho(x, t + \Delta t) = \rho(x - u\Delta t, t).$$

If Δt is sufficiently small, one can expand both sides of the equation by Taylor series expansion and retain only up to the linear term

$$\rho(x,t) + \Delta t \frac{\partial \rho(x,t)}{\partial t} = \rho(x,t) - u\Delta t \frac{\partial \rho(x,t)}{\partial x}$$

or canceling of Δt on both sides, one gets the one-dimensional advection equation

$$\frac{\partial \rho}{\partial t} + u\frac{\partial \rho}{\partial x} = 0.$$

From this discussion it is clear that the exact solution of Equation (1.10) is given as

$$\rho(x,t) = F(x - ut), \tag{1.12}$$

where the initial condition $\rho(x,0) = F(x)$. Equation (1.12) defines a right-traveling wave that propagates (i.e., convects or advects) the initial property (density) distribution to the right at the convection/advection velocity u. The aforementioned analytical solution indicates that the initial property (density) profile $\rho(x,0) = F(x)$ simply propagates (i.e., convects/advects) to the right with the constant velocity u, its shape and magnitude is unchanged.

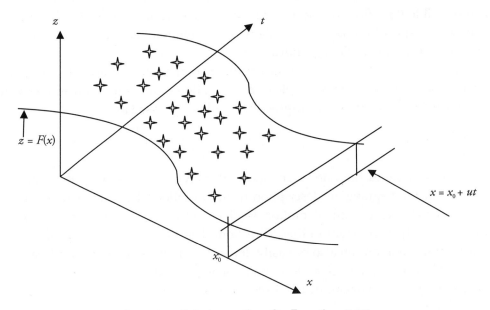

Figure 1.3   Solution surface for Equation (1.10).

If one moves with the solution point $x(t) = x_o + ut$, Equation (1.10) tells us that the rate of change of ρ is zero, i.e., in other words, ρ is a constant along a line $x = x_o + ut$. Figure 1.3 shows the solution surface in the (x,t,z)-space. It is clear from Figure 1.3 that the lines $x = x_o + ut$ are a family of parallel lines in the plane $z = 0$ that intersect the plane $t = 0$ at $x = 0$. The equation says that the height of the solution surface is always the same along such a line, i.e., the intersection of this solution surface with the plane $t = constant$ is a curve that is identical with the curve $z = F(x)$ at $t = 0$, but displaced in the x-direction by a distance ut. Thus, the solution represents a disturbance with arbitrary shape $F(x)$ translating uniformly with speed u in the positive x-direction if $u > 0$, or in the negative x-direction if $u < 0$.

It is clear that "information" about the initial distribution of ρ "propagates" or is "carried along" the lines $x = x_o + ut$ in the plane $z = 0$. These lines are called the characteristic curves, or simply the characteristics of the equation. The characteristic equation is then given as

$$\frac{dx}{dt} = u. \tag{1.13}$$

Integrating Equation (1.13) provides us the characteristic curves. The solution of the ODE Equation (1.13) involves one integration constant that determines where the characteristic curves intersect the x axis. Subsequently one needs to construct the solution surface that has the same value $F(x)$ along each characteristic in the x–t plane as that in the initial plane, $t = 0$.

Consider a general first-order partial differential equation as follows:

$$a\frac{\partial u}{\partial x} + b\frac{\partial u}{\partial y} = c. \tag{1.14}$$

If a, b, and c are functions of x, y, and u, then Equation (1.14) is called a quasilinear PDE. If a and b are functions of x and y while c is a function of x, y and u, then Equation (1.14) is called a semilinear PDE. If a, b, and c are functions of x and y only, then Equation (1.14) is called a linear PDE. Quasilinear PDE is one in which the highest-order terms are linear.

1.4　First-order Equations: Method of Characteristics

Consider the simplest case of the following first-order linear partial differential equation

$$a(x,y)\frac{\partial u}{\partial x} + b(x,y)\frac{\partial u}{\partial y} = c(x,y). \tag{1.15}$$

Assume that one can find a solution $u(x,y)$. Consider the function $S = \{x,y,u(x,y)\}$. If $u(x,y)$ is a solution of Equation (1.15), at each point (x,y), it is possible to express Equation (1.15) as the dot product

$$[a(x,y), b(x,y), c(x,y)] \cdot [u_x(x,y), u_y(x,y), -1] = 0. \tag{1.16}$$

From calculus, the normal to the surface $S = \{x,y,u(x,y)\}$ at the point $[x,y,u(x,y)]$ is given by

$$N(x,y) = [u_x(x,y), u_y(x,y), -1].$$

It is thus clear that if the vector $[a,b,c]$ is perpendicular to $[u_x, u_y, -1]$, then the vector $[a,b,c]$ lies in the tangent plane to S. Hence, to obtain a solution to Equation (1.15), one needs to find a surface S such that at each point (x,y,u) on S, the vector $[a,b,c]$ lies in the tangent plane. To construct such a surface, one first obtains a curve that lies in S. It is clear that the vector $[a,b,c]$ need to lie in the tangent plane to the surface S at each point (x,y,u) on the surface. Let there be a curve C parameterized by s such that at each point on the curve C, the vector $[a,b,c]$ will be tangent to the curve. That is, for a curve C parameterized as $C = \{(x(s),y(s),u(s))\}$, the following three conditions need to be satisfied

$$\frac{dx}{ds} = a(x(s),y(s)) \tag{1.17}$$

$$\frac{dy}{ds} = b(x(s),y(s)) \tag{1.18}$$

$$\frac{du}{ds} = c(x(s),y(s)) \tag{1.19}$$

Such a curve when it exists is called an integral curve for the vector field $[a,b,c]$. For solving a PDE of the form given in Equation (1.15), we need to find the integral curves for the vector field $V = [a(x,y),b(x,y),c(x,y)]$ associated with the PDE. These integral curves are known as *characteristic curves*. The aforementioned characteristic curves are obtained by solving the system of ordinary differential equations (1.17)-(1.19)(ODE).

Once the characteristic curves for Equation (1.15) are obtained, one needs to construct a solution of Equation (1.15) by forming a surface S as a union of these characteristic curves. Such a surface $S = x,y,u$ for which the vector field $V = [a(x,y),b(x,y),c(x,y)]$ lies in the tangent plane to S at each point (x,y,u) on S is known as the integral surface. Through the introduction of characteristic equations, the original PDE (Equation (1.15)) can be reduced to a system of ODEs. The concept is to solve the characteristic equations, obtain an union of the so-called characteristic curves to form a surface that would provide for the solution of the PDE (Equation (1.15)).

1.5 Second-order Quasilinear PDEs: Classification Using Method of Characteristics

The general quasilinear second-order nonhomogeneous PDE in two independent variables x and y are given as

$$a\frac{\partial^2 u}{\partial x^2} + b\frac{\partial^2 u}{\partial x \partial y} + c\frac{\partial^2 u}{\partial y^2} + d\frac{\partial u}{\partial x} + e\frac{\partial u}{\partial y} + fu = g. \tag{1.20}$$

The classification of PDEs to parabolic, elliptic, and hyperbolic PDEs are analogous to the classification of conic section. For example, conics are generally described by the second-order algebraic equation

$$Ax^2 + Bxy + Cy^2 + Dx + Ey + F = 0. \tag{1.21}$$

The conics as described by Equation (1.21) are classified as parabolic, elliptic, and hyperbola based on the sign of the discriminant, $B^2 - 4AC$, ($B^2 - 4AC = 0$ is defined as a parabola, $B^2 - 4AC < 0$ is defined as an elliptic and $B^2 - 4AC > 0$ is defined as a hyperbola). In exactly the same way, the second-order quasilinear PDE (Equation (1.20)) is classified based on the sign of the discriminant $b^2 - 4ac$, where a, b, and c refer to the coefficients of the highest (second-order) derivative; $b^2 - 4ac = 0$ is referred to as a parabolic partial differential equation, $b^2 - 4ac < 0$ is referred to as an elliptic partial differential equation, and $b^2 - 4ac > 0$ is referred to as hyperbolic partial differential equation. In this section, the classification of Equation (1.20) using the characteristics is examined. Earlier it was shown that the characteristic curves for the one-dimensional advection equation [Equation (1.10)] are the lines $x = x_o + ut$ in the plane $z = 0$. The solution for Equation (1.10) represents a disturbance with arbitrary shape $F(x)$ translating uniformly with speed u in the positive x-direction if $u > 0$, or in the negative x-direction if $u < 0$, i.e., "information" about the initial distribution of ρ "propagates" or is "carried along" the characteristic curves.

As discontinuities in the derivatives of the solution, if they exist, must propagate along the characteristics, it is possible to utilize the characteristics themselves to classify the second-order quasilinear PDEs. The following question is posed. Are there any curves in the solution domain passing through a general point P along which the highest-order derivatives [in the case of Equation (1.20)], the second-order derivatives of $u(x, y)$, i.e., u_{xx}, u_{xy}, and u_{yy}, are multi-valued or discontinuous? Such curves, if they exist, are the paths of information propagation. One equation that relates the three second-order derivatives of $u(x, y)$ is the PDE [Equation (1.20)] itself. One can obtain two more such equations as follows:

$$d\left(\frac{\partial u}{\partial x}\right) = \frac{\partial^2 u}{\partial x^2}dx + \frac{\partial^2 u}{\partial y \partial x}dy; \tag{1.22}$$

$$d\left(\frac{\partial u}{\partial y}\right) = \frac{\partial^2 u}{\partial x \partial y}dx + \frac{\partial^2 u}{\partial y^2}dy. \tag{1.23}$$

Equations (1.20), (1.22) and (1.23) can be written in matrix form with the second-order derivatives as unknown. If the determinant of the coefficient matrix vanishes, the second-order derivatives of $u(x, y)$ are indeterminate and thus, multi-valued or discontinuous. Setting the determinant of the coefficient matrix to zero, yields $a(dy)^2 - b(dy)(dx) + c(dx)^2 = 0$, whose solution is obtained from the quadratic formula

$$\frac{dy}{dx} = \frac{-b \pm \sqrt{b^2 - 4ac}}{2a}.$$

(1.24)

Equation (1.24) is the ordinary differential equation for the two families of characteristic curves in the x, y plane, corresponding to the $\pm$ signs. The two families of characteristic curves, if they exist, may either be real and repeated, complex, or real and distinct. This requirement is equivalent to the discriminant $b^2 - 4ac = 0$, $b^2 - 4ac < 0$, and $b^2 - 4ac > 0$, i.e., the original PDE being either parabolic PDE, elliptic PDE, or hyperbolic PDE. Hence, while elliptic PDEs do not have any real characteristics, parabolic PDEs have one real and repeated characteristic, and hyperbolic PDEs have two real and distinct characteristic curves.

The existence of characteristic curves in the solution domain provides for the introduction of concepts such as domain of dependence and range of influence. The domain of dependence of a point $P(x, y)$ in the solution domain is defined as the region of the solution domain upon which the solution at point $P(x, y)$ depends. In other words, the solution at any point P depends on the solution over the domain of dependence. The range of influence of a point $P(x, y)$ in the solution domain is defined as the region of the solution domain in which the solution is influenced by the solution at point $P(x, y)$. That is, the solution at a point P influences the solution over the range of influence. As parabolic and hyperbolic PDEs have real characteristic curves, they will have a definite domain of dependence and range of influence in the real domain. However, elliptic PDEs do not have real characteristic curves. Hence, elliptic PDEs do not have a definite domain of dependence and range of influence in the real domain; thus, the entire solution domain of an elliptic PDE is both its domain of dependence and range of influence of every point in the solution domain. In order to further understand the concept of domain of dependence and range of influence, let us consider specific examples of (i) parabolic, (ii) hyperbolic, and (iii) elliptic PDEs.

The one-dimensional linear heat conduction equation is an example of a parabolic PDE,

$$\frac{\partial T}{\partial t} = \alpha \frac{\partial^2 T}{\partial x^2}.$$

(1.25)

Two other equations that can relate the second-order derivatives are as follows:

$$d\left(\frac{\partial T}{\partial x}\right) = \frac{\partial^2 T}{\partial x^2}dx + \frac{\partial^2 T}{\partial x \partial t}dt \tag{1.26}$$

$$d\left(\frac{\partial T}{\partial t}\right) = \frac{\partial^2 T}{\partial t \partial x}dx + \frac{\partial^2 T}{\partial t^2}dt \tag{1.27}$$

The characteristic differential equation is found by setting the determinant of the coefficient matrix to zero. This yields $\alpha[dt]^2 = 0$, which when integrated provides for time being equal to a constant for the characteristic paths. An example of a hyperbolic PDE is the second-order linear wave equation given by

$$\frac{\partial^2 u}{\partial t^2} = c^2 \frac{\partial^2 u}{\partial x^2}, \tag{1.28}$$

Two other equations that can relate the second-order derivatives are as follows:

$$d\left(\frac{\partial u}{\partial x}\right) = \frac{\partial^2 u}{\partial x^2}dx + \frac{\partial^2 u}{\partial x \partial t}dt \tag{1.29}$$

$$d\left(\frac{\partial u}{\partial t}\right) = \frac{\partial^2 u}{\partial t \partial x}dx + \frac{\partial^2 u}{\partial t^2}dt. \tag{1.30}$$

The characteristic differential equation is found by setting the determinant of the coefficient matrix to zero. This yields $c^2[dt]^2 = [dx]^2$, which consequently yields

$$\frac{dt}{dx} = \pm\frac{1}{c},$$

indicating that two distinct and real families of characteristic paths exist for a hyperbolic PDE.

An example of an elliptic PDE is the second-order linear Laplace equation given by

$$\frac{\partial^2 \phi}{\partial x^2} + \frac{\partial^2 \phi}{\partial y^2} = 0. \tag{1.31}$$

Two other equations that can relate the second-order derivatives are as follows:

$$d\left(\frac{\partial \phi}{\partial x}\right) = \frac{\partial^2 \phi}{\partial x^2}dx + \frac{\partial^2 \phi}{\partial x \partial y}dy \tag{1.32}$$

$$d\left(\frac{\partial \phi}{\partial y}\right) = \frac{\partial^2 \phi}{\partial y \partial x}dx + \frac{\partial^2 \phi}{\partial y^2}dy \tag{1.33}$$

The characteristic differential equation is found by setting the determinant of the coefficient matrix to zero. This yields $[dy]^2 = -[dx]^2$, which subsequently yields

$$\frac{dy}{dx} = \pm i$$

indicating that the characteristic curves for an elliptic PDE do not lie in the real domain. The aforementioned concepts of domain of dependence and range of influence are illustrated for each of the aforementioned second-order linear PDE in Figure 1.4 (a) for a parabolic PDE, Figure 1.4(b) for a hyperbolic PDE and Figure 1.4 (c) for an elliptic PDE.

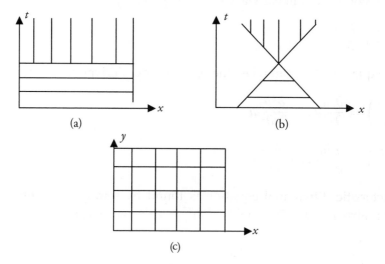

Figure 1.4 Domain of dependence (horizontal hatching) and range of influence (vertical hatching) for (a) parabolic, (b) hyperbolic, and (c) elliptic PDEs.

One can employ a similar strategy to classify the first-order quasilinear PDE. Consider classification of the first-order quasilinear PDE

$$a\frac{\partial u}{\partial t} + b\frac{\partial u}{\partial x} = c. \tag{1.34}$$

One additional equation that relates the first-order derivative is as follows:

$$d(u) = \frac{\partial u}{\partial t}dt + \frac{\partial u}{\partial x}dx \tag{1.35}$$

The characteristic differential equation is found by setting the determinant of the coefficient matrix to zero. This yields $adx - bdt = 0$. Solving for dx/dt gives

$$\frac{dx}{dt} = \frac{b}{a}. \tag{1.36}$$

Equation (1.36) is the differential equation for a family of characteristic curves in the solution domain along which the first derivatives of u may be discontinuous or multi-valued. As a and b are real functions, the characteristic curves always exist and they are real characteristic curves. Hence, a single quasilinear first-order PDE is always a hyperbolic PDE. The one-dimensional first-order advection equation is an example of a hyperbolic PDE.

1.6   Wave Equation

In this section, we derive the one-dimensional wave equation, which is the simplest form of the wave equation for an idealized string. The following assumptions on the physical string are presumed to hold. Assume that a flexible string of length L is tightly stretched along the x-axis with one of its end point at $x = 0$ and the other end point at $x = L$. It is further assumed that the tension force on the string is the only dominant force, whereas all other forces acting on the string are negligible. Moreover, it is assumed that no external forces are applied to the string. Furthermore, it is assumed that the weight of the string is negligible and that the damping forces can also be neglected. Considering the string to be flexible, it follows that at each point, the tension force has constant magnitude; moreover, it has the direction of the tangent line to the string. It is also assumed that each point of the string moves only vertically. Let $u(x,t)$ denote the vertical displacement at time t of the point x on the string. At a fixed initial time, $t = t_o$, the shape of the string is given by the known function $u(x,t_o)$. The objective is to find the shape of the string at all points x and at time t, i.e., $u(x,t)$. To find the shape of the string at all points at a later time, one needs to solve the one-dimensional wave equation with associated initial and boundary conditions.

Consider a small element of the string between the points x and $x + \Delta x$ ($\Delta x > 0$ is assumed small; moreover, it is assumed that this element moves vertically). The total force to which this element is subject to is the tension force exerted at the left end $T(x,t)$ and the tension force exerted at the right end $T(x+\Delta x,t)$ by the rest of the string. These forces have the same constant magnitude T. Let $\theta(x,t)$ be the angle between $T(x,t)$ and the x-axis and $\theta(x+\Delta x,t)$ be the angle between $T(x+\Delta x,t)$ and the x-axis. It is assumed that these angles are between 0 and π. As we are assuming that we are dealing with

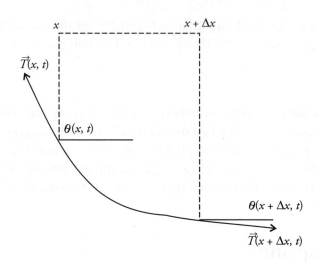

Figure 1.5 String element at time t subject to tension forces.

small vibrations, then either θ is close to 0 (at location $(x + \Delta x, t)$) or close to π (at location (x,t)).

The total vertical force acting on the element is given by F = vertical component of tension force at (x,t) plus vertical component of tension force at $(x + \Delta x, t)$, i.e.,

$$F = T(x+\Delta x,t)\sin[\theta(x+\Delta x,t)] + T(x,t)\sin[\theta(x,t)] = T\{\sin[\theta(x+\Delta x,t)] + \sin\theta(x,t)\}.$$
(1.37)

For θ close to zero, $\sin\theta \sim \tan\theta \sim \theta$, whereas for θ close to π, $\sin\theta \sim -\tan\theta \sim \pi - \theta$. Moreover, the shape of the string at a fixed time t is given as the graph of the function $u(x,t)$ (t fixed and x varies); slope of the tangent line at location x_o is given by $\tan\theta_o$, where θ_o is the inclination angle. It follows then that

$$\tan\theta(x,t) = \frac{\partial u(x,t)}{\partial x} \qquad \text{and} \qquad \tan\theta(x+\Delta x,t) = \frac{\partial u(x+\Delta x,t)}{\partial x}.$$
(1.38)

Substituting Equation (1.38) in Equation (1.37), one obtains the total vertical force acting on the element as

$$F = T(x,t)\left(\frac{\partial u(x+\Delta x,t)}{\partial x} - \frac{\partial u(x,t)}{\partial x}\right).$$
(1.39)

Using Newton's second law of motion, $F = \Delta m\, a$, where Δm is the mass of the element and a is the acceleration of the element at time t. The mass of the element $\Delta m = \rho\Delta x$, where ρ is the density of the string material. The acceleration of the element at time t can be written as

$$a = \frac{\partial^2 u}{\partial t^2}. \tag{1.40}$$

Using the expressions for Δm and a, Newton's second law of motion becomes

$$\rho \frac{\partial^2 u}{\partial t^2} = \frac{T}{\Delta x}\left(\frac{\partial u(x+\Delta x,t)}{\partial x} - \frac{\partial u(x,t)}{\partial x}\right) \tag{1.41}$$

In the limit when $\Delta x \to 0$, Equation (1.41) becomes

$$\frac{\partial^2 u}{\partial t^2} = c^2 \frac{\partial^2 u}{\partial x^2}, \tag{1.42}$$

where $c^2 = T/\rho$ is the square of the speed of the wave. Equation (1.42) is known as the one-dimensional wave equation.

1.7   Linear Advection Equation

The wave equation is closely related to the so-called *advection equation*, which in one dimension takes the form

$$\frac{\partial u}{\partial t} + v\frac{\partial u}{\partial x} = 0. \tag{1.43}$$

The aforementioned equation describes the passive advection of some scalar field u carried along by a flow of constant speed v. Let the initial condition be $u(x,0) = u_o$. Based on the method of characteristics discussed in Section 1.4, it follows that the characteristic equations are

$$\frac{dt}{ds} = 1; \quad \text{with } t(0) = 0;$$

this implies $t = s$;

$$\frac{dx}{ds} = v; \quad \text{with } x(0) = x_o;$$

this implies that $x = x_o + vs$ and $x = x_o + vt \to x_o = x - vt$. Furthermore,

$$\frac{du}{ds} = 0; \quad \text{with } s(0) = u_o(x_o).$$

The unique solution of the advection equation is

$$u(x,t) = u_o(x - vt). \tag{1.44}$$

The solution (Equation (1.44)) is just an initial condition u_o shifted by vt to the right (for $v > 0$) or to the left ($v < 0$), which remains constant along the characteristic curves, $du/ds = 0$.

1.8   Laplace Equation

Consider a thin plate having some width w and some length l; it also has a very small thickness t. The faces of this plate are insulated to ensure that no heat flows in the direction of the thickness t.   Assume that the top edge of the plate is maintained at a higher temperature while the other three edges are maintained at a same lower temperature. In this situation, heat flows into the plate through the top edge and out of the plate through the other three edges. Assume that there are no mechanism/processes for generation of internal energy within the plate. In the aforementioned circumstance, one is interested in obtaining the temperature (T) distribution within the plate. The temperature within the plate will vary within the horizontal plane in terms of x (width-wise coordinate) and y (length-wise coordinate) and the temperature distribution within the plate $T(x,y)$ will be governed by the following two-dimensional Laplace equation:

$$\frac{\partial^2 T}{\partial x^2} + \frac{\partial^2 T}{\partial y^2} = 0 \tag{1.45}$$

The non-homogeneous (right-hand side is a known function of space) form of the Laplace equation is called Poisson equation. Solving Equation (1.45) subject to the boundary conditions (specified temperature on all the four edges) will determine the temperature distribution $T(x,y)$ within the plate. The Laplace equation arises in several problems in ideal fluid flow, heat diffusion, mass diffusion, and in electrostatics. As time does not appear in the Laplace equation, and the prescribed temperature on all the four edges is also independent of time, the solution of the Laplace equation (temperature distribution within the plate) will also not depend on time. Such problems in which time does not appear in the governing equations are known as equilibrium problems.

Equation (1.45) can be solved by the *method of separation of variables* in which it is assumed that the equation has a solution of the form

$$T(x,y) = X(x)Y(y). \tag{1.46}$$

Substituting Equation (1.46) in Equation (1.45), one obtains after dividing by XY, the following

$$\frac{1}{X}\frac{d^2X}{dx^2} = -\frac{1}{Y}\frac{d^2Y}{dy^2} = -k^2. \tag{1.47}$$

As the first term and second terms of Equation (1.47) depend only on x and y, respectively, each of them should depend on a constant, say $-k^2$. The solution is then a product of $X(x) = c_1\sin(kx) + c_2\cos(kx)$ and $Y(y) = c_3\sinh(ky) + c_4\cosh(ky)$. It is possible to reduce the number of constants from 5 to 4. For example, if $c_1 c_3 \neq 0$, it is possible to redefine $c_1 c_3 = A$; $c_2/c_1 = B$; $c_4/c_3 = C$; and write the solution as

$$T(x,y) = A\left[\sin(kx) + B\cos(kx)\right]\left[\sinh(ky) + C\cosh(ky)\right]. \tag{1.48}$$

The constants A, B, C, and k are to be determined from the given initial and boundary conditions.

1.9 Method of Separation of Variables for the One-dimensional Heat Equation

In this section, the method of separation of variables is used to solve the one-dimensional heat equation

$$\frac{\partial u}{\partial t} = \alpha\frac{\partial^2 u}{\partial x^2}. \tag{1.49}$$

In this method, the solution is assumed to be of the form

$$u(x,t) = X(x)T(t). \tag{1.50}$$

Substituting Equation (1.50) in Equation (1.49) and dividing by XT, one gets

$$\frac{1}{X}\frac{d^2X}{dx^2} = \frac{1}{\alpha T}\frac{dT}{dt} = -k^2 \tag{1.51}$$

As the first term and second terms of Equation (1.51) depend only on x and t, respectively, each of them should depend on a constant, say $-k^2$. The solution is then a product of

$$X(x) = c_1\cos(kx) + c_2\sin(kx) \quad \text{and} \quad T(t) = c_3 e^{-\alpha k^2 t} \quad \text{given by}$$

$$u(x,t) = e^{-\alpha k^2 t}\left[c_1\cos(kx) + c_2\sin(kx)\right], \tag{1.52}$$

where c_3 is assumed to be unity without any loss of generality. With boundary conditions $u(x=0,t)=0$ and $u(x=L,t)=0$, one gets $c_1=0$ and $k=n(\pi/L)$; the solution is as follows:

$$u_n(x,t) = b_n e^{-\alpha\left(\frac{n\pi}{L}\right)^2 t} \sin\frac{n\pi x}{L} \qquad \text{for} \quad n=1,2,\cdots \tag{1.53}$$

Satisfying the initial condition $u(x,t=0)=f(x)$, one gets the solution of the heat equation as

$$u(x,t) = \sum_{n=1}^{\infty} b_n e^{-\alpha\left(\frac{n\pi}{L}\right)^2 t} \sin\frac{n\pi x}{L} \tag{1.54}$$

where

$$b_n = \frac{2}{L}\int_0^L f(x) \sin\frac{n\pi x}{L}\, dx.$$

1.10   Method of Separation of Variables for the One-dimensional Wave Equation

In this section, the method of separation of variables is used to solve the one-dimensional wave equation

$$\frac{\partial^2 u}{\partial t^2} = c^2\frac{\partial^2 u}{\partial x^2}. \tag{1.55}$$

As in the case of the heat equation, the solution is assumed to be of the form

$$u(x,t) = X(x)\,T(t). \tag{1.56}$$

Substituting Equation (1.56) in Equation (1.55) and dividing by XT, one obtains

$$\frac{1}{X}\frac{d^2 X}{dx^2} = \frac{1}{c^2 T}\frac{d^2 T}{dt^2} = -k^2. \tag{1.57}$$

As the first term and second terms of Equation (1.57) depend only on x and t, respectively, each of them should depend on a constant, say $-k^2$. The solution is then a product of

$$X(x) = c_1\cos(kx)+c_2\sin(kx) \quad \text{and} \quad T(t)=c_3\cos(kct)+c_4\sin(kct) \quad \text{given by}$$

$$u(x,t) = [c_1\cos(kx)+c_2\sin(kx)]\,[c_3\cos(kct)+c_4\sin(kct)]. \tag{1.58}$$

With boundary conditions $u(x = 0, t) = 0$ and $u(x = L, t) = 0$, one gets $c_1 = 0$ and $k = n(\pi/L)$; solution is as follows:

$$u_n(x, t) = \left(\alpha_n \cos \frac{n\pi ct}{L} + \beta_n \sin \frac{n\pi ct}{L} \right) \sin \frac{n\pi x}{L} \qquad \text{for} \quad n = 1, 2, \cdots \tag{1.59}$$

Satisfying the initial conditions $u(x, 0) = f(x)$ and $u_t(x, 0) = g(x)$; one gets the solution of the wave equation

$$u(x, t) = \sum_{n=1}^{\infty} (\alpha_n \cos \omega_n t + \beta_n \sin \omega_n t) \sin k_n x, \tag{1.60}$$

where

$$\omega_n = \frac{n\pi c}{L}, \qquad k_n = \frac{n\pi}{L},$$

$$\alpha_n = \frac{2}{L} \int_0^L f(x) \sin k_n x \, dx, \quad \text{and}$$

$$\beta_n = \frac{2}{n\pi c} \int_0^L g(x) \sin k_n x \, dx.$$

Exercises 1a (Question and answer)

1. Find the type (linear, semilinear, quasilinear, or nonlinear) of the following partial differential equations:

(a) $x\dfrac{\partial z}{\partial x} + y\dfrac{\partial z}{\partial y} = z$

(b) $x\dfrac{\partial z}{\partial x} + y\dfrac{\partial z}{\partial y} = z^3$

(c) $(x+y)\dfrac{\partial z}{\partial x} + (x-y)\dfrac{\partial z}{\partial y} = xy$

(d) $z\dfrac{\partial z}{\partial x} + y\dfrac{\partial z}{\partial y} = z$

(e) $x\left(\dfrac{\partial z}{\partial x}\right)^2 + y\left(\dfrac{\partial z}{\partial y}\right)^2 = z$

Answer: (a) Linear (b) Semilinear (c) Linear (d) Quasilinear (e) Nonlinear

Exercises 1b (Questions only)

1. Find the general integral of the first-order partial differential equation

$$x\frac{\partial z}{\partial x} + y\frac{\partial z}{\partial y} = z.$$

   Answer: $F(x/y, z/y) = 0$

2. Given the first-order partial differential equation

$$y\frac{\partial z}{\partial x} - x\frac{\partial z}{\partial y} = 0,$$

   find the nature of the characteristic curves.

   Answer: The characteristic curves are a family of circles passing through the origin.

3. Given the first-order partial differential equation

$$a\frac{\partial z}{\partial x} + b\frac{\partial z}{\partial y} = 0,$$

   where a and b are constants. Find the general solution.

   Answer: $z = f(ay - bx)$

4. Given the first-order partial differential equation

$$a\frac{\partial z}{\partial x} + b\frac{\partial z}{\partial y} = c,$$

   where a, b, and c are constants, find the general solution.

   Answer: $z = f(ay - bx) + (c/a)x$

5. Solve the first-order partial differential equation

$$x\frac{\partial z}{\partial x} - y\frac{\partial z}{\partial y} = z,$$

   with initial conditions $z = x^2$ on $y = x$; $1 \leq y \leq 2$.

   Answer: $z(x,y) = x\sqrt{xy}$

6. Solve the first-order partial differential equation

$$x(y-z)\frac{\partial z}{\partial x} + y(x+z)\frac{\partial z}{\partial y} = (x+y)z,$$

   with initial conditions $z = x^2 + 1$ on $y = x$.

Answer: $\dfrac{xy}{u} = \dfrac{x+u-y-1}{x+u-y}$

7. Show that the following second-order partial differential equation of the form

$$A\frac{\partial^2 u}{\partial x^2} + B\frac{\partial^2 u}{\partial x \partial y} + C\frac{\partial^2 u}{\partial y^2} + D\frac{\partial u}{\partial x} + E\frac{\partial u}{\partial y} + Fu + G = 0, \tag{E1.1}$$

subject to the following transformation of independent variables from x and y to ξ and η, where A, B, C, D, E, F, and G are functions of x and y only, which can be put in the canonical or normal form. Show that the transformed equations are of the following form

$$\overline{A}(\xi_x, \xi_y)\frac{\partial^2 u}{\partial \xi^2} + \overline{B}(\xi_x, \xi_y, \eta_x, \eta_y)\frac{\partial^2 u}{\partial \xi \partial \eta} + \overline{C}(\eta_x, \eta_y)\frac{\partial^2 u}{\partial \eta^2} = F[\xi, \eta, u(\xi, \eta), u_\xi(\xi, \eta), u_\eta(\xi, \eta)] \tag{E1.2}$$

where

$$\begin{aligned}
\overline{A}(\xi_x, \xi_y) &= A\xi_x^2 + B\xi_x\xi_y + C\xi_y^2, \\
\overline{B}(\xi_x, \xi_y, \eta_x, \eta_y) &= 2A\xi_x\eta_x + B(\xi_x\eta_y + \xi_y\eta_x) + 2C\xi_y\eta_y, \\
\overline{C}(\eta_x, \eta_y) &= A\eta_x^2 + B\eta_x\eta_y + C\eta_y^2.
\end{aligned}$$

In the aforementioned set of equations, the subscripts indicate partial derivatives.

8. Moreover, show that for the aforementioned second-order partial differential equation (E1.1), the following relation can be obtained

$$\overline{B}^2 - 4\overline{A}\,\overline{C} = (\xi_x\eta_y - \xi_y\eta_x)^2(B^2 - 4AC).$$

9. For the hyperbolic case, where $B^2 - 4AC > 0$, show that Equation (E1.2) will be transformed and result in the following simple, canonical form given by

$$\frac{\partial^2 u}{\partial \xi \partial \eta} = \phi(\xi, \eta, u, u_\xi, u_\eta)$$

10. For the parabolic case, where $B^2 - 4AC = 0$, show that Equation (E1.2) will be transformed and result in the following simple, canonical form given by

$$\frac{\partial^2 u}{\partial \eta^2} = \phi(\xi, \eta, u, u_\xi, u_\eta)$$

11. For the elliptical case, where $B^2 - 4AC < 0$, show that Equation (E1.2) will be transformed and result in the following simple, real canonical form given by

$$\frac{\partial^2 u}{\partial \alpha^2} + \frac{\partial^2 u}{\partial \beta^2} = \psi[\alpha, \beta, u, u_\alpha(\alpha, \beta), u_\beta(\alpha, \beta)],$$

where

$$\alpha = \frac{1}{2}(\xi + \eta) \qquad \text{and} \qquad \beta = \frac{1}{2}(\xi - \eta).$$

12. Reduce the following second-order partial differential equation

$$\frac{\partial^2 u}{\partial x^2} = x^2 \frac{\partial^2 u}{\partial y^2}$$

to its canonical form.

Answer: $u_{\xi\eta} = \dfrac{1}{4(\xi - \eta)}(u_\xi - u_\eta)$

13. Reduce the following second-order partial differential equation

$$\frac{\partial^2 u}{\partial x^2} + 2\frac{\partial^2 u}{\partial x \partial y} + \frac{\partial^2 u}{\partial y^2} = 0$$

to its canonical form.

Answer: $u_{\eta\eta} = 0$

14. Reduce the following second-order partial differential equation

$$\frac{\partial^2 u}{\partial x^2} + x^2 \frac{\partial^2 u}{\partial y^2} = 0$$

to its canonical form.

Answer: $u_{\alpha\alpha} + u_{\beta\beta} = -\dfrac{1}{2\alpha} u_\alpha$

2

Equations of Fluid Motion

2.1   Introduction

A fluid is a substance that yields to applied shear stress; i.e., it is a substance that continuously deforms under an applied shear stress. Fluids comprise both liquids and gases. While in the macroscopic scale, fluids are made up of molecules, it is possible to disregard the molecular viewpoint while discussing fluid properties by invoking the concept of "continuum." The "continuum" hypothesis assumes that fluid is continuous and made up of a very large number of fluid elements. The advantage of the "continuum" hypothesis is that it provides for continuously distributed body of matter in which fluid properties vary smoothly. A fluid element is said to be composed of several millions of molecules within it and has a scale for which macroscopic fluid properties such as density and temperature can be prescribed. Hence in a "continuum" hypothesis approach, the fluid element is the smallest scale of analysis. The scale of the fluid element cannot be too small as such an element would not provide a robust average of fluid properties. Moreover, the scale of fluid element cannot be too large, as this would contribute to smoothing over and ignoring relevant and important scales of variability.

An obvious length scale for gases is the "mean free path" (λ), which is defined as the average distance traversed by a gaseous molecule between two successive collisions. The mean free path for liquid molecules assume typically much smaller values than that for gas. For the atmosphere, the "mean free path" at sea level is 10^{-7}m, at 50 km, it is 10^{-4}m, and at 150 km, it is 1 m. The "characteristic length scale" (L) for a typical fluid situation can be defined as the most important length scale component of the fluid motion. For example, for a fluid flowing in a long tube of circular cross section, the radius or diameter of the circular cross section is the "characteristic length scale." The "continuum" hypothesis is applicable for flow situations in which the "characteristic length scale" L of the flow is much larger than the "mean free path"

λ. For the atmosphere, the "continuum" hypothesis is applicable for fluid motion up to the mesospheric height of 90 km. It is clear that the fluid velocity at a point is nothing but the velocity of the fluid element present at that point at that particular instant of time.

2.2   Lagrangian and Eulerian Description of Fluid Motion

There are inherently two approaches to describe fluid motion. In the Lagrangian description, any single fluid element is marked and identified and the fluid flow characteristics such as velocity, pressure, acceleration and temperature, are determined during the movement of the fluid through space and time. The position of the marked fluid element will change with time in the Lagrangian description; joining the position of the marked fluid element with time gives the "pathline."

In the Eulerian description , one deals with any fixed point in the space occupied by the flow. Here, fluid observations are carried out to depict changes in fluid properties and/or flow characteristics at that particular fixed point, i.e., in the Eulerian description, fluid flow properties such as pressure, velocity, temperature etc., are expressed as functions of space and time. Streamlines are lines that are joined by connecting tangents to the velocity field associated with various fluid elements at different locations at the same time. In general, the streamline flow pattern will change with time.

2.2.1   Substantive or total derivative

It is important to understand the physical concept of a "material derivative" or "substantive derivative" before one seeks to derive the governing equations of fluid motion. The model that one invokes is that of an infinitesimally small fluid element moving with the flow with a velocity

$$\vec{V} = u\hat{i} + v\hat{j} + w\hat{k}, \tag{2.1}$$

where u, v and w are the x, y and z components of fluid velocity in Cartesian space and are given by

$$u = u(x, y, z, t); \tag{2.2}$$

$$v = v(x, y, z, t); \tag{2.3}$$

$$w = w(x, y, z, t). \tag{2.4}$$

Consider, in general, the scalar temperature field,

$$T = T(x,y,z,t). \tag{2.5}$$

Let the fluid element be located at position 1 at an initial time t_1; hence, the temperature of the fluid element at the position 1 is given by

$$T_1 = T(x_1,y_1,z_1,t_1). \tag{2.6}$$

The same fluid element moves to a new point 2 at a later time, t_2. Hence, the temperature of the aforementioned fluid element, at time t_2, is

$$T_2 = T(x_2,y_2,z_2,t_2). \tag{2.7}$$

One can expand the function T = T(x, y, z, t) in a Taylor's series about point 1 as indicated in Equation (2.8)

$$T_2 = T_1 + \left\{\frac{\partial T}{\partial x}\right\}_1 (x_2-x_1) + \left\{\frac{\partial T}{\partial y}\right\}_1 (y_2-y_1) + \left\{\frac{\partial T}{\partial z}\right\}_1 (z_2-z_1) + \left\{\frac{\partial T}{\partial t}\right\}_1 (t_2-t_1)$$
$$+ (\text{quadratic and higher order terms}) \tag{2.8}$$

Dividing Equation (2.8) by (t_2-t_1) and neglecting the quadratic and higher order terms, one obtains

$$\frac{T_2-T_1}{t_2-t_1} = \left\{\frac{\partial T}{\partial x}\right\}_1 \left[\frac{x_2-x_1}{t_2-t_1}\right] + \left\{\frac{\partial T}{\partial y}\right\}_1 \left[\frac{y_2-y_1}{t_2-t_1}\right] + \left\{\frac{\partial T}{\partial z}\right\}_1 \left[\frac{z_2-z_1}{t_2-t_1}\right] + \left\{\frac{\partial T}{\partial t}\right\}_1 \tag{2.9}$$

The left hand side (LHS) of Equation (2.9) determines the average time rate of change of temperature of the fluid element as it moves from point 1 to point 2. In the limit, as time t_2 approaches time t_1, the LHS of Equation (2.9) is called the "substantive" or the "total derivative" DT/Dt and provides the instantaneous time rate of change of temperature of the fluid element as it moves through point 1. Clearly, DT/Dt provides for the time rate of change of temperature of a fluid element as it moves in space, an example of Lagrangian description of fluid motion. The term inside the square bracket in the first term in the right hand side (RHS) of Equation (2.9) is nothing but u, the x component of fluid velocity, whereas the terms inside the square bracket in the second and third terms in the RHS of Equation (2.9), are nothing but v and w, the y and z components of fluid velocity. The last term in the RHS of Equation (2.9) refers physically to the local time rate of change of temperature at the fixed point 1 and is indicated as a "local derivative." The first three terms in the RHS of Equation (2.9) are known as "advective derivative," and indicated as $\vec{V} \cdot \nabla T$, which is physically defined as the time rate of change of temperature due to the movement of the fluid element from one location to another in the flow field where the fluid flow properties do not have the same value. Clearly, the local derivative term and the advective derivative term together make up the total derivative and Equation (2.9) can be written as

$$\frac{DT}{Dt} = \frac{\partial T}{\partial t} + \vec{V} \cdot \nabla T = \frac{\partial T}{\partial t} + u\frac{\partial T}{\partial x} + v\frac{\partial T}{\partial y} + w\frac{\partial T}{\partial z}. \tag{2.10}$$

For a marked fluid element, A, the position in three-dimensional space $x_A(t)$, $y_A(t)$, and $z_A(t)$ describe the motion of that fluid element A. Hence,

$$\vec{V}_A = V_A(x_A(t), y_A(t), z_A(t)). \tag{2.11}$$

The total derivative of the velocity of the marked fluid element, A is then given by

$$\vec{a}_A = \frac{d\vec{V}_A}{dt} = \frac{\partial \vec{V}_A}{\partial t} + \frac{\partial \vec{V}_A}{\partial x}\frac{dx_A}{dt} + \frac{\partial \vec{V}_A}{\partial y}\frac{dy_A}{dt} + \frac{\partial \vec{V}_A}{\partial z}\frac{dz_A}{dt} = \frac{\partial \vec{V}_A}{\partial t} + \frac{\partial \vec{V}_A}{\partial x}u_A(t) + \frac{\partial \vec{V}_A}{\partial y}v_A(t) + \frac{\partial \vec{V}_A}{\partial z}w_A(t). \tag{2.12}$$

It is to be noted that the acceleration of the fluid element is the time rate of change of its fluid velocity. Hence, the component of acceleration in the x-direction, denoted by a_x, is simply the time rate of change of the x-component of fluid velocity u; as one is following the motion of a fluid element, this time rate of change of fluid velocity is given by the total derivative. Thus,

$$a_x = \frac{Du}{Dt}; \ a_y = \frac{Dv}{Dt}; \ a_y = \frac{Dw}{Dt}. \tag{2.13}$$

The RHS of the three relations of Equation (2.13) can in turn be expanded to local derivatives and advective derivative terms, respectively.

2.2.2   Conservation of mass principle: Continuity equation

The conservation of mass principle can be very easily applied to a fluid element. Let the mass of a fluid element be δm at a certain position and at a certain instant of time. Let ρ and δV be the density and the volume of the aforementioned fluid element, i.e., $\delta m = \rho \delta V$. The conservation of mass principle states that in the absence of mass sources and sinks, the time rate of change of mass δm of the fluid element as it moves in space shall be zero, following the motion of the fluid element, i.e., the total derivative of the mass of the fluid element is zero.

$$\frac{D\{\delta m\}}{Dt} = \frac{D\{\rho \delta V\}}{Dt} = \rho\frac{D\{\delta V\}}{Dt} + \delta V\frac{D\{\rho\}}{Dt} = 0; \tag{2.14}$$

$$\frac{D\rho}{Dt} + \rho \left\{ \frac{1}{\delta V} \frac{D[\delta V]}{Dt} \right\} = 0. \tag{2.15}$$

The second term in the LHS of Equation (2.15) shows the product of fluid density and the fractional time rate of change of volume of the fluid element following the motion of the fluid element; physically, the latter is known as the divergence of the fluid velocity and, hence, Equation (2.15) is written as

$$\frac{D\rho}{Dt} + \rho \nabla \cdot \vec{V} = 0. \tag{2.16}$$

Equation (2.16) is a statement of conservation of mass and is called the continuity equation. It is possible to write the total derivative of density (first term in the LHS of Equation (2.16) as the sum of the local derivative of density and the advective derivative of density. Hence, using the vector identity, Equation (2.16) can be written in another equivalent form as follows:

$$\frac{\partial \rho}{\partial t} + \nabla \cdot \left\{ \rho \vec{V} \right\} = 0 \tag{2.17}$$

2.2.3   Conservation of momentum principle: Momentum equation

The underlying physical principle of conservation of momentum is based on Newton's second law of motion as applied to a moving fluid element. The forces that can act on a fluid element can be broadly characterized as (i) body forces and (ii) surface forces. Body forces are forces that while acting on a fluid element are not dependent on the presence or absence of neighboring fluid elements, whereas surface forces are those that are dependent on the presence or absence of neighboring fluid elements. Examples of body forces are gravitational, electric and magnetic forces whereas examples of surface forces are shear and pressure gradient forces. The conservation of momentum principle when applied to a fluid element requires that the acceleration experienced by the aforementioned fluid element would equal to the net force per unit mass that is acting on the fluid element. Let ρ be the density having lengths dx, dy, and dz of a fluid element. Let the component of the total force (surface and body) in the x-direction, be given as F_x whose expression is as follows:

$$F_x = \left\{ \frac{-\partial p}{\partial x} + \frac{\partial \tau_{xx}}{\partial x} + \frac{\partial \tau_{yx}}{\partial y} + \frac{\partial \tau_{zx}}{\partial z} \right\} dxdydz + \rho f_x dxdydz, \tag{2.18}$$

where f_x is the x component of the body force per unit mass and τ_{xx}, τ_{yx}, and τ_{zx} are the normal and shear stresses that act in the x direction. The first term in the RHS is the x component of the pressure gradient force. Similar expressions hold for the y and z components of the total force acting on the fluid element. Considering the mass of the fluid element, m = ρ dx dy dz, and invoking Newton's second law of motion, one obtains the x component of the momentum equation to be

$$\rho \frac{Du}{Dt} = \frac{-\partial p}{\partial x} + \frac{\partial \tau_{xx}}{\partial x} + \frac{\partial \tau_{yx}}{\partial y} + \frac{\partial \tau_{zx}}{\partial z} + \rho f_x. \tag{2.19}$$

Similarly, the y and z components of the momentum equations are given as follows:

$$\rho \frac{Dv}{Dt} = \frac{-\partial p}{\partial y} + \frac{\partial \tau_{xy}}{\partial x} + \frac{\partial \tau_{yy}}{\partial y} + \frac{\partial \tau_{zy}}{\partial z} + \rho f_y; \tag{2.20}$$

$$\rho \frac{Dw}{Dt} = \frac{-\partial p}{\partial z} + \frac{\partial \tau_{xz}}{\partial x} + \frac{\partial \tau_{yz}}{\partial y} + \frac{\partial \tau_{zz}}{\partial z} + \rho f_z. \tag{2.21}$$

Equations (2.19, 2.20 and 2.21) are the components of the momentum equation in x, y and z directions and are derived directly by applying the conservation of momentum principle to a moving fluid element. Equations (2.19, 2.20 and 2.21) are momentum equations in non-conservation form and are known together with the continuity equation as the Navier–Stokes equations, derived independently by Navier and Stokes in the first half of the nineteenth century. The momentum equations in conservation form are given as

$$\frac{\partial (\rho u)}{\partial t} + \nabla \cdot (\rho u \vec{V}) = \frac{-\partial p}{\partial x} + \frac{\partial \tau_{xx}}{\partial x} + \frac{\partial \tau_{yx}}{\partial y} + \frac{\partial \tau_{zx}}{\partial z} + \rho f_x; \tag{2.22}$$

$$\frac{\partial (\rho v)}{\partial t} + \nabla \cdot (\rho v \vec{V}) = \frac{-\partial p}{\partial y} + \frac{\partial \tau_{xy}}{\partial x} + \frac{\partial \tau_{yy}}{\partial y} + \frac{\partial \tau_{zy}}{\partial z} + \rho f_y; \tag{2.23}$$

$$\frac{\partial (\rho w)}{\partial t} + \nabla \cdot (\rho w \vec{V}) = \frac{-\partial p}{\partial z} + \frac{\partial \tau_{xz}}{\partial x} + \frac{\partial \tau_{yz}}{\partial y} + \frac{\partial \tau_{zz}}{\partial z} + \rho f_z. \tag{2.24}$$

Newtonian fluids are those for which the shear stress is proportional to the strain rate, i.e., velocity gradients. The two most important fluids that manifest in Earth's atmosphere and hydrosphere are Newtonian fluids. For Newtonian fluids, Stokes proposed the following relations

$$\tau_{xx} = \lambda \nabla \cdot \vec{V} + 2\mu \frac{\partial u}{\partial x}, \tag{2.25}$$

$$\tau_{yy} = \lambda \nabla \cdot \vec{V} + 2\mu \frac{\partial v}{\partial y}, \tag{2.26}$$

$$\tau_{zz} = \lambda \nabla \cdot \vec{V} + 2\mu \frac{\partial w}{\partial z}, \tag{2.27}$$

$$\tau_{xy} = \tau_{yx} = \mu \left\{ \frac{\partial v}{\partial x} + \frac{\partial u}{\partial y} \right\}, \tag{2.28}$$

$$\tau_{xz} = \tau_{zx} = \mu \left\{ \frac{\partial u}{\partial z} + \frac{\partial w}{\partial x} \right\}, \tag{2.29}$$

$$\tau_{yz} = \tau_{zy} = \mu \left\{ \frac{\partial w}{\partial y} + \frac{\partial v}{\partial z} \right\}. \tag{2.30}$$

where λ is the bulk viscosity coefficient and μ is the molecular viscosity coefficient. Stokes proposed the following relation between λ and μ, $\lambda = -(2/3)\mu$. Substituting the expressions for normal and shear stresses and the relation between λ and μ, one obtains for the incompressible fluid flow with constant μ, the Navier–Stokes equation in non-conservation form to be

$$\frac{\partial u}{\partial x} + \frac{\partial v}{\partial y} + \frac{\partial w}{\partial z} = 0, \tag{2.31}$$

$$\left\{ \frac{\partial u}{\partial t} + u\frac{\partial u}{\partial x} + v\frac{\partial u}{\partial y} + w\frac{\partial u}{\partial z} \right\} = \frac{-1}{\rho}\frac{\partial p}{\partial x} + \nu \left\{ \frac{\partial^2 u}{\partial x^2} + \frac{\partial^2 u}{\partial y^2} + \frac{\partial^2 u}{\partial z^2} \right\} + f_x, \tag{2.32}$$

$$\left\{ \frac{\partial v}{\partial t} + u\frac{\partial v}{\partial x} + v\frac{\partial v}{\partial y} + w\frac{\partial v}{\partial z} \right\} = \frac{-1}{\rho}\frac{\partial p}{\partial y} + \nu \left\{ \frac{\partial^2 v}{\partial x^2} + \frac{\partial^2 v}{\partial y^2} + \frac{\partial^2 v}{\partial z^2} \right\} + f_y, \tag{2.33}$$

$$\left\{ \frac{\partial w}{\partial t} + u\frac{\partial w}{\partial x} + v\frac{\partial w}{\partial y} + w\frac{\partial w}{\partial z} \right\} = \frac{-1}{\rho}\frac{\partial p}{\partial z} + \nu \left\{ \frac{\partial^2 w}{\partial x^2} + \frac{\partial^2 w}{\partial y^2} + \frac{\partial^2 w}{\partial z^2} \right\} + f_z, \tag{2.34}$$

where ν is the kinematic coefficient of viscosity.

2.2.4   Euler's equation of motion for an ideal fluid

Ideal fluids are fluids that are incompressible and non-viscous. The governing equations that describe the flow of an ideal fluid are known as Euler's equation of motion and are given as follows:

$$\frac{\partial u}{\partial x} + \frac{\partial v}{\partial y} + \frac{\partial w}{\partial z} = 0, \tag{2.35}$$

$$\left\{ \frac{\partial u}{\partial t} + u\frac{\partial u}{\partial x} + v\frac{\partial u}{\partial y} + w\frac{\partial u}{\partial z} \right\} = \frac{-1}{\rho}\frac{\partial p}{\partial x} + f_x, \tag{2.36}$$

$$\left\{ \frac{\partial v}{\partial t} + u\frac{\partial v}{\partial x} + v\frac{\partial v}{\partial y} + w\frac{\partial v}{\partial z} \right\} = \frac{-1}{\rho}\frac{\partial p}{\partial y} + f_y, \tag{2.37}$$

$$\left\{ \frac{\partial w}{\partial t} + u\frac{\partial w}{\partial x} + v\frac{\partial w}{\partial y} + w\frac{\partial w}{\partial z} \right\} = \frac{-1}{\rho}\frac{\partial p}{\partial z} + f_z. \tag{2.38}$$

2.2.5 Conservation of energy principle: Thermodynamic energy equation

The principle of conservation of energy is based on the first law of thermodynamics, which states that for a thermodynamic system, the amount of heat added would partly go into increasing the internal energy and partly in doing external work, i.e., Change in internal energy = Heat added - Work done by the system, dU = dq - $pd\alpha$, where dU is the change in internal energy, dq is the amount of heat added and $pd\alpha$ is the work done by the system. For an ideal gas, dU = c_v dT, where c_v is the specific heat at constant volume and, hence, the first law of thermodynamics is expressed as

$$c_v \frac{dT}{dt} = \frac{dq}{dt} - p\frac{d\alpha}{dt}. \tag{2.39}$$

A question that naturally arises is whether Equation (2.39) is applicable for a moving fluid element. Total thermodynamic energy E of fluid element of density ρ and volume δV is given by $E = \left\{ e + \frac{1}{2}\vec{V}\cdot\vec{V} \right\}\rho\delta V$, where e is the internal energy per unit mass and the second term in the RHS within brackets is the kinetic energy per unit mass. To compute the work done onto the fluid element, one takes the product of the resultant of external forces and velocity vector. Recall that the forces that are important are pressure gradient force, gravity, and viscous forces.

$$\text{Work done by the gravity force} = \rho g \cdot \vec{V}\delta V = -\rho g w \delta V, \tag{2.40}$$

where w is the vertical component of the fluid velocity To obtain the work done by the pressure gradient force, consider a parallelepiped volume δV of sides δx, δy, and δz. The work done by the pressure gradient force on the fluid element in the x direction is given by $-\left\{ \frac{\partial(pu)}{\partial x} \right\}\delta x \delta y \delta z$. Considering similar expressions in the y and z directions, the total rate at which work is done to the system by pressure gradient force equals $-\nabla\cdot\left\{ p\vec{V} \right\}\delta V$. The statement of conservation of energy is written as

$$\frac{D}{Dt}\left\{\rho\left[e+\frac{1}{2}\vec{V}\cdot\vec{V}\right]\delta V\right\}=-\nabla\cdot(p\vec{V})\delta V+\rho\vec{g}\cdot\vec{V}\delta V+\rho\dot{J}\delta V, \tag{2.41}$$

where it is assumed that rate at which the work is done by the viscous forces are ignored and $\dot{J}$ is the rate of heating per unit mass due to radiation, conduction, and latent heat release; D/Dt is the total or the substantive derivative. Expanding Equation(2.41), one obtains

$$\rho\delta V\frac{D}{Dt}\left\{e+\frac{1}{2}\vec{V}\cdot\vec{V}\right\}+\left\{e+\frac{1}{2}\vec{V}\cdot\vec{V}\right\}\frac{D(\rho\delta V)}{Dt}=-\vec{V}\cdot\nabla p\delta V-p\nabla\cdot\vec{V}\delta V-\rho gw\delta V+\rho\dot{J}\delta V. \tag{2.42}$$

Conservation of mass principle states that the rate of change of mass of a fluid element is zero following the motion of the fluid element and, hence, the second term in the LHS becomes identically zero. Dividing throughout by δV, one obtains

$$\rho\frac{D}{Dt}\left\{e+\frac{1}{2}\vec{V}\cdot\vec{V}\right\}=-\vec{V}\cdot\nabla p-p\nabla\cdot\vec{V}-\rho gw+\rho\dot{J}. \tag{2.43}$$

Considering the momentum equation in vector form with pressure gradient force and gravitational forces but without the viscous forces,

$$\frac{D\vec{V}}{Dt}=-\frac{1}{\rho}\nabla p+\vec{g}. \tag{2.44}$$

Taking inner product of $\rho\vec{V}$ to Equation (2.44), one obtains the mechanical energy equation

$$\rho\frac{D}{Dt}\left\{\frac{1}{2}\vec{V}\cdot\vec{V}\right\}=-\vec{V}\cdot\nabla p-\rho gw. \tag{2.45}$$

Subtracting Equation (2.45) from Equation (2.43), one obtains the thermal energy equation, given by

$$\rho\frac{De}{Dt}=-p\nabla\cdot\vec{V}+\rho\dot{J}. \tag{2.46}$$

From the continuity equation, $\nabla\cdot\vec{V}=-\frac{1}{\rho}\frac{D\rho}{Dt}$ Hence, after dividing Equation(2.46) throughout by ρ and shifting the first term in the RHS to LHS, it becomes

$$\frac{p}{\rho}\nabla\cdot V=\frac{p}{\rho}\left\{-\frac{1}{\rho}\frac{D\rho}{Dt}\right\}=-\frac{p}{\rho^2}\left\{\frac{D}{Dt}(\alpha^{-1})\right\}=p\frac{D\alpha}{Dt}. \tag{2.47}$$

Substituting Equation (2.47) in Equation (2.46), one obtains

$$c_v \frac{DT}{Dt} + p \frac{D\alpha}{Dt} = \dot{J}, \tag{2.48}$$

where internal energy e equals c_vT. Thus, the first law of thermodynamics is applicable to a fluid element in motion. The second term on the LHS of Equation (2.48) representing the rate of working by the fluid system, provides a conversion between thermal and mechanical energy. It is this conversion process that enables the solar heat energy to drive the motion of the atmosphere.

2.3   Equations Governing Atmospheric Motion

The continuity equation, which is statement of conservation of mass, remains the same when it is applied to the atmosphere. However, the momentum equations that govern the evolution of atmospheric motion are not exactly the same as the momentum equations that govern fluid motion. It is evident that the most appropriate and convenient manner of describing the motion in the atmosphere would be from a frame of reference that is fixed on the Earth. However, a frame of reference fixed on the Earth is actually a rotating frame as the Earth rotates with constant angular velocity. Moreover, a rotating frame of reference is an accelerating frame of reference and, hence, is non-inertial. The principle of conservation of momentum based on Newton's second law of motion cannot then be directly applied to a frame of reference fixed on the Earth, as Newton's laws of motion are applicable only to an inertial frame of reference. To apply Newton's second law of motion to a rotating frame of reference, we need to identify the acceleration terms that are responsible for the departure from the inertial behaviour and treat these terms as pseudo forces or fictitious forces .

2.3.1   Rotating frame of reference

The Earth rotates around an axis passing through its geographical north pole with a constant angular velocity $\vec{\Omega}$. A reference frame fixed on the Earth with its center coinciding with the center of the Earth, its vertical axis coinciding with the Earth's polar axis and the horizontal axes of the frame lying on the equatorial plane would provide a rotating frame.

Consider an air parcel that is at a fixed position with respect to the Earth's surface. As the Earth rotates with an angular velocity $\vec{\Omega}$, the position of the air parcel does not change in the frame of reference that is fixed on the Earth. However, the position

vector of the air parcel when observed from an inertial frame of reference will change
with time; the same is given by

$$\left\{\frac{D\vec{r}}{Dt}\right\}_I = \vec{\Omega} \times \vec{r}, \tag{2.49}$$

where the suffix I indicates the inertial frame of reference and $\vec{r}$ is the position vector.
If, however, the air parcel has a relative velocity $\left\{\frac{D\vec{r}}{Dt}\right\}_R$ with respect to the fixed Earth
surface, then the velocity of the air parcel as observed from an inertial frame of
reference is given by

$$\left\{\frac{D\vec{r}}{Dt}\right\}_I = \left\{\frac{D\vec{r}}{Dt}\right\}_R + \vec{\Omega} \times \vec{r}. \tag{2.50}$$

Equation (2.50) indicates that the inertial velocity of an air parcel (LHS) is the sum of
the relative velocity of the air parcel (the first term in the RHS of the equation) and the
velocity (the second term in the RHS of the equation) associated with the movement
of the reference frame

$$\{\vec{v}\}_I = \{\vec{v}\}_R + \vec{\Omega} \times \vec{r}. \tag{2.51}$$

Newton's second law of motion relates the inertial acceleration to the net force per unit
mass acting on a fluid element. It can be observed that the inertial acceleration (rate of
change of inertial velocity in the inertial reference frame) is related to the rate of change
of inertial velocity in the rotating frame of reference by the same transformation as
indicated in Equation (2.50). Hence,

$$\left\{\frac{D\vec{v}_I}{Dt}\right\}_I = \left\{\frac{D\vec{v}_I}{Dt}\right\}_R + \vec{\Omega} \times \vec{v}_I. \tag{2.52}$$

Substituting Equation (2.51) in Equation (2.52), one obtains

$$\left\{\frac{D\vec{v}_I}{Dt}\right\}_I = \left\{\frac{D\vec{v}_R}{Dt}\right\}_R + 2\vec{\Omega} \times \vec{v}_R + \vec{\Omega} \times (\vec{\Omega} \times \vec{r}). \tag{2.53}$$

The LHS of Equation (2.53) is the inertial acceleration and is related to the net force
per unit mass acting on a fluid element, as follows:

$$\left\{\frac{D\vec{v}_I}{Dt}\right\}_I = -\frac{1}{\rho}\nabla p + \vec{g} + \nu\nabla^2\vec{v}. \tag{2.54}$$

Relating the LHS of Equations (2.53) and (2.54), one obtains

$$\left\{\frac{D\vec{v_R}}{Dt}\right\}_R = -\frac{1}{\rho}\nabla p + \vec{g} + \nu\nabla^2\vec{v} - 2\vec{\Omega}\times\vec{v_R} - \vec{\Omega}\times(\vec{\Omega}\times\vec{r}). \tag{2.55}$$

The last term in the RHS of Equation (2.55) is called the centrifugal force and is usually combined with the Newtonian gravitational force $\vec{g}$ to form the effective gravity force $\vec{g}_{eff}$. The last but one term in the RHS of Equation (2.55) is known as the Coriolis force and is the additional term that manifests in the momentum equation while describing the evolution of the atmosphere. The aforementioned arguments are exactly applicable for describing the motion of the oceans. Hence, the momentum equations that describe the atmospheric and oceanic motions are

$$\left\{\frac{D\vec{v_R}}{Dt}\right\}_R = -\frac{1}{\rho}\nabla p - \nabla\phi + \nu\nabla^2\vec{v} - 2\vec{\Omega}\times\vec{v_R}, \tag{2.56}$$

where the effective gravity force is expressed as a gradient of a function ϕ, where ϕ is known as "geopotential."

2.3.2 Conservation of energy: Thermodynamic energy equation for atmosphere

For deriving the thermodynamic energy equation for atmosphere, one invokes the first law of thermodynamics and applies the same to an air parcel of fixed mass δM. Denoting the internal energy per unit mass as e, the entropy per unit mass as s. and the volume per unit mass as α, one can write expressions for the total internal energy δU, for the total entropy δS, and for the total volume δV as

$$\delta U = e\delta M; \; \delta S = s\delta M; \; \delta V = \alpha\delta M. \tag{2.57}$$

The first law of thermodynamics can then be written as

$$dU = TdS - pdV, \tag{2.58}$$

where the LHS represents infinitesimal change in internal energy and the first and second terms in the RHS represent infinitesimal changes in heat added and infinitesimal change in the work done. Equation (2.58) when written in terms of each quantity expressed per unit mass, is given as

$$d(e\delta M) = Td(s\delta M) - pd(\alpha\delta M). \tag{2.59}$$

Equation (2.59) when applied over a time interval δt, becomes

$$\frac{d(e\delta M)}{\delta t} = T\frac{d(s\delta M)}{\delta t} - p\frac{d(\alpha\delta M)}{\delta t}. \tag{2.60}$$

Taking the limit of the time interval δt as tending to zero, each term of Equation (2.60) represents the rate of change following the air parcel (the total or substantive derivative) and one obtains

$$\frac{D(e\delta M)}{Dt} = T\frac{D(s\delta M)}{Dt} - p\frac{D(\alpha\delta M)}{Dt}. \tag{2.61}$$

However, from the conservation of mass principle, δM, the mass of the air parcel is conserved following the motion of the air parcel. Hence, Equation (2.61) becomes

$$\frac{De}{Dt} = T\frac{Ds}{Dt} - p\frac{D\alpha}{Dt}. \tag{2.62}$$

Equation (2.62) is the expression of the first law of thermodynamics. The atmosphere is assumed to be an ideal case and, hence, one can use the following relations

$$p\alpha = RT;\ e = c_v T;\ c_p - c_v = R, \tag{2.63}$$

where R is the specific gas constant of air and c_p is the specific heat at constant pressure. Substituting Equation (2.63) in Equation (2.62), one obtains

$$c_v\frac{DT}{Dt} = T\frac{Ds}{Dt} - R\frac{dT}{dt} + \left\{\frac{RT}{p}\right\}\frac{Dp}{Dt}. \tag{2.64}$$

Equation (2.64) can be expressed as

$$T\frac{Ds}{Dt} = c_p\frac{dT}{dt} - \left\{\frac{RT}{p}\right\}\frac{Dp}{Dt} \tag{2.65}$$

While considering adiabatic motion in the atmosphere, an important quantity known as "potential temperature" is often invoked. Potential temperature (θ) is defined as the temperature that an air parcel would have if it were brought dry adiabatically to the standard reference pressure of p_o (p_o = 1000 hPa). It is related to temperature (T) and entropy per unit mass (s) by the following relations

$$\theta = T\left\{\frac{p_o}{p}\right\}^{(R/c_p)};\ s = c_p ln(\theta). \tag{2.66}$$

Comparing Equations (2.48), (2.62) and (2.66), one obtains the expression for the thermodynamic energy equation

$$c_p \frac{D\theta}{Dt} = \frac{\theta}{T} J. \tag{2.67}$$

In the absence of diabatic heating processes, J would be zero and specific entropy would be conserved following the motion.

2.3.3   Geostrophic balance equations

The momentum equations that describe the atmospheric and oceanic motions are given by Equation (2.56). In the absence of viscous forces and restricting the atmospheric and oceanic motion to the horizontal, Equation (2.56) becomes

$$\frac{D\vec{v}}{Dt} = -\frac{1}{\rho}\nabla p - 2\vec{\Omega} \times \vec{v}. \tag{2.68}$$

Furthermore, assuming that the acceleration term (term in the LHS) of Equation (2.68) is quite small and can be neglected in comparison with the two terms in RHS, one obtains

$$2\vec{\Omega} \times \vec{v} = -\frac{1}{\rho}\nabla p. \tag{2.69}$$

Equation (2.69) is known as the "geostrophic balance equations" and is a result of a balance between the horizontal pressure gradient force and the Coriolis force. The geostrophic balance is often a reasonable approximation for large-scale motion of the atmosphere and oceans away from the equator. These equations provide a very useful diagnostic relationship to obtain the velocity (wind in the atmosphere and current in the ocean) from the mass fields (pressure fields). However, the assumption of geostrophic balance cannot provide for the evolution of the velocity field (wind in the atmosphere and current in the ocean) with time. The absence of time evolution of the velocity field is an inherent limitation of the assumption of geostrophic balance. It is to be noted that invoking of the geostrophic balance and the associated geostrophic winds and geostrophic currents correspond to non-accelerated motion in the horizontal direction in the atmosphere and oceans.

2.3.4   Hydrostatic balance equation

The momentum equations that describe the atmospheric and oceanic motions are given by Equation (2.56). In the absence of viscous forces and restricting the atmospheric and oceanic motion to the vertical direction, it is noted that the two most dominant force terms correspond to the vertical component of the pressure gradient force and the gravity force. Retaining the aforementioned two most dominant force terms, Equation (2.56) becomes

$$0 = -\frac{1}{\rho}\frac{\partial p}{\partial z} - g; \quad \frac{\partial p}{\partial z} = -g\rho. \tag{2.70}$$

Equation (2.70) is known as the hydrostatic equation and the balance between the vertical component of the pressure gradient force that is directed upward and the gravity force that is directed downward is known as the "hydrostatic balance." The implication of the applicability of the "hydrostatic balance" is that the atmosphere or ocean has no acceleration in the vertical direction. The hydrostatic equation is, however, not applicable for small-scale atmospheric motion such as in a "tornado" for which the vertical acceleration cannot be ignored.

2.3.5 Governing equations of motion of atmosphere with pressure as a vertical coordinate

The hydrostatic balance equation provides for a monotonic relationship between pressure and geometric height and allows one to utilize pressure as an equally valid vertical coordinate. It turns out that it is advantageous to formulate the governing equations of motion of the atmosphere with pressure as the vertical coordinate as the continuity equation, thermodynamic energy, and the expression for horizontal pressure gradient force assume simpler forms. The governing equations of atmospheric motion with pressure as a vertical coordinate are given as

$$\frac{D\vec{v}_h}{Dt} + fk \times \vec{v}_h = -\nabla_p \phi \text{ [horizontal momentum equation]}, \tag{2.71}$$

$$\frac{\partial \phi}{\partial p} = -\frac{RT}{p} \text{ [hydrostatic equation]}, \tag{2.72}$$

$$\left\{\frac{\partial u}{\partial x} + \frac{\partial v}{\partial y}\right\}_p + \frac{\partial \omega}{\partial p} = 0 \text{ [continuity equation]}, \tag{2.73}$$

$$\frac{\partial T}{\partial t} + u\frac{\partial T}{\partial x} + v\frac{\partial T}{\partial y} - S_p\omega = \frac{j}{c_p} \text{ [thermodynamic energy equation]}, \tag{2.74}$$

$$p = \rho RT \text{ [equation of state]}, \tag{2.75}$$

$$\frac{D}{Dt} = \frac{\partial}{\partial t} + u\frac{\partial}{\partial x} + v\frac{\partial}{\partial y} + \omega\frac{\partial}{\partial p}, \tag{2.76}$$

where $f = 2\Omega\sin\phi$ is the Coriolis parameter, v_h is the horizontal velocity vector, 'ω' is the vertical velocity component in pressure coordinates, and S_p is the static stability parameter for the isobaric coordinate and

$$S_p = \frac{RT}{c_p p} - \frac{\partial T}{\partial p} = -\frac{T}{\theta}\frac{\partial\theta}{\partial p}. \tag{2.77}$$

2.3.6 Quasi-geostrophic equations of motion of atmosphere with pressure as a vertical coordinate

For mid-latitude synoptic-scale motions in the atmosphere, it is generally observed that the horizontal velocity components assume values that are close to their geostrophic values and the atmosphere is approximately in hydrostatic balance. Synoptic scale motions are those that have a horizontal extent of about 1000 km. These synoptic-scale motions over mid-latitudes provide for atmospheric motion that have horizontal velocities that are approximately geostrophic (quasi-geostrophic). Quasi-geostrophic equations of motion of the atmosphere with pressure as a vertical coordinate are useful to model and forecast the dynamics of weather systems in mid-latitudes. The quasi-geostrophic model approach provides for the simplification of the three-dimensional equations of motion of the atmosphere without losing the prognostic (forecast) terms in the governing equations. Hence, the aforementioned approach allows the governing quasi-geostrophic equations of motion to be integrated with time to forecast the dynamics of weather systems in mid-latitude regions. The following approximations are assumed while deriving the quasi-geostrophic equations of motion:

1. The horizontal wind is separated into the geostrophic wind component and a departure from the geostrophic wind component (the ageostrophic wind component); $\vec{v}_h = \vec{v}_g + \vec{v}_a$ and advection is dominated by the geostrophic component.

2. The mid-latitude beta plane approximation is assumed to hold, i.e., the Coriolis parameter dependence on latitude ϕ is taken into account and $f = f_0 + \beta y$; where $\beta = df/dy$ is the meridional variation of the Coriolis parameter and β is assumed constant. Here, $f_o = 2\Omega\sin\phi_o$; $\beta = 2\Omega\cos\phi_o/a$, where a is the radius of the Earth.

3. Small vertical temperature perturbations are assumed. The temperature T is expanded into a background temperature T_o, which is a function of pressure

only and a perturbation temperature T′, i.e., T(x,y,p,t) = T$_o$(p) +T′(x,y,p,t) with the following holding good

$$\left|\frac{\partial T_o}{\partial p}\right| >> \left|\frac{\partial T'}{\partial p}\right|.$$

Applying the aforementioned assumptions, one obtains the quasi-geostrophic equations of motion in the pressure coordinate system as follows:

$$\vec{v}_g = \frac{1}{f_o} k \times \nabla \phi \text{ [geostrophic wind],} \tag{2.78}$$

$$\frac{D\vec{v}_g}{Dt} = -f_o k \times \vec{v}_a - \beta y k \times \vec{v}_g \text{ [horizontal momentum equation],} \tag{2.79}$$

$$\frac{\partial u_a}{\partial x} + \frac{\partial v_a}{\partial y} + \frac{\partial \omega}{\partial p} = 0 \text{ [continiuty equation],} \tag{2.80}$$

$$\left\{\frac{\partial}{\partial t} + u\frac{\partial}{\partial x} + v\frac{\partial}{\partial y}\right\}\left\{-\frac{\partial \phi}{\partial p}\right\} - \sigma\omega = \frac{R\dot{J}}{pc_p} \text{ [thermodynamic energy equation],} \tag{2.81}$$

where $\sigma = -\frac{RT_o}{p}\frac{\partial[ln\theta_o]}{\partial p}$

2.3.7 Shallow water equations

About 99 % of the total atmospheric mass is concentrated in the first 32 km above the Earth, whereas the mean depth of oceans is 4 km. Comparing the aforementioned vertical extent of both the atmosphere and the oceans with their respective horizontal extents, the latter, of the order of several tens of thousands of kilometres, it is abundantly clear that both the atmosphere and the oceans are essentially thin and shallow fluids. The shallow water equations describe the evolution of a hydrostatic constant density (homogeneous) incompressible fluid flow and are given by

$$\frac{\partial u}{\partial t} + u\frac{\partial u}{\partial x} + v\frac{\partial u}{\partial y} - fv = -g\frac{\partial h}{\partial x}, \tag{2.82}$$

$$\frac{\partial v}{\partial t} + u\frac{\partial v}{\partial x} + v\frac{\partial v}{\partial y} + fu = -g\frac{\partial h}{\partial y}, \tag{2.83}$$

$$\frac{\partial h}{\partial t} + u\frac{\partial h}{\partial x} + v\frac{\partial h}{\partial y} + h\left\{\frac{\partial u}{\partial x} + \frac{\partial v}{\partial y}\right\} = 0, \tag{2.84}$$

where u and v are x and y components of fluid velocity, f is the Coriolis parameter, g is the acceleration due to gravity, and h is the vertical extent of the fluid layer. The advantage of the shallow water equations is that the essential nonlinearity of the atmospheric and oceanic flows is retained while considerable simplifications in terms of the smaller number of dependent variables are achieved.

2.3.8 Vorticity equation for incompressible fluid: Curl of the Navier–Stokes equation

Vorticity is defined as the curl of the velocity field. It is possible to obtain the vorticity equation by taking the curl of the momentum equation for incompressible fluid, the latter appearing in the Navier–Stokes equation. The momentum equation in vector form for a viscous incompressible fluid is

$$\frac{\partial \vec{v}}{\partial t} + \vec{v} \cdot \nabla \vec{v} = -\frac{1}{\rho} \nabla p - \nabla \phi + \nu \nabla^2 \vec{v}. \tag{2.85}$$

The following vector identity holds

$$\vec{v} \cdot \nabla \vec{v} = \frac{1}{2} \nabla (\vec{v} \cdot \vec{v}) - \vec{v} \times (\nabla \times \vec{v}). \tag{2.86}$$

Defining the vorticity vector $\vec{w} = \nabla \times \vec{v}$, one obtains, by taking the curl of Equation 2.86

$$\nabla \times \{\vec{v} \cdot \nabla \vec{v}\} = \frac{1}{2} \nabla \times \nabla (\vec{v} \cdot \vec{v}) - \nabla \times (\vec{v} \times \vec{w}). \tag{2.87}$$

As the curl of a gradient of scalar function is always zero, the first term in the RHS of Equation 2.87 is zero. The second term in the RHS of Equation 2.87 becomes

$$\nabla \times (\vec{w} \times \vec{v}) = (\vec{v} \cdot \nabla)\vec{w} - (\vec{w} \cdot \nabla)\vec{v} + \vec{w}(\nabla \cdot \vec{v}) - \vec{v}(\nabla \cdot \vec{w}). \tag{2.88}$$

The third term in the RHS of Equation (2.88) is zero as the fluid is incompressible whereas the fourth term in the RHS of Equation (2.88) is zero as the divergence of curl of a vector is always zero. Taking the curl of the momentum equation Equation (2.85) and assuming constant density and using the fact that the curl of gradient of a scalar function is always zero, one obtains the vorticity equation for incompressible fluid

$$\frac{D\vec{w}}{Dt} = (\vec{w} \cdot \nabla)\vec{v} + \nu \nabla^2 \vec{w}. \tag{2.89}$$

2.3.9   Vorticity equation for atmospheric and oceanic flows

For the large-scale motions in the atmosphere and the oceans, only the vertical component of relative vorticity is considered important. To derive an equation for the vertical component of relative vorticity, the inner product of unit vertical vector ($\vec{k}$) is applied to the curl of the momentum equation (i.e., Equation (2.56) without the viscous force term). The aforementioned application leads to the following vorticity equation

$$\frac{D(\zeta+f)}{Dt} = -(\zeta+f)\left\{\frac{\partial u}{\partial x}+\frac{\partial v}{\partial y}\right\} + \frac{1}{\rho^2}\left\{\frac{\partial p}{\partial y}\frac{\partial \rho}{\partial x}-\frac{\partial p}{\partial x}\frac{\partial \rho}{\partial y}\right\} - \left\{\frac{\partial w}{\partial x}\frac{\partial v}{\partial z}-\frac{\partial w}{\partial y}\frac{\partial u}{\partial z}\right\}, \quad (2.90)$$

where ζ is the vertical component of the relative vorticity vector, f is the Coriolis parameter, and $\zeta+f$ is called the absolute vorticity. f is also known as planetary vorticity .

$$\zeta = \frac{\partial v}{\partial x}-\frac{\partial u}{\partial y}. \quad (2.91)$$

The three terms in the RHS of Equation (2.90) are known as the "horizontal velocity divergence term," "solenoidal term," and "tilting term."

2.3.10   Non-divergent vorticity equation for atmospheric and oceanic flows

For large-scale systems in the atmosphere and oceans, all the three terms in the RHS of Equation (2.90) have typically small magnitudes as compared to the magnitude of terms in the LHS of Equations (2.90) and (2.90) becomes

$$\frac{\partial \zeta}{\partial t}+u\frac{\partial \zeta}{\partial x}+v\frac{\partial \zeta}{\partial y}+\beta v=0, \quad (2.92)$$

where β is the latitudinal variation of the Coriolis parameter. The fourth term in the LHS of Equation (2.92) denotes the advection of planetary vorticity, whereas the second and third terms in the LHS of Equation (2.92) refer to the horizontal advection of relative vorticity.

2.3.11   Boussinesq approximation

It is to be noted that in the atmosphere, density fluctuations are essential for representing the buoyancy force and cannot be ignored totally. The density variations in the atmosphere are retained only in the buoyancy term that appears in the vertical momentum equation; density appearing in all other terms is assumed to have a constant mean value ρ_o. The aforementioned simplifications are known as the Boussinesq approximation. In the Boussinesq approximation, density ρ is written as the sum of constant density ρ_o and the density fluctuations $\delta\rho$, i.e,

$$\rho = \rho_o + \delta\rho(x, y, z, t), \tag{2.93}$$

where $|\delta\rho| < \rho_o$ The aforementioned reference density ρ_o is associated with a reference pressure p_o, the latter being a function of z

$$p = p_o(z) + \delta p(x, y, z, t). \tag{2.94}$$

The reference pressure p_o is related to the reference density ρ_o by the hydrostatic equation

$$\frac{dp_o}{dz} = -g\rho_o. \tag{2.95}$$

The governing equations of motion that assume Boussinesq approximation are given as follows:

$$\frac{Du}{Dt} = -\frac{1}{\rho_o}\frac{\partial p}{\partial x} + fv \text{ [x component of momentum equation]}, \tag{2.96}$$

$$\frac{Dv}{Dt} = -\frac{1}{\rho_o}\frac{\partial p}{\partial y} - fu \text{ [y component of momentum equation]},, \tag{2.97}$$

$$\frac{Dw}{Dt} = -\frac{1}{\rho_o}\frac{\partial p}{\partial z} + g\frac{\delta T}{T_o} \text{ [vertical component of momentum equation]}, \tag{2.98}$$

where δT is the departure of temperature from its basic state value T_o (z), where the total temperature is equal to the basic state temperature and the departure from the basic state.

$$\frac{D}{Dt}\left\{\frac{\delta T}{T_o}\right\} + \frac{w}{T_o}\left\{\frac{dT_o}{dz} + \frac{g}{c_p}\right\} = \frac{\dot{q}}{c_p T_o} \text{ [thermodynamic energy equation]}, \tag{2.99}$$

$$\frac{\partial u}{\partial x} + \frac{\partial v}{\partial y} + \frac{\partial w}{\partial z} = 0 \text{ [continuity equation]}. \tag{2.100}$$

2.3.12   Anelestic approximation

In the atmosphere, the density decreases appreciably in the vertical direction. In the anelastic approximation, the density ρ and pressure p are defined in terms of reference values and the associated differences. The difference between the Boussinesq approximation and the anelastic approximation is that whereas in the former, the reference density is considered a constant, the reference density in the anelastic approximation is a function of z, and is given by

$$\rho = \tilde{\rho}(z) + \delta\rho(x,y,z,t); \ p = \tilde{p}(z) + \delta p(x,y,z,t). \tag{2.101}$$

The reference pressure and reference density are related to the hydrostatic equation as follows:

$$\frac{\partial \tilde{p}}{\partial z} = -g\tilde{\rho}(z) \tag{2.102}$$

The governing equations of motion that assume anelastic approximation are as follows:

$$\frac{Du}{Dt} = -\frac{\partial}{\partial x}\left\{\frac{\delta p}{\tilde{\rho}}\right\} + fv \ [\text{x component of momentum equation}], \tag{2.103}$$

$$\frac{Dv}{Dt} = -\frac{\partial}{\partial y}\left\{\frac{\delta p}{\tilde{\rho}}\right\} - fu \ [\text{y component of momentum equation}], \tag{2.104}$$

$$\frac{Dw}{Dt} = -\frac{\partial}{\partial z}\left\{\frac{\delta p}{\tilde{\rho}}\right\} + g\frac{\delta\theta}{\tilde{\theta}}[\text{vertical component of momentum equation}] \tag{2.105}$$

$$\frac{D}{Dt}\left\{\frac{\delta\theta}{\tilde{\theta}}\right\} + w\frac{d}{dz}\left\{ln\tilde{\theta}\right\} = \frac{\dot{q}}{c_p T_o} \ [\text{thermodynamic energy equation}], \tag{2.106}$$

$$\frac{\partial u}{\partial x} + \frac{\partial v}{\partial y} + \frac{1}{\tilde{\rho}}\frac{\partial}{\partial z}\left\{\tilde{\rho}w\right\} = 0 \ [\text{continuity equation}] \tag{2.107}$$

.

2.3.13   Conservation of water vapour mixing ratio equation

One of the important measures of water vapour in the atmosphere is the water vapour mixing ratio (q) that is defined as the ratio of mass of water vapour to the mass of dry air. The statement of conservation of water vapour mixing ratio q simply states that the total amount of water vapor in an air parcel is conserved following the motion, except for the presence of moisture sources such as evaporation E, and moisture sinks such as condensation C, and is given by

$$\frac{\partial q}{\partial t} + \vec{v} \cdot \nabla q = E - C. \tag{2.108}$$

Multiplying Equation (2.108) by density ρ and the continuity equation

$$\frac{\partial \rho}{\partial t} = -\nabla \cdot (\rho \vec{v}). \tag{2.109}$$

by the water vapour mixing ratio q and adding the aforementioned two equations, one obtains the conservation of the water vapour mixing ratio in flux form assuming non-divergence of velocity field

$$\frac{\partial (\rho q)}{\partial t} = -\nabla \cdot (\rho \vec{v} q) + \rho (E - C). \tag{2.110}$$

Whereas the LHS term in Equation (2.110) refers to the local rate of change of moisture flux with time, the first term in the RHS of Equation (2.110) refers to the convergence of the moisture flux and the last term in RHS refers to the source/sink term of moisture flux. It is to be noted that it is possible to include similar additional conservation equations for additional variables such as liquid water, etc., provided the corresponding sources and sinks are duly included.

2.3.14   Mean equations of turbulent flow in the atmosphere

Turbulent flow, is a highly irregular and fluctuating fluid flow that manifests at large Reynolds number. Atmospheric boundary layer is the lowest portion of the atmosphere where viscous forces cannot be neglected and where the fluid flow is strongly influenced by interaction with the surface of the Earth. The flows in the atmospheric and oceanic boundary layers are turbulent in nature. In a turbulent flow, field variables such as temperature, velocity, moisture, and pressure that are measured at a point generally fluctuate rapidly in time; providing the essential signature of turbulent flow. It is clear that while dealing with turbulent motions in the atmosphere or oceans, it would be more meaningful to treat fluid variables such as velocity and temperature that do not fluctuate very rapidly with time. To realise this, and for the measurements to be truly representative of the large-scale flow, an averaging is done (known as Reynolds averaging) on fluid field variables over a time interval. The time over which averaging is performed should be sufficiently long to average out small-scale eddy fluctuations, but at the same time, sufficiently short to preserve trends in the large-scale flow field and not smooth out the large-scale flow fields. The flux form of the horizontal momentum equation is written as

$$\frac{\partial}{\partial t}(\rho u) + \frac{\partial}{\partial x}(\rho u^2) + \frac{\partial}{\partial y}(\rho uv) + \frac{\partial}{\partial z}(\rho uw) - f\rho v = -\frac{\partial p}{\partial x}, \tag{2.111}$$

$$\frac{\partial}{\partial t}(\rho v) + \frac{\partial}{\partial x}(\rho uv) + \frac{\partial}{\partial y}(\rho v^2) + \frac{\partial}{\partial z}(\rho vw) + f\rho u = -\frac{\partial p}{\partial y}. \tag{2.112}$$

We can define each dependent variable into mean and fluctuating parts, as follows:

$$u = \bar{u} + u'; \; v = \bar{v} + v'; \; w = \bar{w} + w'; \; p = \bar{p} + p'; \; \rho = \bar{\rho}; \; \rho' = 0. \tag{2.113}$$

The time averaged forms of the momentum equation, is then given by substituting Equation (2.113) in the governing equations of momentum, energy, and continuity and employing the rule of Reynolds averaging that directs that the mean of a fluctuating quantity is zero whereas the mean of a product of two fluctuating quantities need not be zero.

$$\frac{\partial \bar{u}}{\partial t} + \bar{u}\frac{\partial \bar{u}}{\partial x} + \bar{v}\frac{\partial \bar{u}}{\partial y} + \bar{w}\frac{\partial \bar{u}}{\partial z} - f\bar{v} = -\frac{1}{\rho}\frac{\partial \bar{p}}{\partial x} - \frac{1}{\rho}\left\{ \frac{\partial(\rho\overline{u'u'})}{\partial x} + \frac{\partial(\rho\overline{u'v'})}{\partial y} + \frac{\partial(\rho\overline{u'w'})}{\partial z} \right\}, \tag{2.114}$$

$$\frac{\partial \bar{v}}{\partial t} + \bar{u}\frac{\partial \bar{v}}{\partial x} + \bar{v}\frac{\partial \bar{v}}{\partial y} + \bar{w}\frac{\partial \bar{v}}{\partial z} + f\bar{u} = -\frac{1}{\rho}\frac{\partial \bar{p}}{\partial y} - \frac{1}{\rho}\left\{ \frac{\partial(\rho\overline{u'v'})}{\partial x} + \frac{\partial(\rho\overline{v'v'})}{\partial y} + \frac{\partial(\rho\overline{v'w'})}{\partial z} \right\}, \tag{2.115}$$

$$\frac{\partial \bar{w}}{\partial t} + \bar{u}\frac{\partial \bar{w}}{\partial x} + \bar{v}\frac{\partial \bar{w}}{\partial y} + \bar{w}\frac{\partial \bar{w}}{\partial z} = -\frac{1}{\rho}\frac{\partial \bar{p}}{\partial z} + g\frac{\bar{\theta}}{\theta_o} - \frac{1}{\rho}\left\{ \frac{\partial(\rho\overline{u'w'})}{\partial x} + \frac{\partial(\rho\overline{v'w'})}{\partial y} + \frac{\partial(\rho\overline{w'w'})}{\partial z} \right\}. \tag{2.116}$$

The time averaged thermodynamic energy equation is

$$\frac{\partial \bar{\theta}}{\partial t} + \bar{u}\frac{\partial \bar{\theta}}{\partial x} + \bar{v}\frac{\partial \bar{\theta}}{\partial y} + \bar{w}\frac{\partial \bar{\theta}}{\partial z} = g\frac{d\theta_o}{dz} - \frac{1}{\rho}\left\{ \frac{\partial(\rho\overline{u'\theta'})}{\partial x} + \frac{\partial(\rho\overline{v'\theta'})}{\partial y} + \frac{\partial(\rho\overline{w'\theta'})}{\partial z} \right\}. \tag{2.117}$$

The time averaged continuity equation is

$$\frac{\partial \bar{u}}{\partial x} + \frac{\partial \bar{v}}{\partial y} + \frac{\partial \bar{w}}{\partial x} = 0, \tag{2.118}$$

where θ_o is base state potential temperature. The last three quantities that arise in the RHS of Equations (2.114) to (2.116) [$\rho\overline{u'u'}$, $\rho\overline{u'v'}$, $\rho\overline{u'w'}$, $\rho\overline{v'u'}$, $\rho\overline{v'v'}$, $\rho\overline{v'w'}$, $\rho\overline{w'u'}$, $\rho\overline{w'v'}$, $\rho\overline{w'w'}$] are new terms that have appeared and were not part of the original momentum equations. Each of the aforementioned new terms is known as Reynolds stress terms or components of the turbulent kinematic momentum flux. For atmospheric and oceanic flows, the molecular viscous term of the form

$\nu\nabla^2 u, \nu\nabla^2 v, \nu\nabla^2 w$, that are normally present in the x, y, and z components of momentum equation are several orders of magnitude smaller than the Reynolds stress terms and, hence, the molecular viscous terms can be neglected. The last three quantities that arise in the RHS of Equation (2.117) $[\rho\overline{u'\theta'}, \rho\overline{v'\theta'}, \rho\overline{w'\theta'}]$ are new terms and are known as heat flux terms or components of the turbulent kinematic heat flux. A similar argument in terms of conservation equation of water vapour mixing ratio will yield three additional new quantities such as $[\overline{u'q'}, \overline{v'q'}, \overline{w'q'}]$; these are known as moisture flux terms or components of the turbulent kinematic flux of water vapour.

Equations (2.114) to (2.118) are five equations; however, they involve dependent variables that are very much larger than five. Each of the three momentum equations and the thermodynamic energy equations introduced additional dependent variables (nine Reynolds stresses and three heat fluxes) when the time averaged equations of turbulent motion are derived, ensuring that the system of governing mean equations of turbulent motion is not a closed system. In order to provide for a closed system of governing mean equations of turbulent motion, the simplest workaround solution is to write the new dependent variables that arise in Equations (2.114) to (2.117) in terms of gradients of velocity components or gradients of temperature or gradients of mixing ratio. This closure theory is known as the K theory, K-closure, or flux-gradient theory and uses for example the following relations

$$(\overline{u'w'}) = -K_m\left\{\frac{\partial \bar{u}}{\partial z}\right\}; (\overline{v'w'}) = -K_m\left\{\frac{\partial \bar{v}}{\partial z}\right\}; (\overline{\theta'w'}) = -K_h\left\{\frac{\partial \bar{\theta}}{\partial z}\right\}; (\overline{w'q'}) = -K_e\left\{\frac{\partial \bar{q}}{\partial z}\right\},$$

(2.119)

where K_m, K_h, and K_e are known respectively as the eddy viscosity for momentum, eddy diffusivity for heat, and eddy diffusivity for moisture. In the K theory, these new quantities are expressed in terms of gradients of velocity, temperature or moisture; however, one has to prescribe the K terms. In the simplest K scheme, the eddy coefficients (sometimes called transfer coefficients) are taken as constant. In more complex K schemes, K_m is parameterized in terms of the mean flow, static stability, and/or turbulence kinetic energy, whereas K_h and K_e are prescribed as a function of K_m.

2.3.15 RANS, LES, and DNS approaches

Investigating and simulating the turbulent flow behaviour is indeed a challenging assignment for the following reasons: (i) fluid properties exhibit random spatial variations in three-dimensions in addition to being unsteady, (ii) flow is aperiodic, and (iii) flow contains a wide range of scales (eddies). In some cases, especially those that pertain to industrial applications, one is mainly interested in the steady-state fluid flow. In the steady-state case, it is not necessary to simulate the detailed instantaneous fluid flow, leading to a great reduction of computational time. This philosophy forms the basis for the Reynolds-averaged Navier–Stokes (RANS) approach in which one solves only for the averaged quantities whereas the effect of all the scales of instantaneous turbulent motion is modeled by a turbulence model. The RANS equations are nothing but mean time-averaged equations of motion for turbulent flow. They are based on the idea proposed by Osborne Reynolds in his well-known Reynolds decomposition, wherein an instantaneous fluid quantity is decomposed into its time-averaged quantity and fluctuating quantity and these equations are primarily used to describe the turbulent flows. The RANS equations solve ensemble-averaged (or time-averaged) Navier–Stokes equations.

Boussinesq proposed the concept of eddy viscosity and succeeded in closing the system of equations by relating the turbulence stresses to the mean flow.Some of the well-known eddy viscosity schemes are (i) zero equation or algebraic equations, (ii) one-equation model, and (iii) two-equation models. An example of a zero equation model is the Prandtl mixing length theory in which the basic philosophy is the extension of the concept of molecular transport through molecules to turbulent transport by turbulent eddies. The zero equation model are purely local in nature and do not have any history. An example of a one-equation model is the k-model, where k is the turbulent kinetic energy per unit mass. An equation for the evolution of k can be derived directly from the Navier–Stokes equation. An important advantage of the k-model with respect to the zero-equation model is the inclusion of history effects in the k-model. It is known that turbulence tends to "remember" its past, atleast for short times. The zero-equation model does not reflect the idea that turbulence briefly remembers its past. An example of the two-equation models is the k–ε model, where ε is the turbulence dissipation rate. The advantage of two-equation models over the one-equation model is the use of the former to model turbulent viscosity transport using two transport equations involving k and ε. RANS approach has been widely utilized in industrial applications for the last few decades due to its relatively modest computing requirement. However, it is important to determine the transient behaviour of the turbulent flow and, hence, the RANS approach is not the ideal approach to invoke while dealing with most turbulent flow situations.

Another approach to investigating and simulating turbulent flow behaviour is the large-eddy simulation (LES), which was proposed by Smagorinsky in 1963 and was first adopted by Deardorff in 1970. The basic philosophy of the LES approach is to obtain resolvable large-scale motions (large eddies) of turbulent flow by subjecting the fluid flow through a low-pass filter. The application of a low pass filter can be envisaged as a temporal and spatial averaging that effectively removes the information related to the small scale-motions from the numerical solution. However, the small-scale information is indeed relevant and significant and, hence, its effect on the turbulent flow field has to be modeled. The aforementioned LES approach has a huge computational cost reduction (as compared with the approach where all scales are computed directly) as it ignores the smallest length scales that are the most computationally expensive to resolve. Rather than resolving the smallest length scales, the LES approach models the small-scale (sub-grid scale (SGS)) motions. Most SGS models use the eddy viscosity concept to model the SGS stress tensor in terms of resolved scale strain rate tensor using the SGS eddy viscosity. Smagorinsky was the first to propose an expression for the SGS eddy viscosity. The results of LES are considered to be more accurate than the results obtained from the RANS approach as the LES captures directly the full details of the large eddies that contain most of the turbulent energy and are responsible for most of the momentum transfer and turbulent mixing. It is to be noted that the large eddies are also modeled in the RANS approach. In addition, the small scales tend to be generally more isotropic and homogeneous than the large ones. Hence, modeling the small-scale motions as in the LES approach, turns out to be easier than modeling all scales within a single model as in the RANS approach. Hence, in the LES approach, one solves the spatially averaged Navier–Stokes equations, and whereas the large-scale motions (large eddies) are directly resolved and determined, the effect of small-scale motions (small eddies) that are smaller than the mesh sizes on the resolved scales are modeled.

The direct numerical simulation (DNS) approach to investigating and simulating turbulent flows rests on the premise that theoretically, all turbulent flows can be simulated by numerically solving the full Navier–Stokes equations. In this approach, the direct solution of the unsteady Navier–Stokes equations is sought in such a way as to resolve all the essential turbulent scales, including resolving even the smallest eddies and time scales of turbulence within a flow. Hence, unlike the RANS or the LES approaches, the DNS approach resolves the whole spectrum of scales. From this, it is clear that the DNS approach does not require any additional closure assumptions and associated equations and, hence, no modeling effort is required. Moreover, in order to obtain accurate flow solutions using the DNS approach, very fine grids and extremely small time steps need to be employed. It is to be noted that the major difficulty in performing turbulence calculations using the DNS approach at Reynolds number values that are required for practical industry applications lies in the extensive amount of computational resources required. In essence, the number of degrees of freedom involved with the solution of the Navier–Stokes equation roughly increase $\approx Re^{9/4}$, where Re is the Reynolds number. It is for this reason that the DNS approach is currently confined to simple flow problems at relatively low Reynolds numbers, making it impossible to apply DNS approach to simulate practical high Reynolds number complex flows. A viable alternative to the DNS approach for the high Reynolds number flows is the LES approach as the latter determines directly only the resolvable large scales unlike the DNS approach where all the scales are computed directly.

2.3.16   Parameterization of physical processes in the atmospheric models

Accurate numerical simulations of the state of the atmosphere, for both short-term weather prediction and or long-term studies on climate, require a correct representation of all the important physical processes that occur in the real atmosphere. The aforementioned physical processes include either processes that are very complex and or too small in scale to represent explicitly in numerical models, and thus need to be represented by physical parameterization schemes . In essence, the parameterization schemes seek to statistically relate the effects of physical processes that cannot be represented directly in a model to variables that are included. Representation of a physical process directly in a model may pose difficulties owing to the following reasons, (i) the physical process is so complex that it is computationally expensive to represent it directly, (ii) the sub-grid scale processes, i.e., those processes that have scales that are smaller than the resolvable grid scale of the model have scales that are so small that it is computationally prohibitive to represent a process directly, and (iii) there is inadequate knowledge about how a physical process works to explicitly represent it mathematically. The present generation of atmospheric models provide for the following physical parameterization schemes applicable for (i) turbulence, (ii) cloud and precipitation microphysics, (iii) radiative transfer, (iv) moist convection, (v) land surface processes, and (vi) gravity wave drag.

2.3.17   Parallel computing

Computer codes were written conventionally for serial computing. Serial codes essentially divide the problem to be solved into smaller instructions which are then executed one after the other on the central processing unit (CPU) of a computer. It is clear that serial computing has the following disadvantages: (i) instructions are executed one by one, and (ii) at a time only one instruction is executed at any moment of time. Clearly, serial computing is impractical for solving large and complex problems. Parallel computing uses multiple processing elements simultaneously for solving any problem and has the following distinct advantages over serial computing: (i) it provides concurrency and saves time and costs as many resources work together thus reducing the time of execution and consequently cutting potential costs, (ii) it can handle complex and larger datasets, (iii) it ensures effective utilization of resources as it can take advantage of non-local resources when the local resources are limited and finite, and (iv) it is guaranteed to more effectively use the hardware. The older generation computers belong to th single instruction single data (SISD) category, an example of which is a serial computer in which only one data stream is used as input and only one instruction stream is acted upon by CPU during any one clock cycle. Parallel computers can either belong to single instruction multiple data (SIMD) or multiple instruction multiple data (MIMD) type. In SIMD machines, each processing unit can operate on a different data element while all the processing units execute the same instruction at any given clock cycle. SIMD machines are ideally suited for specialized problems characterized by a high degree of regularity, for example, in graphics and in image processing. In MIMD machines, every processor may be working with a different data stream and may be executing a different instruction stream. Currently, the modern supercomputers belong to the MIMD category. MIMD machines are capable of running multiple instruction streams simultaneously to cooperatively execute a single program. The overall objective is to achieve high speed in scientific computing by breaking the computation into pieces that are sufficiently independent to be performed in parallel using several processes that run on separate hardware units but are sufficiently cooperative to assist in solving the scientific problem. There are broadly two basic types of MIMD architectures, commonly known as shared memory and distributed memory multiprocessors. Whereas distributed memory refers to a multiprocessor computer system in which each processor has its own private memory, a shared memory multiprocessor offers a single memory space used by all processors.

In the shared-memory architecture, all the CPU-cores can access the same memory; however, this above attribute is a limitation of the shared-memory approach as all the CPU-cores compete for access to memory over a shared bus. The limitation of the shared-memory approach can be minimized by introducing memory caches or by putting several processors together in a (non-uniform memory access - NUMA) architecture. However, there is no way one can attain the hundreds of thousands of CPU-cores that are needed for today's multi-petaflop supercomputers and, hence, systems with shared-memory architecture are not manufactured presently. However, the concept of communicating directly through memory remains a useful programming abstraction. Hence, in many systems, this is handled by the hardware and the programmer does not need to insert any special directives. Some common programming techniques that use these abstractions are OpenMP. Distributed memory architecture has traditionally been associated with processors performing computation on local memory and then using explicit messages to transfer data with remote processors. This makes the computing complex for the programmer; however, it simplifies the hardware implementation as the system no longer has to maintain the fallacy that all memory is actually shared. This type of programming has traditionally been used with supercomputers that have hundreds or thousands of processing elements. A commonly used programming technique that uses the aforementioned approach is MPI.

Exercises 2 (Questions and answers)

1. What is a "streakline"?

   Answer: Assume particles are continuously being introduced at a particular point. Then the "streakline" is the line that is formed due to the release of the aforementioned particles from that particular point. Alternatively, "streakline" is also defined as the locus of fluid particles that have sequentially passed through a prescribed track in the flow.

2. Under what conditions do "streamlines," "streaklines," and "pathlines" coincide?

   Answer: Streamlines, streaklines, and pathlines coincide under steady state conditions.

3. What is the definition of an "incompressible fluid"?

   Answer: If the changes in density of a fluid element associated with changes in pressure following the motion are very small and negligible, then the fluid is defined as an "incompressible fluid."

4. What is the definition of a "steady fluid flow"?

   Answer: In general, fluid properties such as pressure, density, temperature, and velocity can change in space at a fixed time. However, when the fluid properties do not change locally with time at a particular fixed point in space, the fluid is referred to as a "steady fluid flow." A fluid flow that is not in steady state is said to be in "unsteady" state.

5. Does the term "incompressible fluid" refer to a fluid having constant and uniform density in space?

   Answer: From the definition of "incompressible fluid," it is clear that it corresponds to a situation where the total derivative of density is zero, following the motion, i.e., $\frac{D\rho}{Dt} = 0$. Using the relation between the total derivative of density and the local derivative of density, one can conclude that for a special situation of an incompressible fluid flow being in steady state and having non-vanishing fluid velocity, the gradient of fluid density is zero, resulting in fluid having uniform and constant density in space.

6. What are the definitions of "body forces" and "surface forces"?

Answer: Body forces acting on a fluid element do not depend on the presence or absence of neighboring fluid particles whereas surface forces that act on a fluid element depend on the presence or absence of neighboring fluid particles.

7. Give examples of "body forces" and "surface forces"?

Answer: Examples of body forces are gravity forces and electromagnetic forces whereas examples of surface forces are pressure gradient forces and viscous forces.

8. What is the definition of "surface stress" and "stress tensor"?

Answer: One cannot discuss the stress at a point without first defining the particular surface through that point on which the stress acts. A small fluid surface element centered at the point $\vec{r}$ is defined by its area δA and by its outward unit normal vector $\vec{n}$. The stress exerted by the fluid on the side toward which $\vec{n}$ points on the surface element is defined as $\vec{\sigma} = \lim_{\delta A \to 0} \frac{\delta \vec{F}}{\delta A}$, where $\delta \vec{F}$ is the force exerted on the surface by the fluid on that side. In the limit $\delta A \to 0$, the stress is independent of the magnitude of the area, but will in general depend on the orientation of the surface element, i.e., on $\vec{n}$. In the absence of shear forces, the stress is always normal to the surface on which it acts; its magnitude is independent of the surface orientation and is expressed only in terms of a single scalar field called pressure $p(\vec{r}, t)$. Consider "reference stresses," at a given point and time t, and let the values of the stresses that are exerted on a surface oriented in the positive x-direction, a surface oriented in the positive y-direction, and a surface oriented in the positive z-direction, be indicated as $\vec{\sigma}_x$, $\vec{\sigma}_y$, $\vec{\sigma}_z$, the aforementioned reference stresses are in turn related to three scalar components each. The nine scalar components (τ_{ij}) are called components of the "stress tensor," which is a tensor of order two having the three diagonal components called "normal stress" components and the remaining six off-diagonal components called "shear stress" components. As the "stress tensor" is a symmetric tensor, there are only six independent stress components. Thus, τ_{xx}, τ_{yx}, and τ_{zx} represent the x, y, and z components of the stress acting on the surface whose outward normal is oriented in the positive x-direction. Whereas the first subscript on τ_{ij} indicates the direction of the stress, the second subscript shows the outward normal of the surface on which it acts.

9. What is the definition of "strain rate tensor"? Is such a tensor a symmetric tensor?

Answer: The strain rate tensor represents that part of the deformation tensor that changes the shape of the local fluid element. The strain rate tensor is a symmetric tensor.

10. What is the definition of a "Newtonian fluid"?

Answer: Any fluid that satisfies Newton's law of viscosity is said to be a "Newtonian fluid." As Newton's law of viscosity relates the shear stress linearly to the rate of strain, all Newtonian fluids have a linear relationship between the shear stress and the rate of strain; the constant of proportionality is known as the "dynamic coefficient of viscosity."

11. How is viscosity interpreted in terms of momentum transfer between two layers of fluid moving with different velocities?

Answer: Viscosity manifests whenever there is transport of momentum between layers of fluid moving with different velocities due to random molecular motion. The motion of molecules in a flowing fluid is due to both mass motion of the fluid medium together with the random motion of molecules due to thermal agitation. Molecules are transported from a higher momentum layer to another layer having lower momentum, thereby increasing the overall momentum in the latter layers. Similarly, molecules with lower momentum move into the higher momentum layer and contribute to reducing the overall momentum in the latter layer. Both processes tend to reduce the relative velocity between the layers. This momentum exchange generates an effective shear force between the two layers. The mechanism is the principal viscosity mechanism for gases, where forces between molecules are small.

12. How does the coefficient of dynamic viscosity of a gas change with increase in temperature?

Answer: When the temperature of a gas increases, the random molecular motion becomes more marked and the associated molecular momentum exchange increases with increase in gas temperature. As greater momentum exchange in a gas is related to larger shear forces, the coefficient of dynamic viscosity of a gas will increase as the temperature increases.

13. How does the coefficient of dynamic viscosity of a liquid change with increase in temperature?

Answer: As far as liquids are concerned, their molecules are closely packed and, hence, the intermolecular forces also play a part in generation of coefficient of dynamic viscosity in addition to molecular interchange. The increase in temperature causes the kinetic or thermal energy to increase and the liquid molecules become more mobile causing a reduction in the attractive binding energy and a corresponding decrease in the coefficient of dynamic viscosity.

14. What are the advantages and disadvantages of the geostrophic balance approximation?

Answer: The advantage of the geostrophic balance approximation is that it provides a very useful diagnostic relationship to determine the velocity (wind) field from the knowledge of the pressure field or geopotential field. The disadvantage of the geostrophic balance approximation is that being a diagnostic relationship, the time dependent terms of the atmospheric variables are no longer present with the result that one cannot prognose or forecast the atmospheric fields with time.

15. What are the advantages of using pressure as a vertical coordinate rather than height?

Answer: When pressure is used as a vertical coordinate, the continuity equation assumes its simplest (non-divergent) form. Moreover, the geostrophic relationship assumes a simpler form with density no longer appearing when defined using pressure as a vertical coordinate.

16. What are the advantages of employing the shallow water equations?

Answer: The shallow water equations are a coupled nonlinear set of three partial differential equations in three dependent variables, namely the horizontal components of velocity and the geopotential. The vertical advection terms of the horizontal momentum equations identically vanish and results in the shallow water equations having a simpler form. Furthermore, the inherent nonlinearities of the atmospheric and oceanic motion as manifested in the advection terms, an important characteristic of atmospheric and oceanic motion are still present in the shallow water equations.

17. What is the equation of state used in ocean models?

Answer: The equation of state for oceans in its most general form is a diagnostic equation for the density in terms of temperature, salinity, and pressure. Typically, the density of standard sea water (ρ), i.e., at 1 atmospheric pressure, is given in terms of ρ_o, density of pure water corresponding to zero salinity as follows: $\rho = \rho_o + AS + BS^{1.5} + CS^2$, where S is the salinity of sea water in ppt (parts per thousand by volume) and the coefficients A, B and C are prescribed functions of temperature.

18. How are the initial conditions that provide the initial state of the atmospheric system obtained?

Answer: The atmospheric observations that are obtained to provide the initial conditions are collected from a variety of sources such as surface weather stations, upper air weather stations, ship data, buoy data, aircraft data, and satellite data. Whereas typically the number of atmospheric observations at any given point of time is of the order of 10^4, the number of degrees of freedom (number of dependent variables multiplied by the number of grid cells through which the atmosphere is discretized) of the atmospheric model is of the order of 10^7. Hence, atmospheric observations cannot by itself provide for the initial state of the atmosphere. Data assimilation techniques are employed to obtain the initial state of the atmospheric system by combining optimally atmospheric observations and short range forecasts to achieve the best possible atmospheric state, which is used as the initial conditions for the atmospheric model.

3

Finite Difference Method

3.1 Introduction

All models that are utilized in atmospheric science are formulated based on fundamental physical principles, such as conservation of mass, conservation of momentum, and conservation of energy. These basic physical principles embodied in the atmospheric models are expressed mathematically as a coupled system of nonlinear partial differential equations. As these mathematical equations are complex and cannot be solved analytically, one has to resort to numerical methods to solve them. Numerical methods approximate a continuous model (system of partial differential equations) to a discrete model (system of difference equations). The process of approximating a continuous model to a discrete model by employing a numerical method is called "numerical discretization."

3.2 Method of Finite Difference

Consider a function f that depends on a single dependent variable x, say $f = f(x)$. Assume that x spans an interval L. Let the interval be partitioned by $N + 1$ equally spaced grid points (including the two end points at the limits of the interval). The grid length is then defined as $\Delta x = L/N$ and the grid points are located at $x_j = j\Delta x$, where $j = 0, 1, 2, \ldots, N$ are integers. Let the value of f at x_j be represented by f_j. Mathematically, the derivative of a function $f(x)$ is defined as

$$\frac{df}{dx} = \lim_{\Delta x \to 0} \frac{f(x + \Delta x) - f(x)}{\Delta x}, \tag{3.1}$$

which is shown graphically in Figure 3.1. As in the finite difference method, the value of Δx is finite and does not go to zero, the derivative and the finite difference (the

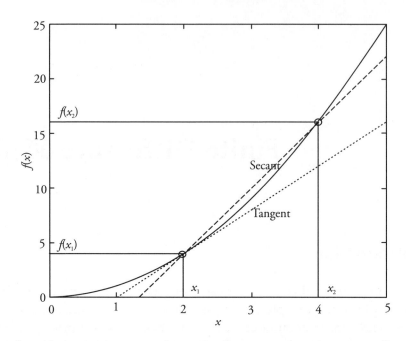

Figure 3.1   As grid size Δx decreases, the secant line approaches the tangent line, the latter being the derivative of f at the point x as $\Delta x \to 0$.

RHS of Equation 3.1) are not exactly identical. Hence, the latter is also called "finite difference approximation" and an error is associated with replacing the derivative of a function with its "finite difference approximation." The value of the derivative of a function $f(x)$ at a grid point x_j can be obtained in three different ways (refer Figure 3.2):

–Forward difference approximation: $\left(\dfrac{df}{dx}\right)_j \approx \dfrac{f_{j+1} - f_j}{\Delta x}$ $\qquad(3.2)$

–Backward difference approximation: $\left(\dfrac{df}{dx}\right)_j \approx \dfrac{f_j - f_{j-1}}{\Delta x}$ $\qquad(3.3)$

–Central difference approximation: $\left(\dfrac{df}{dx}\right)_j \approx \dfrac{f_{j+1} - f_{j-1}}{2\Delta x}$ $\qquad(3.4)$

Each one of the aforementioned finite difference approximations can be obtained from the Taylor series expansion of $f(x + \Delta x)$ about the value of x, which is given by

$$f(x+\Delta x) = f(x) + \Delta x f'(x) + \frac{(\Delta x)^2}{2!} f''(x) + \cdots + \frac{(\Delta x)^n}{n!} f^n(x) + \cdots \qquad(3.5)$$

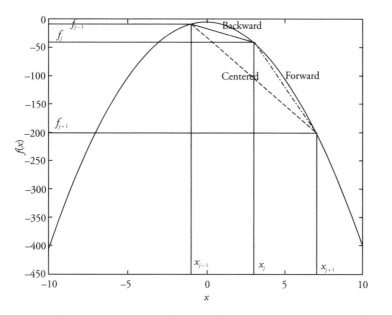

Figure 3.2 Backward, forward, and central finite difference approximations of the derivative of a function $f(x)$ at the point $j\Delta x$. The function f is defined at the grid points $x_j = j\Delta x$ so that $f_j = f(x_j) = f(j\Delta x)$, where Δx is the grid size and $j = 0, 1, 2, \ldots$, are integers.

3.2.1 Forward difference scheme

Using the Taylor series (Equation (3.5)) with $f(x + \Delta x)$ as f_{j+1} and $f(x)$ as f_j, the following equation is obtained:

$$f_{j+1} = f_j + \Delta x \left(\frac{df}{dx}\right)_j + \frac{1}{2}(\Delta x)^2 \left(\frac{d^2 f}{dx^2}\right)_j + \frac{1}{6}(\Delta x)^3 \left(\frac{d^3 f}{dx^3}\right)_j + \cdots \tag{3.6}$$

The forward difference approximation is then given as

$$\frac{f_{j+1} - f_j}{\Delta x} = \left(\frac{df}{dx}\right)_j + \frac{1}{2}\Delta x \left(\frac{d^2 f}{dx^2}\right)_j + \frac{1}{6}(\Delta x)^2 \left(\frac{d^3 f}{dx^3}\right)_j + \cdots \tag{3.7}$$

The difference between Equation (3.7) and the approximated derivative $(df/dx)_j$ is called the truncation error for the forward difference approximation and is given by

$$\varepsilon = \frac{1}{2}\Delta x \left(\frac{d^2 f}{dx^2}\right)_j + \frac{1}{6}(\Delta x)^2 \left(\frac{d^3 f}{dx^3}\right)_j + \cdots \tag{3.8}$$

The truncation error associated with the forward difference approximation of a derivative of a function has an order of accuracy of first-order, which is indicated as $\varepsilon = O(\Delta x)$.

3.2.2 Backward difference scheme

The Taylor series expansion of $f(x - \Delta x)$ about the value of x can be given by

$$f(x - \Delta x) = f(x) - \Delta x f'(x) + \frac{(\Delta x)^2}{2!} f''(x) - \cdots + \frac{(\Delta x)^n}{n!} f^n(x) + \cdots \tag{3.9}$$

The backward difference approximation is then given as

$$\frac{f_j - f_{j-1}}{\Delta x} = \left(\frac{df}{dx}\right)_j - \frac{1}{2} \Delta x \left(\frac{d^2 f}{dx^2}\right)_j + \frac{1}{6}(\Delta x)^2 \left(\frac{d^3 f}{dx^3}\right)_j + \cdots \tag{3.10}$$

The difference between Equation (3.10) and the approximated derivative $(df/dx)_j$ is called the truncation error for the backward difference approximation and is given by

$$\varepsilon = -\frac{1}{2} \Delta x \left(\frac{d^2 f}{dx^2}\right)_j + \frac{1}{6}(\Delta x)^2 \left(\frac{d^3 f}{dx^3}\right)_j + \cdots \tag{3.11}$$

As in the case of forward difference approximation, the truncation error associated with the backward difference approximation of a derivative of a function has an order of accuracy of first-order, which is indicated as $\varepsilon = O(\Delta x)$.

3.2.3 Central difference scheme

Subtracting Equation (3.9) from Equation (3.6) eliminates the first-order term and one gets

$$\frac{f_{j+1} - f_{j-1}}{2\Delta x} = \left(\frac{df}{dx}\right)_j + \frac{1}{6}(\Delta x)^2 \left(\frac{d^3 f}{dx^3}\right)_j + \frac{1}{120}(\Delta x)^4 \left(\frac{d^4 f}{dx^4}\right)_j + \cdots \tag{3.12}$$

The difference between Equation (3.12) and the approximated derivative $(df/dx)_j$ is called the truncation error for the central difference approximation and is given by

$$\varepsilon = \frac{1}{6}(\Delta x)^2 \left(\frac{d^3 f}{dx^3}\right)_j + \frac{1}{120}(\Delta x)^4 \left(\frac{d^4 f}{dx^4}\right)_j + \cdots = O\left[(\Delta x)^2\right]. \tag{3.13}$$

The truncation error associated with the central difference approximation of a derivative of a function has an order of accuracy of second-order, as indicated in Equation (3.13).

3.2.4   Centered fourth-order difference scheme

A finite difference scheme of the fourth-order accuracy for the derivative of a function can be obtained if we expand the Taylor series expansion of $f(x+2\Delta x)$ about x:

$$f_{j+2} = f_j + 2\Delta x \left(\frac{df}{dx}\right)_j + \frac{1}{2}(2\Delta x)^2 \left(\frac{d^2 f}{dx^2}\right)_j + \frac{1}{6}(2\Delta x)^3 \left(\frac{d^3 f}{dx^3}\right)_j + \frac{1}{24}(2\Delta x)^4 \left(\frac{d^4 f}{dx^4}\right)_j$$
$$+ \frac{1}{120}(2\Delta x)^5 \left(\frac{d^5 f}{dx^5}\right)_j + \cdots \quad (3.14)$$

and if we expand the Taylor series expansion of $f(x-2\Delta x)$ about x:

$$f_{j-2} = f_j - 2\Delta x \left(\frac{df}{dx}\right)_j + \frac{1}{2}(2\Delta x)^2 \left(\frac{d^2 f}{dx^2}\right)_j - \frac{1}{6}(2\Delta x)^3 \left(\frac{d^3 f}{dx^3}\right)_j + \frac{1}{24}(2\Delta x)^4 \left(\frac{d^4 f}{dx^4}\right)_j$$
$$- \frac{1}{120}(2\Delta x)^5 \left(\frac{d^5 f}{dx^5}\right)_j + \cdots \quad (3.15)$$

Subtracting Equations (3.15) from Equation (3.14), one gets

$$\frac{f_{j+2} - f_{j-2}}{4\Delta x} = \left(\frac{df}{dx}\right)_j + \frac{4}{6}(\Delta x)^2 \left(\frac{d^3 f}{dx^3}\right)_j + \frac{16}{120}(\Delta x)^4 \left(\frac{d^5 f}{dx^5}\right)_j + \cdots \quad (3.16)$$

Equation (3.16) is second-order accurate as is the central difference scheme, Equation (3.12). Combining Equations (3.12) and (3.16), one gets the following:

$$\frac{4}{3}\frac{f_{j+1} - f_{j-1}}{2\Delta x} - \frac{1}{3}\frac{f_{j+2} - f_{j-2}}{4\Delta x} = \left(\frac{df}{dx}\right)_j - \frac{1}{30}(\Delta x)^4 \left(\frac{d^5 f}{dx^5}\right)_j + \cdots \quad (3.17)$$

The truncation error of the fourth-order central difference approximation is given by

$$\varepsilon = -\frac{1}{30}(\Delta x)^4 \left(\frac{d^5 f}{dx^5}\right)_j + \cdots = O\left[(\Delta x)^4\right]. \quad (3.18)$$

The truncation error associated with the fourth-order central difference approximation of a derivative of a function has an order of accuracy of fourth-order, as indicated in Equation (3.18).

3.2.5   Finite difference scheme for second derivatives and Laplacian

Consider a function f that depends on two dependent variables x and y, say $f = f(x,y)$. Consider a grid in the (x,y) plane as shown in Figure 3.3. Let x and y both span an interval L. Let the interval be partitioned by $M+1$ and $N+1$ equally spaced grid points (including the two end points at the limits of the interval) for x and y, respectively. The grid length in the x and y directions are then defined as $\Delta x = L/M$ and $\Delta y = L/N$; here, the grid points are located in the plane at $x_i = i\Delta x$, where $i = 0, 1, 2, \ldots, M$ and at $y_j = j\Delta y$, where $j = 0, 1, 2, \ldots, N$ and where i, j, M, and N are integers. Let the value of f at (x_i, y_j) be represented by $f_{i,j}$. The grid points in Figure 3.3 are identified by an index i, which increases in the positive x-direction, and an index j, which increases in the positive y-direction. Let (i, j) be the index of a reference grid point in Fig. 3.3, then the grid point immediately to the right of the reference point is designated as $(i + 1, j)$ and the grid point immediately to the left of the reference point is $(i - 1, j)$. Similarly the grid point directly above the reference point is $(i, j + 1)$, and the grid point directly below the reference point is $(i, j - 1)$. The basic idea of the finite difference method is to replace the derivatives of the governing equations with algebraic difference quotients; this results in a system of algebraic equations that can then be solved for the dependant variables at the discrete grid points in the flow field.

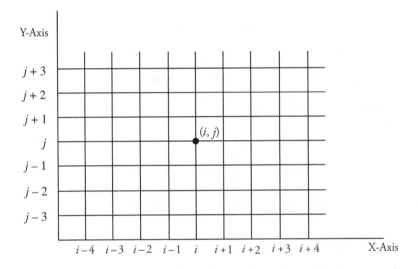

Figure 3.3 Illustration of grid cells in the xy plane using grid sizes Δx and Δy.

The value of the second derivative of a function $f(x, y)$ with respect to say, x, at a grid point (x_i, y_j) using central difference scheme is as follows:

From Figure 3.3 and using the central difference scheme, one gets

$$\left(\frac{\partial f}{\partial x}\right)_{i,j} = \frac{f_{i+1,j} - f_{i-1,j}}{2\Delta x};$$
$$\tag{3.19}$$

$$\left(\frac{\partial^2 f}{\partial x^2}\right)_{i,j} = \frac{\left(\frac{\partial f}{\partial x}\right)_{i+1/2,j} - \left(\frac{\partial f}{\partial x}\right)_{i-1/2,j}}{\Delta x} = \frac{\frac{f_{i+1,j} - f_{i,j}}{\Delta x} - \frac{f_{i,j} - f_{i-1,j}}{\Delta x}}{\Delta x} = \frac{f_{i+1,j} - 2f_{i,j} + f_{i-1,j}}{(\Delta x)^2}.$$

(3.20)

It is possible to obtain Equation (3.20) by simply adding Equations (3.6) and (3.9) to obtain the following relation:

$$\frac{f_{i+1} - 2f_i + f_{i-1}}{(\Delta x)^2} = \left(\frac{d^2 f}{dx^2}\right)_i + \frac{(\Delta x^2)}{12}\left(\frac{d^4 f}{dx^4}\right)_i,$$

(3.21)

with index i replacing j index. The RHS of Equations (3.20) and LHS of Equation (3.21) are identical except for the number of indices and, hence, they represent central difference approximations of the second derivative of a function f with respect to x. The difference between Equation (3.21) and the approximate second derivative $(d^2 f/dx^2)_i$ is called the truncation error of the central difference approximation for the second derivative and is given by

$$\varepsilon = \frac{1}{12}(\Delta x)^2 \left(\frac{d^4 f}{dx^4}\right)_i + \cdots = O\left[(\Delta x)^2\right].$$

(3.22)

Moreover, one can obtain the second-order partial derivative of function f with respect to y as follows:

$$\left(\frac{\partial^2 f}{\partial y^2}\right)_{i,j} = \frac{\left(\frac{\partial f}{\partial y}\right)_{i,j+1/2} - \left(\frac{\partial f}{\partial y}\right)_{i,j-1/2}}{\Delta y} = \frac{\frac{f_{i,j+1} - f_{i,j}}{\Delta y} - \frac{f_{i,j} - f_{i,j-1}}{\Delta y}}{\Delta y} = \frac{f_{i,j+1} - 2f_{i,j} + f_{i,j-1}}{(\Delta y)^2}.$$

(3.23)

Thus, the truncation error associated with the central finite difference approximation of the second derivative of function f with respect to y is second-order accurate.

A second-order mixed derivative of function f with respect to both x and y would be defined as

$$\left(\frac{\partial^2 f}{\partial x \partial y}\right) = \frac{\partial}{\partial x}\left(\frac{\partial f}{\partial y}\right) = \frac{\partial}{\partial y}\left(\frac{\partial f}{\partial x}\right),$$

(3.24)

$$\left(\frac{\partial^2 f}{\partial x \partial y}\right)_{i,j} = \frac{\left(\frac{\partial f}{\partial y}\right)_{i+1,j} - \left(\frac{\partial f}{\partial y}\right)_{i-1,j}}{2\Delta x} = \frac{\frac{f_{i+1,j+1} - f_{i+1,j-1}}{2\Delta y} - \frac{f_{i-1,j+1} - f_{i-1,j-1}}{2\Delta y}}{2\Delta x},$$

(3.25)

which may be simplified to obtain

$$\left(\frac{\partial^2 f}{\partial x \partial y}\right)_{i,j} = \frac{f_{i+1,j+1} - f_{i+1,j-1} - f_{i-1,j+1} + f_{i-1,j-1}}{4\Delta x \Delta y} + O\left[(\Delta x)^2, (\Delta y)^2\right]. \qquad (3.26)$$

3.3   Time Integration Schemes

The time derivative schemes that are used to replace time derivatives in partial differential equations (PDEs) are relatively simple; they, usually are of second-order and sometimes even only first-order accuracy. This fact is in sharp contrast to the general experience with ordinary differential equations, where very accurate methods, such as the Runge–Kutta schemes (which are mostly fourth-order accurate), are extremely successful. There is a basic reason for this behaviour. With an ordinary differential equation of first-order, the equation and a single initial condition is all that is required for an exact solution. Thus, the numerical-solution error is entirely due to the degree of inadequacy of the scheme. With a PDE, the error associated with the numerical solution arises from both the shortcomings of the scheme and the insufficient information about the initial conditions, information regarding which are only known at discrete grid points. Thus, an increase in accuracy of the applied numerical scheme takes care of only one of the aforementioned shortcomings and, hence, the net result of using higher order accurate schemes is not too impressive.

Another important reason for not requiring a scheme of higher order accuracy for approximations of the time derivative terms is that, in order to meet the stability requirement for the applied finite difference scheme, it is usually necessary to choose a time step significantly smaller than that required for adequate accuracy. Once a time step has been specified, other errors, for example those that arise from spatial differencing, are much greater than those that arise owing to time differencing. Hence, it makes better sense to focus efforts to reduce those errors that arise due to spatial differencing, rather than in efforts on increasing the accuracy of the time-differencing schemes. This, of course, does not mean that it is not necessary to carefully consider the properties of various possible time-differencing schemes. It is important to keep in mind that accuracy is only one important consideration when choosing a finite difference scheme.

In order to define some time integration schemes, let us consider a general first-order differential equation

$$\frac{du}{dt} = f(u,t). \qquad (3.27)$$

To discretize Equation (3.27), one divides the time axis into segments of equal length Δt. The approximated value of $u(t)$ at time $t = n\Delta t$ is denoted as u_n. In order to compute u_{n+1}, one needs to determine the value of u at least at u_n and often also at u_{n-1}. A number of time-differencing schemes are available for application to Equation (3.27).

3.3.1   Two-time level schemes

These are schemes that use two different time levels, n and $n + 1$, so that the time integration yields

$$u_{n+1} = u_n + \int_{t_n}^{t_{n+1}} f(u,t)\,dt. \tag{3.28}$$

The problem now is that f only exists as discrete values f_n and f_{n+1} at times $n\Delta t$ and $(n + 1)\Delta t$, respectively.

Forward Euler scheme

If the integral on the RHS of Equation (3.28) is estimated using the value of the integrand at the initial point, we have:

$$u_{n+1} = u_n + \Delta t f_n, \tag{3.29}$$

which is known as the *forward Euler scheme*.

The truncation error for this scheme is $O(\Delta t)$, i.e., the scheme is accurate to first-order. It is said to be uncentered, as the "time derivative" pertains to time level $n + 1/2$ and the function to time level n. Schemes in which the unknown u_{n+1} is explicitly written in terms of known u_n are called explicit schemes. The forward Euler scheme is an example of an explicit finite difference scheme.

Backward Euler scheme

If the integral on the RHS of Equation (3.28) is estimated using the value of the integrand at the final point, we have:

$$u_{n+1} = u_n + \Delta t f_{n+1}, \tag{3.30}$$

which is known as the *backward Euler scheme*.

The backward scheme is uncentered in time and accurate to $O(\Delta t)$. If, as here, a value of f is taken at time level $n + 1$ and f depends on u, i.e., u_{n+1}, the scheme is said to be implicit as the unknown u_{n+1} cannot be explicitly written in terms of known u_n. Implicit finite difference schemes that involve a PDE would require matrix inversion methods for obtaining solutions. For an ordinary differential equation

(ODE), employing the backward scheme, may not be difficult, as there is need to solve only a difference equation for u_{n+1}; however, for a PDE, we would need to solve a set of simultaneous equations, one for each grid point of the computational region. If the value of f_n does not depend on u_{n+1}, then the scheme is still an explicit scheme.

Crank–Nicolson scheme

The Crank–Nicolson scheme can be obtained by taking the average of forward and backward Euler schemes to obtain

$$u_{n+1} = u_n + \frac{\Delta t}{2} \left(f_n + f_{n+1} \right). \tag{3.31}$$

In this scheme, f is assumed to vary linearly between its values at $n\Delta t$ and $(n+1)\Delta t$ so that its mean value is used for the integral. It is clear that this scheme is an implicit scheme, as it requires information from the future time level $n+1$. Note here that the finite-difference approximation is centered at $n+1/2$ between the two time steps n and $n+1$ and, hence, is second-order accurate in time. This scheme is also known as a trapezoidal scheme.

Matsuno forward–backward scheme

To increase the accuracy as obtained from both the Euler forward and backward schemes, respectively an iterative time integration scheme is introduced called the Matsuno scheme, which is initiated by an Euler forward time step:

$$u_{n+1}^* = u_n + \Delta t f_n. \tag{3.32}$$

In this scheme, the value of u_{n+1}^* obtained at the end of Euler forward time step serves as an approximation of f_{n+1}, which hereafter is used to make a backward step to yield a final u_{n+1}:

$$u_{n+1} = u_n + \Delta t f_{n+1}^*, \tag{3.33}$$

where $f_{n+1}^* = f(u_{n+1}^*, t_{n+1})$. The Matsuno scheme is explicit and is of accuracy $O(\Delta t)$.

Heun's scheme

This is similar to the Matsuno scheme and is explicit; however it is of second-order accuracy, as the second step is made using the Crank–Nicolson scheme instead of the Euler backward scheme. The two steps in Heun's scheme are given by

$$u_{n+1}^* = u_n + \Delta t f_n \tag{3.34}$$

$$u_{n+1} = u_n + \frac{\Delta t}{2}\left(f_n + f_{n+1}^*\right) \tag{3.35}$$

3.3.2 Three-time level schemes

These schemes use the time at three levels and the time integration becomes

$$u_{n+1} = u_{n-1} + \int_{t_{n-1}}^{t_{n+1}} f(u,t)\,dt \tag{3.36}$$

Leapfrog scheme

The simplest three-level scheme is to assign f a constant value equal to that at the middle of the time interval of length $2\Delta t$, which yields the leapfrog scheme .

$$u_{n+1} = u_{n-1} + 2\Delta t f_n \tag{3.37}$$

The leapfrog scheme has an accuracy of $O(\Delta t)^2$. It is the most widely used scheme in atmospheric and oceanic models.

Adams–Bashforth scheme

The Adams–Bashforth scheme in the atmospheric sciences is, in fact, a simplified version of the original Adams–Bashforth scheme, which is fourth-order accurate. The simplified version is obtained when f in Equation (3.36) is approximated by a value obtained at the center of the interval Δt by a linear extrapolation using values f_{n-1} and f_n.

$$u_{n+1} = u_n + \frac{\Delta t}{2}\left[3f_n - f_{n-1}\right]. \tag{3.38}$$

The Adams–Bashforth scheme is an explicit, second-order accurate scheme.

Milne–Simpson scheme

In the Milne–Simpson scheme, Simpson's rule is utilized to calculate the integral in Equation (3.36) follows:

$$u_{n+1} = u_{n-1} + \frac{\Delta t}{3}\left[f_{n-1} + 4f_n + f_{n+1}\right]. \tag{3.39}$$

The Milne–Simpson scheme is an implicit scheme.

Exercises 3a (Questions and answers)

1. Find the expression for the first-order derivative that represents diffusive flux having variable coefficients.

   Answer: Let $f(x) = d(x)(\partial u/\partial x)$. Represent the first derivative of f with respect to x, using central derivatives and assuming that i is the index of x.

   $$\left(\frac{\partial f}{\partial x}\right)_i \approx \frac{f_{i+1/2} - f_{i-1/2}}{\Delta x} = \frac{1}{\Delta x}\left[d_{i+1/2}\frac{u_{i+1} - u_i}{\Delta x} - d_{i-1/2}\frac{u_i - u_{i-1}}{\Delta x}\right].$$

   $$\left(\frac{\partial f}{\partial x}\right)_i = \frac{d_{i+1/2}u_{i+1} - (d_{i+1/2} + d_{i-1/2})u_i + d_{i-1/2}u_{i-1}}{(\Delta x)^2}.$$

2. Obtain an expression for the second (mixed) derivative of u with respect to x and y using central difference with i and j being the indices of x and y.

   Answer:

   $$\left(\frac{\partial^2 u}{\partial x \partial y}\right)_{i,j} = \left[\frac{\partial}{\partial x}\left(\frac{\partial u}{\partial y}\right)\right]_{i,j} = \left[\frac{\partial}{\partial y}\left(\frac{\partial u}{\partial x}\right)\right]_{i,j} = \frac{\left(\frac{\partial u}{\partial y}\right)_{i+1,j} - \left(\frac{\partial u}{\partial y}\right)_{i-1,j}}{2\Delta x}$$

   $$= \frac{\left(\frac{u_{i+1,j+1} - u_{i+1,j-1}}{2\Delta y}\right) - \left(\frac{u_{i-1,j+1} - u_{i-1,j-1}}{2\Delta y}\right)}{2\Delta x}$$

   $$= \frac{u_{i+1,j+1} - u_{i+1,j-1} - u_{i-1,j+1} + u_{i-1,j-1}}{4\Delta x \Delta y}.$$

3. Derive the expression for the *Richardson extrapolation* for the first derivative of f with respect to x.

   Answer: Let us assume

   $$f'(x) = \frac{f(x_i + h) - f(x_i - h)}{2h} - \frac{f'''(x_i)}{3!}h^2 + O(h^4)$$

   and

   $$f'(x) = \frac{f(x_i + h/2) - f(x_i - h/2)}{h} - \frac{f'''(x_i)}{4 \times 3!}h^2 + O(h^4).$$

   Multiplying the second expression by 4 and subtracting it from the first expression, one obtains the expression for the Richardson extrapolation for the first derivative of f with respect to x.

   $$f'(x) = \frac{1}{3}\left[4\frac{f(x_i + h/2) - f(x_i - h/2)}{h} - \frac{f(x_i + h) - f(x_i - h)}{2h}\right] + O(h^4).$$

Exercises 3b (Questions only)

1. Obtain the third-order approximation of the first-order derivative of u with respect to x, where i is the index of x.

   Answer:

   $$\left(\frac{\partial u}{\partial x}\right)_i = \frac{2u_{i+1} + 3u_i - 6u_{i-1} + u_{i-2}}{6\Delta x} \quad \text{(backward-biased approximation).}$$

   $$\left(\frac{\partial u}{\partial x}\right)_i = \frac{-u_{i+2} + 6u_{i+1} - 3u_i - 2u_{i-1}}{6\Delta x} \quad \text{(forward-biased approximation).}$$

2. Obtain the fourth-order approximation of the first-order derivative of u with respect to x, where i is the index of x.

   Answer: $\left(\dfrac{\partial u}{\partial x}\right)_i = \dfrac{-u_{i+2} + 8u_{i+1} - 8u_{i-1} + u_{i-2}}{12\Delta x}$.

3. Obtain the fourth-order approximation of the second-order derivative of u with respect to x, where i is the index of x.

   Answer: $\left(\dfrac{\partial^2 u}{\partial x^2}\right)_i = \dfrac{-u_{i+2} + 16u_{i+1} - 30u_i + 16u_{i-1} - u_{i-2}}{12(\Delta x)^2}$.

4. Let $f(x) = A\sin 2\pi x/L$. By utilizing the second-order central finite difference expression for the first derivative of f with respect to x, obtain the finite difference expression for the derivative of f as

   $$\frac{df}{dx} = A\frac{2\pi}{L}\cos\left(\frac{2\pi x}{L}\right)\frac{\sin\left[(2\pi/L)\Delta x\right]}{(2\pi/L)\Delta x}.$$

5. Show that the expression of the first derivative of f with respect to x in Exercise 3b Q 4 approaches the analytical expression of the derivative as $(2\pi/L)\Delta x$ approaches zero.

6. Show that for the case $\Delta x = L/2$, the finite difference expression for the first derivative of f with respect to x in Exercise 3b Q 4 would always be zero regardless of the value of x.

7. Obtain the expression for the third-order Adams–Bashforth method.

   Answer: $u_{n+1} = u_n + \dfrac{\Delta t}{12}(23f_n - 16f_{n-1} + 5f_{n-2})$

8. Obtain the expression for the third-order Adams–Moulton method, which is an implicit scheme and an extension of the backward Euler and the Crank–Nicolson schemes.

   Answer: $u_{n+1} = u_n + \dfrac{\Delta t}{12}(5f_{n+1} + 8f_n - f_{n-1})$

9. Obtain the expression for the second-order Runge–Kutta method, which is of the predictor–corrector type.

Answer:

Predictor step (forward Euler): $u^*_{n+1/2} = u_n + \frac{\Delta t}{2} f_n$

Corrector step (mid-point rule): $u_{n+1} = u_n + \Delta t f(t_{n+1/2}, u^*_{n+1/2})$

10. Obtain the expression for the third-order Runge–Kutta method.

Answer: $u_{n+1} = u_n + \dfrac{\Delta t}{6}(k_1 + 4k_2 + k_3)$, where

$k_1 = f(t_n, u_n)$, $k_2 = f(t_n + \frac{1}{2}\Delta t, u_n + \frac{1}{2}\Delta t k_1)$, and $k_3 = f(t_n + \Delta t, u_n - \Delta t k_1 + 2\Delta t k_2)$

11. Obtain the expression for the fourth-order Runge–Kutta method.

Answer:

$u_{n+1} = u_n + (\Delta t/6)(k_1 + 2k_2 + 2k_3 + k_4)$

$k_1 = f(t_n, u_n)$

$k_2 = f(t_n + \Delta t/2, u_n + k_1 \Delta t/2)$

$k_3 = f(t_n + \Delta t/2, u_n + k_2 \Delta t/2)$

$k_4 = f(t_n + \Delta t, u_n + \Delta t k_3)$

Python examples

1. The following provides the python code for the forward difference operator, backward difference operator, central difference operator, and fourth-order central difference operator to discretize the first-order derivative.

Let $f(x) = 2x^3 + 4x^2 - 5x$.

Exact derivative, $\frac{df}{dx} = 6x^2 + 8x - 5$.

Hence, the derivative at $x = 1$,

$6 + 8 - 5 = 9$

```
# Begin python code
import numpy as np
def forDiff(f,x,h): # Forward difference
    return (f(x+h) - f(x))/h


def  backDiff(f,x,h): # Backward difference
    return (f(x) - f(x-h))/h


def centDiff(f,x,h): # Central difference
```

```python
    return (f(x+h) - f(x-h))/(2*h)


def fourthOrderCeDif(f,x,h): # Fourth order Central difference
    return (4/(6*h))*(f(x+h)-f(x-h))- (1/(12*h))*(f(x+(2*h))
        -f(x-(2*h)))


def myFunc(x):
    return 2*x**3 + 4*x**2 - 5*x


s=np.zeros(2)
s[0]=0.1 #Time step (i)
s[1]=0.01 # Time step (ii)
x = 1
exact = 6*x**2 + 8*x - 5
print("Exact value =", exact)
for i in range(0,2):
    print("Time step h =", s[i])
    h=s[i]
    dydt_f = forDiff(myFunc,x,h)
    dydt_b = backDiff(myFunc,x,h)
    dydt_c = centDiff(myFunc,x,h)
    dydt_Fc = fourthOrderCeDif(myFunc,x,h)

    error_f = abs((dydt_f - exact)/exact)*100
    error_b = abs((dydt_b - exact)/exact)*100
    error_c = abs((dydt_c - exact)/exact)*100
    error_Fc = abs((dydt_Fc - exact)/exact)*100
    print("Forward difference =", dydt_f)
    print("Backward difference =", dydt_b)
    print("Central difference =", dydt_c)
    print("Fourth-order Central difference = " , dydt_Fc)
    print("Percentage error in Forward difference =", error_f)
    print("Percentage error in Backward difference =", error_b)
    print("Percentage error in Central difference =", error_c)
    print("Percentage error in Fourth-order Central difference =",
        error_Fc)
#End python code
```

Output

Exact value = 9

Time step h = 0.1

Forward difference = 10.020000000000016

Backward difference = 8.019999999999996

Central difference = 9.020000000000005

Fourth-order Central difference = 9.000000000000007

Percentage error in Forward difference = 11.333333333333506

Percentage error in Backward difference = 10.888888888888932

Percentage error in Central difference = 0.2222222222222767

Percentage error in Fourth-order Central difference = 7.894919286223335e-14

Time step h = 0.01

Forward difference = 9.100200000000047

Backward difference = 8.90019999999998

Central difference = 9.000200000000014

Fourth-order Central difference = 9.000000000000009

Percentage error in Forward difference = 1.113333333333857

Percentage error in Backward difference = 1.108888888889107

Percentage error in Central difference = 0.0022222222223749416

Percentage error in Fourth-order Central difference = 9.86864910777917e-14

4
Consistency and Stability Analysis

In this chapter, we will introduce the concept of consistency, stability, and convergence of finite difference schemes. Towards this end, various finite difference schemes are introduced and applied to well-known partial differential equations (PDEs) of the parabolic, hyperbolic and elliptic type.

4.1 Consistency and Stability Analysis

Analytical or closed form solutions of PDEs provide closed-form solutions/ expressions that depict the variation of the dependant variables in the domain. The method of finite differences provide the numerical solutions with the values of the dependent variables at discrete grid points in the domain. Consider Figure 3.3, which shows a domain of interest in the xy plane in which the spacing of the grid points in the x-direction is uniform, and is given by Δx. Similarly, the spacing of the grid points in the y-direction is also uniform, and is given by Δy. In reality, it is not mandatory that Δx or Δy be constant (uniform). However, generally, PDEs are numerically solved on a grid that has uniform spacing in each (x and y) direction, as this simplifies the programming, and often yields higher accuracy solutions. In the present chapter, we assume uniform spacing in each coordinate direction. However, in general $\Delta x \neq \Delta y$.

4.2 Basic Aspects of Finite Differences

Consider the following one-dimensional unsteady state linear heat conduction equation. Here, u (temperature) is a function of x and t (time), whereas α is a constant known as coefficient of thermal diffusivity:

$$\frac{\partial u}{\partial t} = \alpha \frac{\partial^2 u}{\partial x^2}. \tag{4.1}$$

Letting the time derivative in Equation (4.1) be replaced by a forward difference approximation, and the spatial derivative by a central difference approximation (this finite difference scheme is called FTCS, forward in time central in space scheme), one gets

$$\frac{u_i^{n+1} - u_i^n}{\Delta t} = \alpha \frac{u_{i+1}^n - 2u_i^n + u_{i-1}^n}{(\Delta x)^2}. \tag{4.2}$$

Equation (4.2) is an explicit finite difference scheme as the unknown u_i^{n+1} is explicitly expressed in terms of known terms u_i^n, u_{i+1}^n, and u_{i-1}^n. Considering the truncation error of the FTCS scheme, one obtains

$$\frac{\partial u}{\partial t} - \alpha \frac{\partial^2 u}{\partial x^2} = \frac{u_i^{n+1} - u_i^n}{\Delta t} - \alpha \frac{u_{i+1}^n - 2u_i^n + u_{i-1}^n}{(\Delta x)^2} + \left[-\frac{\Delta t}{2} \left(\frac{\partial^2 u}{\partial t^2} \right)_i^n + \alpha \frac{(\Delta x)^2}{12} \left(\frac{\partial^4 u}{\partial x^4} \right)_i^n + \cdots \right]. \tag{4.3}$$

The terms in the square brackets in Equation (4.3) are the truncation error terms of the FTCS scheme. Observing Equation (4.3), it can be inferred that as $\Delta x \to 0$ and $\Delta t \to 0$, the truncation error approaches zero. Hence, in the limiting case, the difference equation (Equation (4.2)) approaches the original partial differential equation (Equation (4.1)). Under such situations, the finite difference representation of the partial differential equation is said to be consistent.

4.2.1 Consistency

A finite difference representation of a partial differential equation (PDE) is said to be consistent if it can be shown that the difference between the PDE and its finite difference (FDE) representation vanishes as the mesh is refined, i.e.,

$$\lim_{\text{mesh} \to 0} (\text{PDE} - \text{FDE}) = \lim_{\text{mesh} \to 0} (\text{Truncation Error}) = 0$$

4.2.2 Convergence

A solution of the algebraic equations that approximate a partial differential equation (PDE) is convergent if the approximate difference solution approaches the exact solution of the PDE for each value of the independent variable as the grid spacing tends to zero. The requirement is $u_i^n = \bar{u}(x_i, t^n)$ as $\Delta x, \Delta t \to 0$, where $\bar{u}(x_i, t^n)$ is the exact solution of the system of algebraic equations.

4.2.3   Lax Equivalence Theorem

The Lax equivalence theorem deals with the numerical solution of partial differential equations and was proposed in 1954 by P. D. Lax. The theorem states that a consistent finite difference method for a well-posed linear initial value problem is convergent if and only if it is stable. A given problem is defined as well-posed, if (i) it has a solution, (ii) the solution is unique, and (iii) the solution depends continuously on data and parameters; i.e., "small" changes in initial or boundary functions and in parameter values result in "small" changes in the solution.

4.3   Errors and Stability Analysis

4.3.1   Introduction

There is a formal way of examining the accuracy and stability of linear equations, which is known as the von Neumann method of stability analysis. The method, otherwise also called the Fourier series method is the most frequently used one for establishing the stability of a finite-difference scheme. However, this stability scheme can only be applied to linear PDEs; it cannot be utilized for nonlinear PDEs. The essence of the von Neumann stability method rests on the following premise. It is known that a solution to a linear equation can be expressed in the form of a Fourier series, each harmonic of which is also a solution. The principle behind the von Neumann method is to test for the stability of a single harmonic solution. Then a necessary condition for the stability of the finite difference scheme is that it is stable to all admissible harmonics.

Consider the one-dimensional linear heat conduction equation, a partial differential equation of the parabolic type, such as Equation (4.1). The numerical solution of this PDE is affected by the two sources of error: discretization error and the round-off error.

4.3.2   Discretization error

Discretization error is the difference between the exact analytical solution of the partial differential Equation (4.1) and the exact (round-off free) solution of the corresponding finite-difference equation (for example, Equation (4.2)). This error is simply the truncation error for the finite difference scheme together with any errors introduced by the numerical treatment of the boundary conditions.

4.3.3 Representation of real numbers in a computer: Round-off error

A representation called the "normalized floating point representation" is used to represent real numbers in computers. In this representation, 32 bits are divided into two parts: a part called the "mantissa with its sign" and the other called the "exponent with its sign." The mantissa represents fractions with a nonzero leading bit and the exponent the power of 2 by which the mantissa is multiplied. This method increases the range of numbers that may be represented using 32 bits. In this method, a binary floating point number is represented by

$$(\text{sign}) \times \text{mantissa} \times 2^{\pm\,\text{exponent}}$$

where the sign is one bit, the mantissa is a binary fraction with a nonzero leading bit, and the exponent is a binary integer. To perform numerical computations without too much error, it is found that at least seven significant decimal digits are necessary. The number of bits required to represent seven decimal digits is approximately 23. The remaining 8 bits $(32 - 23 - 1 = 8)$ are allocated for the exponent. The largest floating point number that can be represented as indicated here is $\sim 3.4 \times 10^{38}$, whereas the smallest floating point number that can be represented is $\sim 0.293 \times 10^{-38}$. Representing a real number '0.7' in binary format, yields the following binary number $0.10110011001100110011001100011\ldots$. As the mantissa can accommodate 23 bits only in a 32 bit computer, the floating point representation of 0.7 has to impose either a "chopping" operation or "rounding off" operation to represent 0.7 in a 32 bit computer.

For example, consider a normalized decimal floating-point form: $\pm 0.b_1 b_2 b_3 \ldots b_k \times 10^n$, where $1 \le b_1 \le 9$ and $0 \le b_i \le 9$, for each $i = 2, 3, \ldots, k$.

1. Let a number y be represented by $\pm 0.b_1 b_2 b_3 \ldots b_k b_{k+1} b_{k+2} \cdots \times 10^n$, then in the "chopping" operation, we simply chop off the digits $b_{k+1} b_{k+2} \ldots$ and represent y after chopping as $\pm 0.b_1 b_2 b_3 \ldots b_k \times 10^n$.

2. Let a number y be represented by $\pm 0.b_1 b_2 b_3 \ldots b_k b_{k+1} b_{k+2} \cdots \times 10^n$, then in the "round-off" operation, we simply add $5 \times 10^{n-(k+1)}$ to y and then chop off the digits $b_{k+1} b_{k+2} \ldots$ to represent y after "round-off" operation.

Most computers employ the "round-off" operation rather than the "chopping" operation as the former has smaller error associated with it. In fact it can be shown that the error committed by rounding a normalized number y to n digits (in base 10) is 0.5×10^{-n}, which is much less than the maximum error committed by chopping, the latter being 0.9×10^{-n}. The resultant error introduced by the "rounding off" operation in representing a real number in a 32 bit computer is called "rounding off error." When an arithmetic operation is performed repeatedly, the "rounding off error" may grow in some situations introducing serious errors. This is the numerical error introduced for a repetitive number of calculations in which the computer is constantly rounding the number to some decimal point. Let A be the analytical solution of the partial differential equation, D be the exact solution of the finite difference equation, and N be the numerical solution from a computer with finite accuracy. Then, Discretization error $= A - D =$ Truncation error $+$ error introduced due to treatment of boundary condition, Round-off error $(\varepsilon) = N - D$ or $N = D + \varepsilon$, where ε is the round-off error, which will henceforth be simply called "error." It is obvious that the numerical solution N must satisfy the finite difference equation (Equation (4.2)). Hence, from Equation (4.2).

$$\frac{D_i^{n+1} + \varepsilon_i^{n+1} - D_i^n - \varepsilon_i^n}{\Delta t} = \alpha \frac{D_{i+1}^n + \varepsilon_{i+1}^n - 2D_i^n - 2\varepsilon_i^n + D_{i-1}^n + \varepsilon_{i-1}^n}{(\Delta x)^2}. \tag{4.4}$$

As by definition, D is the exact solution of the finite difference equation, it exactly satisfies

$$\frac{D_i^{n+1} - D_i^n}{\Delta t} = \alpha \frac{D_{i+1}^n - 2D_i^n + D_{i-1}^n}{(\Delta x)^2}. \tag{4.5}$$

Subtracting Equation (4.5) from Equation (4.4), one gets

$$\frac{\varepsilon_i^{n+1} - \varepsilon_i^n}{\Delta t} = \alpha \frac{\varepsilon_{i+1}^n - 2\varepsilon_i^n + \varepsilon_{i-1}^n}{(\Delta x)^2}. \tag{4.6}$$

From Equation (4.6), one notices that the error ε also satisfies the difference equation. If errors ε_i^n are already present at some stage (at time index n of the solution of this equation), then the solution will be stable if the ε_i^n remain "within bounds," as the solution progresses in the marching direction, (i.e., from time t to $t + \Delta t$, or time index n to $n+1$). If the errors ε_i^n assume larger magnitude (grow without bounds) during the progression of the solution from time t to $t + \Delta t$, or time index n to $n+1$, then the solution is unstable. Hence, for a solution to be stable, the required condition to be satisfied is

$$\left| \frac{\varepsilon_i^{n+1}}{\varepsilon_i^n} \right| \leq 1. \tag{4.7}$$

4.3.4 Stability analysis of FTCS scheme as applied to one-dimensional heat conduction equation

In this subsection, the stability of the FTCS scheme is examined using the Von Newmann stability scheme for the one-dimensional linear heat conduction equation (also called as one-dimensional diffusion equation), i.e., Equation (4.2). The Von Neumann stability analysis is the analysis of stability using Fourier methods. Let the distribution of errors along the x-axis be given by a Fourier series in x, and then let the time-dependence of errors be exponential in t, i.e.,

$$\varepsilon(x,t) = e^{at} \sum_m e^{\iota k_m x}, \tag{4.8}$$

where ι is the unit complex number and k_m is the wave number of mth mode. As Equation (4.2) is a linear equation and it is known that linear equations satisfy the principle of superposition, it is adequate to examine the behaviour of one term of the aforementioned series rather than the entire series. One term of the series in Equation (4.8) will appear as

$$\varepsilon(x,t) = e^{at} e^{\iota k_m x}. \tag{4.9}$$

Substituting Equation (4.9) to the Equation (4.6), one gets

$$\frac{e^{a(t+\Delta t)} e^{\iota k_m x} - e^{at} e^{\iota k_m x}}{\Delta t} = \alpha \frac{e^{at} e^{\iota k_m (x+\Delta x)} - 2 e^{at} e^{\iota k_m x} + e^{at} e^{\iota k_m (x-\Delta x)}}{(\Delta x)^2}. \tag{4.10}$$

Dividing Equation (4.10) throughout by $e^{at} e^{\iota k_m x}$ one gets,

$$\frac{e^{a\Delta t} - 1}{\Delta t} = \alpha \frac{e^{\iota k_m \Delta x} - 2 + e^{-\iota k_m \Delta x}}{(\Delta x)^2}. \tag{4.11}$$

$$e^{a\Delta t} = 1 + \frac{\alpha \Delta t}{(\Delta x)^2} \left(e^{\iota k_m \Delta x} + e^{-\iota k_m \Delta x} - 2 \right) = 1 + \frac{\alpha (2\Delta t)}{(\Delta x)^2} \left[\cos(k_m \Delta x) - 1 \right] = 1 - 4 \frac{\alpha \Delta t}{(\Delta x)^2} \left[\sin^2(k_m \Delta x/2) \right] \tag{4.12}$$

The LHS of Equation (4.7) is called the amplification factor and is given by $e^{a\Delta t}$. From Equation (4.12) and Equation (4.7), the condition to be satisfied by the FTCS scheme to be stable is

$$\left| \frac{\varepsilon_i^{n+1}}{\varepsilon_i^n} \right| = \left| e^{a\Delta t} \right| = \left| 1 - 4 \frac{\alpha \Delta t}{(\Delta x)^2} \left[\sin^2(k_m \Delta x/2) \right] \right| \le 1. \tag{4.13}$$

For Equation (4.13) to hold, the following two conditions must be satisfied simultaneously

$$1 - 4 \frac{\alpha \Delta t}{(\Delta x)^2} \left[\sin^2(k_m \Delta x / 2) \right] \leq 1. \tag{4.14}$$

$$1 - 4 \frac{\alpha \Delta t}{(\Delta x)^2} \left[\sin^2(k_m \Delta x / 2) \right] \geq -1. \tag{4.15}$$

Equation (4.14) requires

$$4 \frac{\alpha \Delta t}{(\Delta x)^2} \left[\sin^2(k_m \Delta x / 2) \right] \geq 0.$$

As $\alpha \Delta t / (\Delta x)^2$ is always positive, Equation (4.14) is satisfied. Equation (4.15) is rewritten as

$$4 \frac{\alpha \Delta t}{(\Delta x)^2} \left[\sin^2(k_m \Delta x / 2) \right] \leq 2. \tag{4.16}$$

Equation (4.16) is identically satisfied if the following condition is satisfied

$$\frac{\alpha \Delta t}{(\Delta x)^2} \leq \frac{1}{2}. \tag{4.17}$$

A choice of time step Δt given the values of α and value of Δx that satisfies Equation (4.17) would ensure that the FTCS scheme for the one-dimensional linear heat conduction is "stable." Hence, FTCS scheme is a "conditionally stable finite difference scheme." For a chosen Δt that satisfies Equation (4.17), the error will not grow in subsequent time marching steps in time, and the numerical solution will proceed in a stable manner. However, if the chosen Δt is greater than what is allowed by Equation (4.17), the error will progressively become larger with time and grow without bounds. This condition requires that for a halving of the spatial resolution, Δx, requires a simultaneous reduction in the time-step Δt by a factor of four in order to maintain numerical stability. It is clear that this condition introduces serious constraints as one has to have absurdly small time-steps while dealing with high resolution calculations.

4.3.5   Richardson central in time and central in space (CTCS) finite difference scheme and its stability

As the central difference approximation of a derivative is of the second-order accuracy, it would be logical to replace the time as well as space derivative of Equation (4.1) using central differences; this scheme is known as the "Richardson finite difference scheme":

$$\frac{u_i^{n+1} - u_i^{n-1}}{2\Delta t} = \alpha \frac{u_{i+1}^n - 2u_i^n + u_{i-1}^n}{(\Delta x)^2}. \tag{4.18}$$

Equation (4.18) is rewritten as

$$u_i^{n+1} = u_i^{n-1} + \frac{2\alpha\Delta t}{(\Delta x)^2}\left(u_{i+1}^n - 2u_i^n + u_{i-1}^n\right). \tag{4.19}$$

Equation (4.19) is an explicit finite difference scheme as the unknown u_i^{n+1} is explicitly expressed in terms of known terms $u_i^{n-1}, u_{i+1}^n, u_i^n,$ and u_{i-1}^n. Applying the von Neumann stability analysis to Equation (4.19), one gets the following quadratic equation for the amplification factor G

$$G^2 + 2\beta G - 1 = 0, \quad \text{where} \quad \beta = 4\frac{\alpha\Delta t}{(\Delta x)^2}\left[\sin^2(k\Delta x/2)\right]. \tag{4.20}$$

Solving the quadratic equation, Equation (4.20)) for G, one gets

$$G_{1,2} = -\beta \pm \sqrt{\beta^2 + 1}. \tag{4.21}$$

As $|G_2| > 1$ for all k, the Richardson scheme is "unconditionally unstable" i.e., it is always unstable.

4.3.6 DuFort–Frankel finite difference scheme and its stability

DuFort–Frankel suggested the following modification to the Richardson scheme to overcome the stability problem of the Richardson scheme. Equation (4.18) is modified by replacing u_i^n as the time average of u_i^{n+1} and u_i^{n-1}.

$$\frac{u_i^{n+1} - u_i^{n-1}}{2\Delta t} = \alpha \frac{u_{i+1}^n - 2\frac{u_i^{n-1}+u_i^{n+1}}{2} + u_{i-1}^n}{(\Delta x)^2}. \tag{4.22}$$

Equation (4.22) is an explicit finite difference equation as the unknown u_i^{n+1} can be grouped and expressed explicitly in terms of known terms $u_i^{n-1}, u_{i-1}^n,$ and u_{i+1}^n. The DuFort–Frankel scheme is rewritten as

$$u_i^{n+1} = \frac{1-2\eta}{1+2\eta}u_i^{n-1} + \frac{2\eta}{1+2\eta}\left(u_{i+1}^n + u_{i-1}^n\right), \tag{4.23}$$

where $\eta = \alpha\Delta t/(\Delta x)^2$. When the von Neumann stability analysis is applied to the DuFort–Frankel scheme (Equation (4.23)), the amplification factor $G(k)$ can be found from the quadratic equation

$$(1+2\eta)G^2 - 4G\eta\cos(k\Delta x) + (2\eta - 1) = 0. \tag{4.24}$$

The two roots of the amplification factor $(G_{1,2})$ is written as

$$G_{1,2} = \frac{2\eta \cos(k\Delta x) \pm \sqrt{1 - 4\eta^2 \sin^2(k\Delta x)}}{(1+2\eta)}. \tag{4.25}$$

For real roots, $\{1 - 4\eta^2 \sin^2(k\Delta x)\} \geq 0$, the following inequality is satisfied

$$G_{1,2} = \frac{2\eta \cos(k\Delta x) + \sqrt{1 - 4\eta^2 \sin^2(k\Delta x)}}{(1+2\eta)} \leq \frac{2\eta \cos(k\Delta x) + 1}{(1+2\eta)} \leq \frac{2\eta + 1}{1+2\eta} \leq 1. \tag{4.26}$$

For imaginary roots, $\{1 - 4\eta^2 \sin^2(k\Delta x)\} \leq 0$, the following expression for amplification factor is derived

$$|G_{1,2}|^2 = \frac{4\eta^2 \cos^2(k\Delta x) + (4\eta^2 \sin^2(k\Delta x) - 1)}{(1+2\eta)^2} = \frac{4\eta^2 - 1}{(1+2\eta)^2} = \frac{(2\eta - 1)}{(1+2\eta)} \leq 1. \tag{4.27}$$

Hence, for both real and imaginary roots, the amplification factor has a magnitude less than or equal to one, for all values of η indicating that the DuFort–Frankel scheme is an unconditionally stable scheme.

However, the DuFort–Frankel scheme (4.23) is not consistent as the leading terms of truncated series form the truncation error for the complete equation are

$$\frac{\alpha}{12}\left(\frac{\partial^4 u}{\partial x^4}\right)_i^n (\Delta x)^2 - \alpha\left(\frac{\partial^2 u}{\partial t^2}\right)_i^n \left(\frac{\Delta t}{\Delta x}\right)^2 - \frac{1}{6}\left(\frac{\partial^3 u}{\partial t^3}\right)_i^n (\Delta t)^2 + \cdots \tag{4.28}$$

This expression for truncation for error is meaningful only if $\Delta t / \Delta x \to 0$ together with $\Delta t \to 0$ and $\Delta x \to 0$. However, Δt and Δx may individually approach zero in such a way that $(\Delta t / \Delta x) = \beta$. Then in the limit of mesh size (Δx) and time step (Δt) going to zero, the truncation error would equal to $-\alpha\beta^2(\partial^2 u/\partial t^2)$, which is not zero, making the DuFort–Frankel scheme not a consistent scheme. However, the scheme is consistent if $\Delta t \ll \Delta x$.

4.3.7 Backward in time and central in space (BTCS) scheme and its stability

The simplest example of an implicit finite difference scheme is to replace the time derivatives of Equation (4.1) by the backward difference scheme and the space derivative of Equation (4.1) by the central difference scheme. The BTCS scheme for the one-dimensional linear heat conduction equation is as follows:

$$\frac{u_i^{n+1} - u_i^n}{\Delta t} = \alpha \frac{u_{i+1}^{n+1} - 2u_i^{n+1} + u_{i-1}^{n+1}}{(\Delta x)^2}. \tag{4.29}$$

It is clear that the BTCS scheme (Equation (4.29)) is an implicit scheme as the unknown terms (u_i^{n+1}, u_{i-1}^{n+1}, and u_{i+1}^{n+1}) are not explicitly written in terms of known values of u at the nth time level. Rearranging Equation (4.29) and defining $\eta = \alpha \Delta t / (\Delta x)^2$ one gets

$$\eta u_{i+1}^{n+1} - (1 + 2\eta) u_i^{n+1} + \eta u_{i-1}^{n+1} = -u_i^n \tag{4.30}$$

For this BTCS scheme, the amplification factor G is given by

$$G = \frac{1}{1 + 4\eta \sin^2(k\Delta x/2)}. \tag{4.31}$$

It is clear from Equation (4.31) that the BTCS scheme is an unconditionally stable scheme as $|G| \leq 1$ for all k. The unknown u_i^{n+1} in Equation (4.30) is not only expressed in terms of the known quantities u_i^n at time level n, but also in terms of unknown quantities u_{i-1}^{n+1} and u_{i+1}^{n+1} at time level $(n+1)$. Hence, Equation (4.30) at a given grid point (i, n) cannot itself provide for the solution of u_i^{n+1}. Equation (4.30) has to be written at all grid points, yielding a system of algebraic equations from which the unknowns u_i^{n+1} for all i can be solved simultaneously. All implicit finite difference solutions are obtained as indicated here. Solving the one-dimensional linear heat conduction equation using BTCS finite difference scheme entails invoking matrix inversion methods. Equation (4.30) can be rewritten as a matrix equation with the coefficient matrix being tridiagonal. Very efficient matrix solution algorithms (such as the *Thomas Algorithm*) are available that deal with tridiagonal matrices. A brief introduction to the Thomas algorithm is presented in the Appendix.

4.3.8   Crank–Nicolson Scheme and its stability

In the Crank–Nicolson scheme, Equation (4.1) is centered at $(i, n+1/2)$ with centered time differencing in time and space yielding the following:

$$\frac{u_i^{n+1} - u_i^n}{\Delta t} = \alpha \frac{u_{i+1}^{n+1/2} - 2u_i^{n+1/2} + u_{i-1}^{n+1/2}}{(\Delta x)^2} \tag{4.32}$$

Each of the three terms in the RHS of Equation (4.32) are averaged between the time indices n and $n+1$ to obtain

$$\frac{u_i^{n+1} - u_i^n}{\Delta t} = \frac{\alpha}{2}\left(\frac{u_{i+1}^{n+1} - 2u_i^{n+1} + u_{i-1}^{n+1}}{(\Delta x)^2} + \frac{u_{i+1}^n - 2u_i^n + u_{i-1}^n}{(\Delta x)^2}\right) \tag{4.33}$$

The truncation error for Equation (4.33) can be obtained by expanding all the terms with respect to time index $(n + 1/2)$ and is given by

$$\frac{1}{8} \left(\frac{\partial^2 u}{\partial x^2} \right)_i^{n+1/2} (\Delta t)^2 + \frac{1}{12} \left(\frac{\partial^4 u}{\partial x^4} \right)_i^{n+1/2} (\Delta x)^2 + \frac{1}{24} \left(\frac{\partial^3 u}{\partial t^3} \right)_i^{n+1/2} (\Delta t)^2 + O\left[(\Delta x)^4, (\Delta t)^4, (\Delta x)^2 (\Delta t)^2 \right]$$

Clearly the Crank–Nicolson scheme is a consistent scheme that is second-order accurate in space and time. Moreover, it is an implicit scheme as the unknown quantities at time $(n + 1)$ are seen in both sides of Equation (4.33). Defining $\eta = \alpha \Delta t / (\Delta x)^2$, Equation (4.33) can be rewritten as

$$-\eta u_{i-1}^{n+1} + (2 + 2\eta) u_i^{n+1} - \eta u_{i+1}^{n+1} = \eta u_{i-1}^n + (2 - 2\eta) u_i^n + \eta u_{i+1}^n. \tag{4.34}$$

Dividing Equation (4.34) by η throughout yields

$$-u_{i-1}^{n+1} + \frac{(2 + 2\eta)}{\eta} u_i^{n+1} - u_{i+1}^{n+1} = u_{i-1}^n + \frac{(2 - 2\eta)}{\eta} u_i^n + u_{i+1}^n. \tag{4.35}$$

Assume that the boundary points corresponding to $x = 0$ is $i = 1$ and that corresponding to $x = L$ is $i = M + 1$. Moreover, let the boundary conditions be $u = A$ at $x = 0$ and $u = D$ at $x = L$. Applying Equation (4.35) over all the grid points $i = 2, i = 3, i = 4, \ldots, i = M - 1, i = M$ other than the boundary points at $(i = 1$ and $i = M + 1)$ results in

$$-A + B_1 u_2^{n+1} - u_3^{n+1} = C_1 \quad \text{for } i = 2$$

$$-u_2^{n+1} + B_2 u_3^{n+1} - u_4^{n+1} = C_2 \quad \text{for } i = 3$$

$$-u_3^{n+1} + B_3 u_4^{n+1} - u_5^{n+1} = C_3 \quad \text{for } i = 4$$

$$\vdots \qquad \vdots \qquad \vdots \quad = \quad \vdots$$

$$-u_{M-1}^{n+1} + B_{M-1} u_M^{n+1} - D = C_{M-1} \quad \text{for } i = M$$

where $B_i = (2 + 2\eta)/\eta$ and $C_{i-1} = u_{i-1}^n + [(2 - 2\eta)/\eta] u_i^n + u_{i+1}^n$. This set of algebraic equations can be written in matrix form

$$
\begin{bmatrix}
B_1 & -1 & 0 & 0 & \cdots & 0 \\
-1 & B_2 & -1 & 0 & \cdots & 0 \\
0 & -1 & B_3 & -1 & \cdots & 0 \\
\vdots & \vdots & \vdots & \vdots & \cdots & \vdots \\
0 & 0 & 0 & 0 & -1 & B_{M-1}
\end{bmatrix}
\begin{bmatrix}
u_2^{n+1} \\
u_3^{n+1} \\
u_4^{n+1} \\
\vdots \\
u_M^{n+1}
\end{bmatrix}
=
\begin{bmatrix}
C_1 + A \\
C_2 \\
C_3 \\
\vdots \\
C_{M-1} + D
\end{bmatrix}
\tag{4.36}
$$

The stability analysis of the von Neumann method for the Crank–Nicolson scheme provides for an amplification factor g to be given as

$$G = \frac{1 - \eta[1 - \cos(k\Delta x)]}{1 + \eta[1 - \cos(k\Delta x)]}.$$ (4.37)

As η is always positive, $|G| \leq 1$, the Crank–Nicolson scheme is an unconditionally stable scheme as the amplification factor is bounded for all values of k.

4.4 Two-dimensional Heat Conduction Equation

The two-dimensional linear heat conduction equation (also called the two-dimensional linear diffusion equation) is given as

$$\frac{\partial u}{\partial t} = \alpha \left(\frac{\partial^2 u}{\partial x^2} + \frac{\partial^2 u}{\partial y^2} \right),$$ (4.38)

where $u = u(x, y, t)$ is a function of x, y and time t. If the initial condition of u is given as a function of x and y at $t = 0$ and the boundary conditions of u is given in both x and y directions. As before, time is discretized with uniform time steps Δt and space (both in the x and y directions) are discretized with uniform grid sizes Δx and Δy in the x and y directions.

4.4.1 FTCS scheme and its stability

The FTCS (forward in time and central in space) scheme for two-dimensional linear heat equation with i, j and n indices representing x, y and time t is as follows:

$$\frac{u_{i,j}^{n+1} - u_{i,j}^n}{\Delta t} = \alpha \left(\frac{u_{i+1,j}^n - 2u_{i,j}^n + u_{i-1,j}^n}{(\Delta x)^2} + \frac{u_{i,j+1}^n - 2u_{i,j}^n + u_{i,j-1}^n}{(\Delta y)^2} \right).$$ (4.39)

Defining $\eta = \alpha\Delta t/(\Delta x)^2$ and $\beta = \alpha\Delta t/(\Delta y)^2$, Equation (4.39) can be rewritten as an explicit finite difference scheme with accuracy $O[(\Delta x)^2, (\Delta y)^2, \Delta t]$ as

$$u_{i,j}^{n+1} = \eta \left(u_{i+1,j}^n + u_{i-1,j}^n \right) + \beta \left(u_{i,j+1}^n + u_{i,j-1}^n \right) + (1 - 2\eta - 2\beta)u_{i,j}^n.$$ (4.40)

It is clear that FTCS (Equation (4.40)) is an explicit finite difference scheme. Assuming the error is of the form

$$\varepsilon_{i,j}^n = G^n e^{i(k_x x_i + k_y y_j)},$$ (4.41)

substituting Equation (4.41) to Equation (4.40), one gets the following expression for the amplification factor G as

$$G = 1 - 4\eta \sin^2(k_x \Delta x/2) - 4\eta \sin^2(k_y \Delta y/2). \tag{4.42}$$

For the FTCS scheme, the stability condition requires that

$$\eta + \beta \leq 1/2. \tag{4.43}$$

This imposes a limit on the choice of time step:

$$\Delta t \leq \frac{(\Delta x)^2 (\Delta y)^2}{2\alpha \left[(\Delta x)^2 + (\Delta y)^2\right]} \tag{4.44}$$

For the special case where $\Delta x = \Delta y$, one has the following condition on the time step

$$\Delta t \leq \frac{(\Delta x)^2}{4\alpha}, \tag{4.45}$$

a condition which is even more restrictive as compared to the one-dimensional linear heat equation.

4.4.2 BTCS scheme and its stability

The BTCS (backward in time and central in space) scheme for the two-dimensional linear heat conduction equation with $i, j,$ and n indices representing x, y and time t is as follows:

$$\frac{u_{i,j}^{n+1} - u_{i,j}^n}{\Delta t} = \alpha \left(\frac{u_{i+1,j}^{n+1} - 2u_{i,j}^{n+1} + u_{i-1,j}^{n+1}}{(\Delta x)^2} + \frac{u_{i,j+1}^{n+1} - 2u_{i,j}^{n+1} + u_{i,j-1}^{n+1}}{(\Delta y)^2} \right). \tag{4.46}$$

Defining $\eta = \alpha \Delta t/(\Delta x)^2$ and $\beta = \alpha \Delta t/(\Delta y)^2$, Equation (4.46) can be rewritten as an implicit finite difference scheme with accuracy $O[(\Delta x)^2, (\Delta y)^2, \Delta t]$ as

$$\eta \left(u_{i+1,j}^{n+1} + u_{i-1,j}^{n+1} \right) + \beta \left(u_{i,j+1}^{n+1} + u_{i,j-1}^{n+1} \right) - (1 + 2\eta + 2\beta) u_{i,j}^{n+1} = -u_{i,j}^n. \tag{4.47}$$

It is clear that BTCS (Equation (4.47)) is an implicit finite difference scheme. Moreover, applying the von Neumann stability analysis, it can be shown that the BTCS scheme for two-dimensional linear heat conduction equation is an unconditionally stable scheme. Equation (4.47) when written as a matrix equation gives rise to a coefficient matrix that is a "five-band matrix."

4.4.3   Alternating Direction Implicit (ADI) method

The basic idea of the ADI method (alternating direction implicit) is to alternate direction and thus, solve two one-dimensional problems at each time step. In the first step, space derivatives with respect to the y-direction are treated explicitly whereas space derivatives in the x-direction are treated implicitly.

$$\frac{u_{i,j}^{n+1/2} - u_{i,j}^{n}}{\Delta t/2} = \alpha \left(\frac{u_{i+1,j}^{n+1/2} - 2u_{i,j}^{n+1/2} + u_{i-1,j}^{n+1/2}}{(\Delta x)^2} + \frac{u_{i,j+1}^{n} - 2u_{i,j}^{n} + u_{i,j-1}^{n}}{(\Delta y)^2} \right). \quad (4.48a)$$

In the second step, space derivatives in the x-direction are treated explicitly while space derivatives in the y-direction are treated implicitly.

$$\frac{u_{i,j}^{n+1} - u_{i,j}^{n+1/2}}{\Delta t/2} = \alpha \left(\frac{u_{i+1,j}^{n+1/2} - 2u_{i,j}^{n+1/2} + u_{i-1,j}^{n+1/2}}{(\Delta x)^2} + \frac{u_{i,j+1}^{n+1} - 2u_{i,j}^{n+1} + u_{i,j-1}^{n+1}}{(\Delta y)^2} \right) \quad (4.48b)$$

Both Equations (4.48a) and (4.48b) can be rewritten in terms of $\eta = \alpha \Delta t/(\Delta x)^2$ and $\beta = \alpha \Delta t/(\Delta y)^2$

$$\eta u_{i+1,j}^{n+1/2} - 2(1+\eta)u_{i,j}^{n+1/2} + \eta u_{i-1,j}^{n+1/2} = -\beta u_{i,j+1}^{n} - 2(1-\beta)u_{i,j}^{n} - \beta u_{i,j-1}^{n} \quad (4.49a)$$

$$\beta u_{i,j+1}^{n+1} - 2(1+\beta)u_{i,j}^{n+1} + \beta u_{i,j-1}^{n+1} = -\eta u_{i+1,j}^{n+1/2} - 2(1-\eta)u_{i,j}^{n+1/2} - \eta u_{i-1,j}^{n+1/2} \quad (4.49b)$$

Each of Equations (4.49a) and (4.49b) provides for one sweep and in each sweep, one solves only a tridiagonal matrix and not the five band matrix that one solves using the BTCS scheme. The ADI method is second-order accurate in time and is unconditionally stable.

4.5   Stability Analysis of One-dimensional Linear Advection Equation

The wave equation is closely related to the so-called *advection equation*, which in one dimension takes the form

$$\frac{\partial \rho}{\partial t} + u\frac{\partial \rho}{\partial x} = 0, \quad (4.50)$$

where u is the x-component of velocity and is assumed constant and ρ is the advected quantity.

The objective is to seek a numerical solution of Equation (4.50) over the region, $x_l \leq x \leq x_r$ subject to the simple Dirichlet boundary conditions $\rho(x_l,t) = \rho(x_r,t) = 0$. As before, one needs to discretize in time $t_n = t_o + n\Delta t$, for $n = 0,1,2,\ldots$ and also discretize in the x direction, $x_i = x_l + i\Delta x$ for $i = 0,1,2,\ldots,N+1$, where $\Delta x = (x_r - x_l)/(N+1)$.

4.5.1 Forward in time and central in space (FTCS) scheme

Applying a forward difference in time and a central difference in space (FTCS scheme), one gets

$$\frac{\rho_i^{n+1} - \rho_i^n}{\Delta t} = -u \frac{\rho_{i+1}^n - \rho_{i-1}^n}{2\Delta x} \tag{4.51}$$

It is clear that Equation (4.51) is an explicit finite difference scheme. With $\lambda = u\Delta t/\Delta x$, Equation (4.51) is rewritten as

$$\rho_i^{n+1} = \rho_i^n - \frac{\lambda}{2}\left(\rho_{i+1}^n - \rho_{i-1}^n\right). \tag{4.52}$$

Assume as before (refer to Section 4.3.4) that the error identically satisfies the difference equation (4.52). Assuming that the error is

$$\varepsilon_i^n = B^{n\Delta t} e^{\iota k i \Delta x}$$

and substituting this in the error difference equation, one gets an expression for the amplification factor

$$G \equiv \frac{\varepsilon_i^{n+1}}{\varepsilon_i^n} = 1 - \iota\lambda \sin(k\Delta x). \tag{4.53}$$

Note that from Equation (4.53), one can get the following

$$|G|^2 = G\overline{G} = 1 + \lambda^2 \sin^2(k\Delta x) > 1, \tag{4.54}$$

indicating that the FTCS scheme is unconditionally unstable scheme for the linear advection equation.

4.5.2 Central in time and central in space (CTCS) scheme and its stability

The most general solution of the one-dimensional linear advection equation is $\rho_o(\xi)$, where ρ_o is an arbitrary function and $\xi = x - ut$. When $t = 0$, ρ_o reduces to simply $\rho_o(x)$, which happens to be the initial condition. As the one-dimensional advection equation is closely related to the wave equation, one assumes that at the initial time $t = 0$,

$$\rho(x,0) = \rho_o(x) = Ae^{\iota kx}. \tag{4.55}$$

Then the general solution is

$$\rho(x,t) = \rho_o(x - ut) = Ae^{\iota k(x - ut)}, \tag{4.56}$$

where $k = 2\pi/L$ and Equation (4.56) refers to a single harmonic of Fourier spectrum with wavelength L, phase speed u and a constant amplitude A. The analytical solution as shown in Equation (4.56) can also be obtained by the method of separation of variables.

Applying the central in time and central in space (CTCS) scheme to the one-dimensional linear advection equation, one gets

$$\frac{\rho_i^{n+1} - \rho_i^{n-1}}{2\Delta t} = -u\frac{\rho_{i+1}^n - \rho_{i-1}^n}{2\Delta x}, \tag{4.57}$$

where i and n denote the space and time indices. Equation (4.57) is also an explicit finite difference scheme, which can be expressed in terms of $\lambda (= u\Delta t/\Delta x)$ as follows:

$$\rho_i^{n+1} = \rho_i^{n-1} - \lambda \left(\rho_{i+1}^n - \rho_{i-1}^n\right). \tag{4.58}$$

Assume as before (refer to Section 4.3.4) that the error identically satisfies the difference equation, Equation (4.58). Assuming that the error $\varepsilon_i^n = B^{n\Delta t}e^{\iota ki\Delta x}$ and substituting this error in the error difference equation, one gets an expression for the quadratic equation involving the amplification factor G

$$G^2 + 2\iota\lambda \sin(k\Delta x)G - 1 = 0. \tag{4.59}$$

Solving the quadratic equation, Equation (4.59) in G, one gets

$$G = -\iota\lambda \sin(k\Delta x) \pm \sqrt{1 - \lambda^2 \sin^2(k\Delta x)}. \tag{4.60}$$

Now, consider the case where $\lambda > 1$. For this case, there will always exist wavelengths for which $\lambda^2 \sin^2(k\Delta x) > 1$. Referring to Equation (4.60), for the case $\lambda > 1$, the quantity under the square root will be negative and the two roots of the quadratic equation (Equation (4.60)) will be imaginary with square of magnitudes exceeding unity for one root. Hence for the case $\lambda > 1$, the CTCS scheme is unconditionally unstable.

Next, consider the case $\lambda \leq 1$. Then the quantity under the square root in Equation (4.60) is real and the two roots of Equation (4.60) are complex quantities with the latter written in polar form

$$G_1 = e^{-\iota\alpha} \quad \text{and} \quad G_2 = e^{\iota(\pi+\alpha)}, \tag{4.61}$$

where α is a real angle given by

$$\alpha = \sin^{-1}[\lambda \sin(k\Delta x)]. \tag{4.62}$$

The magnitudes of the amplification factor G from Equation (4.61) clearly show that the CTCS scheme is a stable finite difference scheme provided the value of λ is less than or equal to one. λ is commonly referred to as the Courant number and the condition that the Courant number be less than or equal to one is known as the Courant–Friedrichs–Lewy (CFL) condition. This condition is a necessary condition for stability for explicit finite difference schemes that solve hyperbolic PDEs.

The solution involving the amplification factor is as follows:

$$\rho_i^n = \left[Me^{-\iota\alpha n} + Ee^{\iota(\pi+\alpha)n}\right] e^{\iota k i \Delta x}, \tag{4.63}$$

where M and E are arbitrary constants. For the initial time, $n = 0$, the complete solution, Equation (4.63), must reduce to the initial condition as given by Equation (4.55). Let us substitute Equation (4.63) for $n = 0$ and equate it with Equation (4.56). Then by assigning $M = A - E$ and as $e^{\iota\pi} = -1$, one gets the complete solution as

$$\rho_i^n = (A - E)e^{\iota k(i\Delta x - \alpha n/k)} + (-1)^n E e^{\iota k(i\Delta x + \alpha n/k)}. \tag{4.64}$$

The complete solution Equation (4.64) corresponding to the difference equation, Equation (4.58) for the case $\lambda \leq 1$ consists of two waves rather than the analytical single wave [Equation (4.56)] for the one-dimensional advection equation. The reason for the manifestation of two waves is the use of central difference in time that resulted in a second-order difference equation, yielding the quadratic equation, Equation (4.59) for the amplification factor. The first term in the right-hand side (RHS) of Equation (4.64) is similar to the analytical wave [Equation (4.56)] with $i\Delta x = x_i$ and has the direction of propagation of wave coinciding with the direction of propagation of the analytical wave. However, the phase speed of the wave corresponding to the first term in the RHS of Equation (4.64) is $un\Delta t = \alpha n/k$, from which the phase speed of the first wave c equals $\alpha/(k\Delta t)$. As the first term in the RHS of Equation (4.64) corresponds to a wave traveling in the same direction of propagation as that of the direction of propagation of the analytical wave, this term is referred to as the "physical mode." The second term in the RHS of Equation (4.64), however, propagates in a direction opposite to the direction of propagation of the analytical wave with the magnitude of the phase speed identical to that of the "physical mode." However, the appearance of the factor $(-1)^n$

in the second term of the RHS of Equation (4.64) indicates there is a phase change at every time step. As such a phase change in every time step is not observed in the analytical wave, the second wave corresponding to the second term in the RHS of Equation (4.64) is referred to as the "computational mode." Hence, the "computational mode" is a spurious wave that needs to be controlled during numerical integrations.

To sum up, while the amplitude of the analytical wave is A, the amplitude for the physical (wave) mode and the computation (wave) mode are $A - E$ and E, respectively. Moreover, while the phase speed of the analytical wave is u, the same for the physical (wave) mode and the computation (wave) mode are $\alpha/(k\Delta t)$ and $-\alpha/(k\Delta t)$, respectively. Furthermore, unlike the analytical wave and the physical (wave) mode for which there are no phase change, the computational (wave) mode shows phase change at every time step.

For the special case where $\lambda = 1$, from Equation (4.62), one can infer that $\alpha = k\Delta x$. From the definition, $\lambda = 1$ corresponds to $u\Delta t = \Delta x$. In the special case of $\lambda = 1$, the physical (wave) mode shall travel with a phase speed of $\frac{\alpha}{k\Delta t} = \frac{k\Delta x}{k\Delta t} = \frac{\Delta x}{\Delta t} = u$, i.e., the physical (wave) mode travels with the same phase speed as the analytical wave.

4.5.3   Upwind methods

Consider a finite difference scheme that is forward in both time and space (FTFS) to solve the one-dimensional linear advection equation. Applying the FTFS scheme to Equation (3.49), one gets

$$\frac{\rho_i^{n+1} - \rho_i^n}{\Delta t} = -u \frac{\rho_{i+1}^n - \rho_i^n}{\Delta x}. \tag{4.65}$$

Equation (4.65) is also an explicit finite difference scheme that can be expressed in terms of the Courant number λ as follows:

$$\rho_i^{n+1} = \rho_i^n - \lambda \left(\rho_{i+1}^n - \rho_i^n \right) \tag{4.66}$$

Assume as before (refer to Section 4.3.4) that the error identically satisfies the difference equation, Equation (3.67). Assuming that the error $\varepsilon_i^n = B^{n\Delta t} e^{iki\Delta x}$ and substituting in the error difference equation, Equation (4.66), the expression for the amplification factor G for the FTFS scheme is

$$G = 1 - \lambda \left(e^{ik\Delta x} - 1 \right) \tag{4.67}$$

From Equation (4.67), it can be shown that the expression for the square of the magnitude of G is of the form

$$|G|^2 = 1 + 2\lambda [1 - \cos(k\Delta x)](1 + \lambda), \tag{4.68}$$

For the FTFS scheme to be stable, $|G| \leq 1$ and for it to be unstable, $|G| > 1$. Note that if λ is positive, which implies a positive u, the RHS of Equation (4.68) exceeds 1, and the scheme would be unstable. In this case, for positive u, the space derivative in Equation (4.66) is taken in the direction of u (i.e., in the downstream direction). However, if u is negative and if $-1 < \lambda < 0$, then the amplification factor $|G| < 1$, as can be verified from Equation (4.68), and the scheme is stable. A negative u implies that the space derivative in Equation (4.66) is taken in the opposite direction of u (i.e., in the upstream direction). From this discussion it is clear that "upstream" or "upwind" schemes provide stable solutions. It can be easily verified that for a backward space difference and forward difference in time scheme (FTBS), the amplification factor is bounded yielding stable solutions only for the "upstream" or "upwind" schemes, i.e., for those schemes where u is positive and $0 < \lambda < 1$. Further, for the FTBS scheme, downstream spatial derivatives (with negative u) yield unstable and amplified solutions.

4.5.4 Lax finite difference scheme and its stability

The Lax method is obtained from the FTCS scheme by effecting a simple modification of Equation (4.52) that involves replacing ρ_i^n by averaging ρ between the neighboring spatial points.

$$\rho_i^{n+1} = \frac{1}{2}\left(\rho_{i+1}^n + \rho_{i-1}^n\right) - \frac{\lambda}{2}\left(\rho_{i+1}^n - \rho_{i-1}^n\right) \tag{4.69}$$

Equation (4.69) is an explicit finite difference scheme. Assume as before (refer to Section 4.3.4) that the error identically satisfies the difference equation, Equation (4.69). Assuming that the error $\varepsilon_i^n = B^{n\Delta t}e^{\imath ki\Delta x}$ and substituting this expression in the error difference equation, Equation (4.69), the expression for the amplification factor G for the Lax scheme is

$$G = \cos(k\Delta x) - \imath\lambda \sin(k\Delta x). \tag{4.70}$$

From Equation (4.69), it can be shown that the expression for the square of the amplification factor G is

$$|G|^2 = \cos^2(k\Delta x) + \lambda^2 \sin^2(k\Delta x) = 1 - (1 - \lambda^2)\sin^2(k\Delta x). \tag{4.71}$$

From Equation (4.71), it is clear that the Lax scheme is stable if $1 - \lambda^2 \geq 0$, i.e., if $\lambda \leq 1$, i.e., if the Courant-Friedrichs-Lewy (CFL) condition is satisfied.

4.5.5   Lax–Wendroff scheme and its stability

The Lax–Wendroff scheme can be obtained in different ways. One way to obtain the aforementioned finite difference is by employing the multi-step method, i.e., by calculating ρ_i^{n+1} using the information available on the half time step, as follows:

$$\rho_i^{n+1/2} = \rho_i^n - \frac{\Delta t}{2} \left(u \frac{\partial \rho}{\partial x} \right)_i^n \tag{4.72}$$

$$\rho_i^{n+1} = \rho_i^n - \Delta t \left(u \frac{\partial \rho}{\partial x} \right)_i^{n+1/2}. \tag{4.73}$$

Employing the central difference to replace the space derivative in the RHS of Equation (4.73), one gets

$$\rho_i^{n+1} = \rho_i^n - \frac{u \Delta t}{\Delta x} \left(\rho_{i+1/2}^{n+1/2} - \rho_{i-1/2}^{n+1/2} \right). \tag{4.74}$$

The quantities at $n + 1/2$ in the RHS of Equation (4.74) are obtained using the Lax scheme and are given as

$$\rho_{i-1/2}^{n+1/2} = \frac{1}{2} \left(\rho_i^n + \rho_{i-1}^n \right) - \frac{u \Delta t}{2 \Delta x} \left(\rho_i^n - \rho_{i-1}^n \right). \tag{4.75}$$

$$\rho_{i+1/2}^{n+1/2} = \frac{1}{2} \left(\rho_{i+1}^n + \rho_i^n \right) - \frac{u \Delta t}{2 \Delta x} \left(\rho_{i+1}^n - \rho_i^n \right). \tag{4.76}$$

Equations (4.75), (4.76) and Equation (4.74) together constitute the Lax–Wendroff scheme. It is clear that the Lax–Wendroff scheme is an explicit finite difference scheme and is a second-order accurate scheme in space and time. The second way to derive the Lax-Wendroff scheme is by first using Taylor series expansion in time to the second-order, as follows:

$$\rho(x_i, t_{n+1}) = \rho(x_i, t_n) + \Delta t \left(\frac{\partial \rho}{\partial t} \right)_i^n + \frac{(\Delta t)^2}{2!} \left(\frac{\partial^2 \rho}{\partial t^2} \right)_i^n + \cdots \tag{4.77}$$

In Equation (4.77), replace time derivative $\partial \rho / \partial t$ by space derivative $-u(\partial \rho / \partial x)$ and replace the second derivative in time $\partial^2 \rho / \partial t^2$ by space derivative $u^2(\partial^2 \rho / \partial x^2)$ to get (the latter follows from the one-dimensional wave equation)

$$\rho(x_i, t_{n+1}) = \rho(x_i, t_n) - u\Delta t \left(\frac{\partial \rho}{\partial x}\right)_i^n + \frac{(u^2 \Delta t)^2}{2!} \left(\frac{\partial^2 \rho}{\partial x^2}\right)_i^n + \cdots \tag{4.78}$$

Finally, the Lax–Wendroff scheme is obtained by replacing space derivatives (up to second-order) in Equation (4.78) by central difference

$$\rho_i^{n+1} = \rho_i^n - \frac{\lambda}{2}\left(\rho_{i+1}^n - \rho_{i-1}^n\right) + \frac{\lambda^2}{2}\left(\rho_{i+1}^n - 2\rho_i^n + \rho_{i-1}^n\right), \tag{4.79}$$

where λ is the Courant number. Equation (4.79) is called the one-step Lax–Wendroff scheme. It is an explicit finite difference scheme and is second-order accurate in time and in space. Assume as before (refer to Section 4.3.4) that the error identically satisfies the difference equation, Equation (4.79). Assuming that the error $\varepsilon_i^n = B^{n\Delta t} e^{\imath k i \Delta x}$ and substituting this value in the error difference equation, Equation (4.79), the expression for the amplification factor G for the one step Lax–Wendroff scheme is

$$G = \left[(1 - \lambda^2) + \lambda^2 \cos(k\Delta x)\right] - \imath\lambda \sin(k\Delta x). \tag{4.80}$$

For the one-step Lax–Wendroff scheme to be stable, $\lambda \leq 1$, i.e., the Courant–Friedrichs–Lewy (CFL) condition is to be satisfied since $G^2 = 1 - \lambda^2(1 - \lambda^2)\ (1 - \cos(k\Delta x))^2$ For $G^2 \leq 1, 1 - \lambda^2 \geq 0$ which implies $\lambda \leq 1$.

4.5.6 Backward in time and central in space (BTCS) scheme and its stability

The backward in time and central in space (BTCS) scheme as applied to the one-dimensional linear advection equation as follows:

$$\frac{\rho_i^{n+1} - \rho_i^n}{\Delta t} = -u\frac{\rho_{i+1}^{n+1} - \rho_{i-1}^{n+1}}{2\Delta x}. \tag{4.81}$$

Equation (4.81) is an implicit finite difference scheme and is of the order of accuracy $[O(\Delta t), O(\Delta x)^2]$. The solution of Equation (4.81) involves solving a matrix equation. Assume as before (refer to Section 4.3.4) that the error identically satisfies the difference equation, Equation (4.79). Assuming that the error $\varepsilon_i^n = B^{n\Delta t} e^{\imath k i \Delta x}$ and substituting this expression in the error difference equation, Equation (4.79), the expression for the amplification factor G for the BTCS scheme is

$$G = \frac{1}{1 + \imath\lambda \sin(k\Delta x)}. \tag{4.82}$$

From Equation (4.82), it can be shown that the expression for the square of the amplification factor G is

$$|G|^2 = \frac{1}{1 + \lambda^2 \sin^2(k\Delta x)}. \tag{4.83}$$

It is clear from Equation (4.83) that the BTCS scheme is unconditionally stable for all values of k.

4.5.7 Crank–Nicolson scheme and its stability

The Crank–Nicolson scheme (refer to Section 4.3.8) can also be applied to the one-dimensional linear advection equation as follows:

$$\frac{\rho_i^{n+1} - \rho_i^n}{\Delta t} = -\frac{u}{2} \left(\frac{\rho_{i+1}^{n+1} - \rho_{i-1}^{n+1}}{2\Delta x} + \frac{\rho_{i+1}^n - \rho_{i-1}^n}{2\Delta x} \right). \tag{4.84}$$

Equation (3.85) is an implicit finite difference scheme with second-order accuracy in space and time. Solution of Equation (4.84) involves solving a matrix equation. Assume as before (refer to Section 4.3.4) that the error identically satisfies the difference equation (4.84). Assuming that the error $\varepsilon_i^n = B^{n\Delta t} e^{\iota k i \Delta x}$ and substituting this in the error difference equation, Equation (4.84), the expression for the amplification factor G for the Crank–Nicolson scheme is given by

$$G = \frac{1 - \iota \left[\frac{\lambda}{2} \sin(k\Delta x) \right]}{1 + \iota \left[\frac{\lambda}{2} \sin(k\Delta x) \right]}. \tag{4.85}$$

From Equation (4.85), it is clear that the Crank–Nicolson scheme is unconditionally stable. Moreover, the Crank–Nicolson scheme does not give rise to the computational (wave) mode despite its second-order accuracy in time and space.

4.6   Matrix Method of Stability Analysis

The stability of finite difference schemes as proposed by von Neumann has already been discussed extensively. One limitation of the von Neumann method is that it does not include boundary conditions in the stability analysis. The matrix method is a stability analysis that includes the effect of boundary conditions, It is discussed here for the one-dimensional heat equation.

4.6.1 Matrix method for the one-dimensional heat equation

Consider the one-dimensional heat conduction equation

$$\frac{\partial u}{\partial t} = \alpha \frac{\partial^2 u}{\partial x^2}. \tag{4.86}$$

Applying the FTCS (forward in time and central in space) finite difference scheme, one gets

$$\frac{u_i^{n+1} - u_i^n}{\Delta t} = \alpha \frac{u_{i+1}^n - 2u_i^n + u_{i-1}^n}{(\Delta x)^2}, \tag{4.87}$$

With $\eta = \alpha \Delta t / (\Delta x)^2$, Equation (4.87) becomes

$$u_i^{n+1} = \eta u_{i-1}^n + (1 - 2\eta) u_i^n + \eta u_{i+1}^n. \tag{4.88}$$

For simplicity, if one assumes that the boundary values $u_0^n = u_N^n = 0$; then Equation (4.88) can be written in matrix form as

$$\begin{bmatrix} u_1^{n+1} \\ u_2^{n+1} \\ u_3^{n+1} \\ \vdots \\ u_{N-1}^{n+1} \end{bmatrix} = \begin{bmatrix} 1-2\eta & \eta & 0 & 0 & \cdots & 0 \\ \eta & 1-2\eta & \eta & 0 & \cdots & 0 \\ 0 & \eta & 1-2\eta & \eta & \cdots & 0 \\ \vdots & \vdots & \vdots & \vdots & \cdots & \vdots \\ 0 & 0 & 0 & 0 & \eta & 1-2\eta \end{bmatrix} \begin{bmatrix} u_1^n \\ u_2^n \\ u_3^n \\ \vdots \\ u_{N-1}^n. \end{bmatrix} \tag{4.89}$$

It follows from Equation (4.89) that

$$U_n = AU_{n-1} = A(AU_{n-2}) = A^n U_0. \tag{4.90}$$

It is clear from Equation (4.90) that the solution after n number of time marching will essentially depend on the character of the matrix A itself. Hence, if there are errors at the grid points at the initial time $t = 0$ that ensure that U_0 is given by U_0', then according to Equation (4.90), the value of U after n time steps will be given by

$$U_n' = A^n U_0', \tag{4.91}$$

assuming that no additional errors are introduced in the subsequent calculations. Hence, the error $\varepsilon_n = U_n - U_n'$, after n time steps is found by subtracting Equation (4.91) from Equation (4.90),

$$\varepsilon_n = A^n \varepsilon_0 \tag{4.92}$$

The finite difference scheme is said to be stable when ε_n remains bounded as n increases indefinitely. To derive the stability criteria using the matrix method, it is assumed that it is possible to express the vector ε_0 as a linear combination of a complete set of linearly independent vectors that comprise an orthogonal-normal set. The latter are obtained as the so-called eigenvectors of the matrix A. First, let the eigenvalues of the matrix A be defined as the roots λ_k, $k = 1, 2, \ldots, N-1$, of the equation

$$|A - \lambda I| = 0, \tag{4.93}$$

where I is the identity matrix and the vertical bars denote a determinant. Using the aforementioned definition, the eigenvectors v_k are the solutions of the homogeneous system of equations:

$$(A - \lambda_k I)v_k = 0 \quad \text{or} \quad Av_k = \lambda_k v_k. \tag{4.94}$$

The eigenvectors of matrix A for an orthonormal set under fairly general conditions with error ε_0 is expressible in terms of the orthonormal set:

$$\varepsilon_0 = \sum_{k=1}^{N-1} c_k v_k, \tag{4.95}$$

where c_k are constants. Substituting Equation (4.95) in Equation (4.92), one gets

$$\varepsilon_n = \sum_{k=1}^{N-1} c_k A^n v_k = \sum_{k=1}^{N-1} c_k A^{n-1} A v_k = \sum_{k=1}^{N-1} c_k A^{n-1} \lambda_k v_k.$$

Repeating the aforementioned procedure,

$$\varepsilon_n = \sum_{k=1}^{N-1} c_k \lambda_k^n v_k, \tag{4.96}$$

From Equation (4.96), it can be clearly inferred that the errors will not grow exponentially with n, provided that the largest magnitude of the eigenvalues is less than or equal to unity, i.e.,

$$|\lambda_k| \leq 1, \quad k = 1, 2, \ldots \tag{4.97}$$

Hence, the essence of the matrix method of stability analysis is to determine the eigenvalues, which in general may not be an easy task. However, there are general specific formulas for certain types of matrices and general theorems that provide bounds of the eigenvalues. Moreover, numerical methods exist that calculate the eigenvalues of a matrix using the computer.

Let the matrix A in Equation (4.89) be expressed as the sum of a unit matrix I and a tridiagonal matrix T as follows:

$$A = I + \eta T, \tag{4.98}$$

where T is a tridiagonal matrix with '-2' as diagonal elements and '1' as off-diagonal nonzero elements. It can be shown that the eigenvalues α_k and β_k of two matrices A and B are related by the same rational function that relates the two matrices A and B. It is known that the eigenvalues of a $(N-1) \times (N-1)$ tridiagonal matrix with elements a, b, and c in that order are given by

$$\lambda_k = b + 2\sqrt{ac}\cos(k\pi/N), \quad k = 1, 2, \dots N-1 \tag{4.99}$$

For matrix T, $a = c = 1$ and $b = -2$. Substituting in Equation (4.98), the values of a, b, and c, one obtains the eigenvalues of A:

$$\lambda_k = 1 - 4\eta \sin^2(k\pi/2N), \quad k = 1, 2, \dots N-1 \tag{4.100}$$

It is immediately clear, that the magnitudes of all the λ_k will all be less than unity if $\eta \leq 0.5$; hence, the finite difference scheme will be stable when

$$\eta = \frac{\alpha \Delta t}{(\Delta x)^2} \leq \frac{1}{2}. \tag{4.101}$$

4.7 Energy Method of Stability Analysis

The two stability methods that were introduced deal essentially with linear partial differential equations. The energy method of stability analysis is one that can also be applied for nonlinear partial differential equations. A brief introduction of the energy method is provided in this section. In the energy method, if the true solution is bounded, one investigates the boundedness of the numerical solution by testing whether the summation of the energy over the domain is bounded at a given time. If the summation of the energy over the domain is bounded at a given time, then it can be established that every value of the velocity over the domain of interest is bounded. This ensures the stability of the finite difference scheme.

Consider the one-dimensional linear advection equation:

$$\frac{\partial \rho}{\partial t} + u \frac{\partial \rho}{\partial x} = 0 \tag{4.102}$$

Let u be a constant and $0 \leq x \leq L$. Let cyclic boundary conditions be prescribed at $x = 0$ and $x = L$, i.e., $\rho(0,t) = \rho(L,t)$. Let the initial condition on ρ be $\rho(x,0) = f(x)$ with

$f(x+L) = f(x)$. The analytical solution of Equation (4.102) with the aforementioned initial condition is given as derived earlier

$$\rho(x,t) = f(x-ut). \tag{4.103}$$

The solution function propagates without any change of shape along the positive x axis with a speed (phase speed) given by u. It can be shown that for the case of periodic conditions, and for the solution being sinusoidal in time, the constant λ that is assigned in the separation of variables method can only assume certain (imaginary) values $\lambda = ik$, where k is the wave number with $ku = \omega$, where ω is the frequency. If ρ is to represent a physical field, this field is the real part of the aforementioned solution. As the advection equation is linear, any linear combination of the aforementioned solution is also a solution of the advection equation. As all the component waves of a disturbance travel with the same (advective) speed, there is no dispersion and the disturbance does not change shape with time. It is important to consider the "energy" defined by Equation (4.104)

$$E(t) = \frac{1}{2} \int_{x=0}^{x=L} \rho^2 dx. \tag{4.104}$$

Multiplying Equation (4.102) by ρ on both sides and integrating with respect to x, results in

$$\frac{\partial E}{\partial t} = -\frac{u}{2} \int_0^L \frac{\partial(\rho^2)}{\partial x} dx = -\frac{u}{2} \left[\rho^2\right]_0^L = 0. \tag{4.105}$$

Equation (4.105) shows that the energy is conserved for the linear advection equation with periodic boundary conditions. This is as it should be as there is no change in shape of the disturbance. It would be important to study an analogous quantity E^n given by

$$E^n = \frac{1}{2} \sum_{i=2}^{N+1} (\rho_i^n)^2 \Delta x. \tag{4.106}$$

As an example of applying the energy method of stability analysis, consider the FTFS scheme of Equation (4.102)

$$\rho_i^{n+1} = \rho_i^n - \lambda(\rho_i^n - \rho_{i-1}^n), \tag{4.107}$$

where $\lambda = u\Delta t / \Delta x$. Multiplying Equation (4.107) by $\rho_i^{n+1} + \rho_i^n$, one gets

$$\left(\rho_i^{n+1}\right)^2 - (\rho_i^n)^2 = -\lambda \left(\rho_i^{n+1} + \rho_i^n\right) \left(\rho_i^n - \rho_{i-1}^n\right). \tag{4.108}$$

Substituting for ρ_i^{n+1} from Equation (4.107) into Equation (4.108), one gets after rearranging

$$\left(\rho_i^{n+1}\right)^2 - \left(\rho_i^n\right)^2 = -\lambda \left[\left(\rho_i^n\right)^2 - \left(\rho_{i-1}^n\right)^2\right] - \lambda\left(1 - \lambda\right)\left(\rho_i^n - \rho_{i-1}^n\right)^2. \tag{4.109}$$

Summing the aforementioned equation over all the grid points and using the boundary conditions $\rho_1^n = \rho_{N+1}^n$, one gets

$$E^{n+1} - E^n = -\lambda\left(1 - \lambda\right) \sum_{i=2}^{N+1} \left(\rho_i^n - \rho_{i-1}^n\right)^2 \Delta x. \tag{4.110}$$

In order to prevent the energy from growing from n to $n+1$, one requires

- $\lambda \geq 0$, which requires $u \geq 0$ and

- $(1 - \lambda) \geq 0$, which requires $\lambda \leq 1$ or $\dfrac{u\Delta t}{\Delta x} \leq 1$.

Having chosen the grid length Δx, one can only get a stable solution if the time step Δt is chosen such that $\Delta t \leq \Delta x/u$, which is the CFL condition. Note that once the CFL condition is satisfied, the energy is forced to decay from step n to step $n+1$. The energy method is quite a general approach for analysing finite difference schemes and can be employed even for nonlinear PDEs with more general boundary conditions. However, for most cases, the energy method requires considerable effort and ingenuity to obtain useful and practical stability criteria.

4.8   Aliasing and Nonlinear Computational Instability

Norman A. Philips in 1959 demonstrated that the finite difference solution of the nonlinear barotropic vorticity equation possessed a distinct type of instability that was very different from the one associated with the forecast that utilized a larger time step than allowed by the conditional stability of the finite difference scheme. Philips correctly interpreted that this instability manifests when the grid system could not resolve waves that had wavelengths shorter than two grid intervals. He surmised that when such small wavelengths were formed due to nonlinear interaction of long waves, the grid system wrongly interpreted them as resolvable long waves. In essence, as the aforementioned instability arose only in nonlinear partial differential equations, the instability was called a "nonlinear computational instability."

4.9   Aliasing Error and Instability

To illustrate the *aliasing error*, consider the one-dimensional nonlinear advection equation:

$$\frac{\partial u}{\partial t} + c\frac{\partial u}{\partial x} = 0. \tag{4.111}$$

As discussed earlier, the general solution of Equation (4.111) is given by

$$u = F(x - ct), \tag{4.112}$$

where F is any arbitrary function. Figure 4.1 depicts an example of the aliasing error. In the figure, the continuous line depicted shows a wave that has a wavelength of $(4/3)\Delta x$. It is known that the smallest wavelength that shall be resolved by a grid system with grid intervals Δx is $2\Delta x$. As $(4/3)\Delta x < 2\Delta x$, the continuous line depicted in Figure 4.1 shows a wave whose wavelength is less than the smallest wavelength that can be resolved by the grid system. The grid points in the x axis, however, perceive

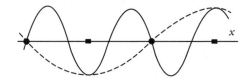

Figure 4.1   A wave (continuous line) of wavelength less than $2\Delta x$, say, $(4/3)\Delta x$, is misrepresented as a resolvable wave (dashed line) of wavelength $4\Delta x$ by the finite difference grid.

and assume that the unresolvable wave of wavelength $(4/3)\Delta x$ is actually a resolvable wave of wavelength $4\Delta x$. Figure 4.1 clearly shows that at the grid points, the wave of wavelength $4\Delta x$ takes exactly the values that the wave of wavelength $(4/3)\Delta x$ would take at those same grid points, if the latter could be represented on the grid system. The aforementioned misrepresentation of a wavelength that is too short to be represented on the grid as a wave having a resolvable wavelength is called "aliasing error."

In Figure 4.1, aliasing manifests as the grid system is too coarse to resolve the wave of wavelength $(4/3)\Delta x$. An alternate way of reiterating the same is that the wave is not adequately sampled by the grid system. Aliasing error is always due to inadequate sampling. It can have a significant impact in observational studies. Observations taken "too far apart" in space (or time) can make a short wave (or high frequency) appear to be a longer wave (or lower frequency). In essence, "aliasing" is due to a high wave number (or frequency) camouflaged as a low wave number (or frequency).

Aliasing error appears while one tries to solve nonlinear partial differential equations as the nonlinear terms in such equations can produce small waves that have wavelengths smaller than the smallest wavelength that can be resolved by the grid system. For example, consider two modes 'A' and 'B' having wavelengths k and l and amplitudes A_o and B_o, respectively, in a one-dimensional grid:

$$A(x_j) = A_o e^{ikj\Delta x} \quad \text{and} \quad B(x_j) = B_o e^{ilj\Delta x}. \tag{4.113}$$

If one combines A and B linearly with α and β as two spatially constant coefficients,

$$\alpha A + \beta B,$$

then no "new" waves are generated with k and l continuing to be the only wave numbers present. However, if one multiplies the two modes A and B together, then one generates a new wave number, $k + l$:

$$AB = A_o B_o e^{i(k+l)j\Delta x}. \tag{4.114}$$

The new wave number $k + l$ can actually have higher values than the highest wave number possible in the given grid system. The highest wave number that can be resolved in a grid system having Δx as grid interval is $\pi/\Delta x$. The new wave number $k + l$ created due to the presence of a nonlinear term may actually satisfy the inequality $(k + l)\Delta x > \pi$, in which case, the new wave number that got created cannot fit in with the given grid system, as it has a wave number larger than the maximum wave number that can be resolved. What, however, happens is that the new mode with the new wave number $k + l$ gets camouflaged and is aliased into a mode that fits in with the given grid system. Let the maximum value of the wave number for a given grid system be represented by k_{max}; then $k_{max} = \pi/\Delta x$. It would be interesting to find out what happens if a new wave of wave number k is generated due to nonlinear interactions such that $k > k_{max}$.

As $2k_{max}\Delta x = 2\pi$, one can assume that $2k_{max} > k > k_{max}$. Writing the expression for $\sin(kj\Delta x)$ as follows:

$$\sin(kj\Delta x) = \sin[(2k_{max} - 2k_{max} + k)j\Delta x] = \sin[2\pi j - (2k_{max} - k)j\Delta x] = \sin[-(2k_{max} - k)j\Delta x]$$

$$= \sin(k^* j\Delta x), \tag{4.115}$$

where

$$k^* = -(2k_{max} - k). \tag{4.116}$$

Similarly, it can be shown that

$$\cos(kj\Delta x) = \cos(k^* j\Delta x). \tag{4.117}$$

It follows from Equations (4.115) and (4.117) and Figure 4.1 that knowing only the values at the grid points, one cannot distinguish the wavenumbers k from the wavenumbers $2k_{max} - k$. Equations (4.115) and (4.117) show that the wave of wavenumber $k > k_{max}$ is misrepresented and misinterpreted by the grid system as a wave of wavenumber $k^* = -(2k_{max} - k)$. The minus sign means that the phase change per Δx is reversed.

For $k > k_{max}$, one gets $k^* < k_{max}$. For $k = k_{max}$, one gets $k^* = k_{max}$. Figure 4.2 shows the misrepresentation of the wavenumber $k > k_{max}$ as shown in Equation (4.116) in which a wave that cannot be resolved actually shows up as a resolvable wave. The wave $k > k_{max}$ gets folded to a wavenumber k^* that is resolvable (i.e., $k^* < k_{max}$).

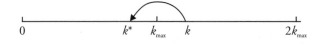

Figure 4.2 Aliasing or misrepresentation of the wavenumber $k > k_{max}$.

According to Figure 4.1, $L = (4/3)\Delta x$, $k = 2\pi/L = 3\pi/(2\Delta x)$. Hence, $k^* = 2k_{max} - k = 2\pi/\Delta x - 3\pi/(2\Delta x) = \pi/(2\Delta x)$. That is the corresponding $L^* = 2\pi/k^* = 4\Delta x$, which was discerned from Figure 4.1. It is easily seen from Equation (4.113) that for situations in which $k > k_{max}$, the change in phase of $A(x_j)$ with increase of j by 1 is more than π. Similarly, for situations in which $k < k_{max}$, the change in phase of $A(x_j)$ with increase of j by 1 is less than π.

The end result of aliasing error in a numerical integration of the governing nonlinear equations of atmospheric motion results in a "cascade" of energy into scales that cannot be resolved. After a certain period of integration time, the aforementioned energy can be expected to grow beyond physically acceptable limits and leads to a catastrophic breakdown. This catastrophic breakdown is referred to as nonlinear (computational) instability and was first discovered by Norman A. Phillips during his early attempts to model and integrate the atmospheric general circulation model using the nonlinear governing equations of atmospheric motion.

4.10   Ways to Prevent Nonlinear Computational Instability

Philips was integrating the nonlinear vorticity equations for one month when he discovered the nonlinear computational instability. He had initially presumed that the nonlinear computational instability can be overcome by choosing smaller time steps for marching the governing equations of atmospheric motions. However, he found that the reduced time step did not help in preventing the nonlinear computational instability. Philips understood the real cause for the instability: the waves that were not resolved by the grid system were getting generated through nonlinear interactions and these waves were erroneously misrepresented as resolvable waves. To overcome this instability, Philips carried out a harmonic analysis of the vorticity fields, eliminating all components with $k > k_{max}/2$ once every two hours of the simulation. This procedure helped Philips to overcome the nonlinear computational instability.

Other possible ways to prevent the manifestation of the nonlinear computational instability are as follows:

1. Employ a finite difference scheme that has a built-in damping of the shortest waves, such as the Lax–Wendroff scheme.

2. Add a dissipative term to a finite difference scheme that is not dissipative and, hence, enable a provision for dissipation to ensure that the amount of dissipation is better controlled

3. Employ schemes based on the Lagrangian/semi-Lagrangian formulation instead of an Eulerian formulation.

4. Employ finite difference schemes that prevent the spurious inflow of energy into the shorter (non-resolvable) waves, instead of adding dissipative finite difference schemes that suppress the amplitude of these short waves.

The best work-around for preventing the nonlinear computational instability is to employ finite difference schemes that prevent the spurious inflow of energy into the shorter (non-resolvable) waves. Arakawa (1966, 1972) was the first to develop such finite difference schemes that prevented the manifestation of nonlinear computational instability.

4.11   Arakawa's Scheme to Prevent Nonlinear Computational Instability

In this section, Arakawa's conservation finite difference scheme to prevent nonlinear computational instability is introduced. Consider a homogeneous, incompressible, inviscid, two-dimensional flow. The vertical component ξ of the relative vorticity vector indicated by ζ is related to the stream function ψ by the following expression

$$\xi = \left(\frac{\partial v}{\partial x} - \frac{\partial u}{\partial y}\right), \qquad u = -\frac{\partial \psi}{\partial y}, \qquad v = \frac{\partial \psi}{\partial x}, \qquad V = \hat{k} \times \nabla \psi, \qquad \xi = \nabla^2 \psi, \quad (4.118)$$

where u and v are the zonal and meridional components of the velocity vector V. With the introduction of stream function ψ, the two-dimensional incompressible continuity equation is identically satisfied. The relative vorticity equation becomes, after applying the scale analysis (divergence term, solenoidal term, and the tilting terms are considered small and, hence, neglected).

$$\frac{\partial \xi}{\partial t} + V \cdot \nabla \xi = 0. \tag{4.119}$$

The continuity equation for an incompressible fluid assumes the following simple form

$$\nabla \cdot V = 0. \tag{4.120}$$

Multiplying Equation (4.120) with ξ and adding the result to Equation (4.119), one obtains

$$\frac{\partial \xi}{\partial t} + \nabla \cdot (\xi V) = 0. \tag{4.121}$$

Applying volume integral to Equation (4.121) over the volume D and using the Leibnitz rule for differentiation under the integral sign, one obtains

$$\frac{d}{dt} \iiint_D \xi \, dV + \iiint_D \nabla \cdot (\xi V) \, dV = 0. \tag{4.122}$$

Utilizing the Gauss divergence theorem, the second volume integral in Equation (4.122) can be transformed into a surface integral, and one obtains

$$\frac{d}{dt} \iiint_D \xi \, dV + \iint_S \xi V \cdot \hat{n} \, dS = 0, \tag{4.123}$$

where $\hat{n}$ is a unit vector normal to dS and pointing from D. Assuming that the fluid flow is contained in a closed domain D so that there is no flow across the domain boundary S, the surface integral in Equation (4.123) becomes zero and one gets

$$\frac{d}{dt} \iiint_D \xi \, dV = 0. \tag{4.124}$$

From Equation (4.124), it follows that the total (or mean) relative vorticity ξ is a conserved quantity.

Multiplying Equation (4.121) by ξ throughout, one obtains

$$\xi\frac{\partial \xi}{\partial t} + \xi\nabla\cdot(\xi V) = 0. \tag{4.125}$$

Using the vector identity,

$$\nabla\cdot(\xi V) = \nabla\xi\cdot V + \xi\nabla\cdot V = \nabla\xi\cdot V,$$

where the vanishing of horizontal velocity divergence due to continuity equation for an incompressible fluid, Equation (4.120) is utilized. Moreover, the second term in LHS of Equation (4.125) becomes

$$\xi(\nabla\xi\cdot V) = \frac{1}{2}\nabla\cdot(\xi^2 V), \tag{4.126}$$

where we again used the vanishing of horizontal velocity divergence due to continuity equation for an incompressible fluid, Equation (4.120). Moreover,

$$\xi\frac{\partial \xi}{\partial t} = \frac{\partial}{\partial t}\left(\frac{1}{2}\xi^2\right). \tag{4.127}$$

Substitution of Equations (4.126) and (4.127) into Equation (4.125), one obtains

$$\frac{\partial}{\partial t}\left(\frac{1}{2}\xi^2\right) + \frac{1}{2}\nabla\cdot(\xi^2 V) = 0. \tag{4.128}$$

Applying volume integral to Equation (4.128) over the volume D and then applying the Gauss divergence theorem to the resulting equation, one obtains

$$\frac{d}{dt}\iiint_D \frac{1}{2}\xi^2\, dV + \iint_S \frac{1}{2}\xi^2 V\cdot\hat{n}\, dS = 0. \tag{4.129}$$

Assuming that the fluid flow is contained in a closed domain D so that there is no flow across the domain boundary S, the surface integral in Equation (4.129) becomes zero. This leads to the fact that the total or mean square vorticity (commonly referred to as *enstrophy*)

$$\frac{d}{dt}\iiint_D \frac{1}{2}\xi^2\, dV = 0 \tag{4.130}$$

is also conserved.

Finally, multiplying Equation (4.121) by ψ throughout, one obtains

$$\psi\frac{\partial \xi}{\partial t} + \psi\nabla\cdot(\xi V) = 0. \tag{4.131}$$

As $\xi = \nabla^2 \psi = \nabla \cdot \nabla \psi$, the first term in Equation (4.131) may be written as

$$\psi \frac{\partial \xi}{\partial t} = \psi \frac{\partial}{\partial t} (\nabla \cdot \nabla \psi) = \psi \nabla \cdot \nabla \left(\frac{\partial \psi}{\partial t} \right) = \nabla \cdot \left(\psi \nabla \frac{\partial \psi}{\partial t} \right) - \nabla \psi \cdot \nabla \frac{\partial \psi}{\partial t}. \qquad (4.132)$$

Further, the second term in Equation (4.131) may be written as

$$\psi \nabla \cdot (\xi V) = \nabla \cdot (\psi \xi V) - \xi V \cdot \nabla \psi. \qquad (4.133)$$

Substitution of Equations (4.132) and (4.133) into Equation (4.131), one obtains

$$\nabla \cdot \left(\psi \nabla \frac{\partial \psi}{\partial t} \right) - \nabla \psi \cdot \nabla \frac{\partial \psi}{\partial t} + \nabla \cdot (\psi \xi V) - \xi V \cdot \nabla \psi = 0. \qquad (4.134)$$

Rearranging Equation (4.134), we obtain

$$\nabla \psi \cdot \nabla \frac{\partial \psi}{\partial t} = \nabla \cdot \left(\psi \nabla \frac{\partial \psi}{\partial t} \right) + \nabla \cdot (\psi \xi V) - \xi V \cdot \nabla \psi. \qquad (4.135)$$

As $V = \hat{k} \times \nabla \psi$, the last term in the RHS of Equation (4.135) is zero. Moreover, the term in the LHS of Equation (4.135) is nothing but the rate of change of mean kinetic energy as

$$\nabla \psi \cdot \nabla \frac{\partial \psi}{\partial t} = \frac{\partial}{\partial t} \left[\frac{1}{2} (\nabla \psi)^2 \right]. \qquad (4.136)$$

From Equations (4.135) and (4.136), one obtains

$$\frac{\partial}{\partial t} \left[\frac{1}{2} (\nabla \psi)^2 \right] = \nabla \cdot \left(\psi \nabla \frac{\partial \psi}{\partial t} \right) + \nabla \cdot (\psi \xi V). \qquad (4.137)$$

Applying volume integral to Equation (4.137) over the volume D and using the Gauss divergence theorem, we obtain an integral equation for Equation (4.137). Assuming that the fluid flow is contained in a closed domain D so that there is no flow across the domain boundary S, one then obtains

$$\frac{d}{dt} \iiint_D \frac{1}{2} (\nabla \psi)^2 \, dV = 0. \qquad (4.138)$$

Equation (4.138) indicates that the mean kinetic energy is also a conserved quantity in D.

To sum up, the vorticity equation expressed by Equation (4.119) for any closed region, conserves mean kinetic energy, mean vorticity, and mean enstrophy in the continuum case. It goes without saying that one would expect that the aforementioned conservation properties also hold for the discrete case, when the partial differential equation is replaced by the finite difference equation.

One can extend the vorticity equation, Equation (4.119) to allow for the variation in the Coriolis parameter (planetary vorticity) f to rewrite Equation (4.119) as

$$\frac{\partial \xi}{\partial t} + V \cdot \nabla(\xi + f) = 0. \tag{4.139}$$

Evaluating the dot product present in Equation (4.139), one obtains

$$\frac{\partial \xi}{\partial t} + u\frac{\partial(\xi + f)}{\partial x} + v\frac{\partial(\xi + f)}{\partial y} = 0. \tag{4.140}$$

Utilizing the stream function relations as indicated in Equation (4.118), Equation (4.140) becomes

$$\frac{\partial \xi}{\partial t} - \frac{\partial \psi}{\partial y}\frac{\partial(\xi + f)}{\partial x} + \frac{\partial \psi}{\partial x}\frac{\partial(\xi + f)}{\partial y} = 0. \tag{4.141}$$

The nonlinear advective terms (last two terms) in Equation (4.141) can be written in terms of the Jacobian $J[\psi, (\xi + f)]$ operator with the Jacobian $J(p,q)$ defined as

$$J(p,q) = \frac{\partial p}{\partial x}\frac{\partial q}{\partial y} - \frac{\partial p}{\partial y}\frac{\partial q}{\partial x}. \tag{4.142}$$

Equation (4.142) can also be written in the following flux forms

$$J(p,q) = \frac{\partial}{\partial x}\left(p\frac{\partial q}{\partial y}\right) - \frac{\partial}{\partial y}\left(p\frac{\partial q}{\partial x}\right) \tag{4.143}$$

and

$$J(p,q) = \frac{\partial}{\partial y}\left(q\frac{\partial p}{\partial x}\right) - \frac{\partial}{\partial x}\left(q\frac{\partial p}{\partial y}\right). \tag{4.144}$$

Because of these flux forms, if p or q is either zero or constant on the boundary of a domain, then the domain integral of $\iint J(p,q)\,dxdy = 0$. Furthermore,

$$\langle p, J(p,q) \rangle = \langle J\left(\tfrac{1}{2}p^2, q\right) \rangle = 0 \tag{4.145}$$

$$\langle q, J(p,q) \rangle = \langle J\left(p, \tfrac{1}{2}q^2\right) \rangle = 0 \tag{4.146}$$

In the aforementioned case, p is ψ (stream function) and q is $(\xi + f)$. It has been shown that multiplying the vorticity equation by a stream function yielded conservation of mean kinetic energy, while multiplying the vorticity equation by vorticity yielded conservation of mean enstrophy.

To prevent nonlinear computational instability, Arakawa set out to construct a finite difference scheme that could conserve the mean values of the vorticity, kinetic energy, and enstrophy; the same properties that the continuous governing nonlinear barotropic equations had conserved. Arakawa rightly deduced that construction of such a finite difference scheme would ensure that a systematic transfer of energy towards the highest wavenumbers would not occur and the nonlinear computational instability discovered by Phillips would be prevented. The method proposed by Arakawa for doing this is to find a form of the Jacobian term in Equation (4.141) that had the appropriate conservation properties that the continuous governing equations exhibited.

There are three alternate ways in which to write the Jacobian in Equation (4.141).

$$J_o^{++}(\psi, \xi + f) = \frac{\partial \psi}{\partial x} \frac{\partial(\xi + f)}{\partial y} - \frac{\partial \psi}{\partial y} \frac{\partial(\xi + f)}{\partial x} \tag{4.147}$$

$$J_o^{+x}(\psi, \xi + f) = \frac{\partial}{\partial x}\left[\psi \frac{\partial(\xi + f)}{\partial y}\right] - \frac{\partial}{\partial y}\left[\psi \frac{\partial(\xi + f)}{\partial x}\right] \tag{4.148}$$

$$J_o^{x+}(\psi, \xi + f) = \frac{\partial}{\partial y}\left[(\xi + f) \frac{\partial \psi}{\partial x}\right] - \frac{\partial}{\partial x}\left[(\xi + f) \frac{\partial \psi}{\partial y}\right] \tag{4.149}$$

Based on the aforementioned various expressions for Jacobians, one can consider several possible ways to construct second-order accurate finite difference approximations to the Jacobian. With the simplest centered space finite differencing, one requires values of ψ and ξ from a box of nine adjacent grid points to evaluate the Jacobians expressed in Equations (4.147)–(4.149) (refer to Figure 4.3). Let the grid size be equal in both x and y directions and let it equal to d. Let ψ_n and $(\xi + f)_n$ be the values of ψ and $\xi + f$ at the point n.

Figure 4.3 shows the labels associated with a stencil of grid points involved in the finite difference expressions for the Arakawa Jacobian. Using these labels, the corresponding finite difference approximations to Equations (4.147)–(4.149) are as follows:

$$J_o^{++} = \frac{(\psi_1 - \psi_3)[(\xi + f)_2 - (\xi + f)_4] - (\psi_2 - \psi_4)[(\xi + f)_1 - (\xi + f)_3]}{4d^2} \tag{4.150}$$

$$J_o^{+x} = \frac{\psi_1[(\xi + f)_5 - (\xi + f)_8] - \psi_3[(\xi + f)_6 - (\xi + f)_7] - \psi_2[(\xi + f)_5 - (\xi + f)_6] + \psi_4[(\xi + f)_8 - (\xi + f)_7]}{4d^2} \tag{4.151}$$

$$J_o^{x+} = \frac{(\xi+f)_2(\psi_5-\psi_6) - (\xi+f)_4(\psi_8-\psi_7) - (\xi+f)_1(\psi_5-\psi_8) + (\xi+f)_3(\psi_6-\psi_7)}{4d^2}$$

$$(4.152)$$

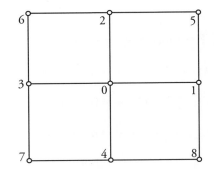

Figure 4.3 Stencil of grid points for evaluating the Arakawa Jacobian.

A more general expression for the approximation to the Jacobian is as follows:

$$J(\psi,\zeta) = \alpha J_o^{++} + \beta J_o^{+x} + \gamma J_o^{x+} \tag{4.153}$$

subject to the consistency requirement that $\alpha + \beta + \gamma = 1$. We need to choose α, β, and γ in such a manner that Equation (4.153) ensures conservation of mean kinetic energy and enstrophy. Consider the gain in enstrophy at point '0' (zero) in Figure 4.3 owing to the value at point '1' and vice versa. The results, are given in Equations (4.154) and (4.155), excluding the factor $4d^2$,

$$(\xi+f)_o J_o^{++} \sim -(\xi+f)_o(\xi+f)_1(\psi_2-\psi_4) + \text{other terms} \tag{4.154}$$

$$(\xi+f)_1 J_1^{++} \sim (\xi+f)_1(\xi+f)_o(\psi_5-\psi_8) + \text{other terms} \tag{4.155}$$

As the aforementioned terms in Equations (4.154) and (4.155) do not cancel with one another, and there is no other situation for the product $\xi_o\xi_1$ to occur over the grid, the sum $\langle \xi J^{++}\rangle$ cannot vanish in general and, hence, $\langle \xi^2\rangle$ will not be conserved [i.e., $\partial(\xi^2)/\partial t \neq 0$]. Similarly, the following terms have

$$(\xi+f)_o J_o^{+x} \sim (\xi+f)_o(\xi+f)_5(\psi_1-\psi_2) + \text{other terms} \tag{4.156}$$

$$(\xi+f)_5 J_5^{+x} \sim (\xi+f)_5(\xi+f)_o(\psi_2-\psi_1) + \text{other terms} \tag{4.157}$$

$$(\xi+f)_o J_0^{x+} \sim -(\xi+f)_o(\xi+f)_1(\psi_5-\psi_8) + \text{other terms} \tag{4.158}$$

$$(\xi+f)_1 J_1^{x+} \sim (\xi+f)_1(\xi+f)_o(\psi_2-\psi_4) + \text{other terms} \tag{4.159}$$

From Equations (4.156) and (4.157), it is clear that J^{+x} conserves enstrophy $\langle \xi^2 \rangle$. Moreover, it is clear from Equations (4.154) and (4.158) as well as Equations (4.155) and (4.159), that

$$\frac{1}{2}\{J^{++} + J^{x+}\}$$

conserves enstrophy $\langle \xi^2 \rangle$. The following terms have

$$\psi_0 J_0^{x+} \sim \psi_0 \psi_5 \left[(\xi + f)_2 - (\xi + f)_1 \right] + \text{other terms} \tag{4.160}$$

$$\psi_5 J_5^{x+} \sim \psi_0 \psi_5 \left[(\xi + f)_1 - (\xi + f)_2 \right] + \text{other terms} \tag{4.161}$$

$$\psi_0 J_0^{++} \sim \psi_0 \psi_1 \left[(\xi + f)_2 - (\xi + f)_4 \right] + \text{other terms} \tag{4.162}$$

$$\psi_1 J_1^{++} \sim \psi_1 \psi_0 \left[(\xi + f)_8 - (\xi + f)_5 \right] + \text{other terms} \tag{4.163}$$

$$\psi_0 J_0^{+x} \sim \psi_0 \psi_1 \left[(\xi + f)_5 - (\xi + f)_8 \right] + \text{other terms} \tag{4.164}$$

$$\psi_1 J_1^{+x} \sim \psi_1 \psi_0 \left[(\xi + f)_4 - (\xi + f)_2 \right] + \text{other terms} \tag{4.165}$$

A similar examination of the finite difference approximation of $\psi J(\psi, \xi)$ shows that both J^{x+} [Equations (4.160) and (4.161)] and $\frac{1}{2}\{J^{++} + J^{+x}\}$ [Equations (4.162) and (4.165)] conserves kinetic energy. Furthermore, both enstrophy and kinetic energy are conserved using the following combination of Jacobians

$$\frac{1}{3}\{J^{++} + J^{x+} + J^{+x}\}.$$

Exercises 4a (Questions only)

1. Show that the backward in time and central in space (BTCS) finite difference scheme is a consistent scheme for the linear one-dimensional heat conduction equation.

2. Show that the central in time and central in space (CTCS) finite difference scheme is a consistent scheme for the linear one-dimensional heat conduction equation.

3. Obtain an expression for the truncation error of the forward in time and central in space (FTCS) finite difference scheme for the linear one-dimensional heat conduction equation

$$\frac{\partial u}{\partial t} = \frac{\partial^2 u}{\partial x^2} \tag{E3.1}$$

Answer: $T_i^n = \left(\dfrac{\partial u}{\partial t} - \dfrac{\partial^2 u}{\partial x^2}\right)_i^n + \left[\dfrac{\Delta t}{2}\dfrac{\partial^2 u}{\partial t^2} - \dfrac{(\Delta x)^2}{12}\dfrac{\partial^4 u}{\partial x^4}\right]_i^n + \left[\dfrac{(\Delta t)^2}{6}\dfrac{\partial^3 u}{\partial t^3}\right]_i^n + O\left[(\Delta t)^3,(\Delta x)^4\right]$

4. Show that by choosing $\Delta t/(\Delta x)^2 = 1/6$, it is possible to obtain a truncation error of the forward in time and central in space (FTCS) finite difference scheme for the linear one-dimensional heat conduction equation (E3.1) to be second-order accurate in time and fourth-order accurate in space, i.e., $O[(\Delta t)^2,(\Delta x)^4]$.

5. Obtain an expression for the truncation error of the DuFort–Frankel finite difference scheme for the linear one-dimensional heat conduction equation (E3.1).

Answer: $T_i^n = \left\{\dfrac{\partial u}{\partial t} - \dfrac{\partial^2 u}{\partial x^2} + \left[\dfrac{(\Delta t)^2}{(\Delta x)^2}\dfrac{\partial^2 u}{\partial t^2}\right]\right\}_i^n + \left[\dfrac{(\Delta t)^2}{6}\dfrac{\partial^3 u}{\partial t^3} - \dfrac{(\Delta x)^2}{12}\dfrac{\partial^4 u}{\partial x^4}\right]_i^n + $

$O\left[\dfrac{(\Delta t)^4}{(\Delta x)^2},(\Delta t)^4,(\Delta x)^4\right]$

6. Obtain an expression for the truncation error of the DuFort–Frankel finite difference scheme for the linear one-dimensional heat conduction equation (E3.1) by setting $r_o = \Delta t/\Delta x$ and letting r_o to be a positive constant.

Answer: $T_i^n = \left(\dfrac{\partial u}{\partial t} - \dfrac{\partial^2 u}{\partial x^2} + r_o^2\dfrac{\partial^2 u}{\partial t^2}\right)_i^n + O\left[(\Delta x)^2\right]$

7. Show from the expression of the truncation error of the DuFort–Frankel finite difference scheme of Exercise 4a Q 6, that with $\Delta x \to 0$, the DuFort–Frankel scheme is consistent with the hyperbolic equation

$$\dfrac{\partial u}{\partial t} + r_o^2\dfrac{\partial^2 u}{\partial t^2} = \dfrac{\partial^2 u}{\partial x^2}.$$

8. Obtain an expression for the truncation error of the DuFort–Frankel finite difference scheme for the linear one-dimensional heat conduction equation (E3.1) by setting $r_o = \Delta t/(\Delta x)^2$ and letting r_o to be a positive constant.

Answer: $T_i^n = \left[r_o^2(\Delta x)^2\dfrac{\partial^2 u}{\partial t^2} + \dfrac{(\Delta t)^2}{6}\dfrac{\partial^3 u}{\partial t^3} - \dfrac{(\Delta x)^2}{12}\dfrac{\partial^4 u}{\partial x^4}\right]_i^n + O\left[r_o^4(\Delta x)^6,(\Delta t)^4,(\Delta x)^4\right].$

9. Show that the DuFort–Frankel finite difference scheme for the linear one-dimensional heat equation is a consistent scheme if $\Delta x \to 0$, $\Delta t \to 0$, and $\Delta t/\Delta x \to 0$.

10. Write down the finite difference expression for the Crank–Nicolson scheme for the linear two-dimensional heat conduction equation, where α is coefficient of thermal conductivity

Answer: $u_{i,j}^{n+1} = u_{i,j}^n + \dfrac{\alpha\Delta t}{(\Delta x)^2}\delta_x^2\left(u_{i,j}^{n+1} + u_{i,j}^n\right)/2 + \dfrac{\alpha\Delta t}{(\Delta y)^2}\delta_y^2\left(u_{i,j}^{n+1} + u_{i,j}^n\right)/2$, where

$\delta_x^2 u_{i,j}^n = u_{i+1,j}^n - 2u_{i,j}^n + u_{i-1,j}^n$ $\qquad$ $\delta_x^2 u_{i,j}^{n+1} = u_{i+1,j}^{n+1} - 2u_{i,j}^{n+1} + u_{i-1,j}^{n+1}$

$\delta_y^2 u_{i,j}^n = u_{i,j+1}^n - 2u_{i,j}^n + u_{i,j-1}^n$ $\qquad$ $\delta_y^2 u_{i,j}^{n+1} = u_{i,j+1}^{n+1} - 2u_{i,j}^{n+1} + u_{i,j-1}^{n+1}$

11. Obtain the amplification factor expression for the Crank–Nicolson scheme for the linear two-dimensional heat conduction equation.

Answer: $\rho = \dfrac{1-2\left[\alpha\Delta t/(\Delta x)^2\right]\sin^2(\xi/2) - 2\left[\alpha\Delta t/(\Delta y)^2\right]\sin^2(\eta/2)}{1+2\left[\alpha\Delta t/(\Delta x)^2\right]\sin^2(\xi/2) + 2\left[\alpha\Delta t/(\Delta y)^2\right]\sin^2(\eta/2)}$,

where ξ and η are the wavenumbers in the x and y directions, respectively.

12. Show that the Crank–Nicolson scheme for the linear two-dimensional heat conduction equation is a stable scheme.

Answer: As the magnitude of the amplification factor ρ is always ≤ 1 (refer to Exercise 4a Q 11), the Crank–Nicolson scheme for the linear two-dimensional heat conduction equation is a stable scheme.

13. Obtain the amplification factor expression for the alternate direction implicit scheme for the linear two-dimensional heat conduction equation.

Answer: $\rho = \dfrac{\left\{1-2\left[\alpha\Delta t/(\Delta x)^2\right]\sin^2(\xi/2)\right\}\left\{1-2\left[\alpha\Delta t/(\Delta y)^2\right]\sin^2(\eta/2)\right\}}{\left\{1+2\left[\alpha\Delta t/(\Delta x)^2\right]\sin^2(\xi/2)\right\}\left\{1+2\left[\alpha\Delta t/(\Delta y)^2\right]\sin^2(\eta/2)\right\}}$

14. Show that the alternate direction implicit scheme for the linear two-dimensional heat conduction equation is a stable scheme.

Answer: As the magnitude of the amplification factor ρ is always ≤ 1 (refer to Exercise 4a Q 13), the alternate direction implicit scheme for the linear two-dimensional heat conduction equation is a stable scheme.

15. Show that the central in time and central in space (CTCS) finite difference scheme is a consistent scheme for the one-dimensional linear advection equation.

16. Write down the CTCS finite difference equation for the one-dimensional linear heat conduction equation. This scheme is unconditionally unstable. If n is the time index and $n-1$ replaces n in the RHS of the CTCS scheme, write down the amplification factor of the aforementioned scheme.

Answer: $\lambda^2 = 1 - 4r\sin^2\left(\dfrac{k\Delta x}{2}\right)$, where $r = \dfrac{\alpha\Delta t}{(\Delta x)^2}$

17. Discuss the stability of the scheme indicated in Exercise 4a Q 16.

Answer: The scheme is a conditionally stable scheme with $r \leq 1/4$ for stability.

18. Show that the forward in time and central in space (FTCS) finite difference scheme is a consistent scheme for the one-dimensional linear advection equation.

19. Show that the backward in time and central in space (BTCS) finite difference scheme is a consistent scheme for the one-dimensional linear advection equation.

20. Show whether the Crank–Nicolson finite difference scheme is a consistent scheme for the one-dimensional linear advection equation.

21. Show whether the Lax finite difference scheme is a consistent scheme for the one-dimensional linear advection equation.

22. Show whether the Lax–Wendroff finite difference scheme is a consistent scheme for the one-dimensional linear advection equation.

23. Show that it is possible to realize the same effects, i.e., effects of nonlinear computational instability, in a linear partial differential equation provided that at some initial time t, the values are same as given in Exercise 4b Q 4, i.e., $U_1 = 0$, let $U_2 > 0$, let $U_3 < 0$, and let $U_4 = 0$.

    Answer: The same effects as in Exercise 4b Q 4 will be realized even if one is dealing with a linear finite difference equation such as

$$\frac{\partial U_j}{\partial t} = -a_j \frac{U_{j+1} - U_{j-1}}{2\Delta x}.$$

Ecercises 4b (Questions and answers)

1. A simple work-around solution was suggested by Norman Phillips to handle "aliasing" or "nonlinear computational instability." Discuss briefly this simple work-around solution.

    Answer: Norman Phillips in 1959, encountered the problem of "aliasing" or "nonlinear computational instability" while solving the nonlinear barotropic vorticity equation. He suggested that if one were to perform Fourier analysis periodically on the grid point numerical model output to successfully eliminate all the waves that have wavelengths less than 4Δ, where Δ is the grid size, it would provide a simple work-around solution.

2. How do we estimate the numerical stability of a nonlinear partial differential equation mapped into finite differences scheme?

    Answer: The energy method is the most appropriate scheme in dealing with matters related to the stability of the solution of nonlinear evolution equations.

3. Why does numerical computational instability develop?

    Answer: Most generally, the instability of the numerical solution starts manifesting as one approaches the boundary. Moreover, it is observed that the instability of the numerical solution can manifest in problems that allow for phase changes, especially in situations as one approaches the phase change.

4. Provide a simple example of nonlinear computational instability.

    Answer: Consider the following partial differential equation:

$$\frac{\partial u}{\partial t} = -u \frac{\partial u}{\partial x}$$

and its corresponding finite difference expression as

$$\frac{\partial U_j}{\partial t} = -U_j \frac{U_{j+1} - U_{j-1}}{2\Delta x},$$

where j is space index and Δx is the grid size. At one particular time t, let $U_1 = 0$, let $U_2 > 0$, let $U_3 < 0$, and let $U_4 = 0$. Then, according to the corresponding finite difference expression, one would obtain the following:

$$\frac{\partial U_1}{\partial t} = 0, \qquad \frac{\partial U_2}{\partial t} > 0, \qquad \frac{\partial U_3}{\partial t} < 0, \qquad \frac{\partial U_4}{\partial t} = 0.$$

With time, the values of U_2 and U_3 would grow without bound and the associated finite difference equation would blow up leading to the simple example of nonlinear computational instability.

5. Can "chopping" of all the high-wave numbers (wavenumbers from $\pi/2$ to π) really help in overcoming the problems related to nonlinear computational instability?

   Answer: For any variable, the wavenumber ranges from 0 to π. However, appearance of the quadratic nonlinear term can have a maximum wavenumber of 2π. Wavenumbers corresponding to 2π cannot, however, be represented in a grid of size Δx. However, if one were to resort to chopping of all the high-wave numbers (wavenumbers from $\pi/2$ to π), this would provide only the wavenumber ranges from 0 to $\pi/2$, for any single variable. Now, with the appearance of the quadratic nonlinear term, one can have a maximum wavenumber of π only, which is possible to represent in a grid of size Δx. However, this method of chopping of all the high-wave numbers (wavenumbers from $\pi/2$ to π), is not a useful one, as one-half of the spectrum is not put to use and, hence, is not a durable solution to overcoming the problems related to nonlinear computational instability.

6. Indicate how a Shapiro filter is utilized to filter the high frequency wavenumbers from a grid point model to provide a means of circumventing the harmful effects of nonlinear computational instability.

   Answer: For grid point models, it has been observed that complete Fourier filtering of the high wave numbers is not necessary. Some models filter high wave numbers but only enough to maintain computational stability. One such filter, the Shapiro filter, is defined as follows:

$$\overline{U}_j^{2n} = [1 - (-D)^n] U_j$$

$$DU_j = \tfrac{1}{4} \left(U_{j+1} - 2U_j + U_{j-1} \right)$$

where the diffusion operator D is applied to the original fields n times. This ensures efficient filtering of the shortest waves (mostly between $2\Delta x$ and $3\Delta x$) without affecting waves of wavelength of $4\Delta x$ or longer. Hence, we obtain an economical and accurate model.

7. What are "quadratically conserving schemes"? How do these schemes help in circumventing the problems related to the nonlinear computational instability?

Answer: Quadratically conserving schemes are a special type of finite difference schemes that conserves both the mean value, of say A and its mean square value A^2 when integrated over a closed domain, i.e., schemes for which

$$\frac{\partial}{\partial t} \sum_{i,j} A = 0 \quad \text{and} \quad \frac{\partial}{\partial t} \sum_{i,j} A^2 = 0$$

A quadratically conserving finite difference scheme will in general not have to deal with the problems related to the nonlinear computational instability.

8. Give an example of a quadratically conserving scheme for a non-divergent atmosphere.

Answer: To obtain a "quadratically conserving scheme," one needs to write a forecast equation for A to be consistent with the continuity equation. Moreover, one obtains an estimate of A at the walls of each cell as a simple average with the neighboring cell. For a non-divergent flow, the continuity equation is a statement of zero divergence, which is given in finite difference form as

$$\frac{u_{i+1/2,j} - u_{i-1/2,j}}{\Delta x} + \frac{v_{i,j+1/2} - v_{i,j-1/2}}{\Delta y} = 0$$

$$\frac{\partial A_{i,j}}{\partial t} = -\frac{u_{i+1/2,j}(A_{i,j} + A_{i+1,j}) - u_{i-1/2,j}(A_{i,j} + A_{i-1,j})}{2\Delta x}$$

$$- \frac{v_{i,j+1/2}(A_{i,j} + A_{i,j+1}) - v_{i,j-1/2}(A_{i,j} + A_{i,j-1})}{2\Delta y}$$

This ensures that $\sum_{i,j} A_{i,j} \frac{\partial A_{i,j}}{\partial t} = 0$ as the fluxes at the walls cancel out. Furthermore, what leaves one cell, will enter the neighboring cell.

9. Indicate one simple and useful way of taking care of the aliasing issue.

Answer: One solution is to systematically filter out or remove waves that have the wavenumber $k > (2/3)k_{max}$.

10. Mention the advantages of having a finite difference scheme that conserves both energy and enstrophy.

Answer: A system in which the finite difference scheme conserves both energy and enstrophy has the characteristic that the fraction of energy flowing to higher wavenumbers is limited; this ensures that the energy cascade in such a system avoids nonlinear computational instability.

Python examples

1. The following program shows the Python implementation of solving the one-dimensional linear heat conduction equation $\frac{\partial u}{\partial t} = \alpha \frac{\partial^2 u}{\partial x^2}$ using the forward in time and central in space (FTCS) scheme, which is a conditionally stable scheme with the following condition to be satisfied, $\frac{\alpha \Delta t}{\Delta x^2} \leq 0.5$. In the program, $\alpha = 4, 0 \leq x \leq L$; $L = 10$; $\Delta x = 0.4$; $\Delta t = 0.016$. The initial conditions on 'u' is $u(x, t = 0) = sin(\pi x/L)$: Boundary conditions are $u(x = 0, t) = 0$; $u(x = L, t) = 0$. The condition to be satisfied for FTCS scheme to be stable is $\frac{\alpha \Delta t}{\Delta x^2} \leq 0.5$. For this case, $\frac{\alpha \Delta t}{\Delta x^2} = 0.42$ is clearly a stable condition. However, with larger time step, $\Delta t = 0.05$, the FTCS scheme is unstable.

```python
#Heat conduction in FTCS
import numpy as np
import matplotlib.pyplot as plt
L = 10                                  # Domain length
alpha = 4                               # Diffusion coefficient
T = 10                                  # Maximum time
dX = float(L)/25
dT = float(T)/600                       # Stable condition
#dT = float(T)/200                          # Unstable condition
u = np.zeros((int(T/dT+1),int(L/dX+1)))
A = np.zeros((int(T/dT+1),int(L/dX+1)))
x = np.linspace(0,L,int(L/dX+1))        #space
t = np.linspace(0,T,int(T/dT+1))        #Time
u[0,:] = np.sin(np.pi*x/float(L))
r = alpha*dT/float(dX**2)
for i in np.linspace(1,T/dT-1,int(T/dT-1)).astype(int):
        # Time
    for j in np.linspace(1,L/dX-1,int(L/dX-1)).astype(int) :
            # Space
        u[i,j] = r*(u[i-1,j+1]-2*u[i-1,j]+u[i-1,j-1])+u[i-1,j]
plt.axis([0,10,0,1])
plt.plot(x,u[0,:],'k.')
plt.plot(x,u[100,:],'k-')
plt.plot(x,u[200,:],'k--')
plt.legend(["Time step = 0","Time step = 100","Time step = 200"])
plt.ylabel('u')
plt.xlabel('x')
plt.title("")
```

```
plt.show()
```

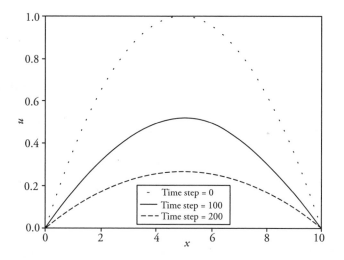

Figure 4.4   Solution of the one-dimensional linear heat conduction equation using the FTCS stable scheme.

2. The following program shows the Python implementation of solving the one-dimensional linear heat conduction equation $\frac{\partial u}{\partial t} = \alpha \frac{\partial^2 u}{\partial x^2}$ using the central in time and central in space (CTCS) scheme, which is an unconditionally unstable scheme. In the program, $\alpha = 4, 0 \leq x \leq L$; $L = 10$; $\Delta x = 0.4$; $\Delta t = 0.02$. The initial conditions on 'u' is $u(x, t = 0) = sin(\pi x/L)$: Boundary conditions are $u(x = 0, t) = 0$; $u(x = L, t) = 0$. For the first time marching, forward in time differencing is used and from $t = \Delta t$ onwards, central differencing in time is utilized.

```
#Heat conduction in CTCS method
import numpy as np
import matplotlib.pyplot as plt
L = 10
alpha = 4
T = 10
dX = float(L)/25
dT = float(T)/501
u = np.zeros((int(T/dT+1),int(L/dX+1)))
x = np.linspace(0,L,int(L/dX+1))          #Space
t = np.linspace(0,T,int(T/dT+1))          #Time
u[0,:] = np.sin(np.pi*x/float(L))
r = alpha*dT/float(dX**2)
```

```
#FTCS
for j in np.linspace(1,L/dX-1,int(L/dX-1)).astype(int) :
        # Space
    u[1,j] = r*(u[0,j+1]-2*u[0,j]+u[0,j-1])+u[0,j]
r2 = r*2
#CTCS
for i in np.linspace(2,T/dT-1,int(T/dT-2)).astype(int):
    # Time
    for j in np.linspace(1,L/dX-1,int(L/dX-1)).astype(int) :
        # Space
        u[i,j] = r2*(u[i-1,j+1]-2*u[i-1,j]+u[i-1,j-1])+u[i-2,j]
plt.axis([0,10,0,1])
plt.plot(x,u[0,:],'k.')
plt.plot(x,u[10,:],'k-')
plt.plot(x,u[26,:],'k--')
plt.legend(["Time_step_=_0","Time_step_=_10","Time_step_=_26"])
plt.ylabel('u')
plt.xlabel('x')
plt.title("")
plt.show()
#End program CTCS
```

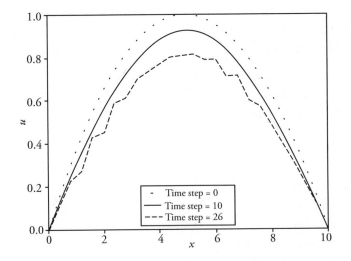

Figure 4.5 Solution of the one-dimensional linear heat conduction equation using the CTCS scheme.

3. The following program shows the Python implementation of solving the one-dimensional linear heat conduction equation $\frac{\partial u}{\partial t} = \alpha \frac{\partial^2 u}{\partial x^2}$ using the backward in time and central in space (BTCS) scheme, which is an unconditionally stable scheme. In the program, $\alpha = 4, 0 \le x \le L$; $L = 10$; $\Delta x = 0.4$; $\Delta t = 0.02$. The initial conditions on 'u' is $u(x, t = 0) = sin(\pi x/L)$: Boundary conditions are $u(x = 0, t) = 0$; $u(x = L, t) = 0$. The Thomas algorithm, which is a very efficient algorithm to solve a system of linear equations where the coefficient matrix is tri-diagonal, is employed in the Python program. The BTCS scheme despite being an unconditionally stable scheme is still a scheme that is only first-order accurate in time

```python
# One D linear heat conduction equation in BTCS method
import numpy as np
import matplotlib.pyplot as plt
L = 10
alpha = 4
T = 10
dX = float(L)/25
dT = float(T)/501
u = np.zeros((int(T/dT+1),int(L/dX)))
A = np.zeros((int(T/dT+1),int(L/dX+1)))
x = np.linspace(0,L,int(L/dX))
t = np.linspace(0,T,int(T/dT+1))
u[0,:] = np.sin(np.pi*x/float(L))
r = alpha*dT/float(dX**2)
a = r
b = -(2*r+1)
c = r
row = np.zeros(int(L/dX-1))
Ad = np.zeros((int(L/dX-1),3))
ia = np.asarray(range(int(L/dX-3)))+1
Ad[0,0] , Ad[0,1] , Ad[0,2] = 0 , b , c
Ad[int(L/dX-2),0] , Ad[int(L/dX-2),1] , Ad[int(L/dX-2),2] = a ,
    b , 0
for i in ia :
    Ad[i,0] , Ad[i,1] , Ad[i,2] = a , b , c
cp = np.zeros(int(L/dX-1))
cp[0]=c/b
ia = np.asarray(range(int(L/dX-2)))+1
for i in ia :
    cp[i] = c/(b-a*cp[i-1])
```

```python
dp = np.zeros(int(L/dX-2))
dp[:] = u[0,1:int(L/dX-1)]
ia = np.asarray(range(int(T/dT-2)))+1
for i in ia :
    tdp = dp
    tdp[0] = tdp[0]/b
    for j in np.asarray(range(np.size(tdp)-1))+1 :
        tdp[j] = (tdp[j]-a*tdp[j-1])/(b-a*cp[j-1])
    nX = tdp
    for j in np.asarray(range(np.size(nX)-1))+1 :
        li = np.size(nX)-j-1
        nX[li] = nX[li] - cp[li]*nX[li+1]
    d = nX
    u[i,1:int(L/dX-1)] = nX
plt.figure(0)
plt.axis([0,L,0,1])
plt.plot(x,u[0,:],'k.')
plt.plot(x,u[100,:],'k-')
plt.plot(x,u[200,:],'k--')
plt.legend(["Time step = 0","Time step = 100","Time step = 200"])
plt.ylabel('u')
plt.xlabel('x')
plt.title("")
plt.show()
# End Program
```

4. The following program shows the Python implementation of solving the one-dimensional linear heat conduction equation $\frac{\partial u}{\partial t} = \alpha \frac{\partial^2 u}{\partial x^2}$ using the Crank–Nicolson scheme, which is an unconditionally stable scheme. In the program, $\alpha = 4, 0 \leq x \leq L;\ L = 1;\ \Delta x = 0.02;\ \Delta t = 0.0033$. The initial conditions on 'u' is $u(x,t=0) = sin(\pi x/L)$: Boundary conditions are $u(x=0,t) = 0;\ u(x=L,t) = 0$. Thomas algorithm which is a very efficient algorithm to solve a system of linear equations where the coefficient matrix is tri-diagonal, is employed in the Python program. This is a scheme that is second-order accurate in time.

```python
#Crank-Nicolson Scheme
import math
import numpy as np
from scipy import sparse
from mpl_toolkits.mplot3d import Axes3D
```

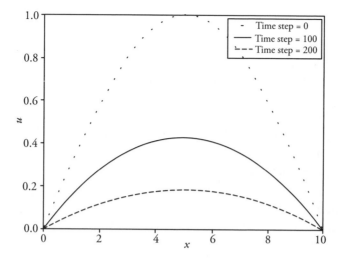

Figure 4.6 Solution of the one-dimensional linear heat conduction equation using the BTCS scheme.

```
import matplotlib.pyplot as plt
from matplotlib import cm
M = 50
N = 60
x0 = 0
L = 1
dx = (L - x0)/(M - 1)
t0 = 0
tF = 0.2
dt = (tF - t0)/(N - 1)
alpha = 4
r = dt*alpha/dx**2
x = np.linspace(x0, L, M)
t = np.linspace(t0, tF, N)
main_diag = ((2 + 2*r)/r)*np.ones((1,M-2))
off_diag = -1*np.ones((1, M-3))
a = main_diag.shape[1]
diagonals = [main_diag, off_diag, off_diag]
A = sparse.diags(diagonals, [0,-1,1], shape=(a,a)).toarray()
U = np.zeros((M, N))
U1 = np.zeros((M, N))
```

```python
print(x)
#----- Initial condition -----
for i in range(0,M):
    U[i,0] = math.sin(math.pi*(x[i])/L)
U[0,:] = 0.0
U[-1,:] = 0.0
print(math.sin(math.pi*0.40816327))
for k in range(1, N):
    for i in range(1,M-2):
        U1[i,k-1]=U[i,k-1]+((2-2*r)/r)*U[i+1,k-1]+U[i+2,k-1]
    c = np.zeros((M-4,1)).ravel()
    b1 = np.asarray([r*U[0,k], r*U[-1,k]])
    b1 = np.insert(b1, 1, c)
    b2 = np.array(U1[1:M-1, k-1])
    b = b1 + b2
    U[1:M-1, k] = np.linalg.solve(A,b)   # Solve x=A\b
X, T = np.meshgrid(t, x)
plt.plot(x,U[:,0],'k.')
plt.plot(x,U[:,10],'k-')
plt.plot(x,U[:,15],'k--')
plt.legend(["Time_step_=_0","Time_step_=_10","Time_step_=_15"])
plt.ylabel('U')
plt.xlabel('x')
plt.title("")
plt.show()
```

5. The following program shows the Python implementation of solving the one-dimensional linear advection equation with constant velocity. $\frac{\partial u}{\partial t} + c\frac{\partial u}{\partial x} = 0$ using the forward in time and central in space (FTCS) scheme, which is an unconditionally unstable scheme. In the program, $c = 0.5, 0 \le x \le L$; $L = 10$; $\Delta x = 0.1$; $\Delta = 0.05$. The initial conditions on 'u' is $u(x, t = 0) = exp^{[-(x-2)^2]}$: Boundary condition is $u(x = 0, t) = 0$.

```python
#Advection in FTCS
import numpy as np
import matplotlib.pyplot as plt
c=0.5
delt=0.05
delx=0.1
lambd=(c*delt)/(2*delx)
```

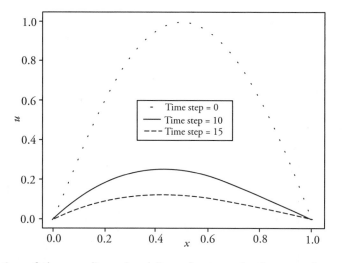

Figure 4.7 Solution of the one-dimensional linear heat conduction equation using the Crank–Nicolson scheme.

```
a=1+(10-0)/delx
b=1+(250-0)/delt
x= np.zeros(int(a))
t= np.zeros(int(a))
x[1]=0
t[1]=0
U = np.zeros((int(b),int(a)))
for i in range(2,int(a)):
    x[i]=x[i-1]+delx
for j in range(2,int(b)):
        t[i]=t[i-1]+delt
for i in range(1,int(a)):
    U[1,i]= np.exp(-(x[i]-2)**2)
U[:,1]=0
U[:,int(a-1)]=0
#FTCS
for i in range(1,int(b-1)):
    for j in range(2,int(a-1)):
        U[i+1,j]=U[i,j]-lambd*(U[i,j+1]-U[i,j-1])
plt.plot(x,U[1,:],'k.')
plt.plot(x,U[100,:],'k-')
```

```
plt.plot(x,U[200,:],'k--')
plt.legend(["Time_step_=_1","Time_step_=_100","Time_step_=_200"])
plt.ylabel('U')
plt.xlabel('x')
plt.title("")
plt.show()
```

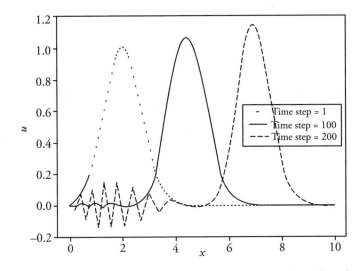

Figure 4.8 Solution of the one-dimensional linear advection equation using the FTCS scheme.

6. The following program shows the Python implementation of solving the one-dimensional linear advection equation with constant velocity $\frac{\partial u}{\partial t} + c\frac{\partial u}{\partial x} = 0$ using the central in time and central in space (CTCS) scheme, which is a conditionally stable scheme with the following condition to be satisfied, the Courant number $\left(\frac{c\Delta t}{\Delta x}\right) \le 1.0$. In the program, $c = 0.5$, $0 \le x \le L$; $L = 10$; $\Delta x = 0.1$; $\Delta t = 0.03$ with the Courant number having a value of 0.15 and, hence, for the aforementioned choice of time step, CTCS scheme is stable. The initial conditions on 'u' is $u(x, t = 0) = exp^{[-(x-2)^2]}$: Boundary condition is $u(x = 0, t) = 0$. If one were to choose a time step that violates the stability condition, CTCS scheme would be unstable.

```
#CTCS Advection
import numpy as np
import matplotlib.pyplot as plt
c=0.5
delt=0.03#Stable condition
#delt=0.3  #Unstable condition
```

```
delx=0.1
lambd=(c*delt)/(2*delx)
CF  =(c*delt)/delx

a=1+(10-0)/delx
b=1+(250-0)/delt
x= np.zeros(int(a))
t= np.zeros(int(a))
x[1]=0
t[1]=0
U = np.zeros((int(b),int(a)))
for i in range(2,int(a)):
    x[i]=x[i-1]+delx
for j in range(2,int(b)):
        t[i]=t[i-1]+delt
for i in range(1,int(a)):
 U[1,i]= np.exp(-(x[i]-2)**2)
 U[:,1]=0
 U[:,int(a-1)]=0
for j in range(2,int(a-1)):
  U[2,j]=U[1,j]-lambd*(U[1,j+1]-U[1,j-1])
for i in range(2,int(b-1)):
 for j in range(2,int(a-1)):
    U[i+1,j]=U[i-1,j]-CF*(U[i,j+1]-U[i,j-1])
plt.plot(x,U[1,:],'k.')
plt.plot(x,U[100,:],'k-')
plt.plot(x,U[200,:],'k--')
plt.legend(["Time_step_=_1","Time_step_=_100","Time_step_=_200"])
plt.ylabel('U')
plt.xlabel('x')
plt.title("")
plt.show()
```

7. The following program shows the Python implementation of solving the one-dimensional linear advection equation with constant velocity $\frac{\partial u}{\partial t} + c\frac{\partial u}{\partial x} = 0$ using the Lax scheme, which is a conditionally stable scheme with the following condition to be satisfied, the Courant number $\left(\frac{c\Delta t}{\Delta x}\right) \leq 1.0$. In the program, $c = 0.5$, $0 \leq x \leq L$; $L = 10$; $\Delta x = 0.1$; $\Delta t = 0.05$ with the Courant number having a value of 0.25 and, hence, for the aforementioned choice of time step, the Lax scheme is stable. The initial conditions on 'u' is $u(x,t=0) = exp^{[-(x-2)^2]}$:

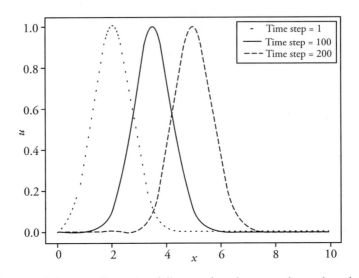

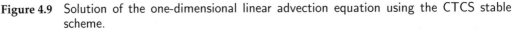

Figure 4.9 Solution of the one-dimensional linear advection equation using the CTCS stable scheme.

Boundary condition is $u(x=0,t)=0$.

```python
#one dimensional advection equation using Lax method
import numpy as np
import matplotlib.pyplot as plt
c=0.5
dt=0.05
dx=0.1
cn=(c*dt)/(2*dx)
a=1+((10-0)/dx)
b=1+((250-0)/dt)
x= np.zeros(int(a))
t= np.zeros(int(b))
u = np.zeros((int(b),int(a)))
x[1]=0
t[1]=0
for i in range(2,int(a)):
    x[i]=x[i-1]+dx
for i in range(2,int(b)):
    t[i]=t[i-1]+dt
for i in range(1,int(a)):
```

```
    u[1,i]=np.exp(-(x[i]-2)**2)
u[:,1]=0
u[:,int(a-1)]=0
for i in range(1,int(b-1)):
    for j in range(2,int(a-1)):

        u[i+1,j]=((u[i,j+1]+u[i,j-1])/2)-cn*(u[i,j+1]-u[i,j-1])
plt.plot(x,u[10,:],'k.')
plt.plot(x,u[100,:],'k-')
plt.plot(x,u[150,:],'k--')
plt.legend(["Time step = 10","Time step = 100","Time step = 150"])
plt.ylabel('u')
plt.xlabel('x')
plt.title("")
plt.show()
```

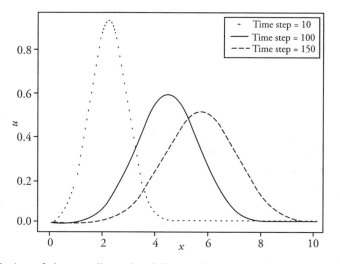

Figure 4.10 Solution of the one-dimensional linear advection equation using the Lax scheme.

8. The following program shows the Python implementation of solving the one-dimensional linear advection equation with constant velocity $\frac{\partial u}{\partial t} + c\frac{\partial u}{\partial x} = 0$ using the Lax–Wendroff scheme, which is a conditionally stable scheme with the following condition to be satisfied, the Courant number $\left(\frac{c\Delta t}{\Delta x}\right) \leq 1.0$. In the program, $c = 0.5$, $0 \leq x \leq L$; $L = 10$; $\Delta x = 0.1$; $\Delta t =$

0.05 with the Courant number having a value of 0.25 and, hence, for the aforementioned choice of time step, the Lax–Wendroff scheme is stable. The initial conditions on 'u' is $u(x, t=0) = exp^{[-(x-2)^2]}$: Boundary condition is $u(x=0,t) = 0$.

```python
#one dimensional advection equation using LAX-WENDROFF method
import numpy as np
import matplotlib.pyplot as plt
c=0.5
dt=0.05
dx=0.1
K=(c*dt)/(2*dx)
a=1+((10-0)/dx)
b=1+((250-0)/dt)
x= np.zeros(int(a))
t= np.zeros(int(b))
u = np.zeros((int(b),int(a)))
x[1]=0
t[1]=0
for i in range(2,int(a)):
    x[i]=x[i-1]+dx
for i in range(2,int(b)):
    t[i]=t[i-1]+dt
for i in range(1,int(a)):
    u[1,i]=np.exp(-(x[i]-2)**2)
u[:,1]=0
u[:,int(a-1)]=0
for i in range(1,int(b-1)):
    for j in range(2,int(a-1)):
        u[i+1,j]=u[i,j]-K*(u[i,j+1]-u[i,j-1])+(K**2)*(u[i,j+1]
            -2*u[i,j]
+u[i,j-1])
plt.plot(x,u[10,:],'k.')
plt.plot(x,u[100,:],'k-')
plt.plot(x,u[150,:],'k--')
plt.legend(["Time step = 10","Time step = 100","Time step = 150"])
plt.ylabel('u')
plt.xlabel('x')
plt.title("")
plt.show()
```

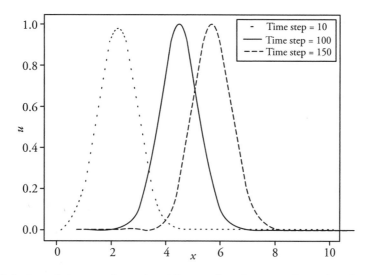

Figure 4.11 Solution of the one-dimensional linear advection equation using the Lax–Wendroff scheme.

9. The following program shows the Python implementation of solving the one-dimensional linear advection equation with constant velocity $\frac{\partial u}{\partial t} + c\frac{\partial u}{\partial x} = 0$ using the upwind or upstream scheme, which is a conditionally stable scheme with the following condition to be satisfied, the Courant number $\left(\frac{c\Delta t}{\Delta x}\right) \leq 1.0$. In the program, $c = 0.5$, $0 \leq x \leq L$; $L = 10$; $\Delta x = 0.1$; $\Delta t = 0.05$ with the Courant number having value of 0.25 and, hence, for the aforementioned choice of time step, the Lax–Wendroff scheme is stable. The initial conditions on 'u' is $u(x, t = 0) = exp^{[-(x-2)^2]}$: Boundary condition is $u(x = 0, t) = 0$.

```
#Upwind method
import numpy as np
import matplotlib.pyplot as plt
c=0.5
dt=0.05
dx=0.1
lamda=(c*dt)/(dx)
a=1+((10-0)/dx)
b=1+((250-0)/dt)
x= np.zeros(int(a))
t= np.zeros(int(b))
u = np.zeros((int(b),int(a)))
x[1]=0
t[1]=0
```

```python
for i in range(2,int(a)):
    x[i]=x[i-1]+dx
for i in range(2,int(b)):
    t[i]=t[i-1]+dt
for i in range(1,int(a)):
    u[1,i]=np.exp(-(x[i]-2)**2)
u[:,1]=0
u[:,int(a-1)]=0
for i in range(1,int(b-1)):
    for j in range(2,int(a-1)):
        u[i+1,j]= u[i,j]*(1-lamda)+lamda*u[i,j-1]
plt.plot(x,u[10,:],'k.')
plt.plot(x,u[100,:],'k-')
plt.plot(x,u[150,:],'k--')
plt.legend(["Time step = 10","Time step = 100","Time step = 150"])
plt.ylabel('u')
plt.xlabel('x')
plt.title("")
plt.show()
```

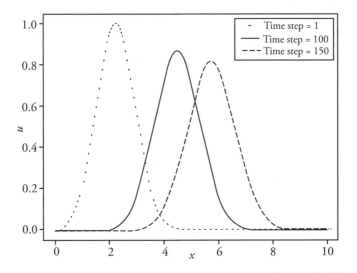

Figure 4.12 Solution of the one-dimensional linear advection equation using the upwind scheme.

5

Oscillation and Decay Equations

5.1 Introduction

The governing equations of the evolution of the atmosphere are time-dependent equations. In this chapter, the problem of time differencing is taken up for detailed discussion without considering issues related to space differencing.

5.2 Properties of Time-differencing Schemes as Applied to the Oscillation Equation

Consider a general first-order differential equation

$$\frac{du}{dt} = f(u,t) \qquad (5.1)$$

The stability and other important properties of the various (two level and three level) time-differencing schemes that were introduced in Section 3.3 depend on the form of the function $f(u,t)$ in Equation (5.1). In order to discuss these properties, one needs to prescribe the form of this function $f(u,t)$. For applications in atmospheric models, it is of interest to consider the case where $f = i\omega u$, i.e., the ordinary differential equation (ODE), where $f = f(u)$:

$$\frac{du}{dt} = i\omega u \qquad (5.2)$$

Equation (5.2) is known as the oscillation equation . It is possible for u to be complex, so that in general, the oscillation equation represents a system of two equations. The parameter ω, the frequency of the system, is taken as real. The exact solution of Equation (5.2) is of the form

$$u(t) = u_o e^{i\omega t},$$ (5.3)

where u_o is the value of u at initial time $t = 0$. The amplitude u_o is an invariant of the system, i.e.,

$$\frac{du_o}{dt} = 0.$$ (5.4)

The following are examples of more familiar equations that reduce to Equation (5.2).
(i) One-dimensional linear advection equation:

$$\frac{\partial u}{\partial t} + c\frac{\partial u}{\partial x} = 0.$$ (5.5)

If $u = u_o e^{-ikx}$, then $\dfrac{du}{dt} = ikcu = i\omega u$, where $\omega = kc$.

(ii) Rotational motion. Pure inertial motion is described by the following equations

$$\frac{du}{dt} - fv = 0.$$ (5.6a)

$$\frac{dv}{dt} + fu = 0.$$ (5.6b)

Multiplying Equation (5.6b) by i and adding it to Equation (5.6a), one gets

$$\frac{d}{dt}(u+iv) + f(-v+iu) = 0.$$ (5.7)

Defining $U = u + iv$, Equation (5.7) becomes

$$\frac{dU}{dt} + ifU = 0.$$ (5.8)

Equation (5.8) has the same form as Equation (5.2) but with $\omega = -f$.

The general solution of Equation (5.2) is $u(t) = u_o e^{i\omega t}$, or, for discrete values $t = n\Delta t$, the solution is $u(n\Delta t) = u_o e^{in\omega \Delta t}$.

If we consider the solution in the complex plane, its argument rotates by $\omega \Delta t$ each time step and there is no change in amplitude. One can utilize the von Neumann method to analyze the properties of the various two time level and three time level difference schemes.

Let $u_{n+1} = \lambda u_n$, where λ is the amplification factor and let $\lambda = |\lambda| e^{i\theta}$. Then the numerical solution can be written as

$$u_n = |\lambda|^n u_o e^{in\theta},$$ (5.9)

where θ represents the phase change of the numerical solution during each time step. As the amplitude of the true solution does not change, one requires that the condition $\lambda \leq 1$ be satisfied for the purposes of stability and classify a scheme as unstable, neutral, or damping, based on whether $|\lambda| > 1, = 1$, or < 1, respectively.

Moreover, a scheme is defined as accelerating, has no effect on the phase speed, or decelerating, based on whether $\theta / (\omega \Delta t) > 1, = 1$, or < 1.

It is always desirable for accuracy purposes to have both the amplification and the relative speed, $\dfrac{\theta}{\omega \Delta t}$, close to unity.

5.3 Properties of Various Two-time Level Differencing Schemes

The three (forward Euler, backward Euler, and the trapezoidal) two-level schemes (refer to Section 3.3) can all be described by a single finite difference equation of the form

$$u_{n+1} = u_n + \Delta t \left(\alpha f_n + \beta f_{n+1} \right), \tag{5.10}$$

with a consistency requirement that $\alpha + \beta = 1$. The forward Euler scheme has $\alpha = 1, \beta = 0$; the backward Euler scheme has $\alpha = 0, \beta = 1$; and the trapezoidal scheme has $\alpha = 0.5, \beta = 0.5$.

Equation (5.10) as applied to the oscillation equation, Equation (5.2), gives

$$u_{n+1} = u_n + i\omega \Delta t \left(\alpha f_n + \beta f_{n+1} \right). \tag{5.11}$$

As $u_{n+1} = \lambda u_n$, it follows from Equation (5.11), that

$$\lambda = \frac{1 + i\alpha p}{1 - i\beta p} = \frac{1 - \alpha \beta p^2 + i(\alpha + \beta)p}{1 + \beta^2 p^2}, \tag{5.12}$$

where $p = \omega \Delta t$.

5.3.1 Forward Euler scheme

For the forward Euler scheme,

$$\lambda = 1 + ip \quad \Rightarrow \quad |\lambda| = \sqrt{1 + p^2}. \tag{5.13}$$

The forward Euler scheme is always unstable for Equation (5.2). If $\Delta t \ll \frac{1}{|\omega|}$, then $p \ll 1$ and $|\lambda| = 1 + \frac{1}{2}p^2 + \cdots$, i.e., $|\lambda| = 1 + O[(\Delta t)^2]$.

Hence, $|\lambda| - 1$ is an order of magnitude less than the maximum allowed by the von Neumann stability condition. However, indiscriminate use of the forward Euler scheme for the solution of the atmospheric equations leads to amplification at a rate that is unacceptable.

5.3.2   Backward Euler scheme

For the backward Euler scheme,

$$\lambda = \frac{1+ip}{1+p^2} \quad \Rightarrow \quad |\lambda| = \frac{1}{\sqrt{1+p^2}}. \tag{5.14}$$

The backward Euler scheme is an unconditionally stable scheme, i.e., it is stable irrespective of the size of Δt. However, as $|\lambda| < 1$, it is a damping scheme with the amount of damping increasing as the frequency (and, hence, p) increases. This property of increasing damping with increase in frequency is a desirable characteristic of a scheme, as it leads to the selective removal of undesirable high-frequency "noise."

5.3.3   Trapezoidal scheme

For the trapezoidal scheme,

$$\lambda = \frac{1-\frac{1}{4}p^2+ip}{1+\frac{1}{4}p^2} \quad \Rightarrow \quad |\lambda| = 1. \tag{5.15}$$

The trapezoidal scheme is a neutral scheme for Equation (5.2). Both the backward Euler scheme and trapezoidal schemes are stable irrespective of the choice of Δt, i.e., both are unconditionally stable schemes for Equation (5.2).

5.3.4   Iterative two-time level scheme

The iterative two-time level schemes in two steps can be described as follows:

$$u_{n+1}^* = u_n + \Delta t f_n \tag{5.16a}$$
$$u_{n+1} = u_n + \Delta t \left(\alpha f_n + \beta f_{n+1}^* \right) \tag{5.16b}$$

with $\alpha + \beta = 1$. For the Matsuno scheme, $\alpha = 0$ and $\beta = 1$ whereas for the Heun scheme, $\alpha = \beta = 0.5$.

When Equations (5.16a) and (5.16b) are applied to the oscillation, Equation (5.2), one gets

$$u_{n+1}^* = u_n + i\omega\Delta t u_n, \tag{5.17a}$$

$$u_{n+1} = u_n + i\omega\Delta t \left(\alpha u_n + \beta u_{n+1}^*\right), \tag{5.17b}$$

where $u_{n+1} = \lambda u_n$, with λ being the amplification factor,

$$\lambda = 1 - \beta p^2 + ip. \tag{5.18}$$

5.3.5   Matsuno scheme

For the Matsuno scheme,

$$\lambda = 1 - p^2 + ip \quad \Rightarrow \quad |\lambda| = \sqrt{1 - p^2 + p^4}. \tag{5.19}$$

The Matsuno scheme is stable if $|p| \leq 1$; hence, to achieve stability, one must ensure that $\Delta t \leq \dfrac{1}{|\omega|}$.

Hence, the Matsuno scheme is conditionally stable and for this scheme, the higher the frequency, the more restrictive the stability condition. Moreover, from Equation (5.19), it can be seen that by differentiating λ with respect to p and assigning it to zero, the amplification factor for the Matsuno scheme has a minimum value at $p = 1/\sqrt{2}$.

Therefore, if we can choose Δt so that $0 < p < 1/\sqrt{2}$ for all the frequencies present, the Matsuno scheme will reduce the relative amplitudes of all the higher frequencies.

5.3.6   Heun scheme

For the Heun scheme,

$$\lambda = 1 - \frac{1}{2}p^2 + ip \quad \Rightarrow \quad |\lambda| = \sqrt{1 + \tfrac{1}{4}p^4}. \tag{5.20}$$

The amplification factor for the Heun scheme is always > 1 and, hence, the Heun scheme is always unstable for Equation (5.2). However, for small p,

$$|\lambda| = 1 + \frac{1}{8}p^4 + \cdots, \quad \text{i.e., } |\lambda| = 1 + O[(\Delta t)^4]. \tag{5.21}$$

The deviation of the amplification factor from 1 is insignificant for small p and, hence, the resulting instability is comparatively weak. It is quite possible to keep the instability within tolerable limits by choosing Δt to be sufficiently small.

The results of the five two-level schemes (forward Euler, backward Euler, trapezoidal, Matsuno, and Heun), applied to Equation (5.2), are summarized in Figure 5.1. As the amplification factors are all even functions of p, it is only necessary to show the curves for $p \geq 0$.

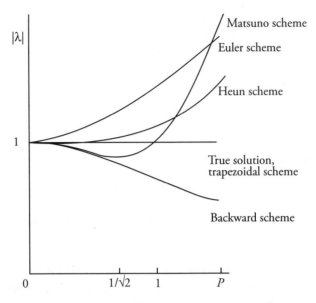

Figure 5.1 Amplification factors as a function of $p = \omega \Delta t$ for the five two-time level schemes discussed in this text and for the true solution.

5.3.7 Phase change of the various two-level schemes

It is of interest to consider the phase change per time step, θ, together with the relative phase change per time step, θ/p, θ being defined as

$$\theta = \tan^{-1}\left[\frac{\Im(\lambda)}{\Re(\lambda)}\right]. \tag{5.22}$$

Forward Euler and backward Euler schemes

From Equations (5.13) and (5.14), one gets

$$\frac{\theta}{p} = \frac{1}{p}\tan^{-1}(p). \tag{5.23}$$

As $\frac{1}{p}\tan^{-1}(p) < 1$, we have $\theta/p < 1$ and both the forward Euler and backward Euler schemes decelerate. For $p = 1$, $\theta = \pi/4$; $\theta/p = \pi/(4p)$.

Matsuno scheme

From Equations (5.19), one gets

$$\frac{\theta}{p} = \frac{1}{p}\tan^{-1}\left(\frac{p}{1-p^2}\right). \tag{5.24}$$

For stability, $|p| \leq 1$ and for relatively small frequencies ($p \ll 1$),

$$\frac{\theta}{p} = 1 + \frac{2}{3}p^2 + \cdots \tag{5.25}$$

Hence, the Matsuno scheme accelerates for Equation (5.2). For $p = 1$, $\theta = \pi/2$; $\theta/p = \pi/(2p)$. The analysis of phase errors of schemes applied to the oscillation equation is not as vital as an analysis of the amplification factor as phase errors do not affect the stability of a scheme. Moreover, the discussion here is confined to time differencing alone, whereas additional phase errors occur as a result of space differencing when solving the governing PDEs that determine the evolution of the atmosphere.

5.4 Properties of Various Three-time Level Differencing Schemes

5.4.1 Leapfrog scheme

When the leapfrog scheme is applied to the oscillation Equation, Equation (5.2), one gets

$$u_{n+1} = u_{n-1} + 2i\omega\Delta t u_n. \tag{5.26}$$

Three-time level schemes such as the leapfrog scheme require two initial conditions to initialize the calculation (i.e., u at $t = (n-1)\Delta t$ and u at $t = n\Delta t$). A single initial condition should have been adequate. However, besides the (physical) initial condition described in the previous line, three-time level schemes also require a computational initial condition u_1. This computational initial condition will have to be determined using a two-time level scheme for the first step. Using Equation (5.26), one gets after using $u_{n+1} = \lambda u_n$ and hence $u_{n+1} = \lambda^2 u_{n-1}$,

$$\lambda^2 - 2ip\lambda - 1 = 0. \tag{5.27}$$

Equation (5.27) is a quadratic equation in λ and, hence, has the following two roots:

$$\lambda_1 = \sqrt{1-p^2} + ip \quad \text{and} \quad \lambda_2 = -\sqrt{1-p^2} + ip. \tag{5.28}$$

Owing to the quadratic nature of Equation (5.27), there are two solutions λ_1 and λ_2, respectively; each such solution is called a mode. If such a solution represents an approximation to the true solution, it is clear that the amplification factor (λ) must tend to unity as $\Delta t \to 0$, i.e., as $p \to 0$. In Equation (5.28), whereas $\lambda_1 \to 1$ as $p \to 0$, $\lambda_2 \to -1$ as $p \to 0$. Thus, the solution with $\lambda = \lambda_1$ is called the physical mode whereas that with $\lambda = \lambda_2$ is not an approximation to the solution and is called the computational mode.

The stability of the leapfrog scheme is indicated as follows. For the aforementioned case, $1 - p^2 > 0$ and $|\lambda_1| = |\lambda_2| = 1$. In this range of p, both the physical and the computational modes are stable and neutral. For the phase change, using Equations (5.22) and (5.28),

$$\theta_1 = \tan^{-1}\left(\frac{p}{\sqrt{1-p^2}}\right) \quad \text{and} \quad \theta_2 = \tan^{-1}\left(\frac{-p}{\sqrt{1-p^2}}\right), \tag{5.29}$$

for $0 \le p < 1$, $\theta_2 = \pi - \theta_1$. As $p \to 0$, $\theta_1 \to p$ whereas $\theta_2 \to (\pi - p)$ so that for small Δt, the physical mode approximates the true solution and the behaviour of the computational mode is quite different. For $p < 0$, $\theta_2 = -(\pi + \theta_1)$.

The phase change is plotted in Figure 5.2 as a function of $x = \frac{p}{\sqrt{1-p^2}}$. For accuracy of the physical mode, θ_1 should approximate closely the phase change of the true solution p. For $|p| \ll 1$, Equation (5.29) reduces to

$$\theta_1 = p + \frac{1}{6}p^3 + \cdots \tag{5.30}$$

The leapfrog scheme is an accelerating scheme for Equation (5.2). However, the acceleration is four times less than that of the Matsuno scheme.

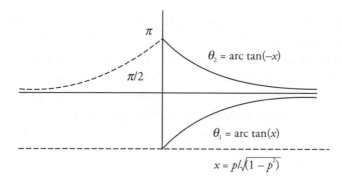

Figure 5.2 Phase change of the physical and the computational mode for the leapfrog scheme.

5.4.2 Adams–Bashforth scheme

Applying the Adams–Bashforth scheme to the oscillation equation, Equation (5.2), one gets

$$u_{n+1} = u_n + \Delta t \left[\frac{3}{2} u_n - \frac{1}{2} u_{n-1} \right]. \tag{5.31}$$

Substituting $u_{n+1} = \lambda^2 u_{n-1}$ to Equation (5.31), one gets a quadratic equation

$$\lambda^2 - \left(1 + \frac{3}{2} ip \right) \lambda + \frac{1}{2} ip = 0. \tag{5.32}$$

For small p, the expressions of the two roots of Equation (5.32) are given as follows:

$$\lambda_1 = \left(1 - \frac{1}{2} p^2 - \frac{1}{8} p^4 - \cdots \right) + i \left(p + \frac{1}{4} p^3 + \cdots \right). \tag{5.33a}$$

$$\lambda_2 = \left(\frac{1}{2} p^2 + \frac{1}{8} p^4 - \cdots \right) + i \left(\frac{1}{2} p - \frac{1}{4} p^3 + \cdots \right). \tag{5.33b}$$

The amplification factors for the Adams–Bashforth scheme are

$$|\lambda_1|^2 = 1 + \frac{1}{4} p^4 + \cdots \qquad \text{and} \qquad |\lambda_2| = \frac{1}{2} p + \cdots \tag{5.34}$$

It can be observed that the physical mode of the Adams–Bashforth scheme is always unstable. However, like the Heun scheme, the amplification is only by a fourth-order term and, hence, the instability can be tolerated when Δt is sufficiently small. Moreover, the Adams–Bashforth scheme has a useful property that the computational mode is damped.

5.5 Properties of Various Schemes as Applied to the Friction Equation

The friction equation naturally arises when one applies the separation of variables approach to the one-dimensional linear heat conduction equation:

$$\frac{\partial T}{\partial t} = \sigma \frac{\partial T}{\partial x^2}, \tag{5.35}$$

where σ is the thermal diffusivity and T is the temperature. Substituting $T(x,t) = \Re\left[u(t)e^{ikx}\right]$ to Equation (5.35), one obtains

$$\frac{du}{dt} = -\sigma k^2 u \tag{5.36}$$

As the thermal diffusivity σ is always positive, $\kappa = \sigma k^2$ is also positive and the friction equation is given by

$$\frac{du}{dt} = -\kappa u, \text{ with } u = u(t) \text{ and } \kappa > 0. \tag{5.37}$$

The general solution of friction equation

$$u(t) = u(t=0)e^{-\kappa t}, \tag{5.38}$$

which indicates that the solution decreases exponentially with time. Equation (5.37) is also known as Decay equation.

5.5.1 Application of various two-time level schemes to the friction equation

The stability of the various two-time level schemes as applied to the friction equation can be studied using the von Neumann method. The three (Euler, backward, and the trapezoidal) two-time level schemes (refer to Section 3.3) can all be described by a single finite difference equation of the form

$$u_{n+1} = u_n + \Delta t \left(\alpha f_n + \beta f_{n+1}\right). \tag{5.39}$$

Applying Equation (5.39) to the friction equation, Equation (5.37), one gets

$$u_{n+1} = u_n - \kappa \Delta t \left(\alpha u_n + \beta u_{n+1}\right). \tag{5.40}$$

Letting $K = \kappa \Delta t$, and applying it to Equation (5.40) one obtains the following expression:

$$u_{n+1} = \frac{1 - \alpha K}{1 + \beta K} u_n. \tag{5.41}$$

Forward Euler scheme

For the forward Euler scheme, applying $\alpha = 1$ and $\beta = 0$, the amplification factor $\lambda = 1 - K$. Hence, the stability requires that $|1 - K| \leq 1$ or $0 < K \leq 2$. It is immediately clear that the stability criteria of particular schemes do not have to meet the same condition when the schemes are applied to different equations. For example, the Euler scheme is always unstable when applied to the oscillation equation whereas it is conditionally stable when applied to the friction equation. For the case of Equation (5.41), one would need to be more particular in the choice of Δt by choosing $K < 1$; this choice prevents the solution represented by Equation (5.41) from oscillating from time step to time step.

Backward Euler scheme

For the backward Euler scheme, applying $\alpha = 0$ and $\beta = 1$, the amplification factor $\lambda = 1/(1+K)$. Hence, the stability requires that $K > 0$. In addition, the solution for the backward Euler scheme does not oscillate in sign.

Trapezoidal scheme

For the trapezoidal scheme, applying $\alpha = \beta = 0.5$, the amplification factor $\lambda = (1 - K)/(1 + K)$. Hence, the stability requires that $K > 0$.

Iterative two-time level scheme

The iterative two-time level scheme as given in Equations (5.16) when applied to Equation (5.37) gives

$$u_{n+1} = (1 - K + \beta K^2)u_n. \tag{5.42}$$

Hence, from Equation (5.42), it is clear that both the Matsuno ($\alpha = 0$ and $\beta = 1$) and Heun ($\alpha = \beta = 0.5$) schemes are stable for sufficiently small values of K.

Three-time level schemes

Examples of three-time level schemes are the leapfrog scheme and Adams–Bàshforth scheme. When the leapfrog scheme is applied to the friction equation, Equation (5.37), it gives

$$u_{n+1} = u_{n-1} - 2\kappa \Delta t\, u_n, \tag{5.43}$$

with the complex amplification factor having the following solutions

$$\lambda_1 = -K + \sqrt{1 + K^2}, \tag{5.44a}$$

$$\lambda_2 = -K - \sqrt{1 + K^2}, \tag{5.44b}$$

where $K = \kappa \Delta t$. The solution associated with λ_1 corresponds to the physical mode whereas the solution associated with λ_2 corresponds to the computational mode. Moreover, the solution for the computational mode is always unstable. Therefore, the leapfrog scheme is unsuitable for obtaining a solution of the friction equation.

The Adams–Bashforth scheme as applied to the friction equation, Equation (5.37), gives the following expression for the (complex) amplification factor:

$$\lambda = \frac{1}{2}\left(1 - \frac{3}{2}K \pm \sqrt{1 - K + \frac{9}{4}K^2}\right). \tag{5.45}$$

Hence, the Adams–Bashforth scheme is stable for sufficiently small values of K; moreover, the computational mode is damped.

Exercises 5a (Questions only)

1. The simple oscillation equation in one dimension may be expanded to include terms such as damping $f(u')$, a possible restoring force term $s(u)$ together with the presence of externally forced excitation $F(t)$, where t is time, u' is the derivative of u with respect to time t, and u is the displacement. Show that the simple oscillation equation can be written as

   $$m\frac{d^2u}{dt^2} + f\frac{du}{dt} + s(u) = F(t),$$

   subject to $u(t = 0) = a$ and $u'(t = 0) = b$.

2. Assuming linear damping with $f(u') = bu'$, write down the finite difference approximation using central differences for the oscillation equation described in Exercise 5a Q 1.

   Answer: $m\dfrac{u_{i+1} - 2u_i + u_{i-1}}{(\Delta t)^2} + b\dfrac{u_{i+1} - u_{i-1}}{2\Delta t} + s(u_i) = F(t_i)$

3. Show that the approximation in Exercise 5a Q 2 reduces to a matrix equation with the coefficient matrix becoming a tridiagonal system of equations that can be solved conveniently using the Thomas algorithm.

4. Assuming nonlinear (quadratic) damping with $f(u') = bu'|u'|$, write down the finite difference approximation using central differences for the oscillation equation shown in Exercise 5a Q 1.

   Answer: Introducing the following approximation:

   $$(u'|u'|)_i = u'(t_{i+1/2})|u'(t_{i-1/2})| = \frac{u_{i+1} - u_i}{\Delta t}\frac{|u_i - u_{i-1}|}{\Delta t}, \text{ we have}$$

   $$m\frac{u_{i+1} - 2u_i + u_{i-1}}{(\Delta t)^2} + b\frac{u_{i+1} - u_i}{\Delta t}\frac{|u_i - u_{i-1}|}{\Delta t} + s(u_i) = F(t_i).$$

5. Obtain a solution of Exercise 5a Q 4 for the quadratic damping.

Answer: Although the finite difference approximation using central differences for the oscillation equation shown in Exercise 5a Q 1 for the quadratic damping results in u_{i+1} appearing on both the LHS and RHS, it turns out that it is possible to combine both the u_{i+1} terms and move it to the LHS thereby obtaining an explicit finite difference equation that can then be solved using standard algorithms.

6. What is the type of PDE that embodies the unsteady one-dimensional convection–diffusion equation of the following type:

$$\frac{\partial f}{\partial t} + u\frac{\partial f}{\partial x} = \alpha\frac{\partial^2 f}{\partial x^2}. \tag{E5.1}$$

Answer: The aforementioned PDE is a parabolic PDE.

7. What is the type of PDE that embodies the steady two-dimensional convection–diffusion equation of the following type:

$$u\frac{\partial f}{\partial x} + v\frac{\partial f}{\partial y} = \alpha\left(\frac{\partial^2 f}{\partial x^2} + \frac{\partial^2 f}{\partial y^2}\right).$$

Answer: The aforementioned PDE is an elliptic PDE.

8. For the unsteady one-dimensional convection–diffusion equation, Equation (E5.1), obtain the forward in time and central in space finite difference scheme.

Answer: $f_i^{n+1} = f_i^n - \dfrac{c}{2}\left(f_{i+1}^n - f_{i-1}^n\right) + d\left(f_{i+1}^n - 2f_i^n + f_{i-1}^n\right)$,

where $c = u\Delta t/\Delta x$, $d = \alpha\Delta t/(\Delta x)^2$.

9. Show that the unsteady one-dimensional convection–diffusion equation, Equation (E5.1), using the forward in time and central in space (FTCS) finite difference scheme is a consistent finite difference scheme.

10. Show that the unsteady one-dimensional convection–diffusion equation, Equation (E5.1), using the forward in time and central in space finite difference scheme is a conditionally stable scheme and obtain the condition to be satisfied for the stability of the FTCS scheme.

Answer: Condition to be satisfied for the FTCS scheme to be a stable scheme $0 \le \lambda \le 0.5$ and $0 \le \gamma \le 2(1-\lambda)$ where $\lambda = \dfrac{\alpha\Delta t}{(\Delta x)^2}$ and $\gamma = \dfrac{u\Delta t}{\Delta x}$.

11. For the unsteady one-dimensional convection–diffusion equation, Equation (E5.1), obtain the backward in time and central in space finite difference scheme.

Answer: $f_i^{n+1} = f_i^n - \dfrac{c}{2}\left(f_{i+1}^{n+1} - f_{i-1}^{n+1}\right) + d\left(f_{i+1}^{n+1} - 2f_i^{n+1} + f_{i-1}^{n+1}\right)$,

where $c = u\Delta t/\Delta x$, $d = \alpha\Delta t/(\Delta x)^2$.

12. Show that the unsteady one-dimensional convection–diffusion equation, Equation (E5.1), using backward in time and central in space finite difference scheme is a consistent finite difference scheme.

13. Show that the unsteady one-dimensional convection–diffusion equation, Equation (E5.1), using backward in time and central in space finite difference scheme is an unconditionally stable scheme.

14. Show that solving the friction equation

$$\frac{du}{dt} = -ku \qquad (k > 0),$$

using the central difference scheme results in one of the roots having a magnitude always greater than 1 and, hence, resulting in unconditionally unstable scheme.

Answer: The central finite difference scheme is written as

$$\frac{u^{n+1} - u^{n-1}}{2\Delta t} = -ku^n.$$

The roots (of the amplification factor) of the central difference scheme are

$$\lambda_1 = -k\Delta t + \sqrt{1 + (k\Delta t)^2}, \quad \lambda_2 = -k\Delta t - \sqrt{1 + (k\Delta t)^2}.$$

λ_2 will always have a magnitude greater than 1 for any $k\Delta t > 0$ and, hence, the central scheme is unconditionally unstable.

15. If the RHS in the finite difference equation of Exercise 5a Q 14 is replaced by $-ku^{n-1}$, find the nature of the stability for such a finite difference scheme.

Answer: If the RHS in the finite difference equation of Exercise 5a Q 14 is replaced by $-ku^{n-1}$, the roots (of the amplification factor) are given by $\lambda_2 = 1 - 2k\Delta t$. The aforementioned scheme is conditionally stable as $\lambda \leq 1$ for $k\Delta t \leq 1$.

16. If the RHS in the finite difference equation of Exercise 5a Q 14 is replaced by $-ku^{n-1}$, find the nature of the solution as in Exercise 5a Q 15 if $1/2 \leq k\Delta t \leq 1$.

Answer: For Exercise 5a Q 15 with $1/2 \leq k\Delta t \leq 1$, $\lambda_2 \leq 0$. Hence, the nature of the solution will exhibit oscillations and change sign for every second time step, considering that the roots (of the amplification factor) are complex.

17. If the RHS in the finite difference equation of Exercise 5a Q 14 is replaced by $-ku^{n-1}$, find the nature of the solution as in Exercise 5a Q 15, if $k\Delta t \leq 1/2$.

Answer: The roots (of the amplification factor) in such a case are real and u will have a more realistic evolution with no numerical oscillations.

Exercises 5b (Questions and answers)

1. Write down the Crank–Nicolson finite difference scheme for the friction equation

$$\frac{du}{dt} = -ku \qquad (k > 0).$$

Answer: $\dfrac{u^{n+1} - u^{n-1}}{2\Delta t} = -\dfrac{k}{2}\left(u^{n+1} + u^n\right)$

2. What is the expression for the root of the amplification factor for the Crank–Nicolson finite difference scheme for the friction equation

$$\frac{du}{dt} = -ku \qquad (k > 0).$$

Indicate whether the Crank–Nicolson finite difference scheme is a stable scheme.

Answer: $\lambda = \dfrac{1 - k\Delta t}{1 + k\Delta t}$. The Crank–Nicolson finite difference scheme is an absolutely stable scheme. However, one requires that $k\Delta t < 1$ for a more realistic evolution with the Crank–Nicolson finite difference scheme with no numerical oscillation.

3. Is it possible for one to employ different finite difference schemes for different terms in the one-dimensional oscillation friction equation?

Answer: Yes, it is possible to employ different schemes for different terms in the one-dimensional oscillation friction equation.

$$\frac{dU}{dt} = i\omega U - \kappa U$$

For example, one can employ the leapfrog scheme for the oscillation term and a forward scheme for the friction term. The finite difference equations for the aforementioned equation using a combination of schemes becomes

$$U^{n+1} = U^{n-1} + 2\Delta t \left[i\omega U^n - \kappa U^{n-1} \right].$$

6

Linear Advection Equation

6.1 Introduction

The advection equation is a very important equation to investigate as this equation conserves the quantity that gets advected following a motion. In this chapter, the various time and space differencing schemes as applied to the advection equation is taken up for detailed discussion.

6.2 Centered Time and Space Differencing Schemes for Linear Advection Equation

Consider a linear one-dimensional advection equation

$$\frac{\partial u}{\partial t} + c\frac{\partial u}{\partial x} = 0, \tag{6.1}$$

where c is a constant and $u = u(x,t)$, and its general solution is given by $u(x,t) = f(x - ct)$, where f is an arbitrary function. If the space derivative in Equation (6.1) is approximated by a central finite difference, one obtains

$$\frac{\partial u_j}{\partial t} = -c\frac{u_{j+1} - u_{j-1}}{2\Delta x}. \tag{6.2}$$

Applying the leapfrog scheme to Equation (6.2) gives

$$\frac{u_j^{n+1} - u_j^{n-1}}{2\Delta t} = -c\frac{u_{j+1}^n - u_{j-1}^n}{2\Delta x}. \tag{6.3}$$

Substituting in Equation (6.2), a tentative solution of the form

$$u_j = \Re\left[U(t)e^{-ikj\Delta x}\right], \tag{6.4}$$

one obtains an equation of the form

$$\frac{dU}{dt} = i\left[\frac{c}{\Delta x}\sin(k\Delta x)\right]U. \tag{6.5}$$

Equation (6.5) is of the form of oscillation equation with $\omega = \frac{c}{\Delta x}\sin(k\Delta x)$

$$\frac{dU}{dt} = i\omega U. \tag{6.6}$$

Figure 6.1 depicts the schematic of the solution of the one-dimensional linear advection equation moving along the characteristic. Here, the solution $u(x,t) = u(x-ct,0)$. i.e., the initial shape of u translates without change of shape with constant speed c. Equation (6.1) is the linear one-dimensional advection equation as expressed in the Eulerian description of fluid motion in which u is the advected quantity and c is the advection speed. It can be written in terms of the Lagrangian description of fluid motion as

$$\frac{du}{dt} = 0. \tag{6.7}$$

The meaning of this Lagrangian description of fluid motion is that the value of the advected quantity (u in this case) does not change following the motion of a fluid element, i.e., the advected quantity u is "conserved" following the motion of the fluid element. The definition of a fluid element follows the continuum hypothesis where a fluid is assumed to be continuous; the fluid element is assumed to be made up of a very large number of molecules such that one can ascribe macroscopic properties such as density, temperature, etc. to the fluid element. In fluid dynamics, one considers an infinite collection of fluid elements. Equation (6.7) indicates that each fluid element maintains its value of the advected quantity u as it moves in time. Assuming that advection is the only process occurring in a fluid system, exactly the same values of the advected quantity u are still in the fluid system at any two different times. Of course, the locations of the fluid elements presumably will have changed with time; however, the maximum value of the advected quantity u over the entire population of fluid elements remains unchanged by advection. Furthermore, if advection is the only physical process occurring in the system, the minimum value of u is unchanged, the average value of u is also unchanged, and in fact all the statistics of the distribution of u over the mass of the fluid are completely unchanged by the advective process. This is an important characteristic of advection. For example, assuming that the probability density function (pdf) of u at a certain time is Gaussian (satisfies the Gaussian distribution), then the pdf of u will still be Gaussian at a later time (having the same mean and standard deviation) if the only physical process affecting the fluid system is "advection" and if no mass enters or leaves the fluid system.

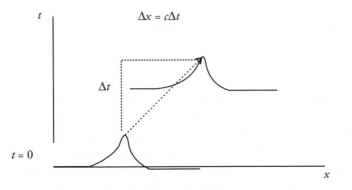

Figure 6.1   Depiction of the schematic of the solution of the one-dimensional linear advection equation moving along the characteristics.

Consider a simple function of u, say u^2. As u is unchanged during advection process, for each element, u^2 will also be unchanged for each element. Furthermore, the aforementioned argument extends to any other function of u. Hence, in an advection process, any function of u will also remain unchanged with time. It follows then that the pdf for any function of u remains unchanged by the advection process. In many situations, the advected quantity u is a positive (non-negative) quantity by definition. For example, the water vapour mixing ratio has to be positive as it is the ratio of two masses, i.e., the ratio of mass of water vapour to the mass of dry air; a negative water vapour mixing ratio is not physically meaningful. However, other advective quantities, such as the zonal component of the atmospheric wind vector, can have either sign; i.e., can be either positive or negative. By convention, positive (negative) zonal wind refers to winds coming from the west (east) or westerlies (easterlies).

Let u be conserved under advection, following motion of each fluid element. Based on the aforementioned discussions, if advection is the only physical process at work, it is abundantly clear that if there are no negative values of u at some initial time, then there will be no negative values of u at any later time either. This deduction is true irrespective of whether the advected quantity under consideration is non-negative by definition (say, the water vapour mixing ratio) or not (say, the zonal component of the atmospheric wind vector).

However, in general, advection is not the only physical process at work and, hence, these advective quantities, such as u, are not really conserved following the motion of the fluid elements. Moreover, many times, various sources and sinks are present within the fluid system that causes the value of u to change with time as the fluid element moves. For example, if u is air temperature, one possible source of heat can be "radiative heating." To incorporate more general processes that include not only advection but also sources (of heat) and sinks, one replaces zero in the RHS of Equation (6.7) with S, where S is the sources and sinks of u per unit time. Equation (6.7) with the nonhomogeneous term S in the RHS is still referred to as a "conservation" equation; it only means that u is still conserved except to the extent that sources or sinks (of u) are to be considered wherever applicable.

If one were to integrate the equation of continuity (expressing conservation of mass) over a closed domain and assume that the closed domain has no mass sources and no mass sinks, and apply Gauss's theorem, the aforementioned fact leads to an important result which is, mass is conserved within the closed domain. In the presence of mass sources and/or mass sinks, one would obtain the result: the mass-weighted average value of u is conserved within the closed domain, except for the effects of mass sources and mass sinks.

While designing finite difference schemes to represent advection (i.e., solve for the advection equation), in addition to considering simplicity, accuracy, stability, and computational ease, one has to also bear in mind the need to have a finite difference scheme that provides for the "conservation property." It is to be noted that as the continuous partial differential equation provides for this conservation property, the discrete finite difference scheme should also have the same "conservation property." If u were the atmospheric wind vector and it is conserved (principle of conservation of momentum), then conservation of u^2 would imply conservation of kinetic energy. Hence, while proposing finite difference schemes that conserve momentum, the same scheme should also ideally provide for conservation of kinetic energy.

Furthermore, it is hoped and expected that a finite difference scheme that represents advection process (in the absence of mass sources and mass sinks of atmospheric humidity), should always provide for a non-negative value, of say, the water vapour mixing ratio at all times. An advection finite difference scheme that ensures this property is called "positive-definite" or "sign-preserving" scheme. Positive definite schemes are desirable, as negative values, of say, the water vapour mixing ratio, that arise through truncation errors associated with space differencing will have to be eliminated before any moist physics process can be considered. It has been observed that the methods used to eliminate the negative values, of say, the water vapour mixing ratio, are inevitably artificial to some extent. Although most of the older advection finite difference schemes do not provide for "positive-definiteness," many of the newer schemes do satisfy it. Besides positive definiteness, one would be interested in imposing additional requirements on the advective finite difference scheme, such as the requirement that the advective finite difference operator should not change the pdf of u over the mass. Unfortunately, this requirement cannot be guaranteed with Eulerian finite difference methods, although one can try to minimize the effects of advection on the pdf, for the case where the shape of the pdf is known *a priori*. However, for numerical models based on Lagrangian methods, the advection process does not change the pdf of the advected quantity u.

6.3   Conservative Finite Difference Methods

The conservation of mass principle (equation of continuity) expressed in terms of the Eulerian description of fluid motion is given by

$$\frac{\partial \rho}{\partial t} = -\nabla \cdot (\rho V), \tag{6.8}$$

where ρ is the fluid density (mass per unit volume) and V is the velocity vector. The conservation equation of u in the presence of sources/sinks expressed using the Lagrangian description of fluid motion is given by

$$\frac{du}{dt} = S \tag{6.9}$$

Writing the LHS of Equation (6.9) as in the Eulerian description of fluid motion, one gets

$$\frac{\partial u}{\partial t} = -(V \cdot \nabla)u + S. \tag{6.10}$$

Multiplying Equations (6.8) by u and multiplying Equation (6.10) by ρ and adding, one gets

$$\frac{\partial (\rho u)}{\partial t} = -\nabla \cdot (\rho V u) + \rho S. \tag{6.11}$$

Equation (6.11) is called the flux form of the conservation equation for the quantity u. It is important to note that if one assigns $u = 1$ and $S = 0$ in Equation (6.11), it reduces to Equation (6.8). This has to be kept in mind while designing finite difference advection schemes.

If one integrates Equation (6.8) over a closed domain R (wherein R experiences no mass sources or mass sinks), then on applying Gauss's theorem, one gets

$$\frac{d}{dt}\int_R \rho dR = 0. \tag{6.12}$$

Equation (6.12) simply states that mass is conserved within the domain R. If one integrates Equation (6.11) over the closed domain R, one gets

$$\frac{d}{dt}\int_R \rho u dR = \int_R \rho S dR. \tag{6.13}$$

Equation (6.13) states that the mass-weighted average value of u is conserved within the domain R, except for the effects of sources and sinks. Both Equations (6.12) and (6.13) are integral forms of the conservation equations for mass and u, respectively. As the continuous PDEs (Equations (6.12) and (6.13)) satisfy the conservation property, it is important to devise finite difference advection schemes that provide for the same conservation property in the difference equation also, i.e., difference approximations of Equations (6.12) and (6.13) are expected to satisfy the following discrete equations:

$$\sum_j \rho_j^{n+1} dR_j = \sum_j \rho_j^n dR_j. \tag{6.14}$$

$$\sum_j (\rho u)_j^{n+1} dR_j = \sum_j (\rho u)_j^n dR_j + \Delta t \sum_j (\rho S)_j^n dR_j. \tag{6.15}$$

Equations (6.14) and (6.15) are finite difference analogs to the integral forms (Equations (6.12) and (6.13)). The effect of source/sink term S in Equation (6.13) is evaluated for simplicity using forward differencing in time, although this need not always be the case.

Consider m to be a mass variable, say density of air, and let ϕ be a "conservative" variable, satisfying the following one-dimensional conservation law:

$$\frac{\partial(m\phi)}{\partial t} + \frac{\partial(mu\phi)}{\partial x} = 0, \tag{6.16}$$

where u is the zonal component of the atmospheric wind velocity and mu is the mass flux. Assigning $\phi = 1$ in Equation (6.16) gives the mass conservation equation

$$\frac{\partial m}{\partial t} + \frac{\partial(mu)}{\partial x} = 0. \tag{6.17}$$

Approximating Equations (6.16) and (6.17) as follows:

$$\frac{d(m_j \phi_j)}{dt} + \frac{\phi_{j+1/2}(mu)_{j+1/2} - \phi_{j-1/2}(mu)_{j-1/2}}{\Delta x_j} = 0, \tag{6.18}$$

$$\frac{dm_j}{dt} + \frac{(mu)_{j+1/2} - (mu)_{j-1/2}}{\Delta x_j} = 0, \tag{6.19}$$

Equations (6.18) and (6.19) retain time derivatives while replacing space derivatives with finite difference schemes. If the "conservative" variable ϕ is uniform everywhere, say $\phi = 1$, then Equation (6.18) reduces identically to Equation (6.19). This is an important point that will be taken up for discussion later. The variables m and ϕ are defined at integer points, whereas u and mu are defined at half-integer points. Such a manner of discretization is called "staggered grid" discretization; here all the dependent variables are not entirely specified at integer points with some dependent variables specified at half-integer points.

Multiplying Equations (6.18) and (6.19) by Δx_j, and summing over the entire domain, one gets

$$\frac{d}{dt} \sum_{j=0}^{J}(m_j \phi_j \Delta x_j) + (mu)_{J+1/2}\phi_{J+1/2} - (mu)_{-1/2}\phi_{-1/2} = 0, \tag{6.20}$$

$$\frac{d}{dt} \sum_{j=0}^{J}(m_j \Delta x_j) + (mu)_{J+1/2} - (mu)_{-1/2} = 0. \tag{6.21}$$

If

$$(mu)_{J+1/2}\phi_{J+1/2} = (mu)_{-1/2}\phi_{-1/2}, \quad \text{and} \quad (mu)_{J+1/2} = (mu)_{-1/2},$$

then Equations (6.20) and (6.21) reduce to

$$\frac{d}{dt} \sum_{j=0}^{J}(m_j \phi_j \Delta x_j) = 0, \tag{6.22}$$

$$\frac{d}{dt} \sum_{j=0}^{J}(m_j \Delta x_j) = 0. \tag{6.23}$$

Equations (6.23) and (6.22) express conservation of mass and of the mass-weighted value of ϕ and are discrete finite difference expressions of the integral forms of the conservation equations for mass and ϕ, respectively (refer to Equations (6.12) and

(6.13)). It is pertinent to note that Equation (6.22) holds irrespective of the form of the interpolation used for $\phi_{j+1/2}$.

By combining Equations (6.16) and (6.17), one obtains the linear one-dimensional advective form of our conservation law:

$$m\frac{\partial\phi}{\partial t} + mu\frac{\partial\phi}{\partial x} = 0. \tag{6.24}$$

From Equations (6.18) and (6.19), one can derive a finite difference "advective form," similar to Equation (6.24).

$$m_j\frac{d\phi_j}{dt} + \frac{(mu)_{j+1/2}\left(\phi_{j+1/2} - \phi_j\right) + (mu)_{j-1/2}\left(\phi_j - \phi_{j-1/2}\right)}{\Delta x_j} = 0. \tag{6.25}$$

As Equation (6.25) is consistent with Equations (6.18) and (6.19), use of Equations (6.25) and (6.19) will provide for the conservation of the mass-weighted value of ϕ and of mass itself. Moreover, notice that if ϕ is uniform over the entire grid, then Equation (6.25) gives $(d\phi_j/dt = 0)$ indicating that ϕ is a conserved variable. This result gives rise to the following important consequence.

If the flux-form advection equation does not reduce to the flux-form continuity equation when ϕ is uniform over the grid, then a uniform tracer field will not remain uniform under advection.

For continuous systems, conservation of 'ϕ' implies conservation of any function of ϕ, such as conservation of ϕ^2, ϕ^n, etc. However, in a finite difference system, it is possible to ensure conservation of at most one non-trivial function of ϕ, say $F(\phi)$, in addition to ϕ, itself. Let F_j denote $F(\phi_j)$, where F is an arbitrary function and let F'_j denote $\frac{d[F(\phi_j)]}{d\phi_j}$. Multiplying Equation (6.25) by F'_j, one gets

$$m_j\frac{dF_j}{dt} + \frac{(mu)_{j+1/2}F'_j(\phi_{j+1/2} - \phi_j) + (mu)_{j-1/2}F'_j(\phi_j - \phi_{j-1/2})}{\Delta x_j} = 0. \tag{6.26}$$

Using Equation (6.19) to rewrite Equation (6.26) in flux form, one gets

$$\frac{d(m_jF_j)}{dt} + \frac{\left\{(mu)_{j+1/2}[F'_j(\phi_{j+1/2} - \phi_j) + F_j] - (mu)_{j-1/2}[-F'_j(\phi_j - \phi_{j-1/2}) + F_j]\right\}}{\Delta x_j} = 0, \tag{6.27}$$

To ensure conservation of $F(\phi)$, one chooses

$$F_{j+1/2} = F'_j(\phi_{j+1/2} - \phi_j) + F_j \tag{6.28}$$

and

$$F_{j-1/2} = -F'_j(\phi_j - \phi_{j-1/2}) + F_j. \tag{6.29}$$

Letting $j \to j+1$ in Equation (6.29), one gets

$$F_{j+1/2} = -F'_{j+1}(\phi_{j+1} - \phi_{j+1/2}) + F_{j+1} \tag{6.30}$$

Eliminating $F_{j+1/2}$ between Equations (6.28) and (6.30), one gets

$$\phi_{j+1/2} = \frac{(F'_{j+1}\phi_{j+1} - F_{j+1}) - (F'_j\phi_j - F_j)}{F'_{j+1} - F'_j}. \tag{6.31}$$

Choosing $\phi_{j+1/2}$ as in Equation (6.31) ensures conservation of both ϕ and $F(\phi)$. Let $F(\phi) = \phi^2$, then $F'(\phi) = 2\phi$ and from Equation (6.31), one gets

$$\phi_{j+1/2} = \frac{(2\phi_{j+1}^2 - \phi_{j+1}^2) - (2\phi_j^2 - \phi_j^2)}{2(\phi_{j+1} - \phi_j)} = \frac{(\phi_{j+1} + \phi_j)}{2}. \tag{6.32}$$

If the grid spacing is uniform, then Equation (6.32) provides second-order accuracy in space.

6.3.1   Leapfrog scheme

When the leapfrog scheme is applied to Equation (6.5), one obtains

$$U^{n+1} = U^{n-1} + 2i\left[\frac{c\Delta t}{\Delta x}\sin(k\Delta x)\right]U^n. \tag{6.33}$$

Assuming $p = \frac{c\Delta t}{\Delta x}\sin(k\Delta x)$, one obtains the quadratic equation for the amplification factor λ

$$\lambda^2 - 2ip\lambda - 1 = 0, \tag{6.34a}$$

whose solution is

$$\lambda = ip \pm \sqrt{1 - p^2}. \tag{6.34b}$$

Based on our earlier analysis, stability requires that the following condition on p be met, $|p| \le 1$, i.e., $\left|\frac{c\Delta t}{\Delta x}\sin(k\Delta x)\right| \le 1$ for any possible k. As $|\sin(kx)| \le 1$, the condition for stability of the leapfrog scheme is $\frac{c\Delta t}{\Delta x} \le 1$, same as the Courant–Friedrichs–Lewy (CFL) condition. Moreover, $\mu = \frac{c\Delta t}{\Delta x}$, where μ is called the Courant number. The CFL condition requires that the Courant number μ be less than or equal to 1.

Figure 6.2 shows a schematic of a finite difference scheme of the one-dimensional linear advection equation where CFL condition holds. A general expression of the finite difference equation (FDE) for a two-time level scheme of the one-dimensional advection equation can be written as

$$u_j^{n+1} = (1 - \mu)u_j^n + \mu u_{j-1}^n,$$

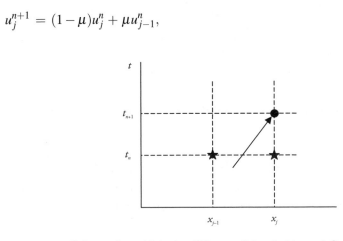

Figure 6.2 Scheme for which the CFL condition holds and Courant number $\mu \leq 1$.

where j and n are the space (x) and time (n) indices. This FDE shows that the solution u at time index $(n+1)$ at a grid point j is obtained as a linear combination of the solution u at the previous time (n) and at grid points j and $j-1$. The solution u_j^n and u_{j-1}^n are assumed to be known at the previous time (n) and are, hence, indicated as stars in Figure 6.2. For $0 < \mu < 1$, the FDE $u_j^{n+1} = (1 - \mu)u_j^n + \mu u_{j-1}^n$ shows that the solution u_j^{n+1} is obtained through interpolation between the known values u_j^n and u_{j-1}^n. As the true solution is known to move along the characteristic and between the two known values (shown as stars) in Figure 6.2, it is clear that for the case $0 < \mu < 1$, the solution (for the previous time n) for u_j^{n+1} is between u_j^n and u_{j-1}^n and the same is interpolated using u_j^n and u_{j-1}^n. In this case, the scheme is stable.

For the scheme in which the Courant number > 1 (refer to Figure 6.3) and where CFL condition fails, it is clear that (the solution moves along the characteristic, which is indicated by the arrow) u_j^{n+1} is outside both known values of u_j^n and u_{j-1}^n (indicated by stars) and the same is extrapolated from the values of u_j^n and u_{j-1}^n.

A third situation where the Courant number is negative ($\mu < 0$) and CFL condition does not hold is shown schematically in Figure 6.4. In this situation also, like Figure 6.3, it is clear that (the solution moves along the characteristic, which is indicated by the arrow) u_j^{n+1} is outside both known values of u_j^n and u_{j-1}^n (indicated by stars) and the same is extrapolated from the values of u_j^n and u_{j-1}^n. In both cases of $\mu > 1$ and $\mu < 0$,

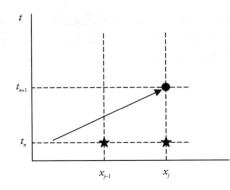

Figure 6.3 Scheme for which the CFL condition does not hold and Courant number $\mu > 1$.

requiring extrapolations of the solutions using u_j^n and u_{j-1}^n, the extrapolations increase the maximum absolute value of u_j^n at each and every time step. Such unrestricted increase of the maximum absolute value of u_j^n at every time step would render the solution unbound and the scheme to be unstable.

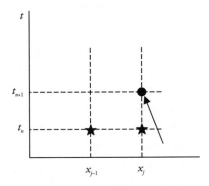

Figure 6.4 Scheme for which the CFL condition does not hold and Courant number $\mu < 0$.

Note that the maximum value of $|p|$, i.e., the minimum stability, is associated with the wave with $k\Delta x = \pi/2$. As the shortest resolvable wavelength on a grid of size Δx is $2\Delta x$, the aforementioned component with wavelength $4\Delta x$, has twice the wavelength as compared to the shortest resolvable wavelength. As Equation (6.34a) is a quadratic equation, there are two solutions for U, the physical (designated as U_1) and the computational (designated as U_2) mode, whose solutions are

$$U_1^{(n)} = \lambda_1^n U_1^{(0)} \qquad \text{and} \qquad U_2^{(n)} = \lambda_2^n U_2^{(0)} \tag{6.35}$$

where λ_1 and λ_2 are solutions of Equation (6.34b). For the stable case, one has the following solution:

$$\lambda_1 = e^{i\theta} \quad \text{and} \quad \lambda_2 = e^{i(\pm\pi-\theta)} = -e^{-i\theta}, \tag{6.36a}$$

where the angle θ is given by

$$\theta = \tan^{-1}\left[\frac{p}{\sqrt{1-p^2}}\right], \tag{6.36b}$$

where the $\pm$ sign is taken according as $p > 0$ or $p < 0$. Combining Equation (6.35) with Equation (6.4), one gets the expressions for the physical mode as

$$u_j^{(n)} = \Re\left[U_1^{(0)}e^{-ik\left(j\Delta x-\frac{\theta}{k\Delta t}n\Delta t\right)}\right] \tag{6.37a}$$

and the computational mode as

$$u_j^{(n)} = \Re\left[(-1)^n U_2^{(0)}e^{-ik\left(j\Delta x+\frac{\theta}{k\Delta t}n\Delta t\right)}\right]. \tag{6.37b}$$

These expressions need to be compared with the true solution of the one-dimensional linear advection equation

$$u(x,t) = \Re\left[U^{(0)}e^{-ik(x-ct)}\right]. \tag{6.38}$$

Comparing Equation (6.37a) with Equation (6.38), it is clear that the phase speed of the physical mode, c_1, is equal to $\theta/(k\Delta t)$.

From the expression of p and Equation (6.36b), it is clear that as $\Delta t \to 0$, $\theta \to p$. Moreover, from the expression of p and as $\Delta x \to 0$, $p \to ck\Delta t$. Hence, from this as $\Delta t \to 0$ and $\Delta x \to 0$, $c_1 \to c$, i.e., the phase speed of the physical mode approaches the phase speed of the solution.

Comparing Equation (6.37b) with Equation (6.38), it is clear that the phase speed of the computational mode, c_2, is equal to $-\theta/(k\Delta t)$. Moreover, from this discussion, it is clear that as $\Delta t \to 0$ and $\Delta x \to 0$, $c_2 \to -c$, i.e., the phase speed of the computational mode approaches minus the phase speed of the solution. Furthermore, the computational mode changes sign at all grid points from each time step to the next time step, (i.e., from n to $n+1$) owing to the presence of the factor $(-1)^n$ in Equation (6.37b).

6.3.2 Matsuno scheme

The Matsuno scheme as applied to the one-dimensional linear advection equation is given by first calculating the approximate values of u_j^{*n+1} using the following forward scheme:

$$\frac{u_j^{*n+1} - u_j^n}{\Delta t} = -c \frac{u_{j+1}^n - u_{j-1}^n}{2\Delta x}. \tag{6.39}$$

The intermediate solution u_{j+1}^{*n+1} and u_{j-1}^{*n+1} are then utilized in the backward scheme as follows:

$$\frac{u_j^{n+1} - u_j^n}{\Delta t} = -c \frac{u_{j+1}^{*n+1} - u_{j-1}^{*n+1}}{2\Delta x}. \tag{6.40}$$

Eliminating the starred quantities in Equation (6.40) from Equation (6.39), one gets

$$\frac{u_j^{n+1} - u_j^n}{\Delta t} = -c \frac{u_{j+1}^n - u_{j-1}^n}{2\Delta x} + c^2 \Delta t \frac{u_{j+2}^n - 2u_j^n + u_{j-2}^n}{(2\Delta x)^2}. \tag{6.41}$$

Equation (6.41) without the second term RHS is just the forward difference time scheme that utilizes central difference in space. The second term in the RHS approaches zero as Δx, $\Delta t \to 0$; however, at fixed Δt, this term approximates $c^2 \Delta t (\partial^2 u / \partial x^2)$ (with centered scheme over twice the grid size), which is essentially a diffusion term. The Matsuno scheme is an explicit scheme; it is first-order accurate in time and second-order accurate in space. It can be shown to be a consistent finite difference scheme. The Matsuno scheme is also a stable scheme if the Courant number is less than or equal to 1. The scheme introduces relatively large dissipation and phase errors.

6.3.3 Lax–Wendroff scheme

For applying the two step Lax–Wendroff scheme to the one-dimensional linear advection equation, first, provisional values of u are calculated at the center of the two rectangular meshes (using centered space and forward time differencing) of the stencil, denoted by stars in Figure 6.5. The values of $u_{j+1/2}^n$ and $u_{j-1/2}^n$ are obtained by averaging the value of u_j^n with neighboring grid points as follows:

$$\frac{u_{j+1/2}^{n+1/2} - \frac{1}{2}\left(u_{j+1}^n + u_j^n\right)}{\Delta t / 2} = -c \frac{u_{j+1}^n - u_j^n}{\Delta x} \tag{6.42a}$$

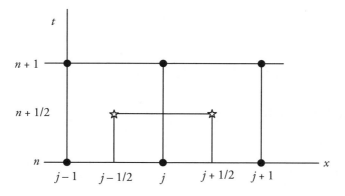

Figure 6.5 Space–time grid stencil arrangements for the two-step Lax–Wendroff scheme.

$$\frac{u_{j-1/2}^{n+1/2} - \frac{1}{2}\left(u_j^n + u_{j-1}^n\right)}{\Delta t/2} = -c\frac{u_j^n - u_{j-1}^n}{\Delta x} \tag{6.42b}$$

Using the aforementioned provisional values (shown as stars in Figure 6.5), a finite difference scheme that is centered in both space and time yields

$$\frac{u_j^{n+1} - u_j^n}{\Delta t} = -c\frac{u_{j+1/2}^{n+1/2} - u_{j-1/2}^{n+1/2}}{\Delta x}. \tag{6.43}$$

Substituting the provisional values from Equations (6.42a) and (6.42b) in Equation (6.43), one gets

$$\frac{u_j^{n+1} - u_j^n}{\Delta t} = -c\frac{u_{j+1}^n - u_{j-1}^n}{2\Delta x} + \frac{1}{2}c^2\Delta t\frac{u_{j+1}^n - 2u_j^n + u_{j-1}^n}{(\Delta x)^2} \tag{6.44}$$

The Lax–Wendroff scheme for the unknown u_j^{n+1} may be written as follows:

$$u_j^{n+1} = u_j^n - \frac{\mu}{2}\left(u_{j+1}^n - u_{j-1}^n\right) + \frac{\mu^2}{2}\left(u_{j+1}^n - 2u_j^n + u_{j-1}^n\right), \tag{6.45}$$

where μ is the Courant number. The form of the Equation (6.44) is similar to the form of the Matsuno scheme, Equation (6.41), except for the second term in the RHS which at constant Δt, reduces to $\frac{1}{2}c^2\Delta t(\partial^2 u/\partial x^2)$.

Unlike the Matsuno scheme, which was first-order accurate in time and second-order accurate in space, the Lax–Wendroff scheme is second-order accurate in both time and in space. Moreover, the term $\frac{1}{2}c^2\Delta t(\partial^2 u/\partial x^2)$ is calculated over an interval $2\Delta x$ and, hence, has its maximum damping effect at $2\Delta x$ wavelength. It has been found that it is indeed desirable to have a finite difference scheme that has damping dependence on wavelength as there are serious problems with finite difference schemes around $2\Delta x$. It is possible to take care of these problems using a dissipative finite difference scheme that preferentially damps the two-grid interval waves.

Stability of the Lax–Wendroff scheme

Substituting

$$u_j^n = \Re\left[U(n)e^{ikj\Delta x}\right]$$

in Equation (6.44), one gets

$$U^{n+1} = \left[1 - \mu^2(1 - \cos(k\Delta x)) - i\mu\sin(k\Delta x)\right]U^n, \tag{6.46}$$

where the quantity under the square bracket in the RHS of Equation (6.46) is the amplification factor λ. From this,

$$|\lambda| = \sqrt{\left[1 - 4\mu^2(1 - \mu^2)\sin^4(k\Delta x/2)\right]}. \tag{6.47}$$

As $|\sin(k\Delta x/2)| \leq 1$, the expression in the square brackets in Equation (6.47) cannot assume a value below $1 - 4\mu^2(1 - \mu^2)$. For $\mu = 1$, $1 - 4\mu^2(1 - \mu^2)$ has a value 1 and for $\mu^2 = 1/2$, $1 - 4\mu^2(1 - \mu^2)$ has a minimum value of zero. Hence, from the aforementioned expression, for $|\lambda| \leq 1$, $(|c|\Delta t/\Delta x) \leq 1$, leading to the CFL condition. The finite differencing scheme has damping property for $|\mu| \leq 1$.

For a given wavelength, the amplification factor λ has a minimum value, given by

$$\lambda_{min} = \sqrt{\left[1 - \sin^4(k\Delta x/2)\right]}, \tag{6.48}$$

which equals zero for $k\Delta x = \pi$, or which is the wavelength of the smallest resolvable wave with grid size Δx. Moreover, for $k \to 0$, (wavelength tending to infinity), $\lambda_{min} \to$ 1. This shows that the damping depends on the wavelength. Figure 6.6 shows the damping of the Lax–Wendroff scheme for various Courant number μ and for two waves with wavelength $2\Delta x$ and $4\Delta x$. From the figure, it is clear that the amount of damping is quite high for the shorter wavelength $2\Delta x$ (continuous line) as compared to the larger wavelength $4\Delta x$ (dashed line). Moreover, for the Lax–Wendroff scheme, the amount of damping depends on the time step and the advection velocity c. This is evidently a disadvantage of the Lax–Wendroff scheme as there are really no reasons for the amount of damping to depend on the time step and the advection velocity. For the Lax–Wendroff scheme, the amount of damping does depend on the time step. It is important to choose the time step Δt in such a way as to ensure that the scheme

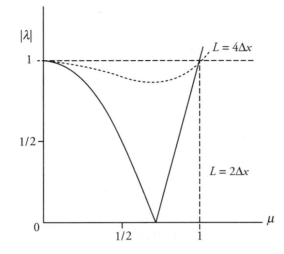

Figure 6.6 Amplification factor of the Lax–Wendroff scheme as a function of the Courant number μ, for waves of wavelength $2\Delta x$ (continuous line) and $4\Delta x$ (dashed line).

is stable. It would not be meaningful to choose the time step for a finite difference scheme on considerations of the amount of damping such a scheme would entail.

The Lax–Wendroff scheme has been broadly used in atmospheric models owing to its attractive characteristics, such as its second-order accuracy in time and space, its explicit nature, and the fact that it is not an unconditionally unstable scheme, and has no existence of the computational mode. The dissipation of the Lax–Wendroff scheme will not be detrimental if its total effect is negligible as compared with the physical dissipation. Moreover, this scheme can be used for controlling the shortest waves. However, if the physical dissipation is very small or nonexistent, it is better not to use the Lax–Wendroff scheme.

6.4 Computational Dispersion: Phase Speed Dependence on Wavelength

The linear one-dimensional advection equation (Equation (6.1)) does not provide for any dispersion whatsoever in the sense that all the wave modes have the same phase speed c. It can be shown that this is generally not the case while considering the finite difference approximation to Equation (6.1).

Consider the linear one-dimensional advection equation (Equation (6.1)) with central differences for space derivatives

$$\frac{\partial u_j}{\partial t} + c\frac{u_{j+1} - u_{j-1}}{2\Delta x} = 0. \tag{6.49}$$

As the time derivative has retained its differential form in Equation (6.49), any error in Equation (6.49) is attributed only to the central space differencing. Equation (6.49) has a solution in the form of a single harmonic component, which is given by

$$u_j = \Re\left[U(t)e^{ikj\Delta x}\right], \tag{6.50}$$

provided that

$$\frac{dU}{dt} + ik\left(c\frac{\sin k\Delta x}{k\Delta x}\right)U = 0. \tag{6.51}$$

Substituting Equation (6.50) on the PDE (Equation (6.1)), one gets

$$\frac{dU}{dt} + ikcU = 0. \tag{6.52}$$

Comparing Equations (6.51) and (6.52), it is clear that the solution of Equation (6.49) propagates with a phase speed c^* that is given by

$$c^* = c\frac{\sin k\Delta x}{k\Delta x}. \tag{6.53}$$

It is clear from Equation (6.53) that the solution of the difference differential equation (Equation (6.49)) propagates with a phase speed c^*, which is a function of the wave number k. Hence, it is clear that the finite (central) differencing in space gives rise to a dispersion of the waves. As $k\Delta x$ increases from zero, the phase speed c^* decreases monotonically from c and becomes zero for the shortest wave having a resolvable wavelength of $2\Delta x$, i.e., when $k\Delta x = \pi$. To sum up, due to space differencing, the solution of the difference differential equation propagates with a phase speed that is less than the true phase speed c. This characteristic of slower phase speed of the solution due to spatial differencing actually increases with the decrease of the wavelength and ultimately for the smallest resolvable wave, the space differencing produces a solution (wave) that is stationary.

Figure 6.7 shows the plot of the solution of Equation (6.49) for a wave of wavelength $2\Delta x$. Looking at the figure the reasons for the stationary behaviour of the two-grid interval wave are obvious. For the wave having wavelength of $2\Delta x$, Figure 6.7 shows that the solution for such a wave has $u_{j+1} = u_{j-1}$ at all grid points. This leads to a situation where $(\partial u_j/\partial t) = 0$ (refer to Equation (6.49)). One of the consequences of the reduction of the true advection speed by the numerical wave is a general retardation of the advection process itself. Furthermore, the dependence of

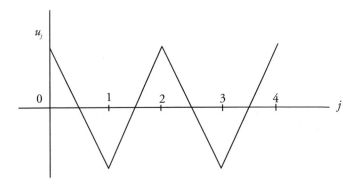

Figure 6.7 Plot of the solution of the difference differential equation for a wave of wavelength $2\Delta x$.

such a reduction of the phase speed on the wave number (wavelength) is quite critical, especially for waves with the smallest wavelength. Moreover, if the pattern that is being advected represents a superposition of more than one wave, the false dispersion arising from the differing reduction of phase speeds for different waves of varying wavelength will ultimately result in a deformation of that pattern. Hence, in the atmospheric evolution of small-scale motions, the latter representing a superposition of many waves, it is important to carefully evaluate the consequences of the effect of such computational dispersion.

6.4.1 Group velocity

The concept of group velocity can be illustrated by considering superposition of two waves, having slightly different wave numbers k_1 and k_2, respectively. Define the following as

$$k = \frac{k_1 + k_2}{2}; \qquad \Delta k = \frac{k_1 - k_2}{2}; \qquad c = \frac{c_1 + c_2}{2}; \qquad \Delta c = \frac{c_1 - c_2}{2}.$$

It is assumed that Δk and Δc are quite small so that the product $\Delta k \Delta c$ can be neglected. From these expressions, it is clear that $k_1 = k + \Delta k$; $k_2 = k - \Delta k$; $c_1 = c + \Delta c$; $c_2 = c - \Delta c$. From this it is clear that $k_1 c_1 = kc + \Delta(kc)$ and $k_2 c_2 = kc - \Delta(kc)$, where term $\Delta k \Delta c$ is neglected, which is reasonable when $k_1 \approx k_2$ and $c_1 \approx c_2$. Using this expression, one can write the sum of two waves having the same amplitude

$$e^{ik_1(x - c_1 t)} + e^{ik_2(x - c_2 t)} = e^{i[(k + \Delta k)x - (kc + \Delta(kc))t]} + e^{i[(k - \Delta k)x - (kc - \Delta(kc))t]}$$

$$= e^{ik(x - ct)}[e^{i[\Delta k x - \Delta(kc)]t} + e^{-i[\Delta k x - \Delta(kc)]t}]$$

$$= 2\cos[\Delta k x - \Delta(kc)t]e^{ik(x-ct)}$$

$$= 2\cos\left[\Delta k\left(x - \tfrac{\Delta(kc)}{\Delta k}t\right)\right]e^{ik(x-ct)}.$$

If Δk is small, the factor $\cos\left[\Delta k\left(x - \tfrac{\Delta(kc)}{\Delta k}t\right)\right]$ may appear as indicated in Figure 6.8 as the outer slowly varying envelope. The envelope "modulates" wave k, the latter being represented by the inner, rapidly varying curve in Figure 6.8. Whereas the short waves move with phase speed c, the envelops of the short waves (wave packets) moves with speed $(\Delta(kc)/\Delta k)$. The differential expression $(d(kc)/dk)$ is known as group velocity c_g, which also equals $(d\omega/dk)$.

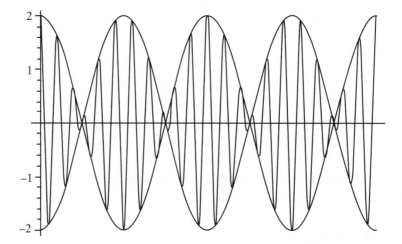

Figure 6.8 Schematic diagram showing the concept of group velocity, which is the velocity of the envelope where the short waves are modulated by the longer waves.

The group velocity is thus defined as

$$c_g = \frac{d\omega}{dk} = \frac{d(kc)}{dk}. \tag{6.54}$$

For the case of the linear one-dimensional advection equation (Equation (6.1)), the analytical phase speed c is a constant. From Equation (6.54), and assuming a constant phase speed c yields $c_g = c$, a situation where the analytical wave is non-dispersive.

However, for the differential difference equation (Equation (6.49)), the group velocity is given by

$$c_g^* = \frac{d(kc^*)}{dk} = c\cos(k\Delta x).$$
(6.55)

From Equation (6.55) it is clear that as $k\Delta x$ increases from zero, the group velocity c_g^* of the solution of the differential difference equation decreases monotonically from c (same as c_g) and becomes equal to $-c$ (same as $-c_g$) for the shortest wave having a wavelength of $2\Delta x$. The plots of the phase velocity (c^* as given in Equation (6.53)) and the group velocity (c_g^* as given in Equation (6.55)) are shown in Figure 6.9. For the

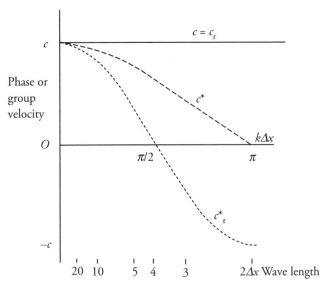

Figure 6.9 Phase speed (c^*) and group velocity (c_g^*) for the linear one-dimensional advection equation and the corresponding differential difference equation with second-order centered space finite difference scheme.

analytical solution of the one-dimensional linear advection equation (Equation (6.1)), as $c = c_g$, both the individual harmonic waves and wave packets propagate at the same constant speed, resulting in a non-dispersive wave. However, replacing the space derivative with the centered finite difference scheme (refer to Equation (6.49)), results in a dispersive wave ($c^* \neq c_g^*$), i.e., the numerical wave shows a decrease of both phase speed and group velocity with the increase of the wave number. This departure from the analytical non-dispersive wave behaviour is especially large for the waves that have small wavelengths. In fact, Figure 6.9 shows that for waves with wavelengths less than $4\Delta x$, group velocity assumes negative values, i.e., the wave packets made up of these waves (with wavelengths less than $4\Delta x$) propagate in the direction opposite to the direction of advection velocity and also opposite to the direction of individual waves.

6.5   Upstream Schemes

The space derivative in the linear one-dimensional advection equation (Equation (6.1)) can also be replaced by the finite difference scheme that is not a centered scheme. This indicates two possibilities; employ either a forward in space difference scheme or a backward in space difference scheme, depending on whether the phase speed c is negative or positive, i.e.,

$$\frac{\partial u_j}{\partial t} + c\frac{u_j - u_{j-1}}{\Delta x} = 0; \quad c > 0 \tag{6.56a}$$

$$\frac{\partial u_j}{\partial t} + c\frac{u_{j+1} - u_j}{\Delta x} = 0; \quad c < 0. \tag{6.56b}$$

The direction of downstream or downwind is in the direction of the advection velocity, whereas the direction of upstream or upwind is in the direction opposite to the direction of the advection velocity. In both Equations (6.56a) and (6.56b), it is clear that for a positive c, the space derivative (in Equation (6.56a)) is replaced by backward differencing in space (i.e., the differencing is in the direction of upstream) and for a negative c, the space derivative (in Equation (6.56b)) is replaced by forward differencing in space (i.e., again the differencing is in the direction of upstream). In short, for both the cases (indicated by Equations (6.56a) and (6.56b)), the space differences are calculated on the side from which the advection velocity reaches the center point j. Owing to this reason, both the schemes that represent Equations (6.56a) and (6.56b) are known as upstream finite difference schemes.

It is possible to apply various time difference schemes for the linear one-dimensional differential difference advection equation. However, the resulting finite difference schemes will only be first-order accurate. Despite their being only first-order accurate, the aforementioned upstream schemes will have one important benefit over the centered finite difference schemes in space when applied to the linear one-dimensional advection equation. When upstream schemes are employed, disturbance cannot propagate in the direction opposite to the direction of the physical advection. This is a very important consideration while dealing with the solution of hyperbolic PDEs, as there will not be any spurious contamination of the numerical solution.

6.5.1 Transportive property

Applying the forward in time and central in space (FTCS) finite difference scheme to the one-dimensional linear advection equation, one obtains

$$\frac{u_j^{n+1} - u_j^n}{\Delta t} + c\frac{u_{j+1}^n - u_{j-1}^n}{2\Delta x} = 0. \tag{6.57}$$

Consider a perturbation δ that is introduced at the mth space location ($j = m$) and at time 'n'. Assuming $c > 0$, one would expect that this perturbation be carried on only along the direction of velocity. At the $(m+1)th$ grid point, downstream of the perturbation, one has for $j = m+1$,

$$\frac{u_{m+1}^{n+1} - u_{m+1}^n}{\Delta t} = -c\frac{0 - \delta}{2\Delta x} = \frac{c\delta}{2\Delta x}. \tag{6.58}$$

Downstream of the grid point where the perturbation is introduced, there is transport of perturbation which is acceptable. However, at the grid point $j = m$, one obtains

$$\frac{u_m^{n+1} - u_m^n}{\Delta t} = -c\frac{0 - 0}{2\Delta x} = 0. \tag{6.59}$$

Equation (6.59) is not a reasonable result as there is no transport of the perturbation at the grid point where the perturbation was introduced. At the upstream grid point ($j = m - 1$), one obtains

$$\frac{u_{m-1}^{n+1} - u_{m-1}^n}{\Delta t} = -c\frac{\delta - 0}{2\Delta x} = -\frac{c\delta}{2\Delta x}. \tag{6.60}$$

Equation (6.60) shows that the transportive property is violated by the FTCS scheme, as for $c > 0$, the perturbations can only be transported downstream.

Applying the upstream scheme to one-dimensional linear advection equation, for $c > 0$, one obtains for the upstream scheme

$$\frac{u_j^{n+1} - u_j^n}{\Delta t} = -c\frac{u_j^n - u_{j-1}^n}{\Delta x}. \tag{6.61}$$

As before, assume that the perturbation is introduced at grid point ($j = m$) and time 'n'. At the downstream grid point location, ($j = m+1$), one obtains

$$\frac{u_{m+1}^{n+1} - u_{m+1}^n}{\Delta t} = -c\frac{0 - \delta}{\Delta x} = \frac{c\delta}{\Delta x}, \tag{6.62}$$

which is again an acceptable result that follows the transportive property.

At grid point $(j = m)$, where the perturbation is introduced, one obtains for the upstream scheme

$$\frac{u_m^{n+1} - u_m^n}{\Delta t} = -c\frac{\delta - 0}{\Delta x} = -\frac{c\delta}{\Delta x}, \tag{6.63}$$

which shows that the perturbation is being transported out of the affected region.

At the upstream grid point location $(j = m - 1)$, one obtains for the upstream scheme

$$\frac{u_{m-1}^{n+1} - u_{m-1}^n}{\Delta t} = -c\frac{0 - 0}{\Delta x} = 0 \tag{6.64}$$

Equation (6.64) indicates that the perturbation is not transported in the upstream direction while employing the upstream scheme showing that the upstream scheme maintains unidirectional flow of information.

Assume that a forward finite difference in time is used to replace the time derivative in Equation (6.56a) and that $c > 0$; hence, clearly the finite difference of the upstream scheme is

$$\frac{u_j^{n+1} - u_j^n}{\Delta t} + c\frac{u_j^n - u_{j-1}^n}{\Delta x} = 0; \quad c > 0 \tag{6.65}$$

The upstream scheme was discussed in Section 4.5.3 where it was shown that the scheme is a stable scheme provided that $-1 < \lambda < 0$, if c is negative and stable for $0 < \lambda < 1$ if c is positive. Moreover, the upstream scheme is found to be a damping scheme with the amount of damping depending on the wavelength. Furthermore, the maximum amount of damping was seen for the shortest wave with $2\Delta x$ wavelength.

Figure 6.10 shows the domains of influence of a grid point for the upstream difference scheme (Equation (6.65)) with $c > 0$ together with schemes having centered differencing and downstream differencing for the true solution. Figure 6.10 clearly brings out the advantage of employing upstream differencing as compared with using centered differencing and downstream differencing. It is known that solution at a grid point value is said to propagate along the characteristics, $x - ct = constant$. Figure 6.10 shows a grid point marked by a circle with the associated characteristic passing through it. Using the upstream finite differencing (refer to Equation (6.65)), the value at the grid point will influence the values at points within the domain shaded by vertical lines as indicated by Figure 6.10. The domains of influence of both the centered finite differencing and downstream finite differencing schemes are also shown in Figure 6.10.

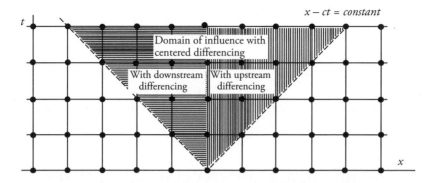

Figure 6.10 Domains of influence of a grid point for the upstream difference scheme (Equation (6.65)) with $c > 0$ together with schemes having centered differencing and downstream differencing for the true solution.

Figure 6.10 clearly shows that the domain of influence of the upstream finite differencing scheme provides the best approximation to the characteristic line representing the domain of influence in the solution. This observation suggests constructing a finite difference scheme for Equation (6.1) by tracing a characteristic back from a point $[j\Delta x, (n+1)\Delta t]$ to intersect the previous time level $t = n\Delta t$ and calculating the value of the solution at the point of intersection by interpolation (refer to Figure 6.11).

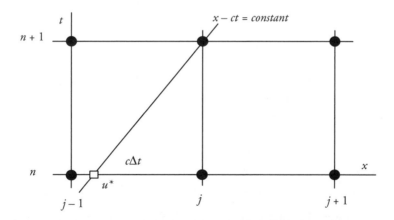

Figure 6.11 Schematic of the construction of a scheme by calculation of a previous value on a characteristic passing through the point $[j\Delta x, (n+1)\Delta t]$.

Setting $u_j^{n+1} = u^*$, and choosing a linear interpolation procedure that utilizes values at two neighboring points at the time $n\Delta t$, one gets

$$u_j^{n+1} = u_{j-1}^n + \frac{u_j^n - u_{j-1}^n}{\Delta x}(\Delta x - c\Delta t). \tag{6.66}$$

Equation (6.66) is identical to Equation (6.61) which is the upstream finite difference scheme for $c > 0$.

6.6   Fourth-order Space Differencing Schemes for Advection Equation

The limitations related to the phase speed error and the computational dispersion described in the previous section arises primarily due to the use of second-order space differencing. Hence, the obvious alternative is to utilize a fourth (higher) order space differencing scheme for the one-dimensional linear advection equation. The central fourth-order finite difference scheme (refer to Section 3.2.4) for the space derivative is given by

$$\left.\frac{\partial u}{\partial x}\right|_j = \frac{4}{3}\frac{u_{j+1} - u_{j-1}}{2\Delta x} - \frac{1}{3}\frac{u_{j+2} - u_{j-2}}{4\Delta x} + O[(\Delta x)^4]. \tag{6.67}$$

Substituting Equation (6.67) into Equation (6.1), one gets

$$\left.\frac{\partial u}{\partial t}\right|_j + c\left(\frac{4}{3}\frac{u_{j+1} - u_{j-1}}{2\Delta x} - \frac{1}{3}\frac{u_{j+2} - u_{j-2}}{4\Delta x}\right) = 0. \tag{6.68}$$

As before, assuming $u_j(t) = \Re\left[U(t)e^{ikj\Delta x}\right]$, one obtains the phase speed for the fourth-order space differencing scheme as

$$c^{**} = c\left[\frac{4}{3}\frac{\sin(k\Delta x)}{k\Delta x} - \frac{1}{3}\frac{\sin(2k\Delta x)}{2k\Delta x}\right]. \tag{6.69}$$

This phase speed expression can be compared with Equation (6.53), the latter providing the phase speed for the second-order space differencing scheme.

Figure 6.12 shows the phase speed for all values of k for the one-dimensional linear advection equation with c being the phase speed of the analytical solution, c^* being the phase speed of the second-order centered space difference solution, and c^{**} being the phase speed of the fourth-order centered space difference solution. It is clear from Figure 6.12 that the phase speed (c^{**}) error for the fourth-order centered space difference solution is substantially reduced especially for small values of $k\Delta x$

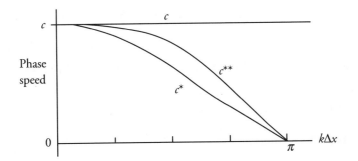

Figure 6.12 Phase speed for the one-dimensional linear advection equation with c being the phase speed of the analytical solution, c^* being the phase speed of the second-order centered space difference solution, and c^{**} being the phase speed of the fourth-order centered space difference solution.

that correspond to large-scale and medium-scale waves. Despite the aforementioned very noticeable increase in accuracy of the phase speed c^{**}, there is still an important limitation in the fourth-order centered space difference solution in terms of the departure from c and the fact that c^{**} is much lower than c as k increases. Furthermore, as $k\Delta x$ approaches π (wavelength λ approaches its smallest value of $2\Delta x$), the increase in the phase speed obtained by fourth-order centered space differencing c^{**} diminishes, until, finally, the wave with wavelength $2\Delta x$ again becomes stationary. Moreover, for the short waves (large value of $k\Delta x$), the slope of the phase speed c^{**} curve for the solution of the fourth-order centered space differencing is greater than that with the phase speed c^* of the second-order centered space differencing indicating that the computational dispersion of the fourth-order centered space differencing scheme waves is greater. Hence, it is clear that the fourth-order centered space difference scheme will distort a short wave disturbance much more rapidly as compared to a second-order centered space difference scheme.

It is known that higher order space differencing would introduce additional grid points; this gives rise to computational modes in space similar to the ones that one encounters for higher (second) order time differencing. Moreover, the formulation of boundary conditions is not straightforward while dealing with fourth-order centered space differencing scheme. Hence, for the case of small-scale disturbances, no higher (third or fourth) order space difference scheme can be more acceptable than the second-order centered space differencing.

6.7   Higher Order Sign Preserving Advection Schemes

It was noted that the upstream scheme satisfies the transportive property. Applying the upstream scheme to a one-dimensional linear advection equation, one gets

$$\frac{u_j^{n+1} - u_j^n}{\Delta t} + c\frac{u_j^n - u_{j-1}^n}{\Delta x} = 0 \text{ for } c > 0. \tag{6.70}$$

Equation (6.70) can be written as

$$u_j^{n+1} = (1-\lambda)u_j^n + \lambda u_{j-1}^n, \tag{6.71}$$

where λ is the Courant number; $\lambda = \frac{c\Delta t}{\Delta x}$. Equation (6.71) has the form of an interpolation. The upstream scheme produces strong damping and is stable when $\lambda \leq 1$. Assuming that the stability criteria is satisfied, Equation (6.71) guarantees the following

$$\text{Min}\left\{u_j^n, u_{j-1}^n\right\} \leq u_j^{n+1} \leq \text{Max}\left\{u_j^n, u_{j-1}^n\right\}. \tag{6.72}$$

Equation (6.72) shows that u_j^{n+1} cannot be smaller than the smallest value of u_j^n or larger than the largest value of u_j^n. Hence, the upstream scheme does not produce any new maxima or minima. If no negative values of 'u' are present initially, real advection cannot produce negative values of 'u' at a later time. The upstream scheme also has this sign preserving property when the stability criteria is satisfied. Such sign preserving schemes are important when the advected quantity is intrinsically a non-negative quantity, such as water vapour mixing ratio or mixing ratio of some trace species. Any scheme that satisfies Equation (6.72) is said to satisfy the monotone property; the upstream scheme is a monotone scheme. All monotone schemes are also sign preserving schemes; however, the opposite is not true. Furthermore, all sign preserving schemes tend to also be stable schemes. This can be easily demonstrated. Assume that one has a linear conservative scheme for a variable 'q', such that

$$\sum_i q_i^n = \sum_i q_i^0, \tag{6.73}$$

where the summation represents a sum over the entire spatial domain and superscripts $n > 0$ and 0 refer to two time levels. Whereas Equation (6.73) shows the relationship only for a single spatial dimension, the argument that follows holds for any dimension. Assume that one applies a scheme that is sign preserving and conserves 'q' while assuming q values vary from q^0 to q^n through n time marchings. Assuming that q_i^0 is of the same sign everywhere, it follows that from Equation (6.73),

$$\sum_i |q_i^n| = \sum_i |q_i^0.|$$ (6.74)

It is to be noted that for any arbitrary variable, ϕ_i, one has

$$\sum_i (\phi_i)^2 \leq \left\{ \sum_i |\phi_i| \right\}^2.$$ (6.75)

From Equations (6.74) and (6.75), it is clear that one can obtain the following inequality

$$\sum_i (q_i^n)^2 \leq \left\{ \sum_i |q_i^0| \right\}^2.$$ (6.76)

It is to be noted that the RHS of Equation (6.76) is a constant; Equation (6.76) shows that $(q_i^n)^2$ summed over the entire spatial domain is bounded for all times demonstrating absolute stability of the scheme in line with the requirements of the energy method of stability analysis. In the aforementioned discussion, it was assumed that q_i^0 is everywhere of the same sign. However, it was observed that the aforementioned assumption is not really necessary as can be discerned from the following arguments. Let

$$q = q^+ + q^-,$$ (6.77)

with the definition that q^+ is positive where q is positive and zero elsewhere. Similarly, one defines that q^- is negative where q is negative and zero elsewhere. With this definition, the total of q is the sum of the two parts (q^+ and q^-) and, hence, advection of q is equivalent to advection of q^+ and q^-, separately. If one were to apply a sign preserving scheme to each part, then each of these two advections are stable and, hence, the advection of q is also stable.

Although the upstream scheme is a sign preserving scheme, it is also only a first-order scheme and produces strong damping. This leads one to seek schemes that are sign preserving as well as having higher order accuracy, say second-order accurate advection schemes. A second-order advection scheme that only occasionally produces a few spurious negative values, i.e., which is by and large sign preserving is given by

$$\frac{d(m_j\phi_j)}{dt} = \frac{(mu)_{j+1/2}\phi_{j+1/2} - (mu)_{j-1/2}\phi_{j-1/2}}{\Delta x_j} = 0,$$ (6.78)

with the option of employing geometric mean for interpolation such as

$$\phi_{j+1/2} = \left\{ \phi_j \phi_{j+1} \right\}^{1/2}$$ (6.79)

or employing harmonic mean for interpolation

$$\phi_{j+1/2} = \frac{2\phi_j\phi_{j+1}}{\phi_j + \phi_{j+1}}. \tag{6.80}$$

Both these interpolations (geometric and harmonic) have the useful property that if either ϕ_{j+1} or ϕ_j is equal to zero, then $\phi_{j+1/2}$ will also be equal to zero. If the time step is sufficientlye small, the aforementioned interpolation along with Equation (6.78) will ensure that the property ϕ will not change sign. It is to be noted that both geometric mean and harmonic mean are nonlinear averages. An even better second-order accurate advective scheme is available where Equation (6.78) is replaced by the following equation,

$$\frac{d(m_j\phi_j)}{dt} + \frac{\left[\left\{(mu)^+_{j+1/2}\phi^+_{j+1/2} + (mu)^-_{j+1/2}\phi^-_{j+1/2}\right\} - \left\{(mu)^+_{j-1/2}\phi^+_{j-1/2} + (mu)^-_{j-1/2}\phi^-_{j-1/2}\right\}\right]}{\Delta x_j} = 0, \tag{6.81}$$

where

$$(mu)^+_{j+1/2} = \frac{(mu)_{j+1/2} + |(mu)_{j+1/2}|}{2} \geq 0, \tag{6.82}$$

$$(mu)^-_{j+1/2} = \frac{(mu)_{j+1/2} - |(mu)_{j+1/2}|}{2} \leq 0, \tag{6.83}$$

$$\phi^+_{j+1/2} = \frac{2\phi_j^{n+1}\phi_{j+1}^n}{\phi_j^n + \phi_{j+1}^n}, \tag{6.84}$$

$$\phi^-_{j+1/2} = \frac{2\phi_j^n\phi_{j+1}^{n+1}}{\phi_j^n + \phi_{j+1}^n}, \tag{6.85}$$

where $(mu)^+_{j+1/2}$ is the mass flux in the positive x direction into cell j+1 from cell j, whereas $(mu)^-_{j+1/2}$ is the mass flux in the negative x direction out of cell j+1 and into cell j. Note from Equations (6.84) and (6.85) that one can associate different interpolated ϕs with the mass flux in the two directions.

6.8   Two-dimensional Linear Advection Equation

The two-dimensional linear advection equation is as follows:

$$\frac{\partial u}{\partial t} + c_x \frac{\partial u}{\partial x} + c_y \frac{\partial u}{\partial y} = 0, \tag{6.86}$$

where $u = u(x,y,t)$ is a function of x, y, and time t whereas c_x and c_y are the x and y components of the advection velocity c. It is assumed that c_x and c_y are constants.

Replacing the space derivatives in Equation (6.86) by second-order centered space differences, one obtains

$$\left.\frac{\partial u}{\partial t}\right|_{j,m} = -c_x \left(\frac{u_{j+1,m} - u_{j-1,m}}{2\Delta x}\right) - c_y \left(\frac{u_{j,m+1} - u_{j,m-1}}{2\Delta y}\right), \tag{6.87}$$

where $u(j\Delta x, m\Delta y)$ indicates the u value at grid points $x = j\Delta x$ and $y = m\Delta y$. Substituting, as before $u_{j,m}(t) = \Re\left[U(t)e^{i(kx+ly)}\right]$ into Equation (6.87), one gets the following oscillation equation

$$\frac{dU}{dt} = i\left[-\frac{c_x}{\Delta x}\sin(k\Delta x) - \frac{c_y}{\Delta y}\sin(l\Delta y)\right]U. \tag{6.88}$$

If the leapfrog scheme (centered) second-order time integration scheme is used for the time derivative, in Equation (6.88), the stability criterion is given by

$$\left|\left(\frac{c_x}{\Delta x}\sin(k\Delta x) + \frac{c_y}{\Delta y}\sin(l\Delta y)\right)\Delta t\right| \leq 1. \tag{6.89}$$

Equation (6.86) has to be identically satisfied for all possible values of the wavenumbers k and l to ensure stability. Consider for simplicity, the special case where $\Delta x = \Delta y = d$. The smallest (minimum resolvable) wavelength in the x and y directions occur when $kd = \pi$ and when $ld = \pi$.

The stability criterion (Equation (6.89)) for the special case $\Delta x = \Delta y = d$ can be shown to assume the form

$$\sqrt{2}c\frac{\Delta t}{\Delta x} \leq 1, \tag{6.90}$$

where the advection velocity c is related to its components c_x and c_y by $c = \sqrt{c_x^2 + c_y^2}$.

It is clear from Equation (6.90) that one must choose a smaller time step ($1/\sqrt{2}$ times smaller) for the two-dimensional linear advection equation than the one employed in the one-dimensional linear advection equation to ensure stability.

6.8.1 Computational dispersion: Phase speed dependence on frequency

In Section 6.4, the resulting computational dispersion due to central space differencing for a one-dimensional linear advection equation was examined. In this section, the computational dispersion due to central time differencing for a one-dimensional linear advection equation will be examined. The effects of the centered time difference scheme is analyzed in the same way as the effects of the centered space differencing space. Replacing the time derivative by central time difference (leapfrog scheme), one gets

$$\frac{u^{n+1} - u^{n-1}}{2\Delta t} + c\left.\frac{\partial u}{\partial x}\right|^n = 0. \tag{6.91}$$

Substituting wave type solutions $u^n = u_o e^{ik(x - c_D n\Delta t)}$ in Equation (6.91), one gets the following expression for the numerical phase speed c_D,

$$c_D = \frac{\sin^{-1}(\omega\Delta t)}{k\Delta t}, \tag{6.92}$$

where c_D is the numerical phase speed and c the true phase speed. The ratio of c_D and c should ideally be as close as possible to one but is given as

$$\frac{c_D}{c} = \frac{\sin^{-1}(\omega\Delta t)}{\omega\Delta t}. \tag{6.93}$$

The computational (c_{D_g}) and physical (c_g) group speeds also differ and their ratio is given as

$$\frac{c_{D_g}}{c_g} = \frac{d\omega_D}{c_g dk} = \frac{d(kc_D)}{c_g dk} = \frac{1}{\sqrt{1 - (\omega\Delta t)^2}}. \tag{6.94}$$

Computational dispersion due to both space and time discretization

Replacing the time derivative by central time difference and the space derivative with central space difference in the one-dimensional linear advection equation, one gets

$$\frac{u_j^{n+1} - u_j^{n-1}}{2\Delta t} + c\frac{u_{j+1}^n - u_{j-1}^n}{2\Delta x} = 0. \tag{6.95}$$

Substituting the wave type solution $u_j^n = u_o e^{ik(j\Delta x - c_D n\Delta t)}$ into Equation (6.95), one gets an expression for the numerical phase speed c_D as

$$c_D = \frac{1}{k\Delta t}\sin^{-1}\left[\frac{c\Delta t}{\Delta x}\sin(k\Delta x)\right], \tag{6.96}$$

where c_D is the numerical phase speed and c the true phase speed. The ratio of c_D and c should ideally be as close as possible to one but is given as

$$\frac{c_D}{c} = \frac{1}{ck\Delta t} \sin^{-1} \left[\frac{c\Delta t}{\Delta x} \sin(k\Delta x) \right]$$

$$= \frac{1}{C_{FL}k\Delta x} \sin^{-1} \left[C_{FL} \sin(k\Delta x) \right], \tag{6.97}$$

where $C_{FL} = (c\Delta t)/\Delta x$ is the Courant number that appears in the CFL condition for stability of advection equation. Equation (6.97) shows that the phase speed is a function of the wave number k. Thus, the finite differencing in space causes a computational dispersion. As $k\Delta x$ increases, the computational phase speed c_D decreases from c to zero when $k\Delta x = \pi$, the latter corresponding to the shortest possible wave length of two grid cells ($\lambda = 2\Delta x$). Thus, all waves propagate at a phase speed c_D that is less than the true phase speed c; this decelerating effect increases as the wave length decreases. Moreover, the two grid waves are stationary. It is clear from Equation (6.97) that if the Courant number (CFL) equals 1, which is the limit of stability for the advection equation, then the computation phase speed is the same as the analytical one ($c_D = c$). In this section, one can draw two important inferences:

- The numerical advection phase speed is slower than the true advection phase speed.

- The numerical advection phase speed changes with wave number.

Exercises 6a (Questions only)

1. Show that the central in time and central in space (CTCS) finite difference scheme is a conditionally stable scheme for the one-dimensional linear advection equation with the requirement that the Courant number is less than or equal to 1.

2. How is the "computational mode" solution handled while utilizing the central in time and central in space (CTCS) finite difference scheme for the one-dimensional linear advection equation.

3. Show that the forward in time and central in space (FTCS) finite difference scheme is an unconditionally unstable scheme for the one-dimensional linear advection equation.

4. Show that the backward in time and central in space (BTCS) finite difference scheme is an unconditionally stable scheme for the one-dimensional linear advection equation.

5. Show that the upwind finite difference scheme is a conditionally stable scheme for the one-dimensional linear advection equation with the requirement that the Courant number has a value less than or equal to 1.

6. Show that the Crank–Nicolson finite difference scheme is an unconditionally stable scheme for the one-dimensional linear advection equation.

7. Show that the Lax finite difference scheme is a conditionally stable scheme for the one-dimensional linear advection equation with the requirement that the Courant number has a value less than or equal to 1.

8. Show that the Lax finite difference scheme is not a consistent finite difference scheme for the one-dimensional linear advection equation.

9. Show that for the Lax finite difference scheme for the one-dimensional linear advection equation, the phase speed of the numerical scheme is exactly the same as the analytical phase speed for the special case where the Courant number equals 1.

10. Write down the finite difference scheme for the one-dimensional linear advection equation, which is second-order accurate in time and fourth-order accurate in space.

    Answer: $\dfrac{u_i^{n+1} - u_i^{n-1}}{2\Delta t} = -c \left(\dfrac{4}{3} \dfrac{u_{i+1}^n - u_{i-1}^n}{2\Delta x} - \dfrac{1}{3} \dfrac{u_{i+2}^n - u_{i-2}^n}{4\Delta x} \right)$

11. Show that the Lax–Wendroff finite difference scheme is second-order accurate in space and time when applied to the one-dimensional linear advection equation.

12. Show that the Lax–Wendroff finite difference scheme is a conditionally stable scheme for the one-dimensional linear advection equation with the requirement that the Courant number has a value less than or equal to 1.

13. Write down the Lax–Wendroff finite difference scheme for the one-dimensional conservation equation

$$\frac{\partial u}{\partial t} + \frac{\partial F(u)}{\partial x} = 0.$$

    Answer:

$$u_{i-1/2}^{n+1/2} = \frac{1}{2}(u_i^n + u_{i-1}^n) - \frac{\Delta t}{2\Delta x}(F_i^n - F_{i-1}^n) \quad \text{where} \quad F_i^n = F(u_i^n)$$

$$u_{i+1/2}^{n+1/2} = \frac{1}{2}(u_{i+1}^n + u_i^n) - \frac{\Delta t}{2\Delta x}(F_{i+1}^n - F_i^n)$$

$$u_i^{n+1} = u_i^n - \frac{\Delta t}{\Delta x}(F_{i+1/2}^{n+1/2} - F_{i-1/2}^{n+1/2})$$

14. Show that the two-step (predictor-corrector) type MacCormack's scheme for the linear one-dimensional advection equation is a consistent finite difference scheme.

15. Show that the two-step (predictor-corrector) type MacCormack's scheme for the linear one-dimensional advection equation is a finite difference scheme with second-order accuracy in space and time.

16. Show that the two-step (predictor-corrector) type MacCormack's scheme for the linear one-dimensional advection equation is a conditionally stable finite difference scheme with the condition that Courant number has to have a value less than or equal to one.

17. Show that the two-step (predictor-corrector) type MacCormack's scheme for the linear one-dimensional advection equation utilizes a forward difference approximation in the predictor step and a backward difference approximation in the second step. This scheme is equivalent to a single step Lax–Wendroff scheme.

18. Write down the finite difference form for the second-order upwind method for the linear one-dimensional advection equation (E6.1).

    Answer: $f_i^{n+1} = f_i^n - \lambda \left(f_i^n - f_{i-1}^n \right) - \dfrac{c(1-c)}{2} \left(f_i^n - 2f_{i-1}^n + f_{i-2}^n \right)$

19. Show that the finite difference scheme mentioned in Exercise 6a Q 18 is a consistent finite difference scheme.

20. Show that the finite difference scheme mentioned in Exercise 6a Q 18 has second-order accuracy in space and time.

21. Show that the finite difference scheme mentioned in Exercise 6a Q 18 is a conditionally stable scheme with the condition that Courant number has to have a value of less than or equal to two.

Exercises 6b (Questions and answers)

1. Show that the Lax finite difference scheme for the one-dimensional linear advection equation

   $$\frac{\partial u}{\partial t} = c \frac{\partial u}{\partial x} \qquad\qquad (E6.1)$$

   has inherent diffusion features.

   Answer: The Lax finite difference scheme for the one-dimensional linear advection equation is given by

   $$u_i^{n+1} = \frac{1}{2} \left(u_{i+1}^n + u_{i-1}^n \right) - \frac{c\Delta t}{2\Delta x} \left(u_{i+1}^n - u_{i-1}^n \right)$$

   We can re-write this as

   $$\left[\left(\frac{1}{2} u_i^{n+1} + \frac{1}{2} u_i^{n+1} \right) - \frac{1}{2} u_i^{n-1} + \frac{1}{2} u_i^{n-1} \right] = \frac{1}{2} \left(u_{i+1}^n + u_{i-1}^n \right) + u_i^n - u_i^n - \frac{c\Delta t}{2\Delta x} \left(u_{i+1}^n - u_{i-1}^n \right)$$

   The third and fourth terms in LHS cancel out while the third and fourth terms in RHS also cancel out. Rearranging the terms in the aforementioned equation, one obtains the following

   $$\frac{1}{2} \left(u_i^{n+1} - u_i^{n-1} \right) = \frac{1}{2} \left(u_{i+1}^n - 2u_i^n + u_{i-1}^n \right) - \frac{c\Delta t}{2\Delta x} \left(u_{i+1}^n - u_{i-1}^n \right) - \frac{1}{2} \left(u_i^{n+1} - 2u_i^n + u_i^{n-1} \right)$$

   or in the continuous limit

   $$\frac{\partial u}{\partial t} = \frac{(\Delta x)^2}{2\Delta t} \frac{\partial^2 u}{\partial x^2} - c \frac{\partial u}{\partial x} - \frac{\Delta t}{2} \frac{\partial^2 u}{\partial t^2}.$$

Although the last term in the RHS tends to zero, as $\Delta t \to 0$, the behaviour of the first term in the right-hand side depends on the behaviour of Δt and Δx, respectively. Taking the derivative of the one-dimensional linear advection with respect to time, one obtains the one-dimensional wave equation

$$\frac{\partial^2 u}{\partial t^2} = c^2 \frac{\partial^2 u}{\partial x^2}.$$

Substitution of the one-dimensional wave equation to the aforementioned equation, one obtains

$$\frac{\partial u}{\partial t} = -c \frac{\partial u}{\partial x} + D \frac{\partial^2 u}{\partial x^2},$$

where

$$D = \frac{(\Delta x)^2}{2\Delta t} - c^2 \frac{\Delta t}{2},$$

where D is a positive definite coefficient, similar to the diffusion coefficient. Hence, application of the Lax finite difference scheme to the one-dimensional linear advection equation has the effect of including additional diffusion term.

2. How is the computational mode that arises naturally in CTCS finite difference scheme for a one-dimensional advection equation suppressed?

Answer: The computational mode that manifests naturally in the CTCS finite difference scheme can be suppressed in two essentially different ways.

(a) The CTCS finite difference scheme can be integrated with a Euler forward or backward at regular intervals of a certain number of time steps (say, 50 time steps). This would eliminate the computational mode.

(b) Another alternate way, which is the most common way in atmospheric models, is to employ a Robert–Asselin filter. First one applies the leapfrog integration to obtain the solution at time level $n+1$,

$$u^{n+1} = (u')^{n-1} + 2\Delta t \left[F(u^n) \right],$$

then the filter is applied as a time smoothing between the time levels $n-1$, n, and $n+1$ so that

$$(u')^n = u^n + \gamma \left[(u')^{n-1} - 2u^n + u^{n+1} \right],$$

where the "primed" quantities indicate the smoothed values and γ is the Asselin coefficient with a typical value around 0.1. The next leapfrog step will be at

$$u^{n+2} = (u')^n + 2\Delta t \left[F(u^{n+1}) \right].$$

3. Write down the two-step (predictor-corrector) type MacCormack's scheme for the linear one-dimensional advection equation. Use f as the advected quantity.

   Answer: The two-step (predictor-corrector) type MacCormack's scheme for the linear one-dimensional advection equation is given as

   First step (predictor):

   $$\overline{f}_i^{n+1} = f_i^n - \lambda \left(f_{i+1}^n - f_i^n\right),$$

   where $\lambda = c\Delta t/\Delta x$ is the convection or Courant number.

   Second step (corrector):

   $$f_i^{n+1} = \frac{1}{2}\left(f_i^n + \overline{f}_i^{n+1}\right) - \lambda\left(\overline{f}_i^{n+1} - \overline{f}_{i-1}^{n+1}\right)/2.$$

Python examples

1. The following python code provides an example of the one-dimensional nonlinear advection equation that has a finite difference scheme which is (i) not energy conserving, and which is (ii) energy conserving.

   Consider the one-dimensional nonlinear advection equation $\frac{\partial u}{\partial t} + u\frac{\partial u}{\partial x} = 0$ Multiplying this equation by u and integrating over the domain (assuming cyclic boundary conditions), one gets $\frac{\partial E}{\partial t} = 0$; $\frac{1}{2}\int_0^L u^2 dx$ where E is the total kinetic energy. The aforementioned equation shows that the total kinetic energy is conserved. It would be desirable to have a finite difference scheme that also demonstrates the conservation of the total kinetic energy. Consider the following finite difference form of the aforementioned equation in which only the advection term has been discretised, as follows:

   $\frac{\partial u_j}{\partial t} = -u_j\frac{u_{j+1}-u_{j-1}}{2\Delta x}$. Multiplying by u_j and summing over all points, one gets $\frac{\partial E'}{\partial t} = -\frac{1}{2}\sum_j(u_j^2 u_{j+1} - u_j^2 u_{j-1}) \neq 0$.

   However, let us consider the following finite difference form of the one-dimensional nonlinear advection equation in which only the advection term has been discretised:

   $\frac{\partial u_j}{\partial t} = -\left\{\frac{u_{j+1}+u_j+u_{j-1}}{3}\right\}\left\{\frac{u_{j+1}+u_{j-1}}{2\Delta x}\right\}.$

   This finite difference form conserves the total kinetic energy leading to $\frac{\partial E'}{\partial t} = 0$:

   ```
   #python code for conservation and non conservational energy
   dt=0.03
   dx=0.1
   NC=0
   C=0
   ```

```python
a=1+(10-0)/dx
b=1+(250-0)/dt
print(int(b))
import numpy as np
import matplotlib.pyplot as plt

x= np.zeros(int(a))
t= np.zeros(int(b))
x[1]=0
t[1]=0
u = np.zeros((int(a),int(b)))
E = np.zeros(int(b))
E1 = np.zeros(int(b))
uc = np.zeros((int(a),int(b)))
Ec = np.zeros(int(b))
E1c = np.zeros(int(b))
for i in range(2,int(a)):
    x[i]=x[i-1]+dx
print(x)
for i in range(2,int(b)):
    t[i]=t[i-1]+dt
print(t)
for i in range(1,int(a)):
    u[i,1]= np.exp(-(x[i]-2)**2)
    uc[i,1]= np.exp(-(x[i]-2)**2)

u[1,:]=0
u[int(a-1),:]=0
uc[1,:]=0
uc[int(a-1),:]=0

for j in range(2,int(a-1)):
    u[j,2]=u[j,1]-u[j,1]*dt/(2*dx)*(u[j+1,1]-u[j-1,1])
    uc[j,2]=uc[j,1]-uc[j,1]*dt/(2*dx)*(uc[j+1,1]-uc[j-1,1])
for i in range(2,int(b-1)):
    for j in range(2,int(a-1)):
        u[j,i+1]=u[j,i-1]-(u[j,i]*dt/dx)*(u[j+1,i]-u[j-1,i])
        uc[j,i+1]=uc[j,i-1]-((uc[j+1,i]+uc[j,i]+uc[j-1,i])*(uc
```

```
                [j+1,i]-uc[j-1,i]))*dt/(3*dx)
for j in range(0,int(b)):
    for i in range(0,int(a)):
        E[j]=u[i,j]*u[i,j]*dx
        E1[j]=E1[j]+E[j]
        Ec[j]=uc[i,j]*uc[i,j]*dx
        E1c[j]=E1c[j]+Ec[j]

plt.plot(t[1:121],(E1[1:121])/2,'k-')
plt.plot(t[1:121],(E1c[1:121])/2,'k--')
plt.legend(["Non_conservation_of_Energy","Conservation_of
Energy"])
plt.ylabel('Energy')
plt.xlabel('Time')
plt.title("")
plt.show()
```

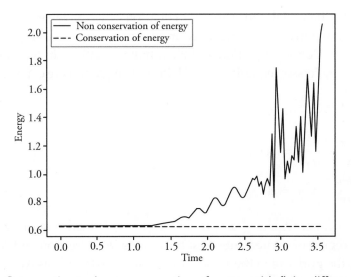

Figure 6.13 Conservation and non-conservation of energy with finite difference methods.

7

Numerical Solution of Elliptic Partial Differential Equations

7.1 Introduction

Elliptic partial differential equations are equations that have second derivatives in space but no time derivatives. The most important examples of elliptic PDEs are the Laplace equation, Poisson equation and Helmholtz equation. In two dimensions, the Laplace equation, Poisson equation, and Helmholtz equation are expressed as follows:

$$\frac{\partial^2 u}{\partial x^2} + \frac{\partial^2 u}{\partial y^2} = 0, \tag{7.1}$$

$$\frac{\partial^2 u}{\partial x^2} + \frac{\partial^2 u}{\partial y^2} = f(x,y), \tag{7.2}$$

$$\frac{\partial^2 u}{\partial x^2} + \frac{\partial^2 u}{\partial y^2} + k^2 u = g(x,y). \tag{7.3}$$

If the region of interest is $a \leq x \leq b$, and $c \leq y \leq d$, then assuming that one can discretize the region of interest into M subdivisions in x direction and N subdivisions in y direction, the grid size in the x and y directions are defined as $\Delta x = (b-a)/M$ and $\Delta y = (d-c)/N$. Considering the Poisson equation and replacing the derivatives with central differences

$$\frac{u_{i+1,j} - 2u_{i,j} + u_{i-1,j}}{(\Delta x)^2} + \frac{u_{i,j+1} - 2u_{i,j} + u_{i,j-1}}{(\Delta y)^2} = f_{i,j}, \tag{7.4}$$

for $1 \leq i \leq M-1$ and $1 \leq j \leq N-1$. Defining $\beta = \Delta x/\Delta y$, Equation (7.4) takes the form

$$\beta^2 u_{i,j-1} + u_{i-1,j} - 2\left(1 + \beta^2\right) u_{i,j} + u_{i+1,j} + \beta^2 u_{i,j+1} = f_{i,j}\left(\Delta x\right)^2. \tag{7.5}$$

The boundary conditions on all the four sides are supposed to be known as shown:

$$u_{0,j} = g_a(a,y_j), \quad u_{M,j} = g_b(b,y_j), \quad u_{i,0} = g_c(x_i,c), \quad u_{i,N} = g_d(x_i,d), \tag{7.6}$$

where g_a, g_b, g_c, and g_d are known prescribed functions. The finite difference scheme is schematically shown in Figure 7.1 where the solution at the grid (i,j), $u_{i,j}$ is related to the solution at the four neighboring grid points $(i+1,j)$, $(i-1,j)$, $(i,j+1)$, and $(i,j-1)$.

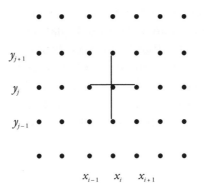

Figure 7.1   Finite difference scheme schematically shown for solving the two-dimensional Poisson equation.

7.1.1   Commonly occurring elliptic problems

Elliptic problems that manifest as elliptic PDEs, arise naturally in steady-state problems, i.e., in problems that are essentially concerned with states of equilibrium. That is, they arise in problems in which time t does not appear as an independent variable and, hence, where the initial values are not relevant. Hence, typical elliptic problems are essentially boundary value problems in which the boundary data are prescribed on given closed boundary curves (or surfaces/hypersurfaces for equations with more than two independent variables). The three most commonly occurring boundary conditions are (i) the Dirichlet problem, in which an elliptic PDE such as the Laplace equation, Equation (7.1), is solved with u being prescribed as a function of position along the whole boundary of the domain, (ii) the Neumann problem , in which an elliptic PDE such as the Laplace equation, Equation (7.1), is solved with $\partial u/\partial n$ being prescribed as a function of position along the whole boundary of the domain, where $\partial/\partial n$ denotes the normal derivative along the boundary, and (iii) the Robin problem, in which an elliptic PDE such as the Laplace equation, Equation (7.1), is solved with $au + b(\partial u/\partial n)$ prescribed along the whole boundary of the domain.

7.2   Direct Methods of Solution

As the dependent variable u [given in Equation (7.6)] is prescribed, and hence known, at the four boundary sides (from boundary values), there are $(M-1) \times (N-1)$ interior grid cells where the solution $u(i,j)$ needs to be determined. It is clear that there are exactly $(M-1) \times (N-1)$ equations in Equation (7.5). Moreover, as the Poisson equation is linear, Equation (7.5) provides a system of linear algebraic equations that could be solved using matrix methods. In order to directly solve the system of linear algebraic equations using matrix methods, one need to express the $u_{i,j}$s as a vector of unknowns. There exists a standard procedure of writing Equation (7.5) as a matrix equation. For this, let $\underset{\sim}{u}$ be the column vector that is obtained by placing one column after another from the columns of $(u_{i,j})$. Thus, one would list $u_{i,1}$ first, then list $u_{i,2}$, etc. Next, one obtains the matrix A that contains the coefficients of Equations (7.4); one also needs to incorporate the boundary conditions given in Equation (7.6) in a vector $\underset{\sim}{b}$. The resulting matrix equation would have the following form;

$$A\underset{\sim}{u} = \underset{\sim}{b} \tag{7.7}$$

Solving the matrix equation, Equation (7.7), is known as the direct method . The advantage of the direct method is that solving Equation (7.7) can be done relatively quickly and accurately. The drawback of the direct method is that one must set up $\underset{\sim}{u}$, A, and $\underset{\sim}{b}$. Furthermore, the matrix A has dimensions $(M-1)(N-1) \times (M-1)(N-1)$, which can be rather large. It is to be noted that although matrix A is large, many of its elements are zero. A matrix having most of its elements zero is called sparse; there are special methods available for efficiently obtaining the solution of matrix equations that involve sparse matrices.

7.3   Iterative Methods of Solution

A usually preferred alternative to the direct method explained in the previous section is to solve the finite difference equations, Equation (7.5), in an iterative manner. For this, one first solves Equation (7.5) for $u_{i,j}$, which yields

$$u_{i,j} = \frac{1}{2(1+\beta^2)} \left[\beta^2 u_{i,j-1} + u_{i-1,j} + u_{i+1,j} + \beta^2 u_{i,j+1} - f_{i,j}(\Delta x)^2 \right]. \tag{7.8}$$

Using Equation (7.8) together with boundary values (Equation (7.6)), one can update $u_{i,j}$ from the values of its neighbors in an iterative manner until convergence is attained for a prescribed tolerance. If this iterative method converges, then the result is an approximate solution of the Poisson equation (7.2). For the iterative solution of the two-dimensional Laplace equation, one solves Equation (7.8) with $f_{i,j}$ being assigned zero. The resulting equation is

$$u_{i,j} = \frac{1}{2(1+\beta^2)}\left[\beta^2 u_{i,j-1} + u_{i-1,j} + u_{i+1,j} + \beta^2 u_{i,j+1}\right]. \tag{7.9}$$

7.3.1 Gauss–Seidel method

The Gauss–Seidel method, as applied to the finite difference approximation of the two-dimensional Laplace equation, is obtained by rearranging as follows, with superscript k assuming integer values $(k = 0, 1, 2, \ldots)$, known as the iteration number:

$$u_{i,j}^{k+1} = \frac{1}{2(1+\beta^2)}\left[\beta^2 u_{i,j-1}^{k+1} + u_{i-1,j}^{k+1} + u_{i+1,j}^{k} + \beta^2 u_{i,j+1}^{k}\right]. \tag{7.10}$$

An initial guess for $u_{i,j}$ at all interior points with $k = 0$ is necessary to start the process. Then, applying Equation (7.10), one can obtain the improved $u_{i,j}$ at all interior points, wherever required boundary values are to be utilized. The iteration process is usually terminated when a convergence criteria as indicated by the following is satisfied for all (i, j). A commonly used convergence criteria is given by

$$\max_{i,j}\left|\Delta u_{i,j}^{k+1}\right| \leq \varepsilon \quad \text{or} \quad \max_{i,j}\left|\frac{\Delta u_{i,j}^{k+1}}{u_{i,j}^{k}}\right| \leq \varepsilon, \tag{7.11}$$

where

$$\Delta u_{i,j}^{k+1} = u_{i,j}^{k+1} - u_{i,j}^{k} \tag{7.12}$$

is called the *correction*. Normally the iteration methods require diagonal dominance to guarantee convergence. The Gauss–Seidel method is also called the "relaxation method."

7.3.2 Successive over relaxation (SOR) method

The convergence rate of the Gauss–Seidel method may be greatly increased by employing the successive over relaxation (SOR) method. In the SOR method, the equation is modified by over relaxing the "correction" by the over relaxation factor ω to obtain

$$u_{i,j}^{k+1} = u_{i,j}^{k} + \omega \Delta u_{i,j}^{k+1}, \tag{7.13}$$

where the correction is as given in Equation (7.12) in which $u_{i,j}^{k+1}$ is obtained using the basic Gauss–Seidel method (7.10). When $\omega = 1.0$, the SOR method reduces to the Gauss–Seidel method. The maximum rate of convergence for the SOR method is obtained for some optimal value of ω that lies between 1.0 and 2.0. The iteration process for the SOR method will continue until the convergence criteria [Equation (7.11)] is satisfied.

7.3.3 Relaxation, sequential relaxation, and successive relaxation methods

In Sections 7.3.1 and 7.3.2, the "relaxation method" and the "successive over relaxation (SOR) method" were briefly introduced as examples of iterative methods to solve elliptic partial differential equations. In this section, a more detailed discussion on the relaxation method and its variants will be presented.

Relaxation method of solving an elliptic partial differential equation

Consider a two-dimensional Helmholtz equation that naturally arises while implementing the semi-implicit method. Clearly, the Helmholtz equation is an example of an elliptic partial differential equation. The Helmholtz equation can be written in schematic form as

$$\nabla^2 T - MT = f(x,y), \tag{7.14}$$

where

$$\nabla^2 \equiv \frac{\partial}{\partial x^2} + \frac{\partial}{\partial y^2}$$

is the two-dimensional *Laplacian operator*. Equation (7.14) with $M = 0$ becomes the *Poisson equation* whereas Equation (7.14) with both $M = 0$ and $f(x,y) = 0$ becomes the *Laplace equation*. Here M and $\hat{f}(x,y)$ are known functions of space. It is assumed that the boundary conditions are prescribed on all the four boundaries that encompass the region of interest. Either the boundary conditions are prescribed on the dependent variable T on all the four boundaries or on the normal derivatives of T on all the four boundaries. The boundary conditions can also be on a linear combination of T and the normal derivatives of T on all the four boundaries.

The simplest finite difference approximation for the Laplacian operator is the 5-point finite difference formula [refer to Equations (3.20) and (3.23)]. Utilizing the aforementioned central difference scheme in space, Equation (7.14) can be written in finite difference form as follows:

$$\frac{T_{i-1,j} + T_{i+1,j} + T_{i,j-1} + T_{i,j+1} - 4T_{i,j}}{h^2} - M_{i,j}T_{i,j} = f_{i,j}, \tag{7.15}$$

where it is assumed that $\Delta x = \Delta y = h$ and $M_{i,j}$ and $f_{i,j}$ are the finite difference representation of M and f at the point (i,j). As the 5-point finite difference formula for the two-dimensional Laplacian operator is an approximation to the second-order, the LHS and RHS of Equation (7.15) are not equal. The difference between the two sides of Equation (7.15) is called the "residual" and is indicated by $R_{i,j}$. Hence, we have

$$\frac{T_{i-1,j} + T_{i+1,j} + T_{i,j-1} + T_{i,j+1} - 4T_{i,j}}{h^2} - M_{i,j}T_{i,j} - f_{i,j} = R_{i,j}$$

or

$$T_{i-1,j} + T_{i+1,j} + T_{i,j-1} + T_{i,j+1} - 4T_{i,j} - M_{i,j}T_{i,j}h^2 - f_{i,j}h^2 = R_{i,j}h^2 = \tilde{R}_{i,j}. \qquad (7.16)$$

As the relaxation method is an iterative method, one need to introduce an index m as a superscript for each of the terms in Equation (7.16), where the index m is an iteration number that gets updated for every iteration. To illustrate the "relaxation method," assume that the mth guess of the dependent variable T at the grid point (x_i, y_j) indicated as (i,j) is $T_{i,j}^m$. Equation (7.16) can then be written as

$$T_{i-1,j}^m + T_{i+1,j}^m + T_{i,j-1}^m + T_{i,j+1}^m - 4T_{i,j}^m - M_{i,j}T_{i,j}^m h^2 - f_{i,j}h^2 = \tilde{R}_{i,j}^m \qquad (7.17)$$

It is to be noted that both $M_{i,j}$ and $f_{i,j}$ do not have superscript m as an iteration number as both M and f are known functions of x and y. $\tilde{R}_{i,j}^m$ is called the mth residual estimate and is a measure of the error of the mth guess. It is clear that if the residuals are to be zero at every grid point over the region of interest, the LHS would be equal to the RHS of Equation (7.15) leading to a solution of the system. However, it is also true that an initial guess estimate of $T_{i,j}^m$ cannot ensure that all the residuals are all zeros at all grid points over the region of interest. Hence, it is realistic to expect that the residuals [LHS of Equation (7.17)] will not be zero at all grid points over the region of interest identically. The aim of the relaxation method is to provide a systematic way of reducing all the residuals to successively smaller values with increase in iterations. Towards this end, let us determine the change in the value of $T_{i,j}^m$ that will reduce the residual $\tilde{R}_{i,j}^m$ to zero. It is assumed that the adjacent values of T ($T_{i-1,j}, T_{i+1,j}, T_{i,j-1}$, and $T_{i,j+1}$) will be left unmodified for now. The aforementioned results in

$$T_{i-1,j}^m + T_{i+1,j}^m + T_{i,j-1}^m + T_{i,j+1}^m - 4T_{i,j}^{m+1} - M_{i,j}T_{i,j}^{m+1}h^2 - f_{i,j}h^2 = 0 \qquad (7.18)$$

Subtracting Equation (7.18) from Equation (7.17), one obtains the improved estimate $T_{i,j}^{m+1}$ from $T_{i,j}^m$ as

$$T_{i,j}^{m+1} = T_{i,j}^m + \frac{\tilde{R}_{i,j}^m}{h^2 M_{i,j} + 4}. \qquad (7.19)$$

Equation (7.19) shows that the residual $\tilde{R}_{i,j}^{m}$ at the grid point (i, j) will reduce to zero in the $(m+1)$th iteration if the $(m+1)$th estimate of T at the same grid point (i, j) is increased over the previous mth estimate of T at the same grid point by $\tilde{R}_{i,j}^{m}/(h^2 M_{i,j} + 4)$. The algorithm of the relaxation method for solving a two-dimensional Helmholtz equation subject to the boundary conditions is as follows. It is assumed that the values of T are given on all the four boundaries as boundary conditions.

1. Make an initial first guess of the dependent variable $T_{i,j}$ at all the interior grid points (other than the boundaries) over the region of interest.

2. Estimate the residual values $\tilde{R}_{i,j}^{m}$ at all the interior grid points (other than the boundaries) over the region of interest using Equation (7.17).

3. Determine the improved estimate $T_{i,j}^{m+1}$ of the dependent variable at all the interior grid points (other than the boundaries) over the region of interest using Equation (7.19).

4. Check for convergence by requiring that

$$\max_{i,j} \left| T_{i,j}^{m+1} - T_{i,j}^{m} \right| \leq \varepsilon,$$

   where ε is the prescribed tolerance. The aforementioned test for convergence need to be satisfied over all the interior grid points over the region of interest. In effect if the maximum modulus difference of T between the $(m+1)$th estimate and the mth estimate over all the interior grid points over the region of interest is less than the prescribed tolerance ε, then the test for convergence is satisfied and convergence can be presumed to be attained.

5. If the aforementioned test for convergence is satisfied over all the interior grid points over the region of interest, the iteration stops. Otherwise, the iteration number is updated by one and the procedure is repeated from steps (2) to (4) till the test for convergence is achieved.

The aforementioned algorithm (method) is also called the "simultaneous relaxation" or just "relaxation" method. As the name indicates, this method does not require any specific sequence (say, increasing x with fixed y or increasing y with fixed x) to be followed in which the residual R values at all interior grid points over the region of interest as well as the updating of the dependent variable T values at all interior grid points over the region of interest are determined. In effect as the name indicates, one can obtain simultaneously the estimates of all the residual values and the next updated values of the dependent variable T at all interior grid points over the region of interest. To sum up, the simultaneous relaxation method provides for an improvement (modification) of the initial guess of the entire field over the region of interest in such a way that the magnitude of the residuals is reduced in a systematic way to successively smaller values with increase of iterations. The "simultaneous relaxation" or just relaxation method is known to converge without fail.

It is to be noted that the corrections to the dependent variable T at a particular grid point [using Equation (7.19)] also changes the residuals at the adjacent grid points over the region of interest. To demonstrate this, consider the grid point $(i, j+1)$ over the region of interest where the mth estimate of residual value has the following expression while solving a two-dimensional Helmholtz equation. Equation (7.17) gets modified to

$$T^m_{i-1,j+1} + T^m_{i+1,j+1} + T^m_{i,j} + T^m_{i,j+2} - 4T^m_{i,j+1} - h^2 M_{i,j+1} T^m_{i,j+1} - h^2 f_{i,j+1} = \tilde{R}^m_{i,j+1} \quad (7.20)$$

Substituting the value of $T^{m+1}_{i,j}$ from Equation (7.19), which is the updated estimate of the dependent variable T at (i, j) into Equation (7.20) for $T^m_{i,j}$, results in a modified estimate of the residual at the $(i, j+1)$th grid point indicated by $\hat{\tilde{R}}^m_{i,j+1}$, as follows:

$$T^m_{i-1,j+1} + T^m_{i+1,j+1} + \left(T^m_{i,j} + \frac{\tilde{R}^m_{i,j}}{h^2 M_{i,j} + 4} \right) + T^m_{i,j+2} - 4T^m_{i,j+1} - h^2 M_{i,j+1} T^m_{i,j+1} - h^2 f_{i,j+1} = \hat{\tilde{R}}^m_{i,j+1}.$$

$$(7.21)$$

Subtracting Equation (7.21) from Equation (7.20) and rearranging terms, one obtains

$$\hat{\tilde{R}}^m_{i,j+1} = \tilde{R}^m_{i,j+1} + \frac{\tilde{R}^m_{i,j}}{(h^2 M_{i,j} + 4)}. \quad (7.22)$$

A little reflection will reveal that an exactly similar expression as in Equation (7.22) will be obtained if one were to substitute the value of $T^{m+1}_{i-1,j+1}$ for $T^m_{i-1,j+1}$ in Equation (7.20) and follow the aforementioned steps. Again similar expressions as in Equation (7.22) will be obtained if one were to substitute the value of $T^{m+1}_{i+1,j+1}$ for $T^m_{i+1,j+1}$ in Equation (7.20) and follow the aforementioned steps as well as while substituting the value of $T^{m+1}_{i,j+2}$ for $T^m_{i,j+2}$ in Equation (7.20). All this indicates that although the $(m+1)$th guess for the (i, j)th grid point given by Equation (7.19) reduces the residual to zero at that grid point, the residuals of the four neighboring grid points $[(i, j), (i, j+2), (i-1, j+1), (i+1, j+1)]$ are increased by precisely the same correction [as can be seen from Equation (7.22)] as was applied to $T^m_{i,j}$. Despite this, each increasing iteration step represents the overall progress and the simultaneous relaxation or relaxation method does converge.

Sequential relaxation method of solving an elliptic partial differential equation

The sequential relaxation method is an improvement over the simultaneous relaxation method and is more widely used as it is usually found to converge more rapidly than the simultaneous relaxation method. It is to be noted that in the simultaneous relaxation method while estimating the $(m+1)$th residual value at any grid point (i,j), only values of the mth estimate of T at the (i,j)th grid point and mth estimate of T at the four neighboring grid points $[(i-1,j),(i+1,j),(i,j-1),(i,j+1)]$ are utilized. If one were to estimate sequentially (say increasing x with fixed y or increasing y with fixed x to span the entire region of interior grid points) the residuals and the improved estimate of all the interior grid points over the region of interest, then while estimating the residual at a particular grid point (i,j), one can invoke and utilize (i) either the improved $(m+1)$th estimate of $T_{i-1,j}^{m+1}$ and/or the improved $(m+1)$th estimate of $T_{i,j-1}^{m+1}$ (as these are available due to sequential marching) in Equation (7.16), or (ii) the previous mth estimates of all the four neighboring points $T_{i-1,j}^m$, $T_{i+1,j}^m$, $T_{i,j-1}^m$, $T_{i,j+1}^m$ are only utilized. The latter procedure is nothing but the "simultaneous relaxation" whereas the former procedure is the improved "sequential relaxation method" or the "successive relaxation method." It is clear that utilizing the most recent estimates of $T_{i-1,j}^{m+1}$ and $T_{i,j-1}^{m+1}$ in Equation (7.16) to estimate the residual would provide for obvious advantages that would lead to faster convergence.

For the sequential relaxation method or the successive relaxation method, Equation (7.17) becomes

$$T_{i-1,j}^{m+1} + T_{i+1,j}^m + T_{i,j-1}^{m+1} + T_{i,j+1}^m - 4T_{i,j}^m - M_{i,j}T_{i,j}^m h^2 - f_{i,j}h^2 = \tilde{R}_{i,j}^m \qquad (7.23)$$

The aim of the sequential relaxation method is to provide a systematic way of reducing all the residuals to successively smaller values with increase in iterations. Towards this end, let us determine the change in the value of $T_{i,j}^m$ that will reduce the residual $\tilde{R}_{i,j}^m$ to zero in Equation (7.23). It is assumed that the adjacent values of $T(T_{i-1,j}, T_{i+1,j}, T_{i,j-1}$, and $T_{i,j+1})$ will be left unmodified for now. The aforementioned results in

$$T_{i-1,j}^{m+1} + T_{i+1,j}^m + T_{i,j-1}^{m+1} + T_{i,j+1}^m - 4T_{i,j}^{m+1} - M_{i,j}T_{i,j}^{m+1}h^2 - f_{i,j}h^2 = 0. \qquad (7.24)$$

Subtracting Equation (7.24) from Equation (7.23), one obtains the improved estimate $T_{i,j}^{m+1}$ from $T_{i,j}^m$ as

$$T_{i,j}^{m+1} = T_{i,j}^m + \frac{\tilde{R}_{i,j}^m}{h^2 M_{i,j}+4}. \qquad (7.25)$$

Equation (7.25) shows that the residual $\tilde{R}_{i,j}^{m+1}$ at the grid point (i, j) will reduce to zero in the $(m+1)$th iteration if the $(m+1)$th estimate of T at the same grid point (i, j) is increased over the previous mth estimate of T at the same grid point by $\tilde{R}_{i,j}^{m}/(h^2 M_{i,j} + 4)$. The algorithm of the sequential relaxation method for solving a two-dimensional Helmholtz equation subject to the boundary conditions is as follows. It is assumed that the values of T are given on all the four boundaries as boundary conditions.

1. Make an initial first guess of the dependent variable $T_{i,j}$ at all the interior grid points (other than the boundaries) over the region of interest.

2. Estimate the residual values $\tilde{R}_{i,j}^{m}$ at all the interior grid points (other than the boundaries) over the region of interest using Equation (7.23).

3. Determine the improved estimate $T_{i,j}^{m+1}$ of the dependent variable at all the interior grid points (other than the boundaries) over the region of interest using Equation (7.25).

4. Check for convergence by requiring that $\max_{i,j} \left| T_{i,j}^{m+1} - T_{i,j}^{m} \right| \leq \varepsilon$, where ε is the prescribed tolerance. This test for convergence needs to be satisfied over all the interior grid points over the region of interest. In effect if the maximum modulus difference of T between the $(m+1)$th estimate and the mth estimate over all the interior grid points over the region of interest is less than the prescribed tolerance ε, then the test for convergence is satisfied and convergence can be presumed to be attained.

5. If the test for convergence is satisfied over all the interior grid points over the region of interest, the iteration stops. Otherwise, the iteration number is updated by one and the procedure is repeated from steps (2) to (4) till the test for convergence is achieved.

It is to be noted that the corrections to the dependent variable T at a particular grid point [using Equation (7.25)], also changes the residuals at the adjacent grid points over the region of interest in the sequential relaxation method as in the simultaneous relaxation method. Despite this, each increasing iteration step represents overall progress and the sequential relaxation method converges.

Successive overrelaxation (SOR) method for solving an elliptic partial differential equation

It is to be noted that the equation as manifested in the Helmholtz equation (or for that manner in any Poisson equation or Laplace equation) has arisen out of a physical situation. Hence, one would expect the solution of such physical problems to possess solutions that have real "nice" properties, such as continuity and differentiability. It goes without saying that one would then expect that residuals in a given area over the region of interest are generally of the same sign for a physical problem. When the sign of the residuals in a given area over the region of interest are of the same sign, it makes sense to "overrelax," that is, to apply a larger correction to the improved estimate of T and in short to overshoot. Equation (7.24) then gets modified to the following expression for the improved estimate

$$T_{i,j}^{m+1} = T_{i,j}^m + \omega \frac{R_{i,j}^m}{h^2 M_{i,j} + 4}, \tag{7.26}$$

where ω is an overrelaxation factor, whose value lies somewhere between 1.0 and 2.0, respectively. The optimum value of ω for a particular elliptic partial differential equation is usually found by numerical trials. The estimate of residual value at grid point (i, j) is obtained from Equation (7.23)

The algorithm of the successive overrelaxation (SOR) method for solving a two-dimensional Helmholtz equation subject to the boundary conditions is as follows. It is assumed that the values of T are given on all the four boundaries as boundary conditions.

1. Make an initial first guess of the dependent variable $T_{i,j}$ at all the interior grid points (other than the boundaries) over the region of interest.

2. Estimate the residual values $\tilde{R}_{i,j}^m$ at all the interior grid points (other than the boundaries) over the region of interest using Equation (7.23).

3. Determine the improved estimate $T_{i,j}^m$ of the dependent variable at all the interior grid points (other than the boundaries) over the region of interest using Equation (7.26).

4. Check for convergence by constraining $\max_{i,j} \left| T_{i,j}^{m+1} - T_{i,j}^m \right| \leq \varepsilon$, where ε is the prescribed tolerance. This test for convergence needs to be satisfied over all the interior grid points over the region of interest. In effect, if the maximum modulus difference of T between the $(m+1)$th estimate and the mth estimate over all the interior grid points over the region of interest is less than the prescribed tolerance ε, then the test for convergence is satisfied and convergence can be presumed to be attained.

5. If the aforementioned test for convergence is satisfied over all the interior grid points over the region of interest, the iteration stops. Otherwise, the iteration number is updated by one and the procedure is repeated from steps (2) to (4) till the condition for convergence is achieved.

It is to be noted that the corrections to the dependent variable T at a particular grid point (using Equation (7.26), also changes the residuals at the adjacent grid points over the region of interest in the successive overrelaxation method as in the simultaneous/sequential relaxation method. Despite this, each increasing iteration step represents overall progress and the "sequential relaxation" method converges. It is found that the SOR method generally converges faster as compared to the simultaneous or sequential relaxation method.

The optimum value, ω_o depends in general on the nature of the mesh, the shape of the domain, and the type of boundary conditions. When the dependent variable is prescribed on all the four boundaries (constant boundary values) for the so-called Dirichlet problem in a rectangular domain of size $(I-1)\Delta x$ by $(J-1)\Delta y$ with uniform Δx and Δy, the optimum value, ω_o is given by

$$\omega_o = 2\left(\frac{1 - \sqrt{1-\xi}}{\xi}\right),\tag{7.27}$$

where ξ is given by

$$\xi = \left\{\frac{1}{1-\beta^2}\left[\cos\left(\frac{\pi}{I-1}\right) + \beta^2\cos\left(\frac{\pi}{J-1}\right)\right]\right\}^2,\tag{7.28}$$

where $\beta = \Delta x/\Delta y$.

With $\omega = \omega_o$, the number of iterations k required to reduce the error to some pre-specified level varies directly with the total number of equations $N = (I-2) \times (J-2)$. For the sequential relaxation method, $k \propto N^2$. Hence, the SOR method with the optimum value of the overrelaxation parameter, ω_o, is considered to perform much better, especially for large problems (large N). Analytic evaluation of the optimum value of ω_o exists only for slightly more general problems.

Neumann boundary condition for relaxation (sequential and SOR) methods

Neumann boundary conditions provide for the normal derivatives on all the four boundaries and not the values of the dependent variable at the four boundaries. The obvious procedure adopted to obtain the solution of an elliptic partial differential equation using the relaxation method with Neumann boundary conditions is to

1. Sweep the grid mesh for the updated $(m+1)$ iterative estimates at all interior grid points over the region of interest.
2. Set the updated iterative $(m+1)$th estimates for the boundary values from the known slope (as given by the Neumann boundary conditions) and the newly calculated adjacent interior grid points.

For example, consider solving the Poisson equation ($M = 0$ in the Helmholtz equation) using the sequential relaxation method. It is assumed as before that $\Delta x = \Delta y = h$. Substituting Equation (7.23) in Equation (7.25) with $M_{i,j} = 0$, one obtains

$$T_{i,j}^{m+1} = T_{i,j}^{m} + \frac{1}{4}\left(T_{i-1,j}^{m+1} + T_{i+1,j}^{m} + T_{i,j-1}^{m+1} + T_{i,j+1}^{m} - 4T_{i,j}^{m} - h^2 f_{i,j} \right). \tag{7.29}$$

Assume (i, jc) is on a boundary, then for a grid point using Equation (7.29), one would have the following.

During interior sweep:

$$T_{i,jc+1}^{m+1} = T_{i,jc+1}^{m} + \frac{1}{4}\left(T_{i-1,jc+1}^{m+1} + T_{i+1,jc+1}^{m} + T_{i,jc}^{m+1} + T_{i,jc+2}^{m} - 4T_{i,jc+1}^{m} - h^2 f_{i,jc+1} \right), \tag{7.30}$$

where $T_{i,jc}^{m}$ is a boundary value that is not known due to the fact that only normal derivatives of the dependent variables are given at the boundary and not the value of the dependent variables. However, from the knowledge of $(\partial T/\partial n)$ at the boundary and knowledge of the LHS of Equation (7.30), one can deduce the boundary value as

$$T_{i,jc}^{m+1} = T_{i,jc+1}^{m+1} - \frac{\partial T}{\partial n}\Delta y. \tag{7.31}$$

A similar procedure can be employed for solving the Poisson equation with Neumann boundary conditions using the SOR method. It is assumed as before that $\Delta x = \Delta y = h$. For the SOR method, Equation (7.29) becomes

$$T_{i,j}^{m+1} = T_{i,j}^{m} + \frac{\omega}{4}\left(T_{i-1,j}^{m+1} + T_{i+1,j}^{m} + T_{i,j-1}^{m+1} + T_{i,j+1}^{m} - 4T_{i,j}^{m} - h^2 f_{i,j} \right). \tag{7.32}$$

Assume (i, jc) is on a boundary, then for a grid point using Equation (7.32), one would have the following.

During interior sweep:

$$T_{i,jc+1}^{m+1} = T_{i,jc+1}^{m} + \frac{\omega}{4}\left(T_{i-1,jc+1}^{m+1} + T_{i+1,jc+1}^{m} + T_{i,jc}^{m+1} + T_{i,jc+2}^{m} - 4T_{i,jc+1}^{m} - h^2 f_{i,jc+1} \right), \tag{7.33}$$

where $T_{i,jc}^{m}$ is a boundary value that is not known as only normal derivatives of the dependent variables are given at the boundary and not the value of the dependent variables. However, from the knowledge of $(\partial T/\partial n)$ at the boundary and LHS of Equation (7.33), one can deduce the boundary value as

$$T_{i,jc}^{m+1} = T_{i,jc+1}^{m+1} - \frac{\partial T}{\partial n}\Delta y. \tag{7.34}$$

Unfortunately, the aforementioned procedure does not converge with the solution drifting, slowly without any end. A study on resolving the aforementioned problem recommends that the derivative boundary condition be incorporated directly into the sequential relaxation and/or SOR method at interior grid points adjacent to the boundaries. Thus, an equation of the form expressed in Equations (7.29) and (7.32) is used only at the interior grid points having more than one node removed from the boundary. At points adjacent to the boundary, for the sequential relaxation method, the boundary term in Equation (7.30) is replaced by the following expression from Equation (7.31) itself resulting in the following expression

$$T_{i,jc+1}^{m+1} = T_{i,jc+1}^{m} + \frac{1}{4}\left[T_{i-1,jc+1}^{m+1} + T_{i+1,jc+1}^{m} + \left(T_{i,jc+1}^{m+1} - \frac{\partial T}{\partial n}\Delta y\right) + T_{i,jc+2}^{m} - 4T_{i,jc+1}^{m} - h^2 f_{i,jc+1}\right],$$
(7.35)

which is solved algebraically after rearrangement of term in LHS of Equation (7.35). After convergence is completed, the final boundary values may be computed from Equation (7.31). Equation (7.35) differs from the cyclic use of (7.30) and (7.31) only in the iteration level of the term $T_{i,jc+1}^{m+1} - (\partial T/\partial n)\Delta y$.

For the SOR method, an equation of the form shown in Equation (7.32) is used only at interior grid points having more than one node removed from the boundary. At points adjacent to the boundary, for the SOR method, the boundary term $T_{i,jc}^{m}$ in Equation (7.33) is replaced by the following expression from Equation (7.34) itself resulting in the following expression

$$T_{i,jc+1}^{m+1} = T_{i,jc+1}^{m} + \frac{\omega}{4}\left[T_{i-1,jc+1}^{m+1} + T_{i+1,jc+1}^{m} + \left(T_{i,jc+1}^{m+1} - \frac{\partial T}{\partial n}\Delta y\right) + T_{i,jc+2}^{m} - 4T_{i,jc+1}^{m} - h^2 f_{i,jc+1}\right],$$
(7.36)

which is solved algebraically after rearrangement of term in LHS of Equation (7.36). After convergence is completed, the final boundary values may be computed from Equation (7.34). Equation (7.36) differs from the cyclic use of Equations (7.33) and (7.34) only in the iteration level of the term $T_{i,jc+1}^{m+1} - (\partial T/\partial n)\Delta y$.

7.4   Multigrid Methods

Multigrid methods for obtaining practical and efficient solutions of elliptic partial differential equations (PDE) were first introduced in the 1970s by Brandt. These methods are capable of solving general elliptic PDE with non-constant coefficients as well as solving nonlinear equations with almost the same speed as solving an elliptic PDE with constant coefficients.

There are two related approaches to the use of multigrid techniques that are, yet, to some extent different. The first approach, called the multigrid method, provides a means to speed up the convergence of a conventional relaxation method, the latter defined on a grid of pre-specified fineness. In this method, one needs to evaluate the source terms only on this grid. The second approach, named the full multigrid (FMG) method, requires one to discretize the same continuous PDE into different sets of discrete finite difference equations using grids of various sizes. It is possible to quit the solution either at a pre-specified fineness, or monitor the truncation error due to the discretization and exit only when the truncation error is tolerably small.

7.4.1   Understanding the two-grid method

Before explaining the concept of a multigrid method, the simpler case of a two-grid method is taken up for consideration. Let the linear elliptic problem whose solution is sought be of the form

$$Ku = f, \tag{7.37}$$

where K is a linear differential operator and f is a source term. If K is the two-dimensional Laplacian operator and $f = \rho(x, y)$ is a known source term, Equation (7.37) becomes the Poisson equation

$$\frac{\partial^2 u}{\partial x^2} + \frac{\partial^2 u}{\partial y^2} = \rho(x, y). \tag{7.38}$$

Discretizing Equation (7.37) on a uniform grid with mesh size h, we can write the resulting set of linear algebraic equations as

$$K_h u_h = f_h. \tag{7.39}$$

Consider the following mesh with equal grid size Δ in both the x and y directions for Equation (7.38):

$$x_j = x_o + j\Delta; \; j = 0, 1, 2, ..J; \; y_k = y_o + k\Delta; \; k = 0, 1, 2, ...K. \tag{7.40}$$

Note that the linear differential operator K is different from maximum value of k. Applying the standard five-point finite difference representation of the two-dimensional Laplacian operators gives

$$u_{j+1,k} + u_{j-1,k} + u_{j,k+1} + u_{j,k-1} - 4u_{j,k} = \Delta^2 \rho_{j,k}. \tag{7.41}$$

In order to write Equation (7.41) in matrix form for unknown vector u, it would be convenient to index the two-dimensional grid points into a single one-dimensional sequence as follows:

$$i = j(K+1) + k; \quad j = 0,1,2,..J; \quad k = 0,1,2...K. \tag{7.42}$$

Equation (7.41) now becomes

$$u_{i+K+1} + u_{i-(K+1)} + u_{i+1} + u_{i-1} - 4u_i = \Delta^2 \rho_i. \tag{7.43}$$

Let $\tilde{u}_h$ represent an approximate solution to Equation (7.43). Denoting u_h to be the exact solution of the difference equation of Equation (7.38), the error or correction in $\tilde{u}_h$ is represented as

$$v_h = u_h - \tilde{u}_h. \tag{7.44}$$

The residual or defect is then defined as

$$d_h = K_h \tilde{u}_h - f_h. \tag{7.45}$$

As K is a linear differential operator, the error satisfies

$$K_h v_h = -d_h. \tag{7.46}$$

Assuming that the difference equation corresponding to Equation (7.38) is being solved by relaxation using one of the well-known iteration methods, such as the Jacobi or Gauss–Seidel method, at each stage, the iteration yields an approximate solution of the equation

$$\hat{K}_h \hat{v}_h = -d_h, \tag{7.47}$$

where $\hat{K}_h$ is a "simpler operator" as compared to K_h; for example, the simpler operator can be the diagonal part of K_h for the Jacobi iteration, or the upper triangle for the Gauss–Seidel iteration. The next improved approximation to the solution is given by

$$\tilde{u}_h^{new} = \tilde{u}_h + \hat{v}_h. \tag{7.48}$$

Consider forming a different approximation $\hat{K}_h$, say, of some appropriate approximation K_H of K_h on a coarser grid with mesh size H, whose value is double that of mesh size h. The residual equation (Equation (7.46)) is now approximated by

$$K_H v_H = -d_H. \tag{7.49}$$

As K_H has smaller dimension, Equation (7.49) will be easier to solve than Equation (7.46). To define the defect d_H on the coarse grid, one requires a restriction operator $\Re$ that restricts d_h to the coarse grid

$$d_H = \Re d_h. \tag{7.50}$$

The restriction operator $\Re$ is also called the fine-to-coarse operator or the injection operator. Once a solution has been obtained for v_H from Equation (7.49), one requires a prolongation operator P that prolongs or interpolates the correction to the fine grid,

$$\tilde{v}_h = P\tilde{v}_H. \tag{7.51}$$

The prolongation operator P is also called the coarse-to-fine operator or the interpolation operator. Both $\Re$ and P are assumed to be linear operators. Finally, the approximation $\tilde{u}_h$ can be updated as follows:

$$\tilde{u}_h^{new} = \tilde{u}_h + \tilde{v}_h. \tag{7.52}$$

The algorithmic steps for advancing one step in the coarse grid correction scheme can be summarized as follows:

1. Compute the defect on the fine grid from the RHS of Equation (7.45).

2. Restrict this defect using Equation (7.50)

3. Solve Equation (7.49) for the correction v_H on the coarse grid

4. Interpolate this correction v_H for the coarse grid onto the fine grid using Equation (7.51)

5. Compute the next approximation of the solution using Equation (7.52)

Comparing the differences between the two methods: (i) relaxation and (ii) coarse grid correction, one can infer that the relaxation scheme converges very slowly in the limit, mesh size tends to zero, i.e., $h \to 0$. The reason for this behaviour is that for the relaxation scheme, the amplitude of the low-frequency smooth components is only slightly modified in each iteration; however, the amplitude of the high-frequency non-smooth components gets reduced by large factors in each iteration, indicating that the relaxation scheme provides good smoothing. It is well known that the smallest wavelength that can be resolved in a mesh with size H is 2H. For the two-grid iteration corresponding to coarse grid correction scheme, components of the error with wavelengths $\leq 2H$ (high frequency components) cannot be represented on the coarse grid of mesh size H and, hence, the errors of the high frequency components cannot be reduced to zero on this coarse grid. However, as indicated earlier, the relaxation method successfully reduces the amplitude of such high frequency non-smooth components by a large factor in each iteration and, hence, it can be utilized to reduce (minimize) the high frequency components on the fine grid. Thus, the contrasting features of the aforementioned two methods leads one to explore the possibility of combining the essential ideas of both relaxation and coarse-grid correction methods to obtain a new "two-grid method" for better results. The algorithmic steps for the two-grid iteration method are as follows:

1. Pre-smoothing step: Compute $\bar{u}_h$ by applying $v_1 \geq 0$ steps of a relaxation method to $\tilde{u}_h$.

2. Coarse grid correction step: Use $\bar{u}_h$ to calculate $\bar{u}_h^{new}$.

3. Post-smoothing step: Compute $\bar{u}_h^{new}$ by applying $v_2 \geq 0$ steps of the relaxation method to $\bar{u}_h^{new}$. The initial guess for the relaxation method using the post-smoothing step is taken from the output of the coarse grid correction step.

The multigrid method is a simple extension from the two-grid method to the aforementioned approach. Instead of solving the coarse-grid defect equation, Equation (7.49) exactly, one can obtain an approximate solution of it by introducing an even coarser grid and using the two-grid iteration method. If the convergence factor of the two-grid method is sufficiently small, one will require only a few steps of this iteration to get a satisfactory approximate solution. Let γ indicate the number of two-grid iterations at each intermediate stage. It is clear that one can apply this idea recursively down to some coarsest grid, where the solution can be found easily, either through affecting the matrix inversion directly or by iterating the relaxation scheme to convergence. The term, "cycle" denotes one iteration of a multigrid method, from finest grid to coarser grids and back to finest grid again. The exact structure of a cycle depends on the value of γ. The cycle is called V-cycle if $\gamma = 1$ whereas it is called W-cycle if $\gamma = 2$. The aforementioned V-cycle and W-cycle are the most important cycles that are employed in practice. Note that for the situation where more than two grids are involved, the pre-smoothing steps after the first one on the finest grid requires an initial approximation for the error v, which is taken to be zero.

7.4.2   Full multigrid (FMG) method

The previous subsection described multigrid as an iterative scheme, where one starts with some initial guess on the finest grid and carries out sufficient cycles (V-cycles, W-cycles,...,) to achieve convergence. In essence, in the aforementioned approach, one applies cycles until some appropriate convergence criterion is satisfied. However, the efficiency of the multigrid method can be improved by invoking the full multigrid algorithm (FMG), the latter also known as nested iteration . Instead of starting with an arbitrary approximation on the finest grid with $u_h = 0$, as in the multigrid method, the first approximation in the FMG method is obtained by interpolating from a coarse grid solution as follows:

$$u_h = Pu_H \tag{7.53}$$

The coarse grid solution itself is found by a similar FMG process from even coarser grids. At the coarsest level, we start with the exact solution.

Note that P in Equation (7.53) need not be the same P used in the multigrid cycles, discussed in Section 7.4.1. It should be at least of the same order as the discretization operator K_h; however, sometimes a higher order operator leads to greater efficiency. One normally requires one or at most two multigrid cycles at each level before proceeding down to the next finer grid. Although there is theoretical guidance on the choice of number of multigrid cycles, one can easily obtain it empirically.

The simple multigrid iteration (cycle) needs the RHS 'f' only at the finest level, whereas the FMG method needs 'f' at all levels. If the boundary conditions are homogeneous, one can use $f_H = \Re f_h$. This suggestion is not always prudent for inhomogeneous boundary conditions, for which it is better to discretize 'f' on each coarse grid. Note that the FMG algorithm produces the solution on all levels and, hence, can be combined with techniques such as the Richardson extrapolation.

7.5   Fast Fourier Transform Methods

The fast Fourier transform can be utilized to solve any elliptic partial differential equation. Let $f(t)$ be the continuous signal that is sampled N times, say, $f[0]$, $f[1]$, $f[2],\ldots f[N-1]$. This signal can be written as a sum of complex exponentials using the discrete Fourier transform (DFT). The DFT is defined for signal $f(t)$, such as that mentioned earlier; these are known only at N instants separated by sample times T, as a sum of complex exponentials $F(k)$, as follows:

$$F(k) = \frac{1}{\sqrt{N}} \sum_{n=0}^{N-1} f(n) \exp[\frac{i2\pi kn}{N}] \tag{7.54}$$

with coefficients $F(k)$ that are given by the inverse DFT formula:

$$f(n) = \frac{1}{\sqrt{N}} \sum_{k=0}^{N-1} F(k) \exp[-\frac{i2\pi kn}{N}] \tag{7.55}$$

For solving an ordinary differential equation in a single dimension, say x, consider a domain $0 \leq x \leq L$ in one (x) dimension and define a lattice of N equally spaced points $x_n = \frac{nL}{N}, n = 0, 1, 2, ..., N-1$. The complex Fourier transform coefficients of $f(x)$ are then defined as

$$g_k = \frac{1}{\sqrt{N}} \sum_{n=0}^{N-1} W^{kn} f_n; \quad W = e^{\frac{2i\pi}{N}}. \tag{7.56}$$

The inverse discrete Fourier transform f_n is given by

$$f_n = \frac{1}{\sqrt{N}} \sum_{k=0}^{N-1} W^{-kn} g_k. \tag{7.57}$$

This will be periodic in x_n with period L and is therefore appropriate for problems that satisfy periodic boundary conditions. If the given problem involves Dirichlet boundary conditions, where the unknown function is prescribed at the boundaries, say for example, $f(0) = f(L) = 0$, then it would be appropriate to use a sine Fourier transform, given by

$$f_n = \sqrt{\frac{2}{N}} \sum_{k=1}^{N-1} \sin\left\{\frac{\pi n k}{N}\right\} g_k. \tag{7.58}$$

If the given problem involves Neumann boundary conditions, i.e., the normal derivatives of the unknown function is prescribed at all the boundaries, then it is appropriate to use the cosine Fourier transform, given by

$$f_n = \sqrt{\frac{2}{N}} \sum_{k=0}^{N-1} \lambda_k \cos\left\{\frac{\pi}{N}\left(n + \frac{1}{2}\right)k\right\} g_k; \text{ with } \lambda_k = \frac{1}{\sqrt{2}} \text{ for } k = 0; \text{ and } \lambda_k = 0 \text{ for } k \neq 0. \tag{7.59}$$

For solving two-dimensional elliptic partial differential equations, such as the Poisson equation, let the discrete form of the Poisson equation be of the form

$$\frac{\partial^2 \phi}{\partial x^2} + \frac{\partial^2 \phi}{\partial y^2} = -\rho \approx \frac{1}{h^2}\left\{\phi_{j+1,k} + \phi_{j-1,k} + \phi_{j,k+1} + \phi_{j,k-1} - 4\phi_{j,k}\right\} = -\rho_{j,k}. \tag{7.60}$$

Let the (x, y) solution domain be discretized on a lattice with N × N grid points in the region $0 \leq x, y \leq 1$, where the function $\rho(x, y)$ is known. Assuming for simplicity that the periodic boundary conditions are prescribed on all the four boundaries, the exponential discrete Fourier transform (DFT) can be used to obtain the solution of the Poisson equation. The discrete Fourier transform is a linear operation and, hence, one can apply the same separately in the x and y directions; it does not matter in which order the transforms are done. The two-dimensional Fourier coefficients are given by

$$\bar{\phi}_{m,n} = \frac{1}{N} \sum_{j=0}^{N-1} \sum_{k=0}^{N-1} W^{mj+nk} \phi_{j,k},$$ (7.61)

$$\bar{\rho}_{m,n} = \frac{1}{N} \sum_{j=0}^{N-1} \sum_{k=0}^{N-1} W^{mj+nk} \rho_{j,k}.$$ (7.62)

The inverse transforms are then given by

$$\phi_{j,k} = \frac{1}{N} \sum_{m=0}^{N-1} \sum_{n=0}^{N-1} W^{-mj-nk} \bar{\phi}_{m,n},$$ (7.63)

$$\rho_{j,k} = \frac{1}{N} \sum_{m=0}^{N-1} \sum_{n=0}^{N-1} W^{-mj-nk} \bar{\rho}_{m,n}.$$ (7.64)

Putting the expressions (Equations (7.63) and (7.64)) into the discretized two-dimensional Poisson equation (Equation (7.60)) and equating coefficients of W^{-mj-nk}, one obtains

$$\frac{1}{h^2} \left\{ W^{-m} + W^m + W^{-n} + W^n - 4 \right\} \bar{\phi}_{m,n} = -\bar{\rho}_{m,n}.$$ (7.65)

Equation (7.65) can be solved for $\bar{\phi}_{m,n}$ as follows:

$$\bar{\phi}_{m,n} = \frac{h^2 \bar{\rho}_{m,n}}{\left\{ 4 - W^{-m} - W^m - W^{-n} - W^n \right\}}.$$ (7.66)

Once $\bar{\phi}_{m,n}$ is known, the inverse Fourier transform given by Equation (7.63) provides the solution.

Although discrete Fourier transform (DFT) is an algorithm that computes the Fourier transform of a digitized (discrete) signal, the fast Fourier transform (FFT) is an optimized implementation of the aforementioned transform. The time complexity of performing a DFT is of the order of N^2, where N is the number of sample points. The FFT, however, achieves the same task as the DFT with much improved efficiency to something of the order of O(NlogN). There are several approaches for FFT to achieve this reduction; one such approach is known as the "decimation in time approach," for sequences whose length is a power of two ($N = 2^r$ for some integer r). This approach divides the N sample points into N/2 sequences, called the odd and even sequences. Once this is done, a property is revealed that is exploited by the algorithm to improve the efficiency. The property is recursive and requires that one computes the first N/2 samples and then uses the recursive formula to update the rest of the samples.

7.6 Cyclic Reduction and Factorization Methods

The cyclic reduction and factorization is a very fast and accurate method for obtaining the numerical solution of elliptic partial differential equations. This method will be illustrated by applying the method to the two-dimensional Poisson equation:

$$\frac{\partial^2 \psi}{\partial x^2} + \frac{\partial^2 \psi}{\partial y^2} = -\zeta, \tag{7.67}$$

where ψ is the stream function and ζ is the relative vorticity. Let the (x, y) domain be discretized with grid sizes Δx and Δy in the x and y directions, with M and N divisions in the x and y directions, respectively. The finite difference approximation of Equation (7.67) is written as

$$\frac{\psi(i+1,j) - 2\psi(i,j) + \psi(i-1,j)}{(\Delta x^2)} + \frac{\psi(i,j+1) - 2\psi(i,j) + \psi(i,j-1)}{(\Delta y^2)} = -\zeta(i,j), \tag{7.68}$$

defined over $i = 1, 2, ..., M$ and $j = 1, 2, ..., N$. Equation (7.68) can be represented as

$$A\psi(i-1,j) + B\psi(i,j) + C\psi(i+1,j) - \psi(i,j+1) - \psi(i,j-1) = G(i,j), \tag{7.69}$$

where

$$A = C = -\left\{\frac{(\Delta y)^2}{(\Delta x)^2}\right\}; \; B = [2 + \left\{2[\frac{(\Delta y)^2}{(\Delta x)^2}]\right\}]; \; G(i,j) = (\Delta y)^2 \zeta(i,j). \tag{7.70}$$

The change in sign in Equation (7.69) can be expressed in a matrix form that has positive values for the main diagonal. Equation (7.69) in matrix form is expressed as

$$\mathbf{D\Psi} = \mathbf{G}, \tag{7.71}$$

Equation (7.71) provides for a system of simultaneous linear equations, where Ψ and $\mathbf{G}$ are column vectors with $M \times N$ elements, whereas $\mathbf{D}$ is a very sparse matrix of $(M \times N)^2$ elements. It is possible to solve Equation (7.71) using a direct method such as Gauss elimination. However, for practical problems, the elements $M \times N$ is far too large to reach a solution using a direct method such as Gauss elimination. The alternate way of solving Equation (7.71) is to expand the aforementioned matrix equation as follows:

$$
D = \begin{bmatrix}
E & -I & 0 & & & & & 0 \\
-I & E & -I & 0 & & & & \\
0 & -I & E & -I & 0 & & & \\
& \ddots & \ddots & \ddots & \ddots & & & \\
& & & 0 & -I & E & -I & 0 \\
& & & & 0 & -I & E & -I \\
0 & & & & & 0 & -I & E
\end{bmatrix} \tag{7.72}
$$

$$
\Psi = \begin{bmatrix}
\psi_{i,1} \\
\psi_{i,2} \\
\cdot \\
\cdot \\
\cdot \\
\psi_{i,N-1} \\
\psi_{i,N}
\end{bmatrix} \tag{7.73}
$$

$$
G = \begin{bmatrix}
g_{i,1} \\
g_{i,2} \\
\cdot \\
\cdot \\
\cdot \\
g_{i,N-1} \\
g_{i,N}
\end{bmatrix} \tag{7.74}
$$

$$
E = \begin{bmatrix}
B_1 & C_1 & 0 & & & & & 0 \\
A_2 & B_2 & C_2 & 0 & & & & \\
0 & A_3 & B_3 & C_3 & & 0 & & \\
& \ddots & \ddots & \ddots & & \ddots & & \\
& & & 0 & A_{M-2} & B_{M-2} & C_{M-2} & \\
& & & & 0 & A_{M-1} & B_{M-1} & C_{M-1} \\
0 & & & & & 0 & A_M & B_M
\end{bmatrix} \tag{7.75}
$$

The individual components of the column vector Ψ and $\mathbf{G}$ given by Equations (7.73) and (7.74) are themselves column vectors of order M. Moreover, E is a matrix having a size $M \times M$. The unit matrix is defined as a $M \times M$ matrix, as follows:

$$I = \begin{bmatrix} 1 & 0 & & & & & & 0 \\ 0 & 1 & 0 & & & & & \\ & 0 & 1 & 0 & & & & \\ & & 0 & 1 & 0 & & & \\ & & & 0 & 1 & 0 & & \\ & & & & 0 & 1 & 0 & \\ & & & & & 0 & 1 & 0 \\ 0 & & & & & & 0 & 1 \end{bmatrix} \tag{7.76}$$

To show that the aforementioned forms combine to yield Equation (7.69), one may expand the matrix equations at one point, say (i=1, j=1). From Equations (7.72) to (7.74), one can write the expansion of the first row of Equation (7.71) as

$$\mathbf{E}\psi(i,1) - \mathbf{I}\psi(i,2) = \mathbf{g}(i,1). \tag{7.77}$$

From Equation (7.77), as we need to expand for i =1, one expands each term of Equation (7.77) for the i index. From Equations (7.75) and (7.76), the aforementioned gives

$$B(1)\psi(1,1) + C(1)\psi(2,1) - \psi(1,2) = \mathbf{g}(1,1). \tag{7.78}$$

If one compares Equation (7.78) to Equation (7.69), one notices that if

$$\mathbf{g}(1,1) = \mathbf{G}(1,1) - A(1)\psi(0,1) + \psi(1,0), \tag{7.79}$$

the aforementioned equations are identical. However, these transferred terms are simply the boundary values that must be provided to solve the Poisson equation with Dirichlet boundary conditions. Hence, Equation (7.72) is the appropriate form of the matrix $\mathbf{D}$ when one has Dirichlet boundary conditions at j = 0 and j = N+1. As the terms of $\mathbf{E}$ are indeed the coefficients of ψ in Equation (7.69), by suitable modification of the coefficients at i =1 and i = M, one can make $\mathbf{E}$ correspond to the matrix of coefficients for any non-periodic boundary conditions at i = 0 and i = M+1.

To illustrate the cyclic reduction and factorization method, consider an example with N = 7 and assume that Dirichlet boundary conditions are specified at the four boundaries. The matrix equation corresponding to Equation (7.71) is given by

$$
\begin{bmatrix}
E & -I & 0 & 0 & 0 & 0 & 0 \\
-I & E & -I & 0 & 0 & 0 & 0 \\
0 & -I & E & -I & 0 & 0 & 0 \\
0 & 0 & -I & E & -I & 0 & 0 \\
0 & 0 & 0 & -I & E & -I & 0 \\
0 & 0 & 0 & 0 & -I & E & -I \\
0 & 0 & 0 & 0 & 0 & -I & E
\end{bmatrix}
\begin{bmatrix}
\psi_{i,1} \\ \psi_{i,2} \\ \psi_{i,3} \\ \psi_{i,4} \cdot \\ \psi_{i,5} \\ \psi_{i,6} \\ \psi_{i,7}
\end{bmatrix}
=
\begin{bmatrix}
g_{i,1} \\ g_{i,2} \\ g_{i,3} \\ g_{i,4} \cdot \\ g_{i,5} \\ g_{i,6} \\ g_{i,7}
\end{bmatrix}
\tag{7.80}
$$

Now multiply the even indexed rows by $\mathbf{E}$ and add to this result the respective rows that are above and below the even indexed rows. The even indexed rows are replaced by the modified rows that were just computed. The new matrix is then given as

$$
\begin{bmatrix}
E & -I & 0 & 0 & 0 & 0 & 0 \\
0 & E^2-2I & 0 & -I & 0 & 0 & 0 \\
0 & -I & E & -I & 0 & 0 & 0 \\
0 & -I & 0 & E^2-2I & 0 & -I & 0 \\
0 & 0 & 0 & -I & E & -I & 0 \\
0 & 0 & 0 & -I & 0 & E^2-2I & 0 \\
0 & 0 & 0 & 0 & 0 & -I & E
\end{bmatrix}
\begin{bmatrix}
\psi_{i,1} \\ \psi_{i,2} \\ \psi_{i,3} \\ \psi_{i,4} \cdot \\ \psi_{i,5} \\ \psi_{i,6} \\ \psi_{i,7}
\end{bmatrix}
=
\begin{bmatrix}
g_{i,1} \\ g_{i,1}+g_{i,3}+Eg_{i,2} \\ g_{i,3} \\ g_{i,3}+g_{i,5}+Eg_{i,4} \\ g_{i,5} \\ g_{i,5}+g_{i,7}+Eg_{i,6} \cdot \\ g_{i,7}
\end{bmatrix}
\tag{7.81}
$$

Expanding the matrix equation, Equation (7.81) for j = 4, one gets the equation

$$
(\mathbf{E}^2-2\mathbf{I})\psi_{i,4} - \mathbf{I}(\psi_{i,2}+\psi_{i,6}) = g_{i,3}+g_{i,5}+\mathbf{E}g_{i,4}
\tag{7.82}
$$

Equation (7.82) are equations that contains only the even indexed ψs. For j = 2 and j = 6, similar expansions occur. Owing to the aforementioned result, one can write a reduced matrix equation for the even indexed ψ's. This equation is given by

$$
\begin{bmatrix}
E^2-2I & -I & 0 \\
-I & E^2-2I & -I \\
0 & -I & E^2-2I
\end{bmatrix}
\begin{bmatrix}
\psi_{i,2} \\ \psi_{i,4} \cdot \\ \psi_{i,6}
\end{bmatrix}
=
\begin{bmatrix}
g_{i,1}+g_{i,3}+Eg_{i,2} \\ g_{i,3}+g_{i,5}+Eg_{i,4} \\ g_{i,5}+g_{i,7}+Eg_{i,6}
\end{bmatrix}
\tag{7.83}
$$

The odd indexed equations make up a system that is referred to as the "eliminated equations." It is immediately clear that Equation (7.83) is similar in form to Equation (7.80) and that the reduction process can be repeated. We can thus write Equation (7.83) as follows:

$$
\begin{bmatrix}
E^{(1)} & -I & 0 \\
-I & E^{(1)} & -I \\
0 & -I & E^{(1)}
\end{bmatrix}
\begin{bmatrix}
\psi_{i,2} \\ \psi_{i,4} \cdot \\ \psi_{i,6}
\end{bmatrix}
=
\begin{bmatrix}
g_{i,2}^{(1)} \\ g_{i,4}^{(1)} \\ g_{i,6}^{(1)} \\ \cdot
\end{bmatrix}
\tag{7.84}
$$

Performing the reduction about the middle row, one has

$$[(\mathbf{E}^{(1)})^2 - 2\mathbf{I}]\,\psi_{i,4} = g_{i,2}^{(1)} + g_{i,6}^{(1)} + Eg_{i,4}^{(1)} \tag{7.85}$$

or

$$\mathbf{E}^{(2)}\,\psi_{i,4} = g_{i,4}^{(2)}. \tag{7.86}$$

To solve Equation(7.86), one must factor the polynomial

$$\mathbf{E}^{(2)} = [(\mathbf{E}^{(1)})^2 - 2\mathbf{I}] = (\mathbf{E}^2 - 2\mathbf{I})^2 - 2\mathbf{I}. \tag{7.87}$$

A fourth-order polynomial in Equation (7.87) can be expressed as

$$\mathbf{E}^{(2)} = (\mathbf{E} - \alpha_1^{(2)}\mathbf{I})(\mathbf{E} - \alpha_2^{(2)}\mathbf{I})(\mathbf{E} - \alpha_3^{(2)}\mathbf{I})(\mathbf{E} - \alpha_4^{(2)}\mathbf{I}). \tag{7.88}$$

To find the roots α_i, we note from Equation (7.88) that

$$\mathbf{E}^{(r+1)} = [(\mathbf{E}^{(r)})^2 - 2\mathbf{I}]. \tag{7.89}$$

This expression is identical in form to the trigonometric identity

$$2cos(2^{r+1}\phi) = [2cos(2^r\phi)]^2 - 2. \tag{7.90}$$

Hence, one can write

$$\mathbf{E}^{(r)} = 2cos(2^r\phi). \tag{7.91}$$

The cosine function expression in the RHS of Equation (7.91) is the Chebyshev polynomial of the first kind, $C_2r(2cos\phi)$, whose zeros are given by

$$\alpha_j^{(r)} = 2cos[\left\{\frac{2j-1}{2^{r+1}}\right\}\pi]; \ j = 1,2,,\ldots.2^r. \tag{7.92}$$

Hence, one has the following expression

$$E^{(r)} = \prod_{j=1}^{2^r}\left\{E - 2cos[\left\{\frac{2j-1}{2^{r+1}}\right\}\pi]I\right\} = 0. \tag{7.93}$$

One can now calculate the values of the $\alpha^{(2)}$ in Equation (7.88) using Equation (7.92). One proceeds to solve Equation (7.86) by noting that it can be written as

$$\left\{E - \alpha_1^{(2)}I\right\}\xi_{i,4} = g_{i,4}, \tag{7.94}$$

where

$$\xi_{i,4} = \left\{ \mathbf{E} - \alpha_2^{(2)} \mathbf{I} \right\} \left\{ \mathbf{E} - \alpha_3^{(2)} \mathbf{I} \right\} \left\{ \mathbf{E} - \alpha_4^{(2)} \mathbf{I} \right\} \psi_{i,4}. \tag{7.95}$$

Equation (7.94) is solved by tri-diagonal Gaussian elimination. One then repeats the Gaussian elimination procedure for the $\left\{ \mathbf{E} - \alpha_2^{(2)} \right\}$ factor in Equation (7.95), obtaining another intermediate solution vector. It is clear from this example that four Gaussian eliminations are necessary to obtain $\psi_{i,4}$. To continue the process of obtaining a solution, one expands Equation (7.84) to get

$$\mathbf{E}^{(1)} \psi_{i,2} = g_{i,2}^{(1)} + \mathbf{I} \psi_{i,4}, \tag{7.96}$$

where $\mathbf{E}^{(1)}$ is the second-order polynomial in E whose roots $\alpha_i^{(1)}$ are given by Equation (7.92) for r = 1. To find $\psi_{i,2}$, two Gaussian eliminations are necessary. $\psi_{i,6}$ is also found with two Gaussian eliminations performed on the equation

$$\mathbf{E}^{(1)} \psi_{i,6} = g_{i,6}^{(1)} + \mathbf{I} \psi_{i,4}. \tag{7.97}$$

With $\psi_{i,2}$, $\psi_{i,4}$, and $\psi_{i,6}$ known, the "eliminated equations," those that have odd indices, can be solved. Each of these equations is a simple matrix equation in E itself, thereby requiring only one Gaussian elimination per equation. Hence, for N = 7, one requires 12 Gaussian eliminations to solve the system (4 for $\psi_{i,4}$, 2 each for $\psi_{i,2}$, and $\psi_{i,6}$ and one each for $\psi_{i,1}$, $\psi_{i,3}$, $\psi_{i,5}$, and $\psi_{i,7}$)

Exercises 7a (Questions only)

1. Write down the fourth-order finite difference approximation for the two-dimensional Laplace equation $\nabla^2 f = 0$. Assume that the grid sizes are different in both the x and y directions, i.e., $\Delta x \neq \Delta y$.

Answer:

$$\frac{f_{i+1,j} - 2f_{i,j} + f_{i-1,j}}{(\Delta x)^2} + \frac{f_{i,j+1} - 2f_{i,j} + f_{i,j-1}}{(\Delta y)^2} + \left[\frac{(\Delta x)^2 + (\Delta y)^2}{12(\Delta y)^2} \right] \left[\frac{f_{i+1,j+1} - 2f_{i,j+1} + f_{i-1,j+1}}{(\Delta x)^2} \right.$$
$$\left. - 2\frac{f_{i+1,j} - 2f_{i,j} + f_{i-1,j}}{(\Delta x)^2} + \frac{f_{i+1,j-1} - 2f_{i,j-1} + f_{i-1,j-1}}{(\Delta x)^2} \right] = 0$$

An alternate expression for this is

$$\left(f_{i+1,j+1} + f_{i+1,j-1} + f_{i-1,j+1} + f_{i-1,j-1} \right) + \frac{(5\beta^2) - 1}{\beta^2 + 1} \left(f_{i+1,j} + f_{i-1,j} \right) + 2(f_{i,j+1} + f_{i,j-1})$$
$$-20f_{i,j} = 0, \quad \text{where} \quad \beta = \Delta x / \Delta y$$

2. Write down the fourth-order finite difference approximation for the two-dimensional Laplace equation $\nabla^2 f = 0$. Assume that the grid sizes are the same in both the x and y directions, i.e., $\Delta x = \Delta y = d$.

Answer:

$$\frac{f_{i+1,j} - 2f_{i,j} + f_{i-1,j}}{d^2} + \frac{f_{i,j+1} - 2f_{i,j} + f_{i,j-1}}{d^2} + \left(\frac{1}{6}\right)\left[\frac{f_{i+1,j+1} - 2f_{i,j+1} + f_{i-1,j+1}}{d^2}\right.$$
$$-2\frac{f_{i+1,j} - 2f_{i,j} + f_{i-1,j}}{d^2} + \left.\frac{f_{i+1,j-1} - 2f_{i,j-1} + f_{i-1,j-1}}{d^2}\right] = 0.$$

An alternate expression for this is

$$\left(f_{i+1,j+1} + f_{i+1,j-1} + f_{i-1,j+1} + f_{i-1,j-1}\right) + \frac{2(5-\beta^2)}{\beta^2+1}[f_{i+1,j} + f_{i-1,j}] + \frac{2(5\beta^2-1)}{\beta^2+1}[f_{i,j+1} + f_{i,j-1}] -$$
$$20f_{i,j} = 0.$$

The aforementioned equation is a well-known 9-point stencil for the finite difference approximation of the two-dimensional Laplace equation.

3. Solve the one-dimensional Poisson equation with the Dirichlet boundary condition of the form

$$\frac{d^2u}{dx^2} + f(x) = 0.$$

Boundary conditions are given as $u(x=0) = a$ and $u(x=1) = b$.

Answer: It is straightforward to integrate the aforementioned second-order ordinary differential equation twice with respect to x and utilize the boundary conditions to obtain the two coefficients. However, if one employs the 5-point central difference approximation for the second derivative of u with respect to x and substitute in the one-dimensional Poisson equation, one obtains a system of linear algebraic equations which is tridiagonal in nature. Incorporating the boundary conditions, the aforementioned tridiagonal system of equations can be solved very efficiently with the Thomas algorithm.

4. Show that the second-order central difference finite difference scheme for the second derivatives of the two-dimensional Laplace equation is a consistent scheme.

5. Find the truncation error of the 5-point and 9-point stencil utilized in the finite difference scheme for the two-dimensional Laplace equation.

Answer: For the 5-point stencil, the truncation error is of the order of $O[(\Delta x)^2, (\Delta y)^2]$, whereas for the 9-point stencil, the truncation error is of the order of $O[(\Delta x)^4, (\Delta y)^4]$.

6. The well-known two-dimensional Poisson equation $\nabla^2 \psi = \zeta$ where ψ is the stream function and $zeta$ is the relative vorticity can also be solved by employing a time-dependent approach. The solution of the elliptic partial differential equation by iteration is analogous to solving

a time dependent problem at an asymptotic steady state, i.e., it is equivalent to solving the following time-dependent partial differential equation

$$\frac{\partial \psi}{\partial t} = \nabla^2 \psi - \zeta.$$

7. For the two-dimensional Laplace equation $\nabla^2 \psi = 0$ utilizing the 5-point stencil and assuming that $\Delta x = \Delta y = d$, one obtains $\psi_{i,j} = \frac{1}{4}(\psi_{i-1,j} + \psi_{i+1,j} + \psi_{i,j-1} + \psi_{i,j+1})$, which indicates that the stream function ψ is the average of all the nearest stream function (to the west, to the east, to the north and to the south).

8. Show that for the sequential or the successive or the successive over relaxation method, the correction to the earlier estimate at a particular grid point also changes the residuals at the adjacent grid points.

Exercises 7b (Questions and answers)

1. By utilizing the 5-point stencil (second-order central difference finite difference scheme for the second derivatives) in the two-dimensional Poisson equation $\nabla^2 f = u$, where u is a known quantity and the boundary condition vanishes on all the four boundaries, show that this system can be written as a block tridiagonal matrix system.

   Answer: Writing the matrix equation as $Af = b$, where f is the vector of unknown quantities, b is the known vector, and A is the coefficient matrix.

$$A = \left(\frac{1}{h^2}\right) \begin{bmatrix} 4 & -1 & & -1 & & & & & \\ -1 & 4 & -1 & & -1 & & & & \\ & -1 & 4 & -1 & & -1 & & & \\ -1 & & -1 & 4 & -1 & & -1 & & \\ & -1 & & -1 & 4 & -1 & & -1 & \\ & & & & & \ddots & & & \\ & & & & & & \ddots & & \\ & & & & -1 & & -1 & 4 & -1 \\ & & & & & -1 & & -1 & 4 \end{bmatrix}$$

The coefficent matrix A can be expressed as a block $(m \times m)$ tridiagonal matrix. Matrix A is a $(nm \times nm)$ matrix

$$
A = \begin{bmatrix}
A_x + 2I_y & -I_y & & & & & \\
-I_y & A_x + 2I_y & -I_y & & & & \\
& -I_y & A_x + 2I_y & -I_y & & & \\
& & & \ddots & & & \\
& & & & \ddots & & \\
& & & & -I_y & A_x + 2I_y & -I_y \\
& & & & & -I_y & A_x + 2I_y
\end{bmatrix}
$$

where A_x is a tridiagonal matrix with diagonal elements having values $2/h^2$; the off-diagonal elements have values $-1/h^2$, whereas I_y is a diagonal unit matrix with diagonal elements having values of $1/h^2$ and all other off-diagonal terms of I_y having zero values.

2. State the conditions under which the "successive over relaxation method" will generally yield better results with faster convergence.

   Answer: It is suggested that in a situation where the residuals in a given area are generally of the same sign over a region that includes many grid points, the "successive overrelaxation method" will generally yield better results with faster convergence.

3. For the sequential or the successive or the successive over relaxation method, is there any order in which the relaxation proceeds.

   Answer: There is no requirement that a certain order be followed. One can sweep across the entire row (consisting of various columns) and then advance to the next row and so on or one can sweep the other way, i.e., one can sweep across the entire column (consisting of various rows) and then advance to the next column.

Python examples

1. Solve the Poisson equation $\frac{\partial^2 \psi}{\partial x^2} + \frac{\partial^2 \psi}{\partial y^2} = \zeta(x,y)$ where $\psi(x,y)$ is the stream function and $\zeta(x,y)$ is the relative vorticity. It is assumed that $\zeta(x,y)$ is known over the region of interest and that the Dirichlet boundary conditions on $\psi(x,y)$ is prescribed at all the four boundaries. The given domain of interest is $0 \leq x \leq 4,000$ km, $0 \leq x \leq 4,000$ km. The relative vorticity $\zeta(x,y)$ corresponds to a cyclonic vortex in the northern hemisphere with the center of the vortex over the center of domain (x_o, y_o). The tangential wind profile associated with this cyclonic vortex is given by

$V(r) = V_{max}[r/r_{max}]$ for $r \leq r_{max}$; $V(r) = V_{max}[r/r_{max}]^{-1/2}$ for $r > r_{max}$, where 'r' is the distance from the center of the cyclonic vortex and r_{max} is the radius corresponding to the largest tangential winds and V_{max} is the magnitude of the maximum tangential winds observed at $r = r_{max}$. Given $r_{max} = 80$ km and $V_{max} = 30$ m/s,

The region of interest is discretized into 100×100 divisions in the east–west and the north–south directions with a grid size equalling $\Delta x = \Delta y = 40$ km. Using the expression for the shear vorticity and the curvature vorticity, the expression for the relative vorticity $\zeta(r)$ is given as follows: $\zeta(r) = 2[V_{max}/r_{max}]$ for $r \leq r_{max}$; $\zeta(r) = -0.5[V/r] + [V/r]$ for $r > r_{max}$. The second-order derivatives of the Poisson equation are replaced by second-order finite differences and the successive over relaxation method is employed to solve the two-dimensional Poisson equation.

```
#SOR method
import math
import numpy as np
import matplotlib.pyplot as plt
n=51
m=101
rmax=80000
Vmax=30
delx=40000
dely=40000
colorinterpolation = 50
colourMap = plt.cm.jet
t=np.zeros(int(m))
x=np.zeros(int(m))
y=np.zeros(int(m))
zeta=np.zeros((int(m),int(m)))
r=np.zeros((int(m),int(m)))
Vr=np.zeros((int(m),int(m)))
for i in range(1,int(m)):
    x[i]=delx*(i-1)
    y[i]=dely*(i-1)
for i in range(1,int(m)):
    for j in range(1,int(m)):
        r[i,j]=((x[i]-x[51])**2+(y[j]-y[51])**2)**(0.5)
        if r[i,j] < rmax:
            Vr[i,j]=Vmax*(r[i,j]/rmax)
            zeta[i,j] =2*Vmax/rmax
```

```
        else:
            Vr[i,j]=Vmax*((rmax/r[i,j])**(0.5))
            zeta[i,j]=((-1/2)*Vr[i,j]/r[i,j])+(Vr[i,j]/r[i,j])
I=1
h=40000
b=1.1**-11
delt=600
res=np.zeros((int(m),int(m)))
psy1=np.zeros((int(m),int(m)))
psy2=np.zeros((int(m),int(m)))
for i in range(1,101):
    for j in range(1,101):
        psy1[i,j]=0
        psy2[i,j]=0
for i in range(2,100):
    for j in range(2,100):
        res[i,j]=(psy1[i+1,j]-4*psy1[i,j]+psy1[i-1,j]+
        psy1[i,j+1]+psy1[i,j-1])/h**2-psy1[i,j]*(I**2)-zeta[i,j]
        psy2[i,j]=psy1[i,j]+1.5*(res[i,j]/(I**2+(4/h**2)))
        print(psy2[3,3])
for i in range(2,100):
    for j in range(2,100):
        t=abs((psy2[i,j]-psy1[i,j])/(psy1[i,j]))
        if t > 0.01:
            psy1[i,j]=psy2[i,j]
        else:
            for i in range(2,100):
                for j in range(2,100):
                    res[i,j]=(psy1[i+1,j]-4*psy1[i,j]+
                    psy1[i-1,j]+psy1[i,j+1]+psy1[i,j-1])/h**2-
                    psy1[i,j]*I**2-zeta[i,j]
                    psy2[i,j]=psy1[i,j]+1.5*(res[i,j]/(I**2+
                    (4/h**2)))
plt.contour(x,y,psy1,20,cmap='RdGy')
plt.ylabel('y')
plt.xlabel('x')
plt.title("")
plt.show()
```

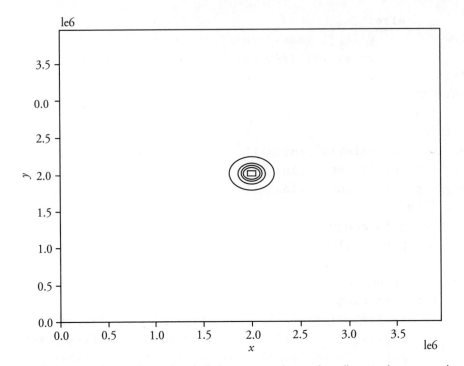

Figure 7.2 Solution of two-dimensional Poisson equation using "successive over-relaxation method."

Shallow Water Equations

8.1 Introduction

The physical process of advection and the various numerical methods that treat advection were discussed in Chapter 6. In this chapter, an important process, "wave propagation," that is observed in the atmosphere will be examined extensively. It is important to note that wave propagation is inherently contained in the governing equations for atmospheric motion. Hence, when the governing equations of atmospheric motion are integrated in time, it is realized as the excitation and propagation of waves in space. It is desirable that the propagation of these waves in the numerical solution is fairly close to the observed wave propagation. To ensure this, it is important to study the various numerical methods that are available to treat the various terms in the governing equations of atmospheric motion that represent wave propagation. The shallow water equations are a simplified set of equations that govern the horizontal propagation of gravity and or inertia–gravity waves.

The shallow water equations are widely employed in the study of atmosphere and oceans. The utility of employing the shallow water equations stems from the fact that both the atmosphere and the oceans are essentially thin fluids, with their vertical extent being very much smaller than their horizontal extent; hence, both the atmosphere and the oceans are inherently "shallow fluids." Furthermore, the inherent nonlinearity of the dynamics of the atmosphere and the oceans are manifest in the shallow water equations.

The shallow water equations describe the evolution of a hydrostatic, constant density (homogeneous) and incompressible fluid flow on the surface of the planet Earth and, hence, these equations are equally applicable for both the atmosphere and the oceans. The applicability of the hydrostatic equation is restricted to situations in which the aspect ratio of the fluid flow, the ratio of the vertical scale to the horizontal scale, is small. There exists a hydrostatic balance when acceleration due to gravity balances the vertically directed pressure gradient force in the vertical momentum equation, resulting in negligible vertical fluid accelerations. The hydrostatic balance is expressed as

$$\frac{\partial p}{\partial z} = -\rho g. \tag{8.1}$$

If the fluid density ρ is assumed constant, and as g is a constant, then Equation (8.1) signifies that the horizontal pressure gradient ($\partial p/\partial x$ and $\partial p/\partial y$) are both independent of the vertical coordinate z.

Differentiating Equation (8.1) with respect to x and y on both sides, one gets

$$\frac{\partial^2 p}{\partial x \partial z} = -\frac{\partial(\rho g)}{\partial x} = 0. \tag{8.2a}$$

$$\frac{\partial^2 p}{\partial y \partial z} = -\frac{\partial(\rho g)}{\partial y} = 0. \tag{8.2b}$$

Equations (8.2a) and (8.2b) can be rewritten as

$$\frac{\partial}{\partial z}\left(\frac{\partial p}{\partial x}\right) = 0. \tag{8.3a}$$

$$\frac{\partial}{\partial z}\left(\frac{\partial p}{\partial y}\right) = 0. \tag{8.3b}$$

As horizontal pressure gradients drive the horizontal fluid flow and the former are independent of height (z), one can look for solutions in which the horizontal fluid flow itself is independent of height. This is the key simplification and chief characteristic that underlies the shallow water system. As the horizontal fluid flow is independent of height, the assumption of incompressibility will imply non-divergent flow ($\nabla \cdot V = 0$) leading to the result that the vertical fluid velocity is linear in height (z). The shallowness of the fluid (thin fluid) with small aspect ratio together with the fact that the fluid flow is incompressible leads to the result that the magnitude of the vertical fluid velocity (w) is much smaller than the magnitude of the horizontal fluid velocity component. The smallness of the vertical fluid velocity component reduces considerably the magnitude of the vertical advection terms in the horizontal momentum equation ($w(\partial u/\partial z)$ and $w(\partial v/\partial z)$). Moreover, due to the smallness of the aspect ratio, the hydrostatic balance is applicable; this hydrostatic balance would replace the vertical component of the fluid momentum equation. Thus, the presence of vertical velocity fluid component is restricted only to the continuity equation and vanishes from the momentum equations.

Figure 8.1 provides a schematic sketch of a vertical section (x–z plane) for a shallow water fluid flow. The statement of conservation of mass (equation of continuity) for a vertical column of fluid, is given by

$$\frac{\partial h}{\partial t} + \nabla \cdot (hV) = 0, \tag{8.4}$$

where h is the vertical extent of the fluid layer, ρ is the constant density of the fluid, and ρh is the mass per unit area of the fluid. The vertically integrated horizontal mass flux of fluid is then given by ρhV, where $V = (u, v)$ is the horizontal velocity vector and $\nabla \equiv (\partial/\partial x, \partial/\partial y)$ is the horizontal gradient operator. Equation (8.4) signifies that the mass of fluid per unit time $[\nabla \cdot (\rho hV)]$ leaving a column of fluid of unit area is equal to minus the time tendency of the mass in this column $-\partial(\rho h)/\partial t$.

$$-\frac{\partial(\rho h)}{\partial t} = \nabla \cdot (\rho hV). \tag{8.5}$$

As the density ρ of the fluid is assumed constant, it cancels out on both sides in Equation (8.5) resulting in Equation (8.4). The horizontal flow velocity V varies with position and time; however, is assumed to be primarily horizontal and not vary significantly with height, the latter due to the homogeneous nature of the fluid and the hydrostatic balance assumption.

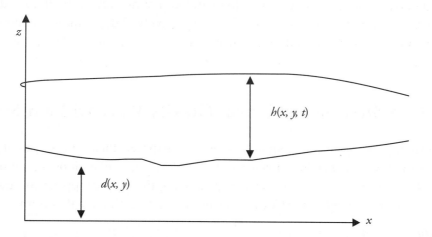

Figure 8.1 Schematic sketch of a vertical section (x–z plane) for derivation of the shallow water equations. The fluid (water) layer has a thickness h, which is a function of position (x,y) and time. The terrain over which the fluid (water) flows has elevation d, which is a function of position (x,y).

The horizontal momentum equation is of the following form in the absence of viscous forces:

$$\frac{\partial V}{\partial t} = -\frac{1}{\rho}\nabla p - f\hat{k} \times V, \tag{8.6}$$

where the Coriolis term with the vertical component of velocity is ignored. Regarding the horizontal pressure gradient term $-\nabla p/\rho$, one can obtain alternative expressions for the same using the following arguments. Assuming that the hydrostatic balance holds, the pressure as a function of height is given by

$$p = \rho g(d + h - z), \tag{8.7}$$

where $d(x,y)$ is the nonuniform terrain elevation and where it is assumed that the pressure at the upper fluid surface is zero. From Equation (8.7), the expression for ∇p becomes equal to $\rho g \nabla(d + h)$. With this replacement, the horizontal momentum equations (Equation (8.6)) become

$$\frac{\partial V}{\partial t} + g\nabla(d + h) + f\hat{k} \times V = 0. \tag{8.8}$$

Equations (8.8) and (8.4) constitute the full set of shallow water equations (two momentum component equations and one continuity equation) having three dependent variables (u, v, and h). It is to be noted that density of the fluid ρ does not appear in either Equations (8.8) and (8.4), respectively. Furthermore, shallow water equations do not require additional thermodynamic equations. For a flat bottom, $d = d(x,y)$ is identically zero in Equation (8.8).

8.2   One-dimensional Linear Gravity Wave without Rotation

For deriving the governing equations of a one-dimensional linear gravity wave without rotation, one needs to linearize the shallow water equations, Equations (8.8) and (8.4), from a state of rest and a constant depth H by dropping the nonlinear advection terms as well as the Coriolis terms and considering a flat bottom:

$$\frac{\partial u}{\partial t} = -g\frac{\partial h}{\partial x}, \tag{8.9a}$$

$$\frac{\partial h}{\partial t} = -H\frac{\partial u}{\partial x}, \tag{8.9b}$$

where $u(x,t)$ and $h(x,t)$ are the perturbation zonal horizontal velocity component and fluid depth, respectively. Equation (8.9a) is the linearized form of the zonal momentum equation while Equation (8.9b) is the linearized depth-integrated form of the continuity equation.

We are looking for solutions of Equations (8.9a) and (8.9b) of the form

$$[u(x,t), h(x,t)] = \Re\left[(\hat{u}, \hat{h}) \, e^{i(kx - \omega t)}\right], \tag{8.10}$$

where $\hat{u}$ and $\hat{h}$ are constants, k is the wavenumber and ω the frequency. Substituting Equation (8.10) into Equations (8.9), one gets

$$\omega\hat{u} = gk\hat{h} \quad \text{and} \quad \omega\hat{h} = Hk\hat{u}. \tag{8.11}$$

One can obtain from Equation (8.11), the following relation between ω and k, known as the dispersion relation

$$\omega^2 = gHk^2. \tag{8.12}$$

The phase speed of the linear one-dimensional gravity wave without rotation is therefore

$$c = \frac{\omega}{k} = \pm\sqrt{gH}. \tag{8.13}$$

It follows that the linear one-dimensional gravity wave without rotation can propagate along the x-axis in either direction with speed $\pm\sqrt{gH}$. It is to be noted that the phase speed is independent of the wave number k and, hence, the linear one-dimensional gravity wave without rotation is a non-dispersive wave.

Replacing the space derivatives in Equations (8.9a) and (8.9b) through the central finite difference scheme, the resulting differential difference equations are

$$\frac{\partial u_j}{\partial t} = -g\frac{h_{j+1} - h_{j-1}}{2\Delta x}, \tag{8.14}$$

$$\frac{\partial h_j}{\partial t} = -H\frac{u_{j+1} - u_{j-1}}{2\Delta x}, \tag{8.15}$$

The solution as in Equation (8.10) takes the following form:

$$[u(x,t), h(x,t)] = \Re\left[(\hat{u}, \hat{h}) \, e^{i(kj\Delta x - \omega t)}\right]. \tag{8.16}$$

Substitution of Equation (8.16) into Equations (8.14) and (8.15) gives

$$\omega\hat{u} = g\frac{\sin k\Delta x}{\Delta x}\hat{h} \quad \text{and} \quad \omega\hat{h} = H\frac{\sin k\Delta x}{\Delta x}\hat{u}. \tag{8.17}$$

Multiplying both sides of both the equations in Equation (8.17) gives the dispersion relation

$$\omega^2 = gH \left(\frac{\sin k\Delta x}{\Delta x} \right)^2. \tag{8.18}$$

Hence, the linear one-dimensional gravity wave without rotation, where the space derivatives are approximated with central differences, propagates with the phase speed c^* given by

$$c^* = \pm\sqrt{gH} \frac{\sin k\Delta x}{k\Delta x} = \pm c \frac{\sin k\Delta x}{k\Delta x}, \tag{8.19}$$

where c is the analytical phase speed of the linear one-dimensional gravity wave without rotation. It is clear from Equation (8.19) that the numerical linear one-dimensional gravity wave without rotation is a dispersive wave, unlike the analytical phase speed. It is noted that space differencing (using central differencing) leads to computational dispersion. In fact, the expression given in Equation (8.19) has the same form as that obtained while considering the advection equation (Equation 6.5). It follows that both the phase speed and group velocity depend on the wavenumber as shown in Figure 6.9. The phase speed c^* decreases as the wavelength decreases; moreover, the wave with wavelength $2\Delta x$ is stationary. However, in the present problem, one is dealing with two dependent variables while the advection equation (Equation 6.1) had only one dependent variable.

Figure 8.2   Grid with two dependent variables (u and h) defined at every grid point.

Figure 8.3   Grid with two dependent variables (u and h) defined at alternate grid points.

While deriving the expression (8.19) for phase speed, it is tacitly assumed that both the dependent variables (u and h) are defined at every grid point as shown in Figure 8.2. However, it is clear from Equations (8.14) and (8.15) that while u_j depends only on h with adjacent space points, (i.e., h_{j+1} and h_{j-1}), h_j depends only on u with adjacent points, (i.e., u_{i+1} and u_{i-1}). This indicates that the grid in Figure 8.2 contains two elementary "sub-grids"; moreover, the solution on one of these sub-grids is completely decoupled from the other. It makes sense to calculate only one of the solutions using the grid shown in Figure 8.3. The grid shown in Figure 8.3 has two dependent variables (u and h) defined at alternate grid points, as the solution of u and h at the next point in time at a given u and h location depends only on the other dependent variable (u and h) and that too only on adjacent points. The arrangement of dependent variables as shown in Figure 8.3 is called a "staggered grid." While the truncation error remains the same for both the non-staggered and the staggered grids, the computational time required to solve Equations (8.14) and (8.15) using the staggered grid is about half the time it takes to solve the same equations in the non-staggered grid. More importantly, waves with $k\Delta x > \pi/2$ are eliminated as the effective grid size is now $2\Delta x$ and the smallest wave that can be resolved has wavelength of $4\Delta x$ (with highest wavenumber being $\pi/(2\Delta x)$). It is clear from Figure 6.9, that the waves that are eliminated in the staggered grid arrangement are the waves ($k\Delta x > \pi/2$) that have largest phase speed errors and are associated with negative group velocities. It is hence clear that the staggered grid arrangement results in significant improvement in the solution through substantial reduction of errors of the phase speed and the group velocity of waves that are retained together; moreover, it reduces by half the computational time without affecting the truncation error. Effectively, the staggered grid arrangement when employed as shown in Figure 8.3, reduces the phase and group velocity diagram as shown in Figure 6.9 to its left half portion only (i.e., considers waves with wavelength larger and equal to $4\Delta x$). If one wishes to consider waves having wavelengths between $2\Delta x$ and $4\Delta x$, one can then utilize the staggered grid arrangement as shown in Figure 8.3 after halving the grid size. The halving of the grid size in the staggered grid arrangement does not increase the computational time as compared to the computational time for the non-staggered grid arrangement.

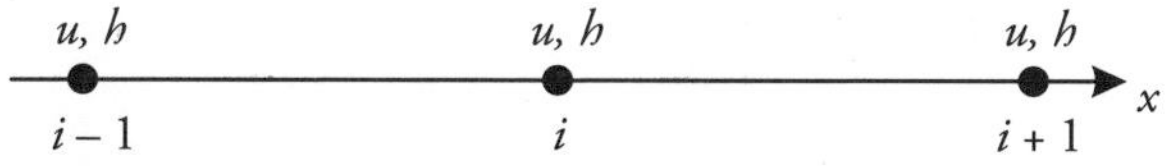

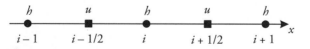

Figure 8.4   Non-staggered arrangement of variables in one-dimensional gravity wave without rotation with both u and h prescribed at all grid points.

Figure 8.5   Staggered arrangement of variables in one-dimensional gravity wave without rotation with h prescribed at all grid points and u prescribed at locations midway between the grid points.

Figure 8.4 depicts the non-staggered arrangement of variables in one-dimensional gravity waves without rotation with both u and h prescribed at all grid points with grid size Δx. The difference operator δ_x on h and u can be defined as

$$\delta_x h = \frac{h(x+\Delta x) - h(x-\Delta x)}{2\Delta x}, \tag{8.20a}$$

$$\delta_x u = \frac{u(x+\Delta x) - u(x-\Delta x)}{2\Delta x}. \tag{8.20b}$$

Eliminating u from Equations (8.9a) and (8.9b), one gets the following second-order wave equation:

$$\frac{\partial^2 h}{\partial t^2} = gH\frac{\partial^2 h}{\partial x^2}. \tag{8.21}$$

Defining the second derivative of h with respect to x by applying δ_x again on each of the two terms in the RHS of Equation (8.20a), one obtains

$$\delta_{xx}h = \frac{h(x+2\Delta x) - 2h(x) + h(x-2\Delta x)}{4(\Delta x)^2}. \tag{8.22}$$

Utilizing Equation (8.22) in the RHS of Equation (8.21), one obtains the following equation:

$$\frac{\partial^2 h}{\partial t^2} = gH\delta_{xx}h. \tag{8.23}$$

Substituting the following wave type solution in Equation (8.23) and keeping time derivative as it is

$$[h(x,t)] = \Re\left[\hat{h}\,e^{i(kx-\omega t)}\right] \tag{8.24}$$

one gets

$$\omega^2 = -\frac{gH}{4(\Delta x)^2}\left(e^{-i2k\Delta x} - 2 + e^{i2k\Delta x}\right)$$

$$= -\frac{2gH}{4(\Delta x)^2}\left[\cos(2k\Delta x) - 1\right]$$

$$= \frac{4gH}{(\Delta x)^2}\left[\sin^2(k\Delta x/2)\cos^2(k\Delta x/2)\right]. \tag{8.25}$$

The dispersion relation (Equation (8.25)) relates frequency to $k\Delta x/2$. Moreover, it is to be noted that the largest discrete wave number corresponds to $k = \pi/\Delta x$ as the smallest wavelength that can be resolved is $2\Delta x$. In order to plot the dispersion relationship (Equation (8.25)), it is convenient to plot the same using nondimensional wavenumber $k' = k\Delta x/\pi$ and nondimensional frequency $\omega' = \omega\Delta x/\sqrt{gH}$ in the x and y axis. Figure 8.6 shows the dispersion relation for the analytical (Equation (8.12)) and the numerical wave (Equation (8.25)) for the linear one-dimensional gravity wave without rotation. It is clear from Equation (8.25) that for the smallest wave (highest wavenumber corresponding to $k\Delta x = \pi$), the phase speed of the numerical wave is stationary as $\omega = 0$ for such a wave. Moreover, it is clear from Equation (8.25) that unlike the analytical wave that is non-dispersive, the numerical wave solution (Equation (8.25)) is dispersive.

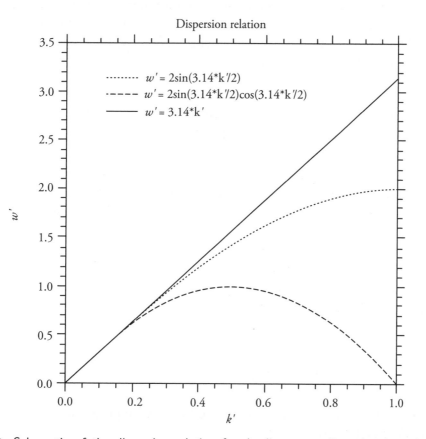

Figure 8.6 Schematic of the dispersion relation for the linear one-dimensional gravity wave without rotation with the straight solid line referring to the analytical wave which is non-dispersive. The symmetric long dashed line line refers to the numerical wave for the non-staggered grid arrangement, whereas the short dashed line refers to the numerical wave for the staggered grid arrangement.

Figure 8.6 shows the schematic diagram of the dispersion relation for the linear one-dimensional gravity wave without rotation with the straight line referring to the analytical wave, which is non-dispersive. The figure uses the non-dimensional wavenumber k' and the non-dimensional frequency ω' in the x and y axis. The symmetric curve (shown as the long dashed line; solution of gravity wave using non-staggered grid) exhibits a false maximum in the frequency ω' at $k = \pi/(2\Delta x)$ or at $k' = 1/2$, which refers to a wavelength of $4\Delta x$. The group velocity has the wrong sign (slope of long dashed line) for all waves that have $k > \pi/(2\Delta x)$ or for those waves whose wavelength is less than $4\Delta x$.

When the group velocity has a wrong sign, the energy is not properly propagated in the adjustment process and manifests as a noise near the sources of energy. The last two characteristics (wrong sign of group velocity for waves having wavelengths smaller than $4\Delta x$ and the smallest wave being stationary) have serious ramifications and, hence, need to be circumvented.

8.3 Staggered Grid Arrangement for Linear One-dimensional Gravity Wave

Figure 8.5 depicts the staggered arrangement of variables in a one-dimensional gravity wave without rotation with h prescribed at all grid points and u prescribed at locations midway between the grid points. In the staggered grid arrangement, the space differences in h in the zonal momentum equation have to be brought to the u location; similarly, the space differences in u in the continuity equation have to be brought to the h location. Writing the space difference for the staggered grid arrangement for Equations (8.9a) and (8.9b), keeping the time derivatives unchanged, one obtains

$$\frac{\partial u_{i+1/2}}{\partial t} = -g\frac{h_{i+1} - h_i}{\Delta x}, \tag{8.26}$$

$$\frac{\partial h_i}{\partial t} = -H\frac{u_{i+1/2} - u_{i-1/2}}{\Delta x}. \tag{8.27}$$

Substituting in Equations (8.26) and (8.27), the wave type solutions for u and h, as given in Equation (8.28)

$$[u(x,t), h(x,t)] = \Re\left[(\hat{u}, \hat{h})\,e^{i(kx-\omega t)}\right], \text{ one gets} \tag{8.28}$$

$$\hat{u}(-i\omega) = -\frac{g}{\Delta x}\hat{h}\left(e^{ik\Delta x} - 1\right)e^{-ik\Delta x/2}. \tag{8.29}$$

$$\hat{h}(-i\omega) = -\frac{H}{\Delta x}\hat{u}\left(e^{ik\Delta x/2} - e^{-ik\Delta x/2}\right). \tag{8.30}$$

Multiplying both sides of Equations (8.29) and (8.30) with one another and canceling common terms, one gets

$$-\omega^2 = \frac{gH}{(\Delta x)^2}\left(e^{ik\Delta x/2} - e^{-ik\Delta x/2}\right)^2. \tag{8.31}$$

Using Euler's formula and after further simplification, one finally obtains

$$\omega^2 = \frac{4gH}{(\Delta x)^2}\sin^2(k\Delta x/2). \tag{8.32}$$

This dispersion relation (in terms of ω' and k') for the staggered grid arrangement (short-dashed line) is plotted in Figure 8.6. The following conclusions can be drawn from Figure 8.6 for the gravity wave for staggered grid arrangement:

1. The gravity wave is still dispersive when compared to the analytical wave, which is non-dispersive. However, the value of phase speed is more closer to the analytical phase speed as compared to non-staggered grid especially for wavelengths greater than $4\Delta x$.

2. There is no false maximum in the frequency ω' as was observed for the non-staggered grid and, hence, no vanishing of group velocity for the staggered grid.

3. The sign of the group velocity (short-dashed line) is correct everywhere although the magnitude is reduced as compared with the analytical wave.

4. The phase speed of the smallest wave ($k\Delta x = \pi$) is not stationary in the staggered grid unlike the non-staggered grid.

The extraordinary enhancement in the characteristics of the gravity wave in terms of its closeness to the analytical wave in the staggered grid arrangement ensures a much improved adjustment process; this is attributed to the fact that the staggered grid arrangement does not utilize averaging operators.

8.4 Linear Inertia–gravity Waves in One-dimension

To study the geostrophic adjustment problem that involves the Coriolis term in the zonal momentum equation, one considers the following linear shallow water equations in one dimension. As the meridional component of velocity v appears through the rotational Coriolis term in the zonal momentum equation, we need to consider the acceleration term in the meridional direction that would balance the

Coriolis term, which is dependent on the zonal velocity component. Considering that the flow is still linear and one-dimensional in nature, the number of independent variables (x and t) are still restricted to two.

$$\frac{\partial u}{\partial t} - fv = -g\frac{\partial h}{\partial x}, \tag{8.33}$$

$$\frac{\partial v}{\partial t} + fu = 0, \tag{8.34}$$

$$\frac{\partial h}{\partial t} = -H\frac{\partial u}{\partial x}. \tag{8.35}$$

Equations (8.33) to (8.35) differ from Equations (8.9a) and (8.9b) essentially due to the appearance of Coriolis terms in the horizontal momentum equations and the appearance of the acceleration term in the meridional momentum equation. This system of equations, Equations (8.33) to (8.35), adds one additional equation and one degree of freedom (additional v) over the previously discussed linear one-dimensional gravity wave problem without rotation.

Substituting the following wave type solution in Equations (8.33) to (8.35),

$$[u(x,t),v(x,t),h(x,t)] = \Re\left[\left(\hat{u},\hat{v},\hat{h}\right)e^{i(kx-\omega t)}\right],\text{one gets} \tag{8.36}$$

a third-order equation in ω, whose solutions are

$$\omega = 0 \quad \text{and} \quad \omega^2 = f^2 + gHk^2. \tag{8.37}$$

Hence, the analytical one-dimensional linear inertia–gravity wave is itself not a non-dispersive wave (depicted as a continuous line in Figures 8.11 and 8.12 for $r = 2.0$ and $r = 0.5$, respectively).

Figures 8.11 and 8.12 depict the schematic diagram of the one-dimensional linear inertia–gravity wave for the continuum (analytical case) and for the non-staggered grid ('A' grid) and the various staggered grids ('B', 'C', and 'D' grids) with the resolution parameter assuming values of 2.0 and 0.5, respectively.

We can define the averaging and differential operator of h with respect to x (j index) as follows:

$$\overline{h}^x = \frac{h_{j+1/2} + h_{j-1/2}}{2}, \tag{8.38}$$

$$\delta_x h = \frac{h_{j+1/2} - h_{j-1/2}}{\Delta x}, \tag{8.39}$$

8.4.1   Non-staggered grid arrangement – Grid 'A'

Utilizing the spatial averaging and differential operators (Equations (8.38) and (8.39)) for Equations (8.33) to (8.35), one gets

$$\frac{\partial u}{\partial t} - fv + g\delta_x \overline{h}^x = 0, \tag{8.40}$$

$$\frac{\partial v}{\partial t} + fu = 0, \tag{8.41}$$

$$\frac{\partial h}{\partial t} + H\delta_x \overline{u}^x = 0. \tag{8.42}$$

We assume the following wave type solution and substitute the same in Equations (8.40) to (8.42)

$$[u(x,t), v(x,t), h(x,t)] = \Re\left[(\hat{u}, \hat{v}, \hat{h})\, e^{i(kx - \omega t)} \right]. \tag{8.43}$$

Figure 8.7 depicts the non-staggered arrangement of variables in one-dimensional inertia gravity wave with rotation with both u, v, and h prescribed at all grid points. The dispersion relation for the non-staggered arrangement of variables in one-

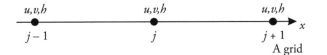

Figure 8.7 Non-staggered arrangement (referred to as 'A' grid) of variables in linear one-dimensional inertia–gravity wave with rotation with all u, v, and h prescribed at all grid points.

dimensional inertia–gravity wave is given in terms of the *Rossby radius of deformation* (L_R) and is

$$\omega^2 = f^2 \left[1 + \frac{4L_R^2}{(\Delta x)^2}\sin^2(k\Delta x/2)\cos^2(k\Delta x/2) \right], \tag{8.44}$$

where $L_R = \sqrt{gH}/f$.

The non-staggered grid arrangement shown in Figure 8.7 for a one-dimensional inertia–gravity wave has a false maximum at $k' = 1/2$ (corresponding wavelength $= 4\Delta x$), irrespective of horizontal resolution Δx as seen in Figure 8.11 (for $r = 2$) and Figure 8.12 (for $r = 0.5$), resulting in zero group velocity at $k' = \frac{1}{2}$ for the non-staggered

'A' grid (refer to long dashed line). The resolution parameter r is defined as $r = 2L_R/\Delta x$. A value of r greater than 1 indicates that the grid can resolve the Rossby radius of deformation whereas a value of r less than 1, indicates that the grid cannot resolve the Rossby radius of deformation. It is clear from Figures 8.11 and 8.12 that the non-staggered grid 'A' suffers from the fact that for wavelengths smaller than $4\Delta x$, the group velocity has the wrong sign as the slope is opposite to the analytical slope for both $r = 2$ and $r = 0.5$. Moreover, for a constant value of f, the frequency has a constant value for the smallest wave ($k' = 1$), yielding zero group velocity for the non-staggered grid 'A' at $k' = 1$.

8.4.2   Staggered grid arrangements – Grid 'B'

Having obtained a significant improvement by utilizing the staggered grid for the linear one-dimensional gravity wave, one is tempted to examine the various possible staggered grids (i.e., arrangement of distribution of variables, not all of them defined at all the grid points) for the linear one-dimensional inertia–gravity wave. Figure 8.8 depicts the staggered arrangement (referred as to 'B' grid) of variables in one-dimensional inertia gravity wave with rotation with both u, v prescribed at center of grid points while h is prescribed at all grid points. Utilizing the spatial averaging and differential operators (Equations (8.38) and (8.39)) to Equations (8.33) to (8.35), one gets for the 'B' grid,

$$\frac{\partial u}{\partial t} - fv + g\delta_x h = 0, \tag{8.45}$$

$$\frac{\partial v}{\partial t} + fu = 0, \tag{8.46}$$

$$\frac{\partial h}{\partial t} + H\delta_x u = 0, \tag{8.47}$$

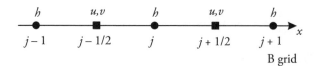

Figure 8.8   Staggered arrangement (referred as to 'B' grid) of variables in linear one-dimensional inertia–gravity wave with rotation with both u, v prescribed at center of grid points while h is prescribed at all grid points.

Assuming the following wave type of solution and substituting the same in Equations (8.45) to (8.47)

$$[u(x,t), v(x,t), h(x,t)] = \Re\left[(\hat{u}, \hat{v}, \hat{h})\, e^{i(kx - \omega t)}\right], \tag{8.48}$$

one obtains the following dispersion relation for the 'B' grid for the linear one-dimensional inertia–gravity wave:

$$\omega^2 = f^2\left[1 + \frac{4L_R^2}{(\Delta x)^2}\sin^2(k\Delta x/2)\right]. \tag{8.49}$$

Unlike the non-staggered grid arrangement ('A' grid) that had a "false maxima" in the dispersion relation corresponding to a wavelength of $4\Delta x$, there are no false maxima present in the dispersion relation for the 'B' grid (shown as short dashed line in Figure 8.11 for $r = 2$ and Figure 8.12 for $r = 0.5$). From the form of Equation (8.49), it is clear that the frequency monotonically increases with increase of the wavenumber, a feature shared by the analytical wave for both $r = 2.0$ and $r = 0.5$. It is clear that the dispersion curve corresponding to grid 'B' is closest to the continuum case for both $r = 2.0$ and $r = 0.5$ as seen in Figures 8.11 and 8.12.

8.4.3   Staggered grid arrangements – Grid 'C'

Figure 8.9 depicts the staggered arrangement (referred to as 'C' grid) of variables in one-dimensional inertia–gravity wave with rotation with both u, prescribed at center of grid points while h and v are prescribed at all grid points. Utilizing the spatial

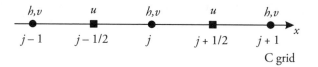

Figure 8.9   Staggered arrangement (referred to as 'C' grid) of variables in linear one-dimensional inertia–gravity wave with rotation with u prescribed at center of grid points while h and v are prescribed at all grid points.

averaging and differential operators (Equations (8.38) and (8.39)) in Equations (8.33) to (8.35), one gets for the 'C' grid

$$\frac{\partial u}{\partial t} - f\bar{v}^x + g\delta_x h = 0, \tag{8.50}$$

$$\frac{\partial v}{\partial t} + f\overline{u}^x = 0,$$

(8.51)

$$\frac{\partial h}{\partial t} + H\delta_x u = 0,$$

(8.52)

Assuming the following wave type of solution and substituting the same in Equations (8.50) to (8.52)

$$[u(x,t), v(x,t), h(x,t)] = \Re\left[(\hat{u}, \hat{v}, \hat{h})\, e^{i(kx - \omega t)}\right],$$

(8.53)

one obtains the following dispersion relation for the 'C' grid for the linear one-dimensional inertia–gravity wave

$$\omega^2 = f^2 \left[\cos^2(k\Delta x/2) + \frac{4L_R^2}{(\Delta x)^2}\sin^2(k\Delta x/2)\right].$$

(8.54)

Unlike the non-staggered grid arrangement ('A' grid) that had a "false maxima" in the dispersion relation corresponding to a wavelength of $4\Delta x$, there are no false maxima present in the dispersion relation for the 'C' grid (shown as a dashed dot line in Figure 8.11 for $r = 2$ and Figure 8.12 for $r = 0.5$). From the form of Equation (8.54), it is clear that the frequency monotonically increases with increase in the wavenumber for a resolution parameter greater than 1.0 (refer to Figure 8.11 for $r = 2.0$), a feature shared by the analytical wave. However, the scheme C has serious problems for resolution parameters less than 1.0 (refer to Figure 8.12 for $r = 0.5$), in which the frequency monotonically decreases with the increase in the wave number. In this case, for $r < 1.0$, the scheme C has the wrong slope (wrong direction of group velocity). Hence, scheme C is not a desirable staggered scheme for situations where the Rossby radius of deformation is not resolved. However, for situations where the Rossby radius of deformation is resolved, scheme C provides a good option of use of the staggered grid.

8.4.4 Staggered grid arrangements - Grid 'D'

Figure 8.10 depicts the staggered arrangement (referred to as 'D' grid) of variables in one-dimensional inertia–gravity wave with rotation with both v, prescribed at the center of grid points while h and u are prescribed at all grid points.

Utilizing the spatial averaging and differential operators (Equations (8.38) and (8.39)) in Equations (8.33) to (8.35), one gets for the 'D' grid

$$\frac{\partial u}{\partial t} - f\overline{v}^x + g\delta_x \overline{h}^x = 0,$$

(8.55)

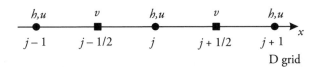

Figure 8.10 Staggered arrangement (referred to as 'D' grid) of variables in linear one-dimensional inertia–gravity wave with rotation with v prescribed at center of grid points while h and u are prescribed at all grid points.

$$\frac{\partial v}{\partial t} + f\overline{u}^x = 0, \tag{8.56}$$

$$\frac{\partial h}{\partial t} + H\delta_x\overline{u}^x = 0. \tag{8.57}$$

Assuming the following wave type of solution and substituting the same in Equations (8.55) to (8.57)

$$[u(x,t), v(x,t), h(x,t)] = \Re\left[\left(\hat{u}, \hat{v}, \hat{h}\right) e^{i(kx-\omega t)}\right], \tag{8.58}$$

one obtains the following dispersion relation for the 'D' grid for the linear one-dimensional inertia gravity wave

$$\omega^2 = f^2\left[\cos^2(k\Delta x/2) + \frac{4L_R^2}{(\Delta x)^2}\sin^2(k\Delta x/2)\cos^2(k\Delta x/2)\right]. \tag{8.59}$$

The staggered 'D' grid (shown as dashed two dots in Figure 8.11 for $r=2$ and in Figure 8.12 for $r=0.5$), also has a false maxima like the non-staggered grid arrangement ('A' grid) at $k'=1/2$ (corresponding wavelength $=4\Delta x$), irrespective of horizontal resolution Δx and for $r=2.0$. However, for $r<1.0$ (refer to Figure 8.12 for $r=0.5$), the staggered 'D' grid does not show a false maxima; instead the 'D' grid shows a monotonic decrease of frequency with increase of wave number, a feature exactly opposite to the analytical wave. From the form of Equation (8.59), it is clear that for the 'D' grid, the wave is stationary, irrespective of horizontal resolution Δx and the value of the resolution parameter r, i.e., the frequency vanishes at $k'=1$ ($k\Delta x=\pi$), corresponding to the smallest resolvable wave of wavelength $2\Delta x$. As even the non-staggered grid ('A' grid) does not have the aforementioned feature of the wave to be stationary, the staggered 'D' grid is worse than even the non-staggered 'A' grid. This characteristic of stationary wave for the 'D' grid is true both for $r=2.0$ and $r=0.5$ (refer to Figures 8.11 and 8.12).

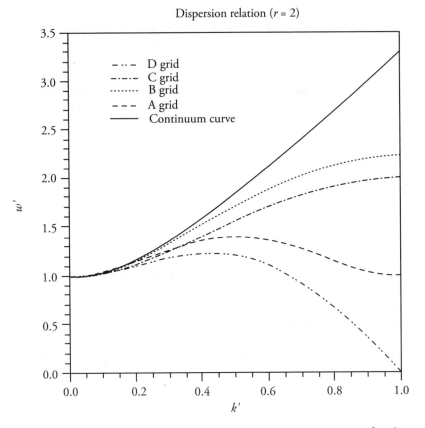

Figure 8.11 Schematic diagram of the one-dimensional inertia–gravity wave for the continuum (analytical case) and for the non-staggered ('A' grid) and the various staggered grids ('B', 'C', and 'D' grids) with the resolution parameter r assuming a value of 2.0.

8.5   Two-dimensional Linear Gravity Wave without Rotation

In two-dimensions, the linearized shallow-water equations that govern the gravity wave without rotation are given by

$$\frac{\partial u}{\partial t} = -g\frac{\partial h}{\partial x}, \tag{8.60a}$$

$$\frac{\partial v}{\partial t} = -g\frac{\partial h}{\partial y}, \tag{8.60b}$$

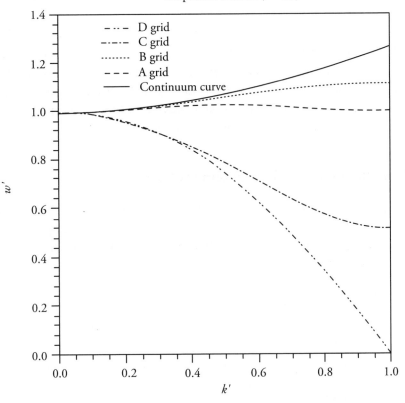

Figure 8.12 Schematic diagram of the one-dimensional inertia–gravity wave for the continuum (analytical case), for the non-staggered ('A' grid), and the various staggered grids ('B', 'C', and 'D' grids) with the resolution parameter r assuming a value of 0.5.

$$\frac{\partial h}{\partial t} = -H\left(\frac{\partial u}{\partial x} + \frac{\partial v}{\partial y}\right). \tag{8.60c}$$

Equations (8.60a) to (8.60c) provide for three dependent variables (u, v, and h) with variations in space (x and y) and time t. Substituting the wave solution

$$[u(x,y,t), v(x,y,t), h(x,y,t)] = \Re\left[(\hat{u}, \hat{v}, \hat{h})\, e^{i(kx+ly-\omega t)}\right], \tag{8.61}$$

one obtains the dispersion relation

$$\omega^2 = gH\left(k^2 + l^2\right) \tag{8.62}$$

The phase speed of the solution wave in two dimensions is defined as $c = \omega/|k|$, where $\vec{k} = (k, l)$ is the wavenumber vector. As $|k| = \sqrt{k^2 + l^2}$, we once again obtain the relationship, $c = \sqrt{gH}$. Figures 8.13 to 8.17 depict the arrangement of all the variables (u, v and h) for both the non-staggered (Grid A) and various staggered grid arrangements (Grid B to Grid E), respectively. Figure 8.18 shows the dispersion surface of the two-dimensional gravity wave without rotation for the continuum case that reveals that the gravity wave is a dispersive wave and that the frequency ω increases with increasing zonal (k') and meridional wave number (l').

8.5.1   Non-staggered grid arrangement (Grid 'A')

Figure 8.13 depicts the arrangement of all the variables (u, v, and h) in a non-staggered grid (Grid 'A'). Let d^* denote the shortest distance between the grid points. Let us also define the spatial averaging and differential operators with respect to x (i index) and y (j index) as follows:

$$\bar{h}^x = \frac{h_{i+1/2,j} + h_{i-1/2,j}}{2}. \tag{8.63}$$

$$\bar{h}^y = \frac{h_{i,j+1/2} + h_{i,j-1/2}}{2}. \tag{8.64}$$

$$\delta_x h = \frac{h_{i+1/2,j} - h_{i-1/2,j}}{d^*}. \tag{8.65}$$

$$\delta_y h = \frac{h_{i,j+1/2} - h_{i,j-1/2,}}{d^*}. \tag{8.66}$$

Applying Equations (8.63) to (8.66) to Equations (8.60a) to (8.60c), for grid 'A', one

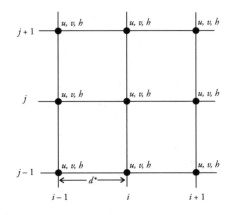

Figure 8.13   Arrangement of variables in a non-staggered grid (Grid 'A').

gets

$$\frac{\partial u}{\partial t} + g\delta_x \overline{h}^x = 0. \tag{8.67}$$

$$\frac{\partial v}{\partial t} + g\delta_y \overline{h}^y = 0. \tag{8.68}$$

$$\frac{\partial h}{\partial t} + H(\delta_x \overline{u}^x + \delta_y \overline{v}^y) = 0. \tag{8.69}$$

Substituting the wave solution, Equation (8.61) to Equations (8.67)–(8.69), one obtains the dispersion relation:

$$\omega^2 = gH \left(\frac{\sin^2 kd^* + \sin^2 ld^*}{d^{*2}} \right) \tag{8.70}$$

The ratio of the group speed c^* (given by Equation (8.70)) to the true phase speed $\sqrt{gH}$ is given by

$$\frac{c^*}{\sqrt{gH}} = \frac{\sqrt{\sin^2 2kd^* + \sin^2 2ld^*}}{2\sqrt{\sin^2 kd^* + \sin^2 ld^*}} \tag{8.71}$$

Figure 8.19 shows the dispersion surface of a two-dimensional gravity wave without rotation for the non-staggered 'A' grid. For $kd^* = \pi$ and $ld^* = \pi$, Equation (8.71) shows that the phase speed c^* is stationary. These wavenumbers correspond to the smallest wavelength that can be resolved. Hence, for the smallest resolvable wave, the phase speed is zero. Figure 8.19 also shows that for grid 'A', for $kd^* = \pi$ and $ld^* = 0$ as well as for $kd^* = 0$ and $ld^* = \pi$, the wave is stationary. This indicates that Grid 'A' is not a desirable grid arrangement. The frequency ω assumes a maximum (refer Figure 8.19) at $kd^* = \pi/2$ and for all ld^* values as well as for $ld^* = \pi/2$ and for all values of kd^*. Thus, the group velocity in the zonal direction is zero for $k = \pi/(2d^*)$ and for all values of ld^* as well as the group velocity in the meridional direction is zero for $l = \pi/(2d^*)$ and for all values of kd^*. Beyond this maximum value of ω, for $\pi/2 < kd^* < \pi$, and for all values of ld^* as well as for $\pi/2 < ld^* < \pi$, and for all values of kd^*, the frequency ω decreases as the wave number k' and l' increases. Thus, for the waves having wave numbers k' and l' in the above range, the group velocity has the wrong sign. As all variables u, v, and h are defined at all the grid points in the 'A' grid, it is relatively easy to construct a higher order accurate scheme. However, the main disadvantage of non-staggered 'A' grid is that the spatial differences are computed over a distance of $2d^*$, and, hence, the adjacent grid points are not coupled for the pressure and convergence terms.

8.5.2   Staggered grid arrangement (Grid 'B')

Figure 8.14 depicts the arrangement of all the variables (u, v, and h) in a staggered grid (Grid 'B'), in which the h variable is centered at grid points while the variables u and v are centered at the box, equidistant from all the h variable locations.

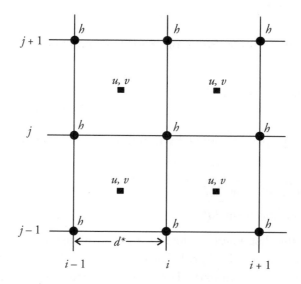

Figure 8.14   Arrangement of variables in a staggered grid (Grid 'B').

Applying Equations (8.63) to (8.66) to Equations (8.60a) to (8.60c), for grid 'B', one gets

$$\frac{\partial u}{\partial t} + g\delta_x \overline{h}^y = 0, \tag{8.72}$$

$$\frac{\partial v}{\partial t} + g\delta_y \overline{h}^x = 0, \tag{8.73}$$

$$\frac{\partial h}{\partial t} + H(\delta_x \overline{u}^y + \delta_y \overline{v}^x) = 0. \tag{8.74}$$

Substituting the wave solution, Equation (8.61) in Equations (8.72) to (8.74), one obtains the dispersion relation

$$\omega^2 = 4gH \left(\frac{\sin^2 \frac{kd^*}{2} \cos^2 \frac{ld^*}{2} + \sin^2 \frac{ld^*}{2} \cos^2 \frac{kd^*}{2}}{d^{*2}} \right). \tag{8.75}$$

In the staggered grid 'B', evaluation of the two sets of variables (wind components u and v and mass h) are evaluated at different points (locations). In grid 'B', the wind components u and v are evaluated at the center of a grid (refer to Figure 8.14) and the masses (h) at the grid corners. As the 'B' grid has both the wind components at the same point (location), grid 'B' is also called a "semi-staggered" grid. Figure 8.20 shows the dispersion surface of a two-dimensional gravity wave without rotation for the staggered 'B' grid. For the 'B' grid, the frequency increases monotonically initially; however, for $kd^* = \pi$ and $ld^* = \pi$, Equation (8.75) shows that the gravity wave is stationary with the phase speed vanishing. For the smallest resolvable wavelength in the x direction, $kd^* = \pi$ and other values of ld^* other than π, Equation (8.75) shows that the phase speed is not stationary. Similarly for the smallest resolvable wavelength in the y direction, $ld^* = \pi$ and other values of kd^* other than π, Equation (8.75) shows that the phase speed is not stationary. But for the case where both $kd^* = \pi$, and $ld^* = \pi$, the frequency vanishes. Moreover, for the 'B' grid, the spatial derivatives with respect to x and y provide for a central difference over d^*, unlike 'A' grid where the spatial derivatives provide for central differences over $2d^*$. Referring to Figure 8.14, it is clear that the spatial derivatives in grid 'B', are implemented after averaging the variable in the "normal" direction, i.e., derivatives with respect to x is effected after averaging the dependent variable in the y direction or vice-versa. Therefore, the staggered 'B' grid is more desirable than the non-staggered 'A' grid.

8.5.3 Staggered grid arrangement (Grid 'C')

Figure 8.15 depicts the arrangement of all the variables (u, v, and h) in a staggered grid (Grid 'C'), in which the h variable is centered at grid points while the variables u are centered at the north and south corners midway between the h locations, whereas the variables v are centered at the west and east corners midway between the h locations.

Applying Equations (8.63) to (8.66) to Equations (8.60a) to (8.60c), for grid 'C', one gets

$$\frac{\partial u}{\partial t} + g\delta_x h = 0, \tag{8.76}$$

$$\frac{\partial v}{\partial t} + g\delta_y h = 0,$$

(8.77)

$$\frac{\partial h}{\partial t} + H(\delta_x u + \delta_y v) = 0.$$

(8.78)

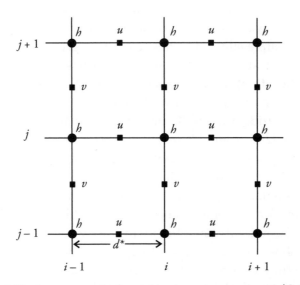

Figure 8.15   Arrangement of variables in a staggered grid (Grid 'C').

Substituting the wave solution, Equation (8.61) to Equations (8.76) to (8.78), one obtains the dispersion relation

$$\omega^2 = 4gH \left(\frac{\sin^2 \frac{kd^*}{2} + \sin^2 \frac{ld^*}{2}}{d^{*2}} \right),$$

(8.79)

Figure 8.21 shows the dispersion surface of the two-dimensional gravity wave without rotation for the staggered 'C' grid. For 'C' grid, unlike the 'B' grid, for $kd^* = \pi$ and $ld^* = \pi$, (the smallest resolvable wavelength in both x and y directions), Equation (8.79) shows that the gravity wave is not stationary with the phase speed not vanishing. From Equations (8.76) to (8.78), it is clear that the spatial difference equations for 'C' grid, do not require any averaging in the spatial directions. Moreover, for the 'C' grid, the spatial derivatives with respect to x and y provide for central difference over d^*, unlike 'A' grid where the spatial derivatives provide for central differences over $2d^*$. Referring to Figure 8.15, it is clear that the spatial derivatives for grid 'C', unlike for grid 'B' are implemented without any averaging the variable in the "normal" direction. This desirable feature of grid 'C', is reflected in Figure 8.21 that provides for increasing values of frequency with increase of k' at constant l' and increasing values of frequency with increase of l' at constant k'. Therefore, the staggered 'C' grid is more desirable than the non-staggered 'A' grid as well as the staggered 'B' grid. At high resolution, the only grid that produces a monotonically increasing frequency (i.e., one with the group velocity vector pointed in the right direction) is the 'C' grid.

8.5.4 Staggered grid arrangement (Grid 'D')

Figure 8.16 depicts the arrangement of all the variables (u, v, and h) in a staggered grid (Grid 'D'), in which the h variable is centered at grid points while the variables v are centered at the north and south corners midway between the h locations, whereas the variables u are centered at the west and east corners midway between the h locations.

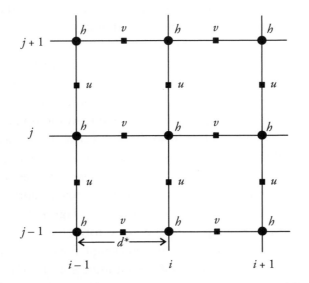

Figure 8.16 Arrangement of variables in a staggered grid (Grid 'D').

Applying Equations (8.63) to (8.66) to Equations (8.60a) to (8.60c), for grid 'D', one gets

$$\frac{\partial u}{\partial t} + g\delta_x \overline{h}^{xy} = 0, \tag{8.80}$$

$$\frac{\partial v}{\partial t} + g\delta_y \overline{h}^{xy} = 0, \tag{8.81}$$

$$\frac{\partial h}{\partial t} + H(\delta_x \overline{u}^{xy} + \delta_y \overline{v}^{xy}) = 0. \tag{8.82}$$

Substituting the wave solution (Equation (8.61)) to Equations (8.80) to (8.82), one obtains the dispersion relation

$$\omega^2 = gH \left(\frac{\sin^2 kd^* \cos^2 \frac{ld^*}{2} + \sin^2 ld^* \cos^2 \frac{kd^*}{2}}{d^{*2}} \right). \tag{8.83}$$

Figure 8.22 shows the dispersion surface of a two-dimensional gravity wave without rotation for the staggered 'D'grid. From Equations (8.67) to (8.69), for 'A' grid, spatial derivatives with respect to x required averaging of the dependent variable in the x direction before the difference could be effected. A similar situation prevailed for 'A' grid with respect to spatial derivatives with respect to y. For grid 'B', however, spatial derivatives required averaging of the dependent variable in the "normal" direction before the difference could be effected. For grid 'C', however, spatial derivatives did not require any averaging of the dependent variable whatsoever. From Equations (8.80) to (8.82), i.e., for grid 'D', it is clear that spatial derivatives required averaging of the dependent variable in both the directions before the difference could be effected. For example, the derivative with respect to x required that the dependent variable be averaged in both the x and y directions before the difference could be effected. Hence, the staggered grid 'D' appears more worse than the non-staggered grid 'A'. For the smallest resolvable wavelength in the x direction ($kd^* = \pi$), Equation (8.83) clearly indicates that the frequency vanishes and the phase speed is stationary for all values of l. Similarly for the smallest resolvable wavelength in the y direction ($ld^* = \pi$), Equation (8.83) clearly indicates that the frequency vanishes and the phase speed is stationary for all values of k. Moreover, it is clear from Figure 8.22 that the frequency actually decreases with increasing wavenumber k' for values greater than 0.5 (for all l'). Similarly, the frequency decreases with increasing l' for values greater than 0.5 (for all values of k'). Hence, over the range of wavenumbers indicated earlier, the sign of the group velocity is wrong. It goes without saying that grid 'D' is the least desirable among all the grids that have been studied, including the non-staggered grid 'A'.

8.5.5   Staggered grid arrangement (Grid 'E')

Figure 8.17 depicts the arrangement of all the variables (u, v, and h) in a staggered grid (Grid 'E'), which is obtained by rotating grid 'B' by 45° in the clockwise direction. For grid 'E', the lines joining the nearest points with the same variable makes an angle of 45° with the grid lines whereas for all the other grids 'A', 'B', 'C', and 'D', these lines are along the grid lines. However, while comparing square grids of the same resolution, i.e., when the distance between neighboring grid points carrying the same variable is the same for each grid, the space (x and y) derivatives for the 'E' grid are taken over a grid interval which is $\sqrt{2}$ times longer than that for the other staggered grids.

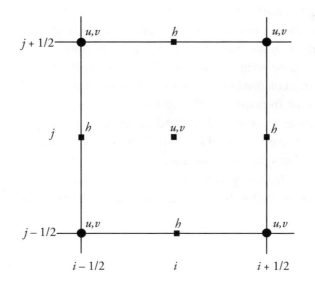

Figure 8.17   Arrangement of variables in a staggered grid (Grid 'E').

Applying Equations (8.63) to (8.66) to Equations (8.60a) to (8.60c), for grid 'E', one gets

$$\frac{\partial u}{\partial t} + g\delta_x h = 0, \tag{8.84}$$

$$\frac{\partial v}{\partial t} + g\delta_y h = 0, \tag{8.85}$$

$$\frac{\partial h}{\partial t} + H(\delta_x u + \delta_y v) = 0. \tag{8.86}$$

Substituting the wave solution, Equation (8.61) to Equations (8.84) to (8.86), one obtains the dispersion relation

$$\omega^2 = gH\left(\frac{\sin^2 \frac{kd^*}{\sqrt{2}} + \sin^2 \frac{ld^*}{\sqrt{2}}}{d^{*2}}\right), \tag{8.87}$$

where d^* is the distance between two "h" grid points

Figure 8.23 shows the dispersion surface of a two-dimensional gravity wave without rotation for the staggered 'E' grid. From Equation (8.87), it is clear that the frequency ω reaches a maximum at $kd^* = \pi/\sqrt{2}$ and at $ld^* = \pi/\sqrt{2}$. Moreover, for the wavenumber in the x and y directions with $kd^* = \sqrt{2}\pi$ and $ld^* = \sqrt{2}\pi$, Equation (8.87) shows that the frequency ω vanishes corresponding to vanishing phase speed. Unlike grids 'A', 'B', and 'D' and similar to grid 'C', grid 'E' does not involve any averaging operation while dealing with spatial derivatives. Figure 8.23 shows that the dispersion surface of a two-dimensional gravity wave without rotation for the staggered 'E' grid indicates increase of frequency with increase of wavenumber k' at a fixed value of l', up to k' assuming a value of $\sqrt{2}/2$ and an increase of frequency with increase of wavenumber l' at a fixed value of k', up to l' assuming a value of $\sqrt{2}/2$. Beyond this range of wavenumbers, the 'E' grid has a wrong sign of group velocity. This shows that the direction of the group velocity is wrong for wavenumbers $k' > \sqrt{2}/2$ and for different value of l' as well as for $l' > \sqrt{2}/2$ and for different values of k.

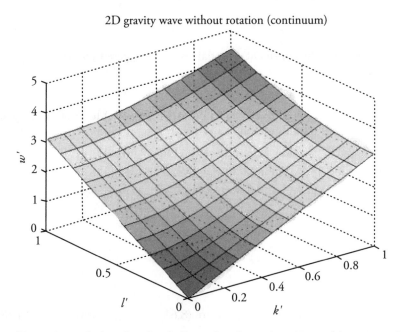

Figure 8.18 Dispersion relation for the 2-dimensional gravity wave without rotation for the continuum case.

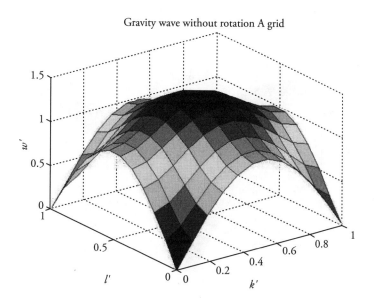

Gravity wave without rotation A grid

Figure 8.19 Dispersion relation for the two-dimensional gravity wave without rotation for the 'A'-grid.

8.6 Two-dimensional Linear Gravity Wave with Rotation

In two dimensions, the linearized shallow-water equations that govern the gravity wave with rotation are given by

$$\frac{\partial u}{\partial t} - fv = -g\frac{\partial h}{\partial x}, \tag{8.88a}$$

$$\frac{\partial v}{\partial t} + fu = -g\frac{\partial h}{\partial y}, \tag{8.88b}$$

$$\frac{\partial h}{\partial t} = -H\left(\frac{\partial u}{\partial x} + \frac{\partial v}{\partial y}\right). \tag{8.88c}$$

Equations (8.88a) to (8.88c) provide for three dependent variables (u, v, and h) with variations in space (x and y) and time t. Substituting the wave solution

$$[u(x,y,t), v(x,y,t), h(x,y,t)] = \Re\left[(\hat{u}, \hat{v}, \hat{h})\, e^{i(kx+ly-\omega t)}\right], \tag{8.89}$$

in Equations (8.88a) to (8.88c), one obtains the dispersion relation

$$\omega^2 = f^2 + gH\left(k^2 + l^2\right). \tag{8.90}$$

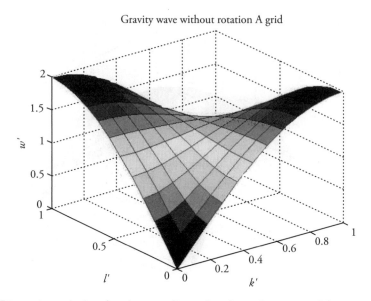

Figure 8.20   Dispersion relation for the two-dimensional gravity wave without rotation for the 'B'-grid.

Comparing Equation (8.62) with Equation (8.90), it is clear that the gravity wave modified by rotation is not a non-dispersive wave. Figure 8.24 and Figure 8.30 depict the dispersion surface for a two-dimensional gravity wave with rotation for the continuum case for $r = 2$ and $r = 0.5$, respectively. In the continuum case, the frequency ω increases with increasing wavenumbers k' and l' respectively. In the continuum case, there is no maximum of frequency for any combination of wavenumber values (Figures 8.24 and 8.30). Moreover, the phase speed is never zero and neither is the group velocity vanishing.

8.6.1   Non-staggered grid arrangement (Grid 'A')

The non-staggered grid arrangement (Grid 'A') is the same as in Figure 8.13. Applying Equations (8.63) to (8.66) to Equations (8.88a) to (8.88c), for grid 'A', one gets

$$\frac{\partial u}{\partial t} - fv + g\delta_x \overline{h}^x = 0, \tag{8.91}$$

$$\frac{\partial v}{\partial t} + fu + g\delta_y \overline{h}^y = 0, \tag{8.92}$$

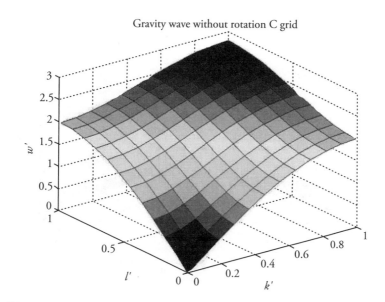

Figure 8.21 Dispersion relation for the two-dimensional gravity wave without rotation for the 'C'-grid.

$$\frac{\partial h}{\partial t} + H\left(\delta_x \overline{u}^x + \delta_y \overline{v}^y\right) = 0. \tag{8.93}$$

Substituting the wave solution (Equation (8.61)) to Equations (8.91) to (8.93), one obtains the dispersion relation

$$\omega^2 = f^2 + gH\left(\frac{\sin^2 kd^* + \sin^2 ld^*}{d^{*2}}\right). \tag{8.94}$$

Figure 8.25 and Figure 8.31 show the dispersion surfaces for the two-dimensional gravity wave with rotation for 'A' grid for $r_x = r_y = 2$ (Figure 8.25) and for $r_x = r_y = 0.5$ (Figure 8.31). Figure 8.25 (for $r_x = r_y = 2$) shows that the frequency ω increases with increasing k' at a fixed l' till $k' = 0.5$; similarly, the frequency ω increases with increasing l' at a fixed k' till $l' = 0.5$. Hence, for wavelengths below $4\Delta x$ and $4\Delta y$, the sign of the group velocity is wrong, a feature clearly at variance with the continuum case.

For $kd^* = \pi$ and $ld^* = \pi$, Equation (8.94) shows that the frequency ω is a constant assuming that the Coriolis parameter is a constant. These wavenumbers correspond to the smallest wavelength that can be resolved. For $kd^* = \pi/2$ and $ld^* = \pi/2$, Equation (8.94) shows that the frequency ω assumes the highest value and is a constant and, hence, for the aforementioned values of wavenumbers, the group velocity is zero

Gravity wave without rotation D grid

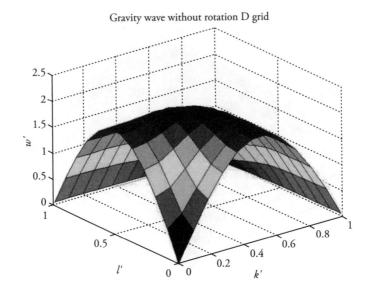

Figure 8.22   Dispersion relation for the two-dimensional gravity wave without rotation for the 'D'-grid.

assuming that the Coriolis parameter is constant. Figure 8.31 ($r_x = 0.5$, $r_y = 0.5$) is very similar to Figure 8.25 ($r_x = 2$, $r_y = 2$) and the former has the same features as that of the latter.

8.6.2   Staggered grid arrangement (Grid 'B')

The staggered grid arrangement (Grid 'B') is the same as in Figure 8.14. Applying Equations (8.63) to (8.66) to Equations (8.88a) to (8.88c), for grid 'B', one gets

$$\frac{\partial u}{\partial t} - fv + g\delta_x \overline{h}^y = 0, \tag{8.95}$$

$$\frac{\partial v}{\partial t} + fu + g\delta_y \overline{h}^x = 0, \tag{8.96}$$

$$\frac{\partial h}{\partial t} + H\left(\delta_x \overline{u}^y + \delta_y \overline{v}^x\right) = 0. \tag{8.97}$$

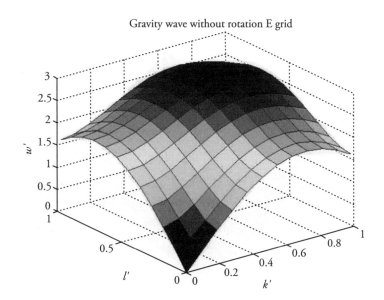

Gravity wave without rotation E grid

Figure 8.23 Dispersion relation for the two-dimensional gravity wave without rotation for the 'E'-grid.

Substituting the wave solution, Equation (8.61) to Equations (8.95) to (8.97), one obtains the dispersion relation

$$\omega^2 = f^2 + 4gH \left(\frac{\sin^2 \frac{kd^*}{2} \cos^2 \frac{ld^*}{2} + \sin^2 \frac{ld^*}{2} \cos^2 \frac{kd^*}{2}}{d^{*2}} \right). \tag{8.98}$$

Figure 8.26 and Figure 8.32 show the dispersion surfaces for the two-dimensional gravity wave with rotation for 'B' grid for $r_x = r_y = 2$ (Figure 8.26) and for $r_x = r_y = 0.5$ (Figure 8.32). Figures 8.26 and 8.32 clearly show an increasing value of frequency with increasing wavenumbers k' and l'. Thus, grid 'B' has the essential feature that characterizes the continuum case. The wavenumbers corresponding to the smallest wavelength that can be resolved assume values of $kd^* = \pi$ and $ld^* = \pi$. For these wavenumbers, the frequency presumes a constant value assuming that the Coriolis parameter is a constant. As seen from Equations (8.95) to (8.97), the spatial derivatives in grid 'B' require spatial averaging in the "normal" direction in both the momentum and continuity equations, However, no spatial averaging is required for the Coriolis terms in the zonal and the meridional momentum equations as both u and v are assigned at the center of the box in grid 'B'. Figure 8.32 ($r_x = r_y = 0.5$) is very similar to Figure 8.26 ($r_x = r_y = 2$) and the former has the same features as that of the latter.

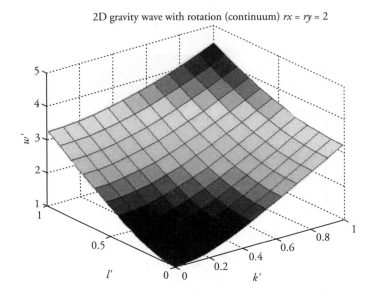

Figure 8.24 Dispersion relation for the two-dimensional gravity wave with rotation for the continuum case for $r_x = r_y = 2$.

8.6.3 Staggered grid arrangement (Grid 'C')

The staggered grid arrangement (Grid 'C') is the same as in Figure 8.15. Applying Equations (8.63) to (8.66) to Equations (8.88a) to (8.88c), for grid 'C', one gets

$$\frac{\partial u}{\partial t} - f\overline{v}^{xy} + g\delta_x h = 0, \tag{8.99}$$

$$\frac{\partial v}{\partial t} + f\overline{u}^{xy} + g\delta_y h = 0, \tag{8.100}$$

$$\frac{\partial h}{\partial t} + H\left(\delta_x u + \delta_y v\right) = 0. \tag{8.101}$$

Substituting the wave solution given in Equation (8.61) to Equations (8.99) to (8.101), one obtains the dispersion relation

$$\omega^2 = f^2\left(\cos^2\frac{kd^*}{2} \times \cos^2\frac{ld^*}{2}\right) + 4gH\left(\frac{\sin^2\frac{kd^*}{2} + \sin^2\frac{ld^*}{2}}{d^{*2}}\right). \tag{8.102}$$

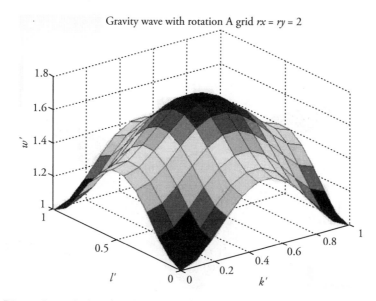

Figure 8.25 Dispersion relation for the two-dimensional gravity wave with rotation for the 'A' grid for $r_x = r_y = 2$.

Figure 8.27 and Figure 8.33 show the dispersion surfaces for the two-dimensional gravity wave with rotation for 'C' grid for $r_x = r_y = 2$ (Figure 8.27) and for $r_x = r_y = 0.5$ (Figure 8.33). The wavenumbers corresponding to the smallest wavelength that can be resolved assume values of $kd^* = \pi$ and $ld^* = \pi$. For these wavenumbers, the frequency is no longer a constant (unlike grid 'A' and grid 'B') but depends on r_x and r_y. As is seen from Equations (8.99) to (8.101), the spatial derivatives with respect to x and y, do not involve any spatial averaging. Hence, it is clear that grid 'C' is better than grid 'A' and grid 'B'. The only terms where spatial averaging is seen is with the Coriolis terms in the zonal and meridional momentum equations. Figure 8.27 (for $r_x = r_y = 2$) shows that the frequency increases with increasing wavenumbers k' and l' over the entire range of wavenumbers indicating that the sign of the group velocity is consistent everywhere for the continuum case. Furthermore, as there is no maximum of frequency in Figure 8.27, there is no indication of the vanishing of the group velocity for 'C' grid for $r_x = r_y = 2$. Values of r_x and $r_y > 1$, indicate that the Rossby radius of deformation is indeed resolvable. However, unlike the other grids (grid 'A' and grid 'B') that had virtually very little differences in the dispersion surface between different values of r_x and r_y (2 and 0.5), grid 'C' for $r_x = r_y = 0.5$ (Figure 8.33) shows very different behaviour to that shown by the same grid for $r_x = r_y = 2$ (Figure 8.27). While there was very little doubt that grid 'C' was the best for $r_x = r_y = 2$, the same is not true for

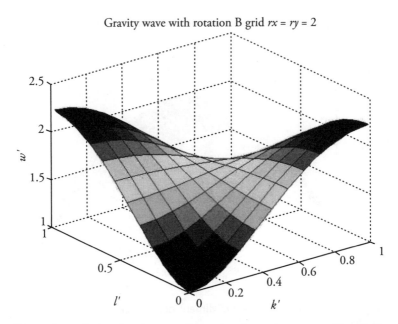

Gravity wave with rotation B grid $rx = ry = 2$

Figure 8.26   Dispersion relation for the two-dimensional gravity wave with rotation for the 'B' grid for $r_x = r_y = 2$.

$r_x = r_y = 0.5$. Figure 8.33 (for $r_x = r_y = 0.5$) shows that for grid 'C', for the smallest wavelength in the x direction ($k' = 1$) and for all values of l' as well as for the smallest wavelength in the y direction ($l' = 1$) and for all values of k', the frequency increases with increasing wavenumbers, in agreement with the continuum case and for grid 'C' for $r_x = r_y = 2$. However, for all other values of k' and l', the frequency decreases with increasing wavenumbers (Figure 8.33) for grid 'C', showing the wrong sign of the group velocity. Hence, grid 'C' is not the best choice of staggered grid for situations in which the Rossby radius of deformation is not resolved. This can be understood as in grid 'C', one adopts horizontal averaging in both directions (x and y) for the Coriolis force term resulting in the "inertia gravity terms" being less accurate than the "Coriolis force terms." The aforementioned effect is most telling in situations where the Rossby radius of deformation is not resolved or in situations where $L_R \ll d^*$, corresponding to a situation where r_x and r_y are both < 1. For the 'C' grid, the minimum distance for horizontal differences is d^*.

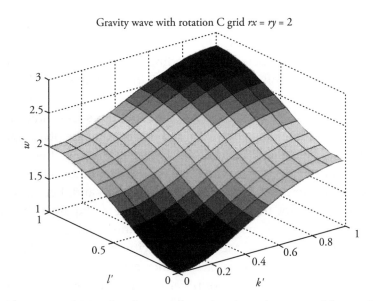

Figure 8.27 Dispersion relation for the two-dimensional gravity wave with rotation for the 'C' grid for $r_x = r_y = 2$.

8.6.4   Staggered grid arrangement (Grid 'D')

The staggered grid arrangement (Grid 'D') is the same as in Figure 8.16. Applying Equations (8.63) to (8.66) to Equations (8.88a) to (8.88c), for grid 'D', one gets

$$\frac{\partial u}{\partial t} - f \overline{v}^{xy} + g \delta_x \overline{h}^{xy} = 0, \tag{8.103}$$

$$\frac{\partial v}{\partial t} + f \overline{u}^{xy} + g \delta_y \overline{h}^{xy} = 0, \tag{8.104}$$

$$\frac{\partial h}{\partial t} + H \left(\delta_x \overline{u}^{xy} + \delta_y \overline{v}^{xy} \right) = 0. \tag{8.105}$$

Substituting the wave solution given in Equation (8.61) to Equations (8.103) to (8.105), one obtains the dispersion relation

$$\omega^2 = f^2 \left(\cos^2 \frac{kd^*}{2} \times \cos^2 \frac{ld^*}{2} \right) + gH \left(\frac{\sin^2 kd^* \cos^2 \frac{ld^*}{2} + \sin^2 ld^* \cos^2 \frac{kd^*}{2}}{d^{*2}} \right). \tag{8.106}$$

Figure 8.28 and Figure 8.34 show the dispersion surfaces for the two-dimensional gravity wave with rotation for 'D' grid for $r_x = r_y = 2$ (Figure 8.28) and for $r_x = r_y = 0.5$ (Figure 8.34). The wavenumbers corresponding to the smallest wavelength in the x direction that can be resolved ($kd^* = \pi$) and for all values of l' as well as the wavenumbers corresponding to the smallest wavelength in the y direction that can be resolved ($ld^* = \pi$) and for all values of k' have the unusual characteristic that the frequency vanishes at these range of wavenumbers, making the phase speeds of these waves zero (refer to Figures 8.28 and 8.34). Figure 8.28 shows that for grid 'D' for $r_x = r_y = 2$, the frequency decreases with increasing wavenumbers ($k' > 0.5$ and for all values of l' as well as for $l' > 0.5$ and for all values of k') and, hence, the sign of the group velocity is wrong for this range of wavenumbers. Equation (8.106) indicates that for $k' = l' = 0.5$, the frequency assumes a maximum value (refer to Figure 8.28) and, hence, at this wavenumber, the group velocity is vanishing. For $r_x = r_y = 0.5$ and for grid 'D', the frequency decreases with increasing wavenumbers k' and l' (refer to Figure 8.34), indicating that the group velocity is wrongly directed for all range of wavenumbers. This indicates that grid 'D' is worse than the non-staggered grid 'A' for both values of $r_x = r_y = 2$ and $r_x = r_y = 0.5$. Moreover, grid 'D' shows its worst characteristic for $r_x = r_y = 0.5$. It is clear from Equations (8.103) to (8.105) that all spatial derivatives require spatial averaging in both the x and y directions, confirming that grid 'D' is the worst grid among all the grids that have been studied.

8.6.5   Staggered grid arrangement (Grid 'E')

The staggered grid arrangement (Grid 'E') is the same as in Figure 8.17. Applying Equations (8.63) to (8.66) to Equations (8.88a) to (8.88c), for grid 'E', one gets

$$\frac{\partial u}{\partial t} - fv + g\delta_x h = 0, \tag{8.107}$$

$$\frac{\partial v}{\partial t} + fu + g\delta_y h = 0, \tag{8.108}$$

$$\frac{\partial h}{\partial t} + H\left(\delta_x u + \delta_y v\right) = 0. \tag{8.109}$$

Substituting the wave solution given in Equation (8.61) to Equations (8.107) to (8.109), one obtains the dispersion relation

$$\omega^2 = f^2 + gH\left(\frac{\sin^2 \frac{kd^*}{\sqrt{2}} + \sin^2 \frac{ld^*}{\sqrt{2}}}{d^{*2}}\right). \tag{8.110}$$

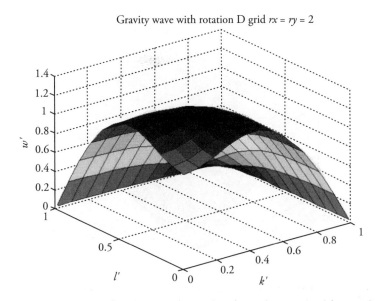

Figure 8.28 Dispersion relation for the two-dimensional gravity wave with rotation for the 'D' grid for $r_x = r_y = 2$.

where d^* is the distance between two "h" grid points

Figure 8.29 and Figure 8.35 show the dispersion surfaces for the two-dimensional gravity wave with rotation for 'E' grid for $r_x = r_y = 2$ (Figure 8.29) and for $r_x = r_y = 0.5$ (Figure 8.35). The grid 'E' shows (refer to Figures 8.29 and 8.35) that the frequency monotonically increases with increasing wavenumbers k' and l', a feature reflected in the continuum case. Furthermore, the dispersion surface as seen for different values of $r_x = r_y = 2$ (Figure 8.29) and $r_x = r_y = 0.5$ (Figure 8.35) are almost similar and exhibit features as seen in the continuum case. From Equations (6.117) to (6.119), it is clear that the spatial derivatives do not involve any spatial averaging unlike grids 'A', 'B' and 'D'. Moreover, unlike grid 'C', where the Coriolis terms required spatial averaging in both the x and y directions, grid 'E' does not require spatial averaging even for the Coriolis terms. For the 'E' grid, the minimum distance for horizontal differences is $\sqrt{2}d^*$ rather than d^* as in grid 'C'.

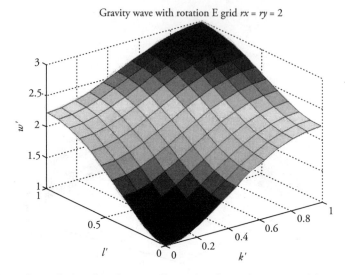

Figure 8.29 Dispersion relation for the two-dimensional gravity wave with rotation for the 'E' grid for $r_x = r_y = 2$.

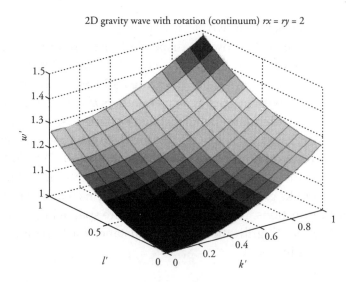

Figure 8.30 Dispersion relation for the two-dimensional gravity wave with rotation for the continuum case for $r_x = r_y = 0.5$.

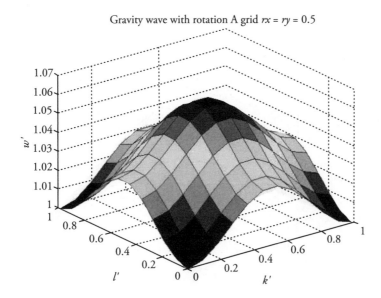

Figure 8.31 Dispersion relation for the two-dimensional gravity wave with rotation for the 'A' grid for $r_x = r_y = 0.5$.

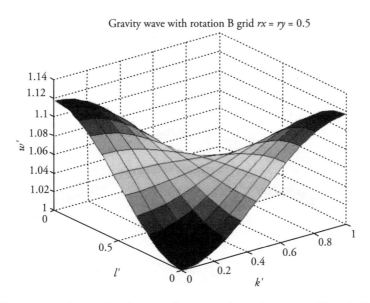

Figure 8.32 Dispersion relation for the two-dimensional gravity wave with rotation for the 'B' grid for $r_x = r_y = 0.5$.

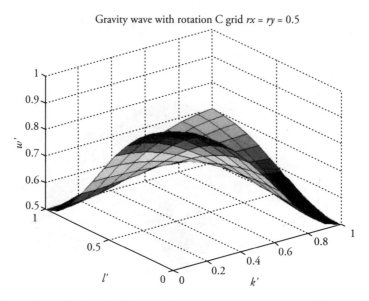

Figure 8.33 Dispersion relation for the two-dimensional gravity wave with rotation for the 'C' grid for $r_x = r_y = 0.5$.

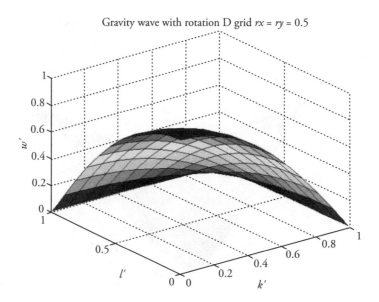

Figure 8.34 Dispersion relation for the two-dimensional gravity wave with rotation for the 'D' grid for $r_x = r_y = 0.5$.

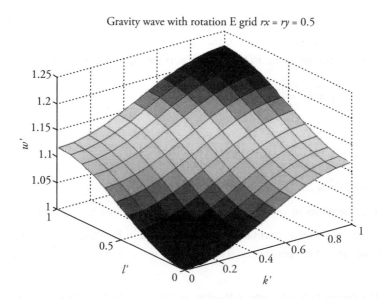

Gravity wave with rotation E grid $rx = ry = 0.5$

Figure 8.35 Dispersion relation for the two-dimensional gravity wave with rotation for the 'E' grid for $r_x = r_y = 0.5$.

Exercises 8a (Questions and answers)

1. Obtain the normal form of the one-dimensional shallow water equations

$$\frac{\partial u}{\partial t} + c\frac{\partial u}{\partial x} + g\frac{\partial h}{\partial x} = 0,$$

$$\frac{\partial h}{\partial t} + c\frac{\partial h}{\partial x} + H\frac{\partial u}{\partial x} = 0,$$

without rotation and interpret the equations in the normal form.

Answer: The one-dimensional shallow water equations without rotation are as follows:

$$\frac{\partial u}{\partial t} + c\frac{\partial u}{\partial x} + g\frac{\partial h}{\partial x} = 0$$

$$\frac{\partial h}{\partial t} + c\frac{\partial h}{\partial x} + H\frac{\partial u}{\partial x} = 0$$

Multiply the aforementioned second equation by an arbitrary parameter λ and add it to the first equation to obtain

$$\frac{\partial}{\partial t}(u + \lambda h) + (c + \lambda H)\frac{\partial u}{\partial x} + (g + \lambda c)\frac{\partial h}{\partial x} = 0.$$

Choose λ such that $\lambda = \frac{g+\lambda c}{c+\lambda H}$ to obtain a partial differential equation with only one dependent variable $(u+\lambda h)$. The two solutions are $\lambda = \pm\sqrt{g/H}$. Substituting this solution into the aforementioned single PDE, one obtains the following two equations:

$$\left[\frac{\partial}{\partial t} + (c + \sqrt{gH})\frac{\partial}{\partial x}\right]\left(u + \sqrt{g/H}\,h\right) = 0,$$

$$\left[\frac{\partial}{\partial t} + (c - \sqrt{gH})\frac{\partial}{\partial x}\right]\left(u - \sqrt{g/H}\,h\right) = 0.$$

These two equations are the equations in normal form.

2. Interpret the equations in the normal form as mentioned in Exercise 8a Q 1.

   Answer: The equations in normal form as mentioned in Exercise 8a Q 1 are equivalent to a system of two advection equations. The quantity $u + \sqrt{g/H}\,h$ is advected with the velocity $c + \sqrt{gH}$.

3. Consider the one-dimensional linear shallow water equations without rotation

$$\frac{\partial u}{\partial t} = -g\frac{\partial h}{\partial x} \quad \text{and} \quad \frac{\partial h}{\partial t} = -H\frac{\partial u}{\partial x}.$$

   Use forward scheme for the first equation and a backward scheme for the second equation resulting in

$$\frac{u_i^{n+1} - u_i^n}{\Delta t} = -g\frac{h_{i+1}^n - h_{i-1}^n}{2\Delta x} \quad \text{and} \quad \frac{h_i^n - h_i^{n-1}}{\Delta t} = -H\frac{u_{i+1}^n - u_{i-1}^n}{2\Delta x}.$$

   Show that after performing certain algebra to this finite difference equation, one will obtain the finite difference equation for the one-dimensional wave equation.

   Answer: Rewrite the finite difference equation centered at $n-1$ instead of n. The equation will become

$$\frac{u_i^n - u_i^{n-1}}{\Delta t} = -g\frac{h_{i+1}^{n-1} - h_{i-1}^{n-1}}{2\Delta x} \quad \text{and} \quad \frac{h_i^{n-1} - h_i^{n-2}}{\Delta t} = -H\frac{u_{i+1}^{n-1} - u_{i-1}^{n-1}}{2\Delta x}$$

   Subtracting the continuity equation from one another (first equation minus second equation) and replacing all the u terms in the continuity equation in terms of h obtained from the zonal momentum equation, one obtains the following equation, which is nothing but a one-dimensional wave equation, centered at $n-1$

$$\frac{h_i^n - 2h_i^{n-1} + h_i^{n-2}}{(\Delta t)^2} = \frac{gH}{(2\Delta x)^2}\left(h_{i+2}^{n-1} - 2h_i^{n-1} + h_{i-2}^{n-1}\right).$$

4. Given the one-dimensional shallow water equations that has both advection and gravity wave terms

$$\frac{\partial u}{\partial t} + c\frac{\partial u}{\partial x} + g\frac{\partial h}{\partial x} = 0,$$

$$\frac{\partial h}{\partial t} + c\frac{\partial h}{\partial x} + H\frac{\partial u}{\partial x} = 0.$$

outline the splitting or the Marchuk method as applied to these equations.

Answer: It is clear that one can employ different finite difference schemes to treat different terms in the governing equation. Even if the individual finite difference schemes employed individually satisfy the stability requirement, one cannot presume that the stability of the combination of different schemes will be stable. In the splitting or the Marchuk method, it is almost certain that the combination of different schemes will be stable, if they individually meet their respective stability requirements.

In the splitting or the Marchuk method, within a given time step, one could first solve the system of advection equations given by

$$\frac{\partial u}{\partial t} + c\frac{\partial u}{\partial x} = 0 \quad \text{and} \quad \frac{\partial h}{\partial t} + c\frac{\partial h}{\partial x} = 0.$$

Denote the provisional values u^{n+1}, h^{n+1}, obtained in this way by u^*, h^*. Use the aforementioned (u^*, h^*) values at the beginning of the time step for solving the remaining subsystem (the terms that govern the gravity wave behaviour) as given here:

$$\frac{\partial u}{\partial t} + g\frac{\partial h}{\partial x} = 0 \quad \text{and} \quad \frac{\partial h}{\partial t} + H\frac{\partial u}{\partial x} = 0.$$

The values u^{n+1}, h^{n+1}, obtained after solving this other subsystem (having the gravity wave term), are now taken as actual approximate values of these variables at the level $n+1$. This procedure is repeated in each following time step to march in the splitting or Marchuk method.

5. Show that the finite difference equations of the splitting or the Marchuk method will be stable if the individual schemes are found to be stable.

Answer: The stability of the splitting or Marchuk method, can be investigated by considering the example in Exercise 8a Q 4. Denote by l_a and l_b, the values of l of the schemes chosen for the numerical solution of subsystems (equation with advection term and equation with the gravity wave term), respectively. Then, one has the following at the end of the first stage (dealing with advection term)

$$u^* = \Re\left(\lambda_a \lambda^n \tilde{u} e^{ikx}\right), \qquad h^* = \Re\left(\lambda_a \lambda^n \tilde{h} e^{ikx}\right)$$

and

$$u^{n+1} = \Re\left(\lambda_b \lambda_a \lambda^n \tilde{u} e^{ikx}\right), \qquad h^{n+1} = \Re\left(\lambda_b \lambda_a \lambda^n \tilde{h} e^{ikx}\right),$$

where λ_a, λ_b, and λ are the amplification factor of the "first finite difference scheme" (say that solved the equation with the advection term), the "second finite difference scheme" (say that solved the equation with the gravity wave term), and the "combined finite difference scheme." From this, one obtains $\lambda = \lambda_a \lambda_b$ and $|\lambda| = |\lambda_a| |\lambda_b|$. Hence, the splitting or the Marchuk method (the combination method) is stable ($|\lambda| \leq 1$), if the individual finite difference schemes that are employed are themselves individually stable.

6. Indicate the advantages of employing the splitting or the Marchuk method.

   Answer: When employing the splitting or the Marchuk method, one does not have to necessarily have the same time steps while integrating separately each one of the subsystems. This would mean that one can utilize a larger value of the time step while solving one of the subsystems, which is known to be a slow process, or utilize a number of smaller steps to calculate the faster process, while solving the subsystem, which is known to be a fast process. This is the chief advantage of the splitting or Marchuk method as one can gain significant computational advantage.

7. Indicate the disadvantages of employing the splitting or the Marchuk method.

   Answer: One of the chief disadvantages of the splitting or the Marchuk method is that if one wishes to investigate the effects of different physical factors one at a time, it usually leads to an increase in the truncation error.

8. What is the advantage of employing the 'C-Grid'?

   Answer: The advantage of employing the 'C-grid' lies in the fact that the velocities are perpendicular on the walls of the grid-box, which makes the differencing straightforward while dealing with (i) the scalar transport in the tracer equation and also while differencing (ii) the continuity equation.

9. Are there situations where the results of employing the 'C-Grid' can be poor while solving the one-dimensional shallow water equations with no rotation?

   Answer: For the 'C-Grid', the frequency increases monotonically with the wavenumber if $\sqrt{gH}/[f(\Delta x)^2] > 1/4$, i.e., so long as the Rossby radius of deformation $\sqrt{gH}/f > \Delta x/2$. If the Rossby radius of deformation is exactly equal to half the grid size Δx, i.e., $\sqrt{gH}/f = \Delta x/2$, the group velocity would vanish identically and for shorter Rossby radii of deformation values, the frequency ω will decrease in an unrealistic way with increasing wavenumber k throughout the range, $0 < k\Delta x < \pi$.

10. Consider the one-dimensional linear shallow water equations without rotation. For a non-staggered grid arrangement, the dispersion relation was obtained as follows:

$$\omega^2 = k^2 gH \frac{\sin^2(k\Delta x)}{(k\Delta x)^2}.$$

Show that in this case, there is a manifestation of additional two computational modes that arise while employing a non-staggered grid arrangement.

Answer: Let $p = k\Delta x$, then the dispersion relation becomes

$$\omega^2 = k^2 gH \frac{\sin^2(p)}{p^2}$$

. Let ω be given and let $p = p_o$ satisfy the aforementioned dispersion relation. It turns out that if $p = p_o$ satisfies this dispersion relation, then $p = -p_o$, $p = \pi - p_o$; $p = -(\pi - p_o)$ also satisfies this dispersion relation. Hence, for each $p = p_o$ satisfying the aforementioned dispersion relation, there are three more variants of p_o that also satisfy the same dispersion relation. Clearly the analytical solution can have only "two physical modes" ($p = p_o$ and $p = -p_o$) whereas the other two additional modes that correspond to [$p = \pi - p_o$ and $p = -(\pi - p_o)$] are computational modes that arise due to the nature of the finite difference scheme.

Exercises 8b (Questions only)

1. Show that a solution obtained by the splitting or the Marchuk method will represent a consistent approximation to the true solution.

2. Write down the nonlinear two-dimensional shallow water equations in f plane using vorticity divergence formulation

3. Write down the nonlinear two-dimensional shallow water equations in β plane using vorticity divergence formulation

4. Write down the nonlinear two-dimensional shallow water equations for no rotation case using vorticity divergence formulation

9

Numerical Methods for Solving Shallow Water Equations

9.1 Introduction

A major concern while performing weather prediction using numerical methods is to appropriately represent the geostrophic adjustment process, which reflects the broad balance between the mass fields and the wind fields. It is known that the gravity inertia waves are closely associated with the geostrophic adjustment process.

9.2 Linear One-dimensional Shallow Water Equations without Rotation

The linear one-dimensional shallow water equations without rotation represent the linear one-dimensional gravity wave and are given as follows:

$$\frac{\partial u}{\partial t} + g\frac{\partial h}{\partial x} = 0, \tag{9.1}$$

$$\frac{\partial h}{\partial t} + H\frac{\partial u}{\partial x} = 0, \tag{9.2}$$

where u is the zonal (east–west)component of fluid velocity, h is the perturbation height of the fluid, g is the acceleration due to gravity, and H is the mean depth of the fluid, which is considered constant. Equations (9.1) and (9.2) correspond to a system of hyperbolic partial differential equations. If one were to eliminate h from Equations (9.1) and (9.2), by differentiating Equation (9.1) with respect to t and differentiating

Equation (9.2) with respect to x, and subtracting one equation from the other, one ends up with the following equation in u:

$$\frac{\partial^2 u}{\partial t^2} + gH\frac{\partial^2 u}{\partial x^2} = 0 \tag{9.3}$$

One would also obtain an equation in h identical to Equation (9.3), if one were to eliminate u from Equations (9.1) and (9.2). If one were to assume, wave type solutions for u and h, such as

$$u = \hat{u}e^{ik(x-ct)} \qquad \text{and} \qquad h = \hat{h}e^{ik(x-ct)}$$

and substitute them in Equations (9.1) and (9.2), one obtains the following expression for the phase speed of the gravity wave

$$c = \pm\sqrt{gH}. \tag{9.4}$$

Equation (9.4) indicates that there are two gravity waves that travel in opposite directions (i.e., in the positive and the negative x axis).

The following sections outline various methods to solve the one-dimensional linear shallow water model using the method of finite differences.

9.3 Solution of Linear One-dimensional Shallow Water Equations without Rotation

Broadly, the methods of solving the linear one-dimensional shallow water equations without rotation can be divided into *explicit* and *implicit* finite difference methods.

9.3.1 Explicit schemes: Leapfrog scheme (non-staggered)

Utilizing central difference finite difference schemes for spatial derivatives and time derivatives, the latter known as the *leapfrog* scheme, for Equations (9.1) and (9.2), one gets

$$\frac{u_j^{n+1} - u_j^{n-1}}{2\Delta t} = -g\frac{h_{j+1}^n - h_{j-1}^n}{2\Delta x}, \tag{9.5}$$

$$\frac{h_j^{n+1} - h_j^{n-1}}{2\Delta t} = -H\frac{u_{j+1}^n - u_{j-1}^n}{2\Delta x}. \tag{9.6}$$

The stability of the leapfrog time integration scheme is determined by substituting the following Equations (9.7) and (9.8) into Equations (9.5) and (9.6).

$$u_j^n = \hat{u}\lambda^n e^{ikx_j}, \tag{9.7}$$

$$h_j^n = \hat{h}\lambda^n e^{ikx_j}. \tag{9.8}$$

The stability is arrived at by seeking a condition under which $|\lambda| \le 1$. Substituting Equations (9.7) and (9.8) in Equations (9.5) and (9.6) ultimately yields the quadratic equation

$$\lambda^2 - 2ip\lambda - 1 = 0, \tag{9.9}$$

where

$$p = \sqrt{gH}\frac{\Delta t}{\Delta x}\sin(k\Delta x). \tag{9.10}$$

The condition for stability is $p^2 \le 1$ since the roots of the quadratic equation are

$$\lambda = ip \pm \sqrt{1 - p^2}, \tag{9.11}$$

which translates into the condition that the time step Δt needs to satisfy for stability of the leapfrog scheme:

$$\Delta t \le \frac{\Delta x}{\sqrt{gH}}. \tag{9.12}$$

The leapfrog scheme [Equations (9.5) and (9.6)] is a three time level scheme as it involves three time levels, $n-1$, n, and $n+1$. It is known that a three time level scheme would contribute to a computational mode,which requires the application of a filter similar to the way the computational mode was handled while dealing with the one-dimensional linear advection equation. It is clear from Equation (9.11) that the leapfrog scheme is a neutral scheme (i.e., the amplification factor is always equal to 1), assuming that the chosen time step satisfies Equation (9.12). Hence, the leapfrog scheme is second-order accurate in time. It is an explicit scheme and a neutral one, which indicates that it is non-damping. However, the leapfrog scheme has phase errors and computational dispersion.

9.3.2   Explicit schemes: FTCS scheme (non-staggered)

Utilizing central difference finite difference schemes for spatial derivatives and forward difference for time derivatives, in Equations (9.1) and (9.2), one gets

$$\frac{u_j^{n+1} - u_j^n}{\Delta t} = -g\frac{h_{j+1}^n - h_{j-1}^n}{2\Delta x}, \tag{9.13}$$

$$\frac{h_j^{n+1} - h_j^n}{\Delta t} = -H\frac{u_{j+1}^n - u_{j-1}^n}{2\Delta x}. \tag{9.14}$$

Assuming, as before, wave type solutions as in Equations (9.7) and (9.8) and following exactly the same procedure to arrive at the stability condition as was done in Section 9.3.1, one gets the following expression for the quadratic equation similar to Equation (9.9)

$$|\lambda|^2 = 1 + gH\left(\frac{\Delta t}{\Delta x}\right)^2 \sin^2(k\Delta x). \tag{9.15}$$

Equation (9.15) shows clearly that the forward in time and central in space finite difference scheme is absolutely unstable as the second term in the right-hand side of Equation (9.15) is always positive and, hence, would violate $|\lambda| \leq 1$.

9.3.3   Fully implicit schemes (non-staggered)

Utilizing central difference finite difference schemes for spatial derivatives and backward difference for time derivatives in Equations (9.1) and (9.2), one gets the fully implicit scheme for one-dimensional linear shallow water equations without rotation

$$\frac{u_j^{n+1} - u_j^n}{\Delta t} = -g\frac{h_{j+1}^{n+1} - h_{j-1}^{n+1}}{2\Delta x}, \tag{9.16}$$

$$\frac{h_j^{n+1} - h_j^n}{\Delta t} = -H\frac{u_{j+1}^{n+1} - u_{j-1}^{n+1}}{2\Delta x}. \tag{9.17}$$

Assuming, as before, wave type solutions as in Equations (9.7) and (9.8) and following exactly the same procedure to arrive at the stability condition as was done in Section 9.3.1, one obtains an expression for the amplification factor λ:

$$\lambda - 1 = -c^2\frac{\lambda^2}{\lambda - 1}\sin^2(k\Delta x), \tag{9.18}$$

where c is the Courant number, $c = \sqrt{gH}(\Delta t / \Delta x)$. Equation (9.18) becomes

$$\lambda^2 \left[1 + c^2 \sin^2(k\Delta x)\right] - 2\lambda + 1 = 0 \tag{9.19}$$

Quadratic equation in λ has the following solutions

$$\begin{aligned}
\lambda &= \frac{1 \pm \sqrt{1 - \left[1 + c^2 \sin^2(k\Delta x)\right]}}{1 + c^2 \sin^2(k\Delta x)} \\
&= \frac{1 \pm ic \sin(k\Delta x)}{1 + c^2 \sin^2(k\Delta x)}.
\end{aligned} \tag{9.20}$$

In order to find the stability of this fully implicit scheme, one needs to obtain the amplitude of the amplification factor λ, which is $|\lambda|^2 = \lambda\lambda^*$, where λ^* is the complex conjugate of λ. From Equation (9.20), one obtains

$$|\lambda|^2 = \frac{1 + c^2 \sin^2(k\Delta x)}{\left[1 + c^2 \sin^2(k\Delta x)\right]^2} = \frac{1}{1 + c^2 \sin^2(k\Delta x)}. \tag{9.21}$$

As for a non-vanishing value of Courant number, $|\lambda|^2 < 1$ for all values of k and all values of c, the fully implicit scheme of the linearized one-dimensional shallow water equations is a unconditionally stable scheme.

9.3.4   Forward–backward scheme (non-staggered)

As Equations (9.1) and (9.2) involve two linear partial differential equations that depend on one another, it is quite natural to solve them using the forward–backward time-stepping scheme, i.e., a forward in time scheme for "u equation" followed by a backward in time scheme for the "h equation." It is also assumed that both variables u and h are colocated, i.e., the scheme is an un-staggered scheme. Moreover, the spatial derivatives are replaced by central space differences. The difference equations of Equations (9.1) and (9.2) are as follows for the forward–backward non-staggered scheme for the linear one-dimensional shallow water equations without rotation:

$$\frac{u_j^{n+1} - u_j^n}{\Delta t} = -g \frac{h_{j+1}^n - h_{j-1}^n}{2\Delta x}, \tag{9.22}$$

$$\frac{h_j^{n+1} - h_j^n}{\Delta t} = -H \frac{u_{j+1}^{n+1} - u_{j-1}^{n+1}}{2\Delta x}. \tag{9.23}$$

Assuming, as before, wave type solutions as in Equations (9.7) and (9.8) and following exactly the same procedure to arrive at the stability condition as was done in Section

9.3.1, one obtains an expression of a quadratic equation for amplification factor λ after some algebra in terms of Courant number c as follows:

$$\lambda^2 + \left[c^2 \sin^2(k\Delta x) - 2\right]\lambda + 1 = 0, \tag{9.24}$$

whose solution after some algebra is

$$\lambda = 1 - \frac{c^2}{2}\sin^2(k\Delta x) \pm \frac{ic}{2}\sin(k\Delta x)\sqrt{4 - c^2 \sin^2(k\Delta x)}. \tag{9.25}$$

Equation (9.25) shows that for $|c| \le 2$, $|\lambda| = 1$. Hence, for $|c| \le 2$, the forward backward scheme is stable. However, for $|c| > 2$, the forward–backward scheme is clearly unstable as can be easily discerned from Equation (9.25). For $|c| > 2$, the square of the amplification factor is given by

$$|\lambda|^2 = \left[1 - \frac{c^2 \sin^2(k\Delta x)}{2} \pm \frac{c}{2}\sin(k\Delta x)\sqrt{c^2 \sin^2(k\Delta x) - 4}\right]^2. \tag{9.26}$$

Equation (9.26) shows that the amplification factor λ can exceed one for $|c| > 2$ and, hence, the forward-backward scheme is unstable for $|c| > 2$. Hence, the forward–backward scheme is conditionally stable. It is clear that the forward–backward scheme allows a time step Δt that is twice the time step allowed by the leapfrog scheme. Moreover, the forward–backward scheme is a neutral scheme in that if it is stable, the amplification factor has a magnitude equal to 1. Although Equation (9.24) (the second equation) appears similar to an implicit scheme, the set of equations, Equations (9.22) and (9.23) are decoupled similar to an explicit scheme and, hence, in the forward–backward scheme, one does not need to solve a coupled system of simultaneous equations, i.e., the forward–backward scheme is indeed an explicit scheme only.

9.3.5   Pressure averaging scheme (non-staggered)

A procedure similar to the forward–backward scheme by some extent is the "pressure averaging" scheme. The name "pressure averaging" scheme arose essentially while employing the primitive (hydrostatic, however, non-geostrophic motion) equations using "height" as the vertical coordinate, where the adjustment (averaging) term for the momentum equations is given by the so-called pressure gradient term. As in this section one is dealing with the shallow-water equations, it would be more appropriate to call this scheme "height" or "geopotential" averaging.

The essential idea of the pressure averaging scheme is to consider the height h in the momentum conservation equations using some average of the height h in the previous present and future time instants and also by employing centered time derivatives only.

For example, the "pressure averaging" scheme as applied to the zonal momentum equation, Equation (9.1) would appear as

$$\frac{u_j^{n+1} - u_j^{n-1}}{2\Delta t} = -\frac{g}{2\Delta x} \left\{ (1 - 2\varepsilon) \left(h_{j+1}^n - h_{j-1}^n \right) + \varepsilon \left[\left(h_{j+1}^{n-1} - h_{j-1}^{n-1} \right) + \left(h_{j+1}^{n+1} - h_{j-1}^{n+1} \right) \right] \right\}.$$
(9.27)

Equation (9.27) reduces to the leapfrog scheme for $\varepsilon = 0$. For $\varepsilon = 1/4$, the von Neumann stability condition reduces to the same conditional stability conditions as for the forward–backward scheme.

9.3.6   Implicit scheme (non-staggered)

An implicit scheme using forward time differences can be constructed using the Crank–Nicolson approach as applied to Equations (9.1) and (9.2):

$$\frac{u_j^{n+1} - u_j^{n-1}}{2\Delta t} = -g \left[\beta \delta h_j^n + (1 - \beta) \delta h_j^{n+1} \right]$$
(9.28)

$$\frac{h_j^{n+1} - h_j^{n-1}}{2\Delta t} = -H \left[\beta \delta u_j^n + (1 - \beta) \delta u_j^{n+1} \right]$$
(9.29)

where β and $(1 - \beta)$ are weights such that their sum $= 1$. Moreover, in Equations (9.28) and (9.29), δ represents a centered finite difference operator corresponding to the first derivative. As a special case, $\beta = 1$, corresponds to forward in time and central in space finite difference scheme, which is unconditionally unstable (refer Section 9.3.2).

9.3.7   Staggered explicit scheme

Equations (9.1) and (9.2) are non-dimensionalized for convenience by employing the velocity scale as $\sqrt{gH}$, length scale as H and time scale as $H/\sqrt{gH}$. Using this, the non-dimensionalized one-dimensional linear shallow water equations without rotation are as follows:

$$\frac{\partial u}{\partial t} + \frac{\partial h}{\partial x} = 0,$$
(9.30)

$$\frac{\partial h}{\partial t} + \frac{\partial u}{\partial x} = 0.$$
(9.31)

Employing the staggered grid (refer to Section 8.3), the finite difference equations for Equations (9.30) and (9.31) become

$$\frac{u_j^{n+1} - u_j^{n-1}}{2\Delta t} = -\frac{h_{j+1/2}^n - h_{j-1/2}^n}{\Delta x}, \tag{9.32}$$

$$\frac{h_{j+1/2}^{n+1} - h_{j+1/2}^{n-1}}{2\Delta t} = -\frac{u_{j+1}^n - u_j^n}{\Delta x}. \tag{9.33}$$

The advantages of such a discretization using the staggered grid is that the centered space derivative in the RHS of Equations (9.32) and (9.33) now uses successive points of the same variable. The dispersion characteristics of this staggered scheme is improved as the effective grid length is halved in the staggered case.

Let us assume discrete traveling wave solutions such as

$$u_j^n = u_o e^{i(kj\Delta x - n\omega\Delta t)}, \tag{9.34}$$

$$h_j^n = h_o e^{i(kj\Delta x - n\omega\Delta t)}. \tag{9.35}$$

Substituting Equations (9.34) and (9.35) into Equations (9.32) and (9.33), one obtains the following relations

$$u_j^n \left(\frac{e^{-i\omega\Delta t} - e^{i\omega\Delta t}}{2\Delta t} \right) + h_j^n \left(\frac{e^{ik\Delta x/2} - e^{-ik\Delta x/2}}{\Delta x} \right) = 0, \tag{9.36}$$

$$h_j^n e^{ik\Delta x/2} \left(\frac{e^{-i\omega\Delta t} - e^{i\omega\Delta t}}{2\Delta t} \right) + u_j^n \left(\frac{e^{ik\Delta x} - 1}{\Delta x} \right) = 0. \tag{9.37}$$

Taking the determinant of Equations (9.36) and (9.37) to zero after canceling common terms, one obtains the dispersion relation as follows:

$$-\sin^2(\omega\Delta t) = -4\lambda^2 \sin^2(k\Delta x/2)$$

or

$$\sin(\omega\Delta t) = \pm 2\lambda \sin(k\Delta x/2), \tag{9.38}$$

where $\lambda = (\Delta t/\Delta x)$. Recall $(\Delta t/\Delta x)$ is the reciprocal of the non-dimensional velocity, c'. It turns out that the non-dimensional velocity is related to the dimensional velocity through the following relation that employs the velocity scale

$$c' = \frac{c}{\sqrt{gH}} = \frac{\Delta x}{\Delta t\sqrt{gH}}, \tag{9.39}$$

$$\lambda = \frac{1}{c'} = \frac{\Delta t\sqrt{gH}}{\Delta x}. \tag{9.40}$$

From Equation (9.38), it is clear that for stability, one requires $0 < \lambda \le 0.5$. Hence, the time step Δt for the staggered scheme must satisfy the following condition that

$$\frac{\Delta t \sqrt{gH}}{\Delta x} \le \frac{1}{2} \tag{9.41}$$

It is clear that the stability requirement for the staggered grid is more stringent than for the non-staggered grid (refer to Equation (9.12) for the explicit non-staggered leapfrog scheme). Assuming Δt and Δx are sufficiently small, one can expand Equation (9.38) by the Taylor series expansion and obtain

$$\omega \Delta t \approx \pm 2\lambda \left[\left(\frac{k\Delta x}{2} \right) - \frac{1}{6} \left(\frac{k\Delta x}{2} \right)^3 + \cdots \right], \tag{9.42}$$

$$\omega \approx \pm k \left[1 - \frac{k^2 (\Delta x)^2}{24} \right]. \tag{9.43}$$

For the staggered grid, Equation (9.43) indicates that the phase error is negative and, hence, the wave is decelerating.

9.3.8 Splitting method

It is noted (refer to Equation (9.12)) that employing an explicit finite difference scheme such as the leapfrog scheme for solving the shallow water equations that is representative of the synoptic scale ($\Delta x = 100$ km) with mean depth equalling 10 km, the required time step is 5 minutes for stability. It can be seen that this time step is much lower than the typical time scale associated with synoptic scale and, hence, utilizing the explicit scheme contributes to larger computational time than what would be desirable. One way of increasing the allowed time step is provided by the splitting method, brief details of the same are provided in this section.

To ensure that the linear one-dimensional shallow water equations are more realistic, one should consider the effects of advection. Hence, Equations (9.1) and (9.2), get modified as given below, after incorporating advection

$$\frac{\partial u}{\partial t} + u_o \frac{\partial u}{\partial x} + g \frac{\partial h}{\partial x} = 0, \tag{9.44}$$

$$\frac{\partial h}{\partial t} + u_o \frac{\partial h}{\partial x} + H \frac{\partial u}{\partial x} = 0. \tag{9.45}$$

Equations (9.44) and (9.45) embody two (advection and gravity) different physical mechanisms. Marchuk suggested that it is better that these two different physical mechanisms be treated separately. Hence, Equations (9.44) and (9.45) are split as follows into the "advection" part and "gravity" part.

$$\frac{\partial u}{\partial t} + u_o \frac{\partial u}{\partial x} = 0 \qquad\qquad \frac{\partial h}{\partial t} + u_o \frac{\partial h}{\partial x} = 0 \qquad\qquad\qquad (9.46)$$

$$\frac{\partial u}{\partial t} + g \frac{\partial h}{\partial x} = 0 \qquad\qquad \frac{\partial h}{\partial t} + H \frac{\partial u}{\partial x} = 0. \qquad\qquad\qquad (9.47)$$

The following procedure is then employed in the spitting method.

1. Use standard finite difference approximations to approximate the derivatives to solve Equation (9.46). Let h^* and u^* denote the respective values after the first time step of solving the advection part of the equations; these can be represented as

    $$h^* = \lambda_{adv} h^n \qquad \text{and} \qquad u^* = \lambda_{adv} u^n,$$

    where λ_{adv} indicates the modifying factor at the end of the first time step after solving the advection part, Equation (9.46).

2. The new values at the end of first time step h^* and u^* are now utilized as the starting point for solving the gravity part of Equation (9.47) to get

    $$h^{n+1} = \lambda_{grav} h^* \qquad \text{and} \qquad u^{n+1} = \lambda_{grav} u^*,$$

    where λ_{grav} indicates the modifying factor at the end of the first time step after solving the gravity part, Equation (9.47).

3. Utilizing steps 1 and 2, one obtains after substitution, the following

    $$h^{n+1} = \lambda h^n \qquad \text{and} \qquad u^{n+1} = \lambda u^n,$$

    where $\lambda = \lambda_{adv} \lambda_{grav}$.

The complete splitting method is stable provided $|\lambda| \leq 1$ and the aforementioned condition will be satisfied only if stability is ensured at each of the separate steps, i.e., if both

$$|\lambda_{adv}| \leq 1 \qquad \text{and} \qquad |\lambda_{grav}| \leq 1.$$

It is to be noted that each of the two physical processes (advection and gravity) have different time scales and, hence, one can exploit this fact in choosing different time steps for the "advection" part and for the "gravity" part. For example, the gravity wave speed is typically larger (by a factor of 3) than the advection speed and, hence, it makes sense to utilize a larger time step (say, Δt) for solving the advection part of

Equation (9.46) while employing a number of smaller time steps (M steps of smaller time step δt such that $\Delta t = M \delta t$) for solving the gravity part of Equation (9.47). The two separate steps will be individually stable provided the following two conditions are satisfied

$$u_o \frac{\Delta t}{\Delta x} \leq 1, \qquad c\frac{\Delta t}{\Delta x} = c\frac{\Delta t}{M \delta x} \leq 1 \tag{9.48}$$

As the gravity wave speed is typically three times the advective speed, it is reasonable to utilize $M = 3$ and take three marchings of the gravity part before each advective step.

9.3.9   Semi-implicit method

It is clear from the discussion in the previous section, that the terms in the shallow water equations that are responsible for the gravity wave (the so-called gravity terms) have a much smaller time step as compared to the non-gravity terms that can possibly have a much larger time step. As the utilization of an explicit finite difference scheme entails a stringent condition on the choice of the time step, it makes sense to treat the so-called gravity terms in an implicit manner and the remaining non-gravity terms to be treated in an explicit fashion. This is the essence of the "semi-implicit" scheme, which has been widely employed in numerical forecasting of the atmosphere and the oceans. It would of course be a "win–win" situation if this scheme has other desirable properties such as "unconditional stability" and relative ease of carrying out the computations without allocating excessive computer time. It has been observed that the semi-implicit scheme possesses the aforementioned desirable properties. It makes sense to introduce the semi-implicit scheme as applied to the linear one-dimensional shallow water equations without rotation. Towards this, one considers Equations (9.30) and (9.31), the non-dimensional form of the linear one-dimensional shallow water equations without rotation as applied to a staggered grid where h variables are defined at the regular grid points while u variables are defined at the mid-point between the regular grid points. A semi-implicit finite difference implementation of Equations (9.30) and (9.31) on a staggered grid is

$$\frac{u_{j+1/2}^{n+1} - u_{j+1/2}^{n}}{\Delta t} = -\frac{1}{2}\left[\frac{h_{j+1}^{n+1} - h_j^{n+1}}{\Delta x} + \frac{h_{j+1}^{n} - h_j^{n}}{\Delta x}\right], \tag{9.49}$$

$$\frac{h_j^{n+1} - h_j^{n}}{\Delta t} = -\frac{1}{2}\left[\frac{u_{j+1/2}^{n+1} - u_{j-1/2}^{n+1}}{\Delta x} + \frac{u_{j+1/2}^{n} - u_{j-1/2}^{n}}{\Delta x}\right]. \tag{9.50}$$

In Equations (9.49) and (9.50), while the time derivative uses central difference scheme centered at $n+1/2$, the RHS involves splitting the gravity terms into two parts, using the "implicit" scheme for the first part and the "explicit" scheme for the second part. Obtaining $u_{j+1/2}^{n+1}$ and $u_{j-1/2}^{n+1}$ from Equation (9.49), one gets

$$u_{j+1/2}^{n+1} = u_{j+1/2}^{n} - \frac{\Delta t}{2\Delta x}\left[(h_{j+1}^{n+1} - h_j^{n+1}) + (h_{j+1}^{n} - h_j^{n})\right], \tag{9.51}$$

$$u_{j-1/2}^{n+1} = u_{j-1/2}^{n} - \frac{\Delta t}{2\Delta x}\left[(h_j^{n+1} - h_{j-1}^{n+1}) + (h_j^{n} - h_{j-1}^{n})\right]. \tag{9.52}$$

Substituting from Equations (9.51) and (9.52) in Equation (9.50) and rearranging terms, one obtains the following equation

$$h_{j+1}^{n+1}\left[-\frac{1}{4}\left(\frac{\Delta t}{\Delta x}\right)^2\right] + h_j^{n+1}\left[1 + \frac{1}{2}\left(\frac{\Delta t}{\Delta x}\right)^2\right] + h_{j-1}^{n+1}\left[-\frac{1}{4}\left(\frac{\Delta t}{\Delta x}\right)^2\right] = -\frac{\Delta t}{\Delta x}\left(u_{j+1/2}^{n} - u_{j-1/2}^{n}\right)$$

$$+ h_{j+1}^{n}\left[\frac{1}{4}\left(\frac{\Delta t}{\Delta x}\right)^2\right] + h_j^{n}\left[1 - \frac{1}{2}\left(\frac{\Delta t}{\Delta x}\right)^2\right] + h_{j-1}^{n}\left[\frac{1}{4}\left(\frac{\Delta t}{\Delta x}\right)^2\right]. \tag{9.53}$$

Equation (9.53) describes a tridiagonal matrix, which is diagonally dominant. Several fast matrix inversion solvers exist for such tridiagonal systems. Once h^{n+1} is determined for all j's, by solving Equation (9.53), the same can be substituted in Equation (9.49), to obtain u^{n+1} for all j's, thereby ensuring the solution of Equations (9.30) and (9.31). For the two-dimensional shallow water equations, a similar procedure would yield a "Helmholtz" type equation instead of Equation (9.53), which can then be solved easily using the "successive over relaxation method."

9.3.10   Stability of the semi-implicit method

In order to discuss the stability of the semi-implicit method, consider the linear one-dimensional shallow water equations in dimensional form (Equations (9.1) and (9.2)). As in Section 9.3.9, a staggered grid is considered where the h variables are defined at the regular grid points while u variables are defined at the mid-point between the regular grid points. The semi-implicit finite difference scheme as applied to Equations (9.1) and (9.2) are given as follows:

$$u_{j+1/2}^{n+1} = u_{j+1/2}^{n} - g\frac{\Delta t}{\Delta x}\left[\theta(h_{j+1}^{n+1} - h_j^{n+1}) + (1-\theta)(h_{j+1}^{n} - h_j^{n})\right], \tag{9.54}$$

$$h_j^{n+1} = h_j^{n} - H\frac{\Delta t}{\Delta x}\left[\theta(u_{j+1/2}^{n+1} - u_{j-1/2}^{n+1}) + (1-\theta)(u_{j+1/2}^{n} - u_{j-1/2}^{n})\right]. \tag{9.55}$$

In Equations (9.54) and (9.55), $\theta = 0$ provides for the explicit scheme, $\theta = 1$ provides for a fully implicit scheme, and $\theta = 0.5$ provides for the semi-implicit scheme. Changing the variable from h to z, using $z = h\sqrt{g/H}$, Equations (9.54) and (9.55) become

$$u_{j+1/2}^{n+1} + \theta\sqrt{gH}\frac{\Delta t}{\Delta x}\left(z_{j+1}^{n+1} - z_j^{n+1}\right) = u_{j+1/2}^n - (1-\theta)\sqrt{gH}\frac{\Delta t}{\Delta x}\left(z_{j+1}^n - z_j^n\right), \quad (9.56)$$

$$z_j^{n+1} + \theta\sqrt{gH}\frac{\Delta t}{\Delta x}\left(u_{j+1/2}^{n+1} - u_{j-1/2}^{n+1}\right) = z_j^n - (1-\theta)\sqrt{gH}\frac{\Delta t}{\Delta x}\left(u_{j+1/2}^n - u_{j-1/2}^n\right). \quad (9.57)$$

Equations (9.56) and (9.57) are investigated for stability by employing the von Neumann method of stability analysis. Towards this end, a Fourier mode is introduced for each field variable u and z. In effect, $u_{j+1/2}^n$ and z_j^n are replaced in Equations (9.56) and (9.57) by $\tilde{u}^n e^{i(j+1/2)\alpha}$ and $\tilde{z}^n e^{ij\alpha}$, where $\tilde{u}^n$ and $\tilde{z}^n$ are the amplitude functions of u and z at time level n and α is the phase angle. After some algebra, Equations (9.56) and (9.57) become

$$\tilde{u}^{n+1} + \theta\Phi\tilde{z}^{n+1}\left(e^{i\alpha/2} - e^{-i\alpha/2}\right) = \tilde{u}^n - (1-\theta)\Phi\tilde{z}^n\left(e^{i\alpha/2} - e^{-i\alpha/2}\right), \quad (9.58)$$

$$\tilde{z}^{n+1} + \theta\Phi\tilde{u}^{n+1}\left(e^{i\alpha/2} - e^{-i\alpha/2}\right) = \tilde{z}^n - (1-\theta)\Phi\tilde{u}^n\left(e^{i\alpha/2} - e^{-i\alpha/2}\right). \quad (9.59)$$

where $\Phi = \sqrt{gH}\frac{\Delta t}{\Delta x}$. Setting $p = 2\Phi\sin(\alpha/2)$, Equations (9.58) and (9.59) can be written as a matrix equation

$$P\tilde{W}^{n+1} = Q\tilde{W}^n \quad (9.60)$$

with

$$\tilde{W}^n = \begin{bmatrix} \tilde{u}^n \\ \tilde{z}^n \end{bmatrix}, \quad P = \begin{bmatrix} 1 & ip\theta \\ ip\theta & 1 \end{bmatrix}, \quad Q = \begin{bmatrix} 1 & -ip(1-\theta) \\ -ip(1-\theta) & 1. \end{bmatrix} \quad (9.61)$$

The amplification matrix is $G = P^{-1}Q$ and a necessary and sufficient condition for stability is that $\|G\|_2 \leq 1$ identically satisfies for every α. As G is a normal matrix, the norm of G is equal to its spectral radius. The eigenvalues of G are given by

$$\lambda_{1,2} = \frac{1 - p^2\theta(1-\theta) \pm ip}{1 + p^2\theta^2}. \quad (9.62)$$

Thus, the condition for the spectral radius of G to be not greater than unity is $1 - 2\theta < 0$, or equivalently, $\theta \geq 1/2$. As for the semi-implicit scheme, $\theta = 1/2$, the semi-implicit scheme is indeed an unconditionally stable scheme.

9.4   Two-dimensional Linear Shallow Water Equations without Rotation

The two-dimensional linear shallow water equations without rotation are as indicated below:

$$\frac{\partial u}{\partial t} + g\frac{\partial h}{\partial x} = 0, \tag{9.63}$$

$$\frac{\partial v}{\partial t} + g\frac{\partial h}{\partial y} = 0, \tag{9.64}$$

$$\frac{\partial h}{\partial t} + H\left(\frac{\partial u}{\partial x} + \frac{\partial v}{\partial y}\right) = 0. \tag{9.65}$$

9.4.1   Leapfrog scheme

Applying the leapfrog scheme with centered space differencing to the two-dimensional linear shallow water equations without rotation, one obtains

$$u_{j,m}^{n+1} = u_{j,m}^{n-1} - 2g\Delta t\, \delta(h_x)_{j,m}^n. \tag{9.66}$$

$$v_{j,m}^{n+1} = v_{j,m}^{n-1} - 2g\Delta t\, \delta(h_y)_{j,m}^n, \tag{9.67}$$

$$h_{j,m}^{n+1} = h_{j,m}^{n-1} - 2H\Delta t\left[\delta(u_x)_{j,m}^n + \delta(v_y)_{j,m}^n\right] \tag{9.68}$$

where δ represents the central finite difference operator corresponding to the first derivative

Let us assume wave type solutions:

$$u_{j,m}^n = Re\left(\lambda^n \tilde{u}\, e^{i(kx+ly)}\right), \tag{9.69}$$

$$v_{j,m}^n = Re\left(\lambda^n \tilde{v}\, e^{i(kx+ly)}\right), \tag{9.70}$$

$$h_{j,m}^n = Re\left(\lambda^n \tilde{h}\, e^{i(kx+ly)}\right). \tag{9.71}$$

Substituting Equations (9.69)–(9.71) to Equations (9.66)–(9.68), one obtains a homogeneous system in λ whose solutions are $\lambda = 1$, $\lambda = -1$ as well as the solution of the following equation:

$$\lambda^2 = 1 - 2A \pm 2\sqrt{A(A-1)}, \tag{9.72}$$

where

$$A = gH\mu^2(\sin^2 X + \sin^2 Y),\qquad(9.73)$$

where $\mu = \Delta t/d^*$, $X = kd^*$, and $Y = ld^*$, where k and l are the zonal and the meridional wave numbers.

The solution $\lambda = -1$ corresponds to a computational mode that arises due to the fact that the leapfrog scheme is a three time level scheme, whereas solution $\lambda = 1$, corresponds to a neutral and stationary solution. For the stability of the leapfrog scheme, the requirement of $\lambda \leq 1$ is met by satisfying the condition $\sqrt{2A} \leq 1$ as can be discerned from Equation (9.72). The aforementioned condition assumes the form:

$$\sqrt{2gH}\,\frac{\Delta t}{d^*} \leq 1.\qquad(9.74)$$

Equation (9.74) shows that the choice of time step for solving linear two-dimensional shallow water equations without rotation is more stringent (factor of $\sqrt{2}$ less than the linear one-dimensional shallow water equations).

9.4.2   Elliassen grid

While using the staggered grid 'D' (refer Section 8.54), it was noticed that the space derivatives in such a grid required extensive averaging (i.e., averaging in both the x and y directions) before the derivatives could be applied. Staggering variables in time as well as space provides a way of avoiding the far-reaching averaging that one has to employ while utilizing the 'D' grid. This technique, proposed by Eliassen, (shown in Figure 9.5), involves representing variables at every second time step on an offset 'D' grid, as illustrated in Figure 9.5. If the right-hand side terms of the shallow water equations are evaluated using variables on the 'D' grid as shown in Figure 8.16, then tendencies are computed for the aforementioned variables at the positions shown on the offset grid (Figure 9.5). On the next time step, the variables are computed back on the original grid 'D'. In essence, Eliassen proposed a regular staggered grid 'D' and alternately, used an offset grid at the next time step to avoid the extensive averaging that is unavoidable with the 'D' grid.

9.4.3   Forward backward scheme

The two-dimensional linear shallow water equations without rotation (Equations (9.63) to (9.65)) are again considered. The forward–backward scheme is obtained by first integrating the gravity wave terms (pressure gradient terms in the momentum equations and the divergence term in the continuity equation) of either the momentum equation or of the continuity equation forward, and then those of the other equation (continuity equation or the momentum equations backward in time). Utilizing centered space differencing as applied to Equations (9.63) to (9.65), the forward-backward scheme finite difference equations are

$$u_{j,m}^{n+1} = u_{j,m}^n - g\Delta t\,\delta(h_x)_{j,m}^n, \tag{9.75}$$

$$v_{j,m}^{n+1} = v_{j,m}^n - g\Delta t\,\delta(h_y)_{j,m}^n, \tag{9.76}$$

$$h_{j,m}^{n+1} = h_{j,m}^n - H\Delta t\left[\delta(u_x)_{j,m}^{n+1} + \delta(v_y)_{j,m}^{n+1}\right]. \tag{9.77}$$

Assuming the wave type solutions for Equations (9.69) to (9.71) and substituting them into Equations (9.75) to (9.77), one obtains a homogeneous system in λ whose solutions are $\lambda = 1$, as well as the remaining two solution of the following equation:

$$\lambda = 1 - \frac{A}{2} \pm \sqrt{\frac{A}{4}(A - 4)}, \tag{9.78}$$

where A is defined as before (Equation (9.73)). The solution $\lambda = 1$ corresponds to a neutral and stationary solution. Solution $\lambda = 1$ and solutions of Equation (9.78) are the same irrespective of whether the momentum equation is first integrated with a

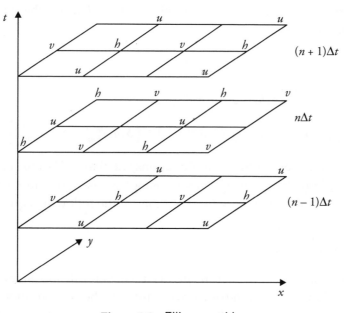

Figure 9.1 Elliassen grid.

forward scheme and the continuity equation with a backward scheme or where the continuity equation is first integrated with a forward scheme and the momentum equations are then integrated with a backward scheme. Equation (9.78) shows that the amplification factors indicate that the forward–backward scheme is stable and neutral for $A \leq 2$. For all admissible waves, the stability condition of the forward–backward scheme is that the time step should satisfy the following condition:

$$\sqrt{\frac{A}{2}} \leq 1 \qquad \text{and} \qquad \sqrt{\frac{gH}{2}} \frac{\Delta t}{d^*} \leq 1. \tag{9.79}$$

Comparison of Equation (9.79) with Equation (9.74) shows that the required time step for meeting the stability requirement for the forward–backward scheme can be twice as large as the required time step for meeting the stability requirement for the leapfrog scheme.

9.4.4   Implicit scheme (trapezoidal method)

The two-dimensional linear shallow water equations without rotation (Equations (9.63) to (9.65)) are again considered. Considering centered finite differences in that centered at $n + 1/2$ and replacing the gravity terms (pressure gradient terms and divergence terms) by their time average (at n and $n + 1$ using the trapezoidal rule), one obtains the following:

$$u_{j,m}^{n+1} = u_{j,m}^n - g\frac{\Delta t}{2} \left[\delta(h_x)_{j,m}^n + \delta(h_x)_{j,m}^{n+1} \right], \tag{9.80}$$

$$v_{j,m}^{n+1} = v_{j,m}^n - g\frac{\Delta t}{2} \left[\delta(h_y)_{j,m}^n + \delta(h_y)_{j,m}^{n+1} \right], \tag{9.81}$$

$$h_{j,m}^{n+1} = h_{j,m}^n - H\frac{\Delta t}{2} \left[\left(\delta(u_x)_{j,m}^n + \delta(v_y)_{j,m}^n \right) + \left(\delta(u_x)_{j,m}^{n+1} + \delta(v_y)_{j,m}^{n+1} \right) \right]. \tag{9.82}$$

Assuming the wave type solutions for Equations (9.69) to (9.71) and substituting them into Equations (9.80)–(9.82), one obtains a homogeneous system in λ whose solutions are $\lambda = 1$, as well as the remaining two solutions of the following equation:

$$\lambda = \frac{1}{1+A/4} \left(1 - \frac{A}{4} \pm i\sqrt{A} \right). \tag{9.83}$$

As before, the solution $\lambda < 1$ corresponds to a neutral and stationary solution. Equation (9.83) shows that for the other two solutions, the amplification factors always satisfy the condition $|\lambda| = 1$, resulting in an unconditionally stable and neutral finite difference scheme.

9.5 Semi-implicit Scheme of Kwizak and Robert

Based on the discussions in earlier sub-sections, it is indeed clear that there is need to utilize an implicit finite difference scheme for the gravity wave terms (pressure gradient terms in the momentum equations and the divergence term in the continuity equation) in the shallow water equations. In effect, this means that there are no distinct advantages in employing an implicit finite difference scheme for advection, Coriolis, and other terms (that are not gravity wave terms) of the shallow water equations. These non-gravity wave terms are associated with slower phase speeds, and, hence, should not require excessively small time steps for linear stability when calculated explicitly. As the implicit scheme with trapezoidal method (refer to Section 9.4.4) is a two-level scheme like the forward–backward scheme, it is convenient to use the Adams–Bashforth scheme for this purpose. Kwizak and Robert in 1971 chose, however, to utilize the leapfrog scheme for time integration and treat the advection and Coriolis terms (non-gravity wave terms) in an explicit finite difference scheme and treat only the gravity wave terms using the implicit scheme with trapezoidal scheme.

Consider the nonlinear two-dimensional shallow water equations with rotation.

$$\frac{\partial u}{\partial t} + u\frac{\partial u}{\partial x} + v\frac{\partial u}{\partial y} - fv = -g\frac{\partial h}{\partial x}, \tag{9.84}$$

$$\frac{\partial v}{\partial t} + u\frac{\partial v}{\partial x} + v\frac{\partial v}{\partial y} + fu = -g\frac{\partial h}{\partial y}, \tag{9.85}$$

$$\frac{\partial h}{\partial t} + \frac{\partial}{\partial x}[(H+h)u] + \frac{\partial}{\partial y}[(H+h)v] = 0, \tag{9.86}$$

where H is the mean depth of the fluid, which is assumed constant. Treating the nonlinear advection terms and the Coriolis terms in Equations (9.84) and (9.85) as being represented as A_u and A_v terms and treating all the nonlinear terms in the continuity equation, Equation (9.86) as being represented as A_h, the aforementioned shallow water equations are rewritten as

$$\frac{\partial u}{\partial t} = -g\frac{\partial h}{\partial x} + A_u, \tag{9.87}$$

$$\frac{\partial v}{\partial t} = -g\frac{\partial h}{\partial y} + A_v, \tag{9.88}$$

$$\frac{\partial h}{\partial t} = -H\left(\frac{\partial u}{\partial x} + \frac{\partial v}{\partial y}\right) + A_h. \tag{9.89}$$

These additional terms A_u, A_v, and A_h are treated explicitly, while the gravity wave terms in Equations (9.87) to (9.89) are treated implicitly over a time interval $2\Delta t$ (leapfrog time integration scheme) giving the following semi-implicit finite difference equations:

$$u_{j,m}^{n+1} = u_{j,m}^{n-1} - g\Delta t \left[\delta(h_x)_{j,m}^{n-1} + \delta(h_x)_{j,m}^{n+1} \right] + 2\Delta t (A_u)_{j,m}^n, \tag{9.90}$$

$$v_{j,m}^{n+1} = v_{j,m}^{n-1} - g\Delta t \left[\delta(h_y)_{j,m}^{n-1} + \delta(h_y)_{j,m}^{n+1} \right] + 2\Delta t (A_v)_{j,m}^n, \tag{9.91}$$

$$h_{j,m}^{n+1} = h_{j,m}^{n-1} - H\Delta t \left[\left(\delta(u_x)_{j,m}^{n-1} + \delta(v_y)_{j,m}^{n-1} \right) + \left(\delta(u_x)_{j,m}^{n+1} + \delta(v_y)_{j,m}^{n+1} \right) \right] + 2\Delta t (A_h)_{j,m}^n. \tag{9.92}$$

Applying the operator δx to Equation (9.90) and the operator δy to Equation (9.91), and adding the resulting equations, one obtains

$$\left[\delta(u_x)_{j,m}^{n+1} + \delta(v_y)_{j,m}^{n+1} \right] = \left[\delta(u_x)_{j,m}^{n-1} + \delta(v_y)_{j,m}^{n-1} \right]$$
$$- g\Delta t \left[\left(\delta_{xx}h_{j,m}^{n-1} + \delta_{yy}h_{j,m}^{n-1} \right) + \left(\delta_{xx}h_{j,m}^{n+1} + \delta_{yy}h_{j,m}^{n+1} \right) \right] + 2\Delta t \left[\delta_x(A_u)_{j,m}^n + \delta_y(A_v)_{j,m}^n \right], \tag{9.93}$$

where

$$\delta_{xx}h = \delta_x(\delta_x h) \qquad \text{and} \qquad \delta_{yy}h = \delta_y(\delta_y h). \tag{9.94}$$

Substituting Equation (9.93) to Equation (9.92), one obtains

$$h_{j,m}^{n+1} = h_{j,m}^{n-1} - 2H\Delta t \left[\delta(u_x)_{j,m}^{n-1} + \delta(v_y)_{j,m}^{n-1} \right] + gH(\Delta t)^2 \left(\nabla_\oplus^2 h_{j,m}^{n-1} + \nabla_\oplus^2 h_{j,m}^{n+1} \right)$$
$$- 2H(\Delta t)^2 \left[\delta_x(A_u)_{j,m}^n + \delta_y(A_v)_{j,m}^n \right] + 2\Delta t (A_h)_{j,m}^n, \tag{9.95}$$

where

$$\nabla_\oplus^2 h = \delta_{xx}h + \delta_{yy}h \tag{9.96}$$

is the finite difference expression for the two-dimensional Laplacian. Equation (9.95) can be rewritten as

$$h_{j,m}^{n+1} - gH(\Delta t)^2 \nabla_\oplus^2 h_{j,m}^{n+1} = F_{j,m}^{n-1} + G_{j,m}^n, \tag{9.97}$$

where the RHS terms are as follows and refer to terms at time level $n-1$ and time levels n:

$$F_{j,m}^{n-1} = h_{j,m}^{n-1} - 2H\Delta t \left[\delta(u_x)_{j,m}^{n-1} + \delta(v_y)_{j,m}^{n-1}\right] + gH(\Delta t)^2 \nabla_\oplus^2 h_{j,m}^{n-1}, \tag{9.98}$$

$$G_{j,m}^n = 2\Delta t \left[(A_h)_{j,m}^n - 2H(\Delta t)^2 \left(\delta_x(A_u)_{j,m}^n + \delta_y(A_v)_{j,m}^n\right)\right]. \tag{9.99}$$

Equation (9.97) is an elliptic partial differential equation of the Helmholtz type and can in principle be solved subject to boundary conditions; $h_{j,m}^{n+1}$ can be thus determined. From Equations (9.90) and (9.91), $u_{j,m}^{n+1}$ and $v_{j,m}^{n+1}$ can also be determined resulting in a solution of the complete two-dimensional shallow water equations with rotation.

Exercises 9a (Questions and answers)

1. Indicate the need for a staggered grid.

   Answer: Consider the simplest shallow water model (linear, one-dimensional and without rotation). It is very clear in the non-staggered grid arrangement to solve these equations that u_i depends only on the adjacent h_i, whereas h_i depends only on the adjacent u_i. Thus, the original non-staggered grid contains two elementary "sub-grids," with the solution on one of these sub-grids being completely decoupled from the other. It would appear in such a situation that it is sufficient to calculate only one of these solutions using the staggered grid (with variables carried at alternate points in space).

2. Indicate the advantages of the staggered grid arrangement.

   Answer: The computation time required (say) to solve the shallow water equations is halved, although the truncation error remains the same.

3. Indicate the various application fields in which the results of shallow water equations are applicable.

   Answer: The results of shallow water equations are applicable in diverse areas such as (i) wave propagation in tsunami forecasting, (ii) storm surge simulations, (iii) river flooding, and (iv) dam break problems. Moreover, the shallow water theory can be successfully applied to resolve a large number of hydrodynamical related questions. The results of shallow water equations are also employed to model landslides and avalanches.

4. Write down the one-dimensional linear shallow water equations without rotation and having a bottom topography given by $\eta(x,y)$.

Answer: The one-dimensional linear shallow water equations without rotation and having a bottom topography $\eta(x,y)$ are

$$\frac{\partial u}{\partial t} = -g\frac{\partial h}{\partial x} \quad \text{and} \quad \frac{\partial h}{\partial t} = -H\frac{\partial u}{\partial x} - \eta\frac{\partial u}{\partial x} - u\frac{\partial \eta}{\partial x}$$

where $h = h(x,y,t)$ is the height of the fluid above the bottom topography whereas H is the mean fluid depth that is assumed to be a constant.

5. Write down the one-dimensional linear shallow water equations with rotation and having a bottom topography given by $\eta(x,y)$.

   Answer: The one-dimensional linear shallow water equations with rotation and having a bottom topography $\eta(x,y)$ are

$$\frac{\partial u}{\partial t} - fv = -g\frac{\partial h}{\partial x}, \quad \frac{\partial v}{\partial t} + fu = 0, \quad \text{and} \quad \frac{\partial h}{\partial t} = -H\frac{\partial u}{\partial x} - \eta\frac{\partial u}{\partial x} - u\frac{\partial \eta}{\partial x},$$

   where $h = h(x,y,t)$ is the height of the fluid above the bottom topography whereas H is the mean fluid depth that is assumed to be a constant and f is the Coriolis parameter.

6. Write down the one-dimensional nonlinear shallow water equations without rotation and having a bottom topography given by $\eta(x,y)$.

   Answer: The one-dimensional nonlinear shallow water equations without rotation and having a bottom topography $\eta(x,y)$ are

$$\frac{\partial u}{\partial t} + u\frac{\partial u}{\partial x} = -g\frac{\partial h}{\partial x} \quad \text{and} \quad \frac{\partial h}{\partial t} + \frac{\partial}{\partial x}[(h+\eta)u] = 0.$$

7. Write down the one-dimensional nonlinear shallow water equations with rotation and having a bottom topography given by $\eta(x,y)$.

   Answer: The one-dimensional nonlinear shallow water equations with rotation and having a bottom topography $\eta(x,y)$ are

$$\frac{\partial u}{\partial t} + u\frac{\partial u}{\partial x} - fv = -g\frac{\partial h}{\partial x}, \quad \frac{\partial v}{\partial t} + u\frac{\partial v}{\partial x} + fu = 0, \quad \text{and} \quad \frac{\partial h}{\partial t} + \frac{\partial}{\partial x}[(h+\eta)u] = 0.$$

8. Write down the two-dimensional linear shallow water equations without rotation and having a bottom topography given by $\eta(x,y)$.

   Answer: The two-dimensional linear shallow water equations without rotation and having a bottom topography $\eta(x,y)$ are

$$\frac{\partial u}{\partial t} = -g\frac{\partial h}{\partial x}, \quad \frac{\partial v}{\partial t} = -g\frac{\partial h}{\partial y}, \quad \text{and} \quad \frac{\partial h}{\partial t} = -H\left(\frac{\partial u}{\partial x} + \frac{\partial v}{\partial y}\right) - \eta\left(\frac{\partial u}{\partial x} + \frac{\partial v}{\partial y}\right) - u\frac{\partial \eta}{\partial x} + v\frac{\partial \eta}{\partial y}.$$

9. Write down the two-dimensional linear shallow water equations with rotation and having a bottom topography given by $\eta(x,y)$.

   Answer: The two-dimensional linear shallow water equations with rotation and having a bottom topography $\eta(x,y)$ are

   $$\frac{\partial u}{\partial t} - fv = -g\frac{\partial h}{\partial x}, \qquad \frac{\partial v}{\partial t} + fu = -g\frac{\partial h}{\partial y}, \quad \text{and}$$

   $$\frac{\partial h}{\partial t} = -H\left(\frac{\partial u}{\partial x} + \frac{\partial v}{\partial y}\right) - \eta\left(\frac{\partial u}{\partial x} + \frac{\partial v}{\partial y}\right) - u\frac{\partial \eta}{\partial x} - v\frac{\partial \eta}{\partial y}.$$

10. Write down the two-dimensional nonlinear shallow water equations without rotation and having a bottom topography given by $\eta(x,y)$.

    Answer: The two-dimensional nonlinear shallow water equations without rotation and having a bottom topography $\eta(x,y)$ are

    $$\frac{\partial u}{\partial t} + u\frac{\partial u}{\partial x} + v\frac{\partial u}{\partial y} = -g\frac{\partial h}{\partial x}, \qquad \frac{\partial v}{\partial t} + u\frac{\partial v}{\partial x} + v\frac{\partial v}{\partial y} = -g\frac{\partial h}{\partial y}, \quad \text{and}$$

    $$\frac{\partial h}{\partial t} + \frac{\partial}{\partial x}[(h+\eta)u] + \frac{\partial}{\partial y}[(h+\eta)v] = 0.$$

11. Write down the two-dimensional nonlinear shallow water equations with rotation and having a bottom topography given by $\eta(x,y)$.

    Answer: The two-dimensional nonlinear shallow water equations with rotation and having a bottom topography $\eta(x,y)$ are

    $$\frac{\partial u}{\partial t} + u\frac{\partial u}{\partial x} + v\frac{\partial u}{\partial y} - fv = -g\frac{\partial h}{\partial x}, \qquad \frac{\partial v}{\partial t} + u\frac{\partial v}{\partial x} + v\frac{\partial v}{\partial y} + fu = -g\frac{\partial h}{\partial y}, \quad \text{and}$$

    $$\frac{\partial h}{\partial t} + \frac{\partial}{\partial x}[(h+\eta)u] + \frac{\partial}{\partial y}[(h+\eta)v] = 0.$$

12. Write down the linear one-dimensional shallow water equations without rotation that also includes the advection terms with constant zonal velocity u_o.

    Answer: The linear one-dimensional shallow water equations without rotation that also includes the advection terms are

    $$\frac{\partial u}{\partial t} + u_o\frac{\partial u}{\partial x} = -g\frac{\partial h}{\partial x} \qquad \text{and} \qquad \frac{\partial h}{\partial t} + u_o\frac{\partial h}{\partial x} = -H\frac{\partial u}{\partial x}.$$

13. Obtain the analytical dispersion relation for question Exercise 9a Q 12.

    Answer: The analytical dispersion relation is given by $c = u_o \pm \sqrt{gH}$.

14. If one employs leapfrog time integration scheme and replaces space derivatives by central differencing for the governing equations indicated in Exercise 9a Q 12, what is the stability of the aforementioned explicit finite difference scheme?

    Answer: The scheme that utilizes leapfrog time integration scheme and central finite differencing for the space derivatives to solve the equations indicated in Exercise 9a Q 12 is a conditionally stable scheme; the condition for the aforementioned scheme to be stable is given by

    $$\Delta t \leq \frac{\Delta x}{u_o + \sqrt{gH}}.$$

15. For the case where $H = 10\,\text{km}$, $u_o = 100\,\text{m/s}$, $\Delta x = 10^5\,\text{m}$, what is the phase speed of the gravity wave? What is the maximum permissible time step that one can use for Exercise 9a Q 14?

    Answer: For $H = 10\,\text{km}$, the phase speed of the gravity wave is $313\,\text{m/s}$. The maximum permissible time step that one can use for Exercise 9a Q 14 is $\Delta t \leq 4$ minutes. It is clear from this case, that the phase speed of the gravity wave mainly determines the maximum permissible time step.

16. For the splitting or the Marchuk method that is invoked to solve the linear one-dimensional shallow water equations without rotation and has an advection term with constant zonal velocity u_o, indicate the stability criteria that needs to be satisfied for each of the individual steps. Moreover, indicate the preferred ratio of the time step of the advection step to the time step of the gravity wave adjustment step.

    Answer: The linear one-dimensional shallow water equations without rotation and with an advection term is given in Exercise 9a Q 12. For stability of the advection step, the following condition needs to be satisfied, $u_o \Delta t / \Delta x \leq 1$, where Δt is the time step for the advective step. As advection is a relatively slower process, one can employ a larger value of the advective time step, i.e., larger value of Δt. For stability of the gravity wave adjustment step, the following condition needs to be satisfied, $c \delta t / \Delta x \leq 1$ where δt is the time step for the gravity wave term adjustment step. As the "gravity wave adjustment" is a faster process, one is forced to employ a smaller value of the gravity wave adjustment time step, i.e., smaller value of δt. Because the phase speed of the gravity wave is typically three times the phase speed of the advection process, it is appropriate to take the ratio of the Δt to δt to be '3' and to have "three gravity adjustment steps" for every advective step.

Exercises 9b (Questions only)

1. Consider the two-dimensional linear shallow water equations without rotation. Assuming forward difference in time, and applying the trapezoidal rule for the gravity wave terms (pressure gradient terms in momentum equation and divergence term in continuity equation), show that while one of the solutions provides for neutral stationery solution ($\lambda = 1$), the amplification factor for the other two solutions are given by

$$\lambda = \frac{1 - B \pm 2i\sqrt{B}}{1 + B},$$

where

$$B = gh(\Delta t)^2 \left[\frac{\sin^2(k\Delta x/2)}{(\Delta x)^2} + \frac{\sin^2(l\Delta y/2)}{(\Delta y)^2} \right].$$

2. For the Exercise 9b Q 1, show that the semi-implicit method provides for stable and neutral solutions.

   Answer: As

   $$|\lambda| = 1 \quad \text{and} \quad |\lambda|^2 = \frac{(1-B)^2}{(1+B)^2} + \frac{4B}{(1+B)^2} = \frac{(1+B)^2}{(1+B)^2} = 1$$

   the semi-implicit method provides for stable and neutral solutions.

3. Write down the shallow water potential vorticity equation?

4. Write down the shallow water equations in spherical coordinates?

5. Write down the shallow water equations in cylindrical coordinates?

Python examples

1. Solve the linear two-dimensional shallow water equations in a f plane

$$\frac{\partial u}{\partial t} - fv = -g\frac{\partial h}{\partial x},$$
$$\frac{\partial v}{\partial t} - fu = -g\frac{\partial h}{\partial y},$$
$$\frac{\partial h}{\partial t} = -H\left\{ \frac{\partial u}{\partial x} + \frac{\partial v}{\partial y} \right\},$$

where u and v are zonal and meridional components of velocity, h is the interface height, H is the mean depth of the fluid, f is the Coriolis parameter, which is considered a constant in the f-plane and g is the acceleration due to gravity.

The python code for solving linear shallow water equations in a beta plane is adapted from the code designed by Prof G Vallis. It was modified to solve the linear shallow water equations in a f plane with the initial condition corresponding to the following. The python code uses Arakawa C grid arrangement.

The initial condition considered is that $h(x,y,t=0) = h_o exp[-(x^2+y^2)/(L_w)^2]$, where $h_o = 1$ km, $L_w = 1/7$. The initial conditions for the zonal and meridional component of velocity are obtained from the geostrophic relationship:

$$v(x,y,t=0) = \frac{g}{f}\frac{\partial h}{\partial x}$$

$$u(x,y,t=0) = -\frac{g}{f}\frac{\partial h}{\partial y}$$

The code uses the following values for the various parameters, $\Delta x = (2 \times 10^4)/128$ km ;$\Delta y = (1 \times 10^4)/129$ km ; $\Delta t = 1000$ s; $f = 1 \times 10^{-5}$ s^{-1}

```python
#python code for Linear Shallow Water equations in f plane


from __future__ import (print_function, division)
import time


import numpy as np
import matplotlib.pyplot as plt


experiment = '2d'            # set to '1d' or '2d'
plot_interval = 20           # plot every n steps


## CONFIGURATION
### Domain
nx = 128
ny = 129


H  = 100.0           # [m]  Average depth of the fluid
Lx = 2.0e7           # [m]  Zonal width of domain
Ly = 1.0e7           # [m]  Meridional height of domain
delx = Lx/128
dely = Ly/129
boundary_condition = 'periodic'  # either 'periodic' or 'walls'
### Coriolis and Gravity
## Change Coriolis to zero and see the difference!
f0 = 1.0e-5    *1.    # [s^-1] f = f0 + beta y
#beta =  2.1*(10**-11)
beta=0
#f0 =0              # [m^-1.s^-1]
```

```python
g = 1.0                    # [m.s^-1]


### Diffusion and Friction
nu = 5.0e4                 # [m^2.s^-1] Coefficient of diffusion
r = 1.0e-4                 # Rayleigh damping at top and bottom of
                                 domain

dt = 1000.0                # Timestep [s]
_u = np.zeros((nx+3, ny+2))
_v = np.zeros((nx+2, ny+3))
_h = np.zeros((nx+2, ny+2))

u = _u[1:-1, 1:-1]                    # (nx+1, ny)
v = _v[1:-1, 1:-1]                    # (nx, ny+1)
h = _h[1:-1, 1:-1]                    # (nx, ny)

state = np.array([u, v, h])


dx = Lx / nx               # [m]
dy = Ly / ny               # [m]

# positions of the value points in [m]
ux = (-Lx/2 + np.arange(nx+1)*dx)[:, np.newaxis]
vx = (-Lx/2 + dx/2.0 + np.arange(nx)*dx)[:, np.newaxis]

vy = (-Ly/2 + np.arange(ny+1)*dy)[np.newaxis, :]
uy = (-Ly/2 + dy/2.0 + np.arange(ny)*dy)[np.newaxis, :]

hx = vx
hy = uy

t = 0.0                    # [s] Time since start of simulation
tc = 0                     # [1] Number of integration steps taken


def update_boundaries():
```

```python
# 1. Periodic Boundaries
#    - Flow cycles from left-right-left
#    - u[0] == u[nx]
if boundary_condition is 'periodic':
    _u[0, :] = _u[-3, :]
    _u[1, :] = _u[-2, :]
    _u[-1, :] = _u[2, :]
    _v[0, :] = _v[-2, :]
    _v[-1, :] = _v[1, :]
    _h[0, :] = _h[-2, :]
    _h[-1, :] = _h[1, :]



# This applied for both boundary cases above
for field in state:
    # Free-slip of all variables at the top and bottom
    field[:, 0] = field[:, 1]
    field[:, -1] = field[:, -2]

    # fix corners to be average of neighbours
    field[0, 0] =  0.5*(field[1, 0] + field[0, 1])
    field[-1, 0] = 0.5*(field[-2, 0] + field[-1, 1])
    field[0, -1] = 0.5*(field[1, -1] + field[0, -2])
    field[-1, -1] = 0.5*(field[-1, -2] + field[-2, -1])


def diffx(psi):
    """Calculate partial/partial x[psi] over a single grid
       square.

    i.e. d/dx(psi)[i,j] = (psi[i+1/2, j] - psi[i-1/2, j]) / dx

    The derivative is returned at x points at the midpoint
       between
    x points of the input array."""
    global dx
```

```
    return (psi[1:,:] - psi[:-1,:]) / dx


def diff2x(psi):
    """Calculate partial2/partial x2[psi] over a single grid
        square.

    i.e. d2/dx2(psi)[i,j] = (psi[i+1, j] - psi[i, j] + psi[i-1,
        j]) / dx^2

    The derivative is returned at the same x points as the
    x points of the input array, with dimension (nx-2, ny)."""
    global dx
    return (psi[:-2, :] - 2*psi[1:-1, :] + psi[2:, :]) / dx**2


def diff2y(psi):
    """Calculate partial 2/partial y2[psi] over a single grid
        square.

    i.e. d2/dy2(psi)[i,j] = (psi[i, j+1] - psi[i, j] + psi[i,
        j-1]) / dy^2

    The derivative is returned at the same y points as the
    y points of the input array, with dimension (nx, ny-2)."""
    global dy
    return (psi[:, :-2] - 2*psi[:, 1:-1] + psi[:, 2:]) / dy**2


def diffy(psi):
    """Calculate partial/partial y[psi] over a single grid
        square.

    i.e. d/dy(psi)[i,j] = (psi[i, j+1/2] - psi[i, j-1/2]) / dy

    The derivative is returned at y points at the midpoint
    between y points of the input array."""
    global dy
    return (psi[:, 1:] - psi[:,:-1]) / dy


def centre_average(phi):
```

```
    """Returns the four-point average at the centres between
        grid points."""
    return 0.25*(phi[:-1,:-1] + phi[:-1,1:] + phi[1:, :-1] +
        phi[1:,1:])


def y_average(phi):
    """Average adjacent values in the y dimension.
    If phi has shape (nx, ny), returns an array of shape (nx,
        ny - 1)."""
    return 0.5*(phi[:,:-1] + phi[:,1:])


def x_average(phi):
    """Average adjacent values in the x dimension.
    If phi has shape (nx, ny), returns an array of shape (nx -
        1, ny)."""
    return 0.5*(phi[:-1,:] + phi[1:,:])


def divergence():
    """Returns the horizontal divergence at h points."""
    return diffx(u) + diffy(v)


def del2(phi):
    """Returns the Laplacian of phi."""
    return diff2x(phi)[:, 1:-1] + diff2y(phi)[1:-1, :]


def uvatuv():
    """Calculate the value of u at v and v at u."""
    global _u, _v
    ubar = centre_average(_u)[1:-1, :]
    vbar = centre_average(_v)[:, 1:-1]
    return ubar, vbar


def uvath():
    global u, v
    ubar = x_average(u)
    vbar = y_average(v)
    return ubar, vbar
```

```python
def absmax(psi):
    return np.max(np.abs(psi))



## DYNAMICS
# These functions calculate the dynamics of the system we are
    interested in
def forcing():
    """Add some external forcing terms to the u, v and h
    equations. This function should return a state array (du,
    dv, dh) that will be added to the RHS of equations (1),
    (2) and (3) when they are numerically integrated."""
    global u, v, h
    du = np.zeros_like(u)
    dv = np.zeros_like(v)
    dh = np.zeros_like(h)
    # Calculate some forcing terms here...
    return np.array([du, dv, dh])


sponge_ny = ny//7
sponge = np.exp(-np.linspace(0, 5, sponge_ny))
def damping(var):
    # sponges are active at the top and bottom of the domain by
        applying Rayleigh friction
    # with exponential decay towards the centre of the domain
    global sponge, sponge_ny
    var_sponge = np.zeros_like(var)
    var_sponge[:, :sponge_ny] = sponge[np.newaxis, :]
    var_sponge[:, -sponge_ny:] = sponge[np.newaxis, ::-1]
    return var_sponge*var


def rhs():
    """Calculate the right hand side of the u, v and h
    equations."""
    u_at_v, v_at_u = uvatuv()    # (nx, ny+1), (nx+1, ny)

    # the height equation
    h_rhs = -H*divergence() + nu*del2(_h) - r*damping(h)
```

```
    # the u equation
    dhdx = diffx(_h)[:, 1:-1]   # (nx+1, ny)
    u_rhs = (f0 + beta*uy)*v_at_u - g*dhdx + nu*del2(_u) -
        r*damping(u)

    # the v equation
    dhdy  = diffy(_h)[1:-1, :]    # (nx, ny+1)
    v_rhs = -(f0 + beta*vy)*u_at_v - g*dhdy + nu*del2(_v) -
        r*damping(v)

    return np.array([u_rhs, v_rhs, h_rhs]) + forcing()

_ppdstate, _pdstate = 0,0
def step():
    global dt, t, tc, _ppdstate, _pdstate

    update_boundaries()

    dstate = rhs()

    # take adams-bashforth step in time
    if tc==0:
        # forward euler
        dt1 = dt
        dt2 = 0.0
        dt3 = 0.0
    elif tc==1:
        # AB2 at step 2
        dt1 = 1.5*dt
        dt2 = -0.5*dt
        dt3 = 0.0
    else:
        # AB3 from step 3 on
        dt1 = 23./12.*dt
        dt2 = -16./12.*dt
        dt3 = 5./12.*dt
```

```
        newstate = state + dt1*dstate + dt2*_pdstate + dt3*_ppdstate
        u[:], v[:], h[:] = newstate
        _ppdstate = _pdstate
        _pdstate = dstate

        t   += dt
        tc += 1


## INITIAL CONDITIONS
f1=1*(10**-4)

g1 = 9.8
L=1/7
x1=np.zeros(int(nx)+1)
y1=np.zeros(int(ny)+1)
for i in range(1,int(nx)+1):
    x1[i]=delx*(i-1)
 #print(x)
for i in range(1,int(ny)+1):
    y1[i]=dely*(i-1)
h_init = np.zeros(((int(nx),int(ny))))
u_init = np.zeros(((int(ny),int(ny))))
v_init = np.zeros(((int(nx),int(130))))
h_o=1000
h_init[64,64] =h_o
u_init[64,64]=0
v_init[64,64]=0
for i in range(0,64):
    for j in range(0,63):
        h_init[i,j] =h_o*np.exp(-(x1[64-i]/L)**2-(y1[64-j]/L)**2)

for i in range(65,127):
    for j in range(0,64):
        h_init[i,j] =h_o*np.exp(-(x1[i-64]/L)**2-(y1[64-j]/L)**2)
for i in range(65,127):
    for j in range(65,128):
        h_init[i,j] =h_o*np.exp(-(x1[i-64]/L)**2-(y1[j-64]/L)**2)
```

```
for i in range(0,64):
    for j in range(65,128):
        h_init[i,j] =h_o*np.exp(-(x1[64-i]/L)**2-(y1[j-64]/L)**2)
#u_init
for i in range(0,64):
    for j in range(0,63):
        u_init[i,j] =(g1/f1)*(-2*x1[64-i]/(L**2))*h_o*np.
            exp(-(x1[64-i]/L)**2-(y1[64-j]/L)**2)


for i in range(65,128):
    for j in range(0,64):
        u_init[i,j] =(g1/f1)*(-2*x1[i-64]/(L**2))*h_o*np.
            exp(-(x1[i-64]/L)**2-(y1[64-j]/L)**2)
for i in range(65,128):
    for j in range(65,128):
        u_init[i,j] =(g1/f1)*(-2*x1[i-64]/(L**2))*h_o*np.
            exp(-(x1[i-64]/L)**2-(y1[j-64]/L)**2)
for i in range(0,64):
    for j in range(65,128):
        u_init[i,j] =(g1/f1)*(-2*x1[64-i]/(L**2))*h_o*np.
            exp(-(x1[64-i]/L)**2-(y1[j-64]/L)**2)
#v_init
for i in range(0,64):
    for j in range(0,63):
        v_init[i,j] =(g1/f1)*(-2*y1[64-j]/(L**2))*h_o*np.
            exp(-(x1[64-i]/L)**2-(y1[64-j]/L)**2)


for i in range(65,127):
    for j in range(0,64):
        v_init[i,j] =(g1/f1)*(-2*y1[64-j]/(L**2))*h_o*np.
            exp(-(x1[i-64]/L)**2-(y1[64-j]/L)**2)
for i in range(65,127):
    for j in range(65,129):
        v_init[i,j] =(g1/f1)*(-2*y1[j-64]/(L**2))*h_o*np.
            exp(-(x1[i-64]/L)**2-(y1[j-64]/L)**2)
for i in range(0,64):
    for j in range(65,129):
        v_init[i,j] =(g1/f1)*(-2*y1[j-64]/(L**2))*h_o*np.
```

```
            exp(-(x1[64-i]/L)**2-(y1[j-64]/L)**2)
# Set the initial state of the model here by assigning to u[:],
    v[:] and h[:].
if experiment is '2d':
    h0 = h_init
    u0 = u_init
    v0 = v_init


# set the variable fields to the initial conditions
u[:] = u0
v[:] = v0
h[:] = h0
plt.ion()
fig = plt.figure(figsize=(8*Lx/Ly, 8))
# create a set of color levels with a slightly larger neutral
    zone about 0
nc = 12
colorlevels = np.concatenate([np.linspace(-1, -.05, nc),
    np.linspace(.05, 1, nc)])


def plot_all(u,v,h):
    hmax = np.max(np.abs(h))
    plt.clf()
    plt.plot()
    X, Y = np.meshgrid(hx, hy)
    plt.contourf(X/Lx, Y/Ly, h.T, cmap=plt.cm.RdBu,
        levels=colorlevels*absmax(h))
    plt.title('h')
    plt.plot()
    plt.savefig('Shallow_water_equation_at_time'+np.str(t)+'dt'
        +np.str(dt)+'.png',dpi=300)
    plt.pause(0.001)
    plt.draw()


im = None
def plot_fast(u,v,h):
    # only plots an imshow of h, much faster than contour maps
    global im
```

```
    if im is None:
        im = plt.imshow(h.T, aspect=Ly/Lx, cmap=plt.cm.RdBu,
            interpolation='bicubic')
        im.set_clim(-absmax(h), absmax(h))
    else:
        im.set_array(h.T)
        im.set_clim(-absmax(h), absmax(h))
    plt.pause(0.001)
    plt.draw()


plot = plot_all



## RUN
# Run the simulation and plot the state
c = time.clock()
nsteps = 1000
for i in range(nsteps):
    step()
    if i % plot_interval == 0:
        plot(*state)
        print('[t={:7.2f} u: [{:.3f}, {:.3f}], v: [{:.3f},
                    {:.3f}], h: [{:.3f}, {:.2f}]'.format(
            t/86400,
            u.min(), u.max(),
            v.min(), v.max(),
            h.min(), h.max()))
        #print('fps: %r' % (tc / (time.clock()-c)))
```

h

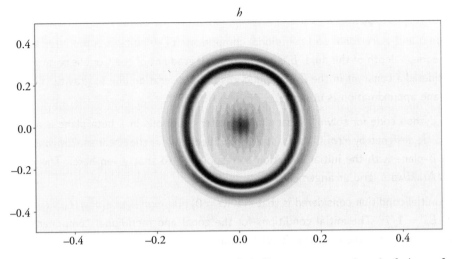

Figure 9.2 Geostrophic height in two-dimensional shallow water equations in f plane after 320 steps.

h

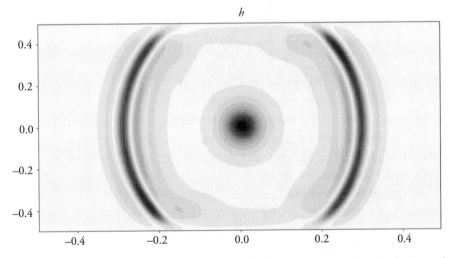

Figure 9.3 Geostrophic height in two-dimensional shallow water equations in f plane after 640 steps.

2. Solve the linear two-dimensional shallow water equations in a beta plane

$$\frac{\partial u}{\partial t} - \beta y v = -g \frac{\partial h}{\partial x},$$

$$\frac{\partial v}{\partial t} + \beta y u = -g \frac{\partial h}{\partial y},$$

$\frac{\partial h}{\partial t} = -H \left\{ \frac{\partial u}{\partial x} + \frac{\partial v}{\partial y} \right\},$

where u and v are zonal and meridional components of velocity, h is the interface height, H is the mean depth of the fluid, β is the meridional gradient of the Coriolis parameter which is considered a constant in the β-plane and g is the acceleration due to gravity. The equatorial β-plane approximation is utilized.

The python code for solving linear shallow water equations in a beta plane is adapted from the code designed by Prof G Vallis. It was modified to solve the linear shallow water equations in a β-plane with the initial condition corresponding to that given here. The python code uses Arakawa C grid arrangement.

The initial condition considered is that $h(x,y,t=0) = h_o exp[-(x^2+y^2)/(L_w)^2]$, where $h_o = 1$ km, $L_w = 1/7$. The initial conditions for the zonal and meridional component of velocity are obtained from the geostrophic relationship

$v(x,y,t=0) = \frac{g}{\beta y} \frac{\partial h}{\partial x},$

$u(x,y,t=0) = -\frac{g}{\beta y} \frac{\partial h}{\partial y}.$

The code uses the following values for the various parameters, $\Delta x = (2 \times 10^4)/128$ km ; $\Delta y = (1 \times 10^4)/129$ km ; $\Delta t = 1000$ s; $\beta = 2.1 \times 10^{-11}$

```
#python code for Linear Shallow Water equations in beta plane


"""Linear Shallow Water Model
-   Typical use: geostrophic adjustment.
-   Original code by James Penn, modified by G. K. Valis
-   Procedural (non-object oriented) version of code
- Two dimensional shallow water in a rotating frame
- Staggered Arakawa-C grid
- fixed boundary conditions in the y-dimension (free slip)
- Linearised about a fluid depth H and u = 0


Dimensions (SI units) are implied via values of constants e.g.
Lx is the width of the domain in metres [m], however, there is
no code dependency on using specific units.  If all input values
are scaled appropriately other units may be used.


eta = H + h


partial/partial t[u] - fv = - g partial/partial x[h] + F      (1)
partial/partial t[v] + fu = - g partial/partial y[h] + F      (2)
```

```
partial/partial t[h] + H(partial/partial x[u] + partial/partial
    y[v]) = F                               (3)


f = f0 + partial y
F is a forcing, default = (0, 0, 0)
"""


from __future__ import (print_function, division)
import time

import numpy as np
import matplotlib.pyplot as plt


experiment = '2d'              # set to '1d' or '2d'
plot_interval = 20             # plot every n steps


## CONFIGURATION
### Domain
nx = 128
ny = 129


H  = 100.0            # [m]  Average depth of the fluid
Lx = 2.0e7            # [m]  Zonal width of domain
Ly = 1.0e7            # [m]  Meridional height of domain
delx = Lx/128
dely = Ly/129
boundary_condition = 'periodic'   # either 'periodic' or 'walls'



### Coriolis and Gravity
## Change Coriolis to zero and see the difference!
#f0 = 1.0e-5    *1.   # [s^-1] f = f0 + beta y
beta =   2.1*(10**-11)
f0 =0                 # [m^-1.s^-1]
g = 1.0               # [m.s^-1]


### Diffusion and Friction
nu = 5.0e4           # [m^2.s^-1] Coefficient of diffusion
```

```
r = 1.0e-4          # Rayleigh damping at top and bottom of domain

dt = 1000.0         # Timestep [s]


_u = np.zeros((nx+3, ny+2))
_v = np.zeros((nx+2, ny+3))
_h = np.zeros((nx+2, ny+2))

u = _u[1:-1, 1:-1]                    # (nx+1, ny)
v = _v[1:-1, 1:-1]                    # (nx, ny+1)
h = _h[1:-1, 1:-1]                    # (nx, ny)

state = np.array([u, v, h])


dx = Lx / nx              # [m]
dy = Ly / ny              # [m]

# positions of the value points in [m]
ux = (-Lx/2 + np.arange(nx+1)*dx)[:, np.newaxis]
vx = (-Lx/2 + dx/2.0 + np.arange(nx)*dx)[:, np.newaxis]

vy = (-Ly/2 + np.arange(ny+1)*dy)[np.newaxis, :]
uy = (-Ly/2 + dy/2.0 + np.arange(ny)*dy)[np.newaxis, :]

hx = vx
hy = uy

t = 0.0                   # [s] Time since start of simulation
tc = 0                    # [1] Number of integration steps taken


## GRID FUNCTIONS
# These functions perform calculations on the grid such as
    calculating
# derivatives of fields or setting boundary conditions
```

```python
def update_boundaries():

    # 1. Periodic Boundaries
    #    - Flow cycles from left-right-left
    #    - u[0] == u[nx]
    if boundary_condition is 'periodic':
        _u[0, :] = _u[-3, :]
        _u[1, :] = _u[-2, :]
        _u[-1, :] = _u[2, :]
        _v[0, :] = _v[-2, :]
        _v[-1, :] = _v[1, :]
        _h[0, :] = _h[-2, :]
        _h[-1, :] = _h[1, :]


    # This applied for both boundary cases above
    for field in state:
        # Free-slip of all variables at the top and bottom
        field[:, 0] = field[:, 1]
        field[:, -1] = field[:, -2]

        # fix corners to be average of neighbours
        field[0, 0] = 0.5*(field[1, 0] + field[0, 1])
        field[-1, 0] = 0.5*(field[-2, 0] + field[-1, 1])
        field[0, -1] = 0.5*(field[1, -1] + field[0, -2])
        field[-1, -1] = 0.5*(field[-1, -2] + field[-2, -1])


def diffx(psi):
    """Calculate partial/partial x[psi] over a single grid
        square.

    i.e. d/dx(psi)[i,j] = (psi[i+1/2, j] - psi[i-1/2, j]) / dx

    The derivative is returned at x points at the midpoint
        between
    x points of the input array."""
    global dx
```

```
    return (psi[1:,:] - psi[:-1,:]) / dx


def diff2x(psi):
    """Calculate partial2/partial x2[psi] over a single grid
       square.

    i.e. d2/dx2(psi)[i,j] = (psi[i+1, j] - psi[i, j] + psi[i-1,
       j]) / dx^2

    The derivative is returned at the same x points as the
    x points of the input array, with dimension (nx-2, ny)."""
    global dx
    return (psi[:-2, :] - 2*psi[1:-1, :] + psi[2:, :]) / dx**2


def diff2y(psi):
    """Calculate partial 2/partial y2[psi] over a single grid
       square.

    i.e. d2/dy2(psi)[i,j] = (psi[i, j+1] - psi[i, j] + psi[i,
       j-1]) / dy^2

    The derivative is returned at the same y points as the
    y points of the input array, with dimension (nx, ny-2)."""
    global dy
    return (psi[:, :-2] - 2*psi[:, 1:-1] + psi[:, 2:]) / dy**2


def diffy(psi):
    """Calculate partial/partial y[psi] over a single grid
       square.

    i.e. d/dy(psi)[i,j] = (psi[i, j+1/2] - psi[i, j-1/2]) / dy

    The derivative is returned at y points at the midpoint
    between y points of the input array."""
    global dy
    return (psi[:, 1:] - psi[:,:-1]) / dy


def centre_average(phi):
```

```python
    """Returns the four-point average at the centres between
        grid points."""
    return 0.25*(phi[:-1,:-1] + phi[:-1,1:] + phi[1:, :-1] +
        phi[1:,1:])


def y_average(phi):
    """Average adjacent values in the y dimension.
    If phi has shape (nx, ny), returns an array of shape (nx,
        ny - 1)."""
    return 0.5*(phi[:,:-1] + phi[:,1:])


def x_average(phi):
    """Average adjacent values in the x dimension.
    If phi has shape (nx, ny), returns an array of shape (nx -
        1, ny)."""
    return 0.5*(phi[:-1,:] + phi[1:,:])


def divergence():
    """Returns the horizontal divergence at h points."""
    return diffx(u) + diffy(v)


def del2(phi):
    """Returns the Laplacian of phi."""
    return diff2x(phi)[:, 1:-1] + diff2y(phi)[1:-1, :]


def uvatuv():
    """Calculate the value of u at v and v at u."""
    global _u, _v
    ubar = centre_average(_u)[1:-1, :]
    vbar = centre_average(_v)[:, 1:-1]
    return ubar, vbar


def uvath():
    global u, v
    ubar = x_average(u)
    vbar = y_average(v)
    return ubar, vbar
```

```python
def absmax(psi):
    return np.max(np.abs(psi))



## DYNAMICS
# These functions calculate the dynamics of the system we are
#     interested in
def forcing():
    """Add some external forcing terms to the u, v and h
    equations. This function should return a state array (du,
    dv, dh) that will be added to the RHS of equations (1),
    (2) and (3) when they are numerically integrated."""
    global u, v, h
    du = np.zeros_like(u)
    dv = np.zeros_like(v)
    dh = np.zeros_like(h)
    # Calculate some forcing terms here...
    return np.array([du, dv, dh])


sponge_ny = ny//7
sponge = np.exp(-np.linspace(0, 5, sponge_ny))
def damping(var):
    # sponges are active at the top and bottom of the domain by
    #     applying Rayleigh friction
    # with exponential decay towards the centre of the domain
    global sponge, sponge_ny
    var_sponge = np.zeros_like(var)
    var_sponge[:, :sponge_ny] = sponge[np.newaxis, :]
    var_sponge[:, -sponge_ny:] = sponge[np.newaxis, ::-1]
    return var_sponge*var


def rhs():
    """Calculate the right hand side of the u, v and h
        equations."""
    u_at_v, v_at_u = uvatuv()    # (nx, ny+1), (nx+1, ny)

    # the height equation
    h_rhs = -H*divergence() + nu*del2(_h) - r*damping(h)
```

```
    # the u equation
    dhdx = diffx(_h)[:, 1:-1]   # (nx+1, ny)
    u_rhs = (f0 + beta*uy)*v_at_u - g*dhdx + nu*del2(_u) -
        r*damping(u)

    # the v equation
    dhdy = diffy(_h)[1:-1, :]    # (nx, ny+1)
    v_rhs = -(f0 + beta*vy)*u_at_v - g*dhdy + nu*del2(_v) -
        r*damping(v)

    return np.array([u_rhs, v_rhs, h_rhs]) + forcing()

_ppdstate, _pdstate = 0,0
def step():
    global dt, t, tc, _ppdstate, _pdstate

    update_boundaries()

    dstate = rhs()

    # take adams-bashforth step in time
    if tc==0:
        # forward euler
        dt1 = dt
        dt2 = 0.0
        dt3 = 0.0
    elif tc==1:
        # AB2 at step 2
        dt1 = 1.5*dt
        dt2 = -0.5*dt
        dt3 = 0.0
    else:
        # AB3 from step 3 on
        dt1 = 23./12.*dt
        dt2 = -16./12.*dt
        dt3 = 5./12.*dt
```

```
        newstate = state + dt1*dstate + dt2*_pdstate + dt3*_ppdstate
        u[:], v[:], h[:] = newstate
        _ppdstate = _pdstate
        _pdstate = dstate

        t   += dt
        tc += 1


## INITIAL CONDITIONS
f1=1*(10**-4)

g1 = 9.8
L=1/7
x1=np.zeros(int(nx)+1)
y1=np.zeros(int(ny)+1)
for i in range(1,int(nx)+1):
    x1[i]=delx*(i-1)
 #print(x)
for i in range(1,int(ny)+1):
    y1[i]=dely*(i-1)
h_init = np.zeros((int(nx),int(ny)))
u_init = np.zeros((int(ny),int(ny)))
v_init = np.zeros((int(nx),int(130)))
h_o=1000
h_init[64,64] =h_o
u_init[64,64]=0
v_init[64,64]=0
for i in range(0,64):
    for j in range(0,63):
        h_init[i,j] =h_o*np.exp(-(x1[64-i]/L)**2-(y1[64-j]/L)**2)

for i in range(65,127):
    for j in range(0,64):
        h_init[i,j] =h_o*np.exp(-(x1[i-64]/L)**2-(y1[64-j]/L)**2)
for i in range(65,127):
    for j in range(65,128):
        h_init[i,j] =h_o*np.exp(-(x1[i-64]/L)**2-(y1[j-64]/L)**2)
```

```
for i in range(0,64):
    for j in range(65,128):
        h_init[i,j] =h_o*np.exp(-(x1[64-i]/L)**2-(y1[j-64]/L)**2)
#u_init
for i in range(0,64):
    for j in range(0,63):
        u_init[i,j] =(g1/f1)*(-2*x1[64-i]/(L**2))*h_o*np.
            exp(-(x1[64-i]/L)**2-(y1[64-j]/L)**2)


for i in range(65,128):
    for j in range(0,64):
        u_init[i,j] =(g1/f1)*(-2*x1[i-64]/(L**2))*h_o*np.
        exp(-(x1[i-64]/L)**2-(y1[64-j]/L)**2)
for i in range(65,128):
    for j in range(65,128):
        u_init[i,j] =(g1/f1)*(-2*x1[i-64]/(L**2))*h_o*np.
            exp(-(x1[i-64]/L)**2-(y1[j-64]/L)**2)
for i in range(0,64):
    for j in range(65,128):
        u_init[i,j] =(g1/f1)*(-2*x1[64-i]/(L**2))*h_o*np.
            exp(-(x1[64-i]/L)**2-(y1[j-64]/L)**2)
#v_init
for i in range(0,64):
    for j in range(0,63):
        v_init[i,j] =(g1/f1)*(-2*y1[64-j]/(L**2))*h_o*np.
            exp(-(x1[64-i]/L)**2-(y1[64-j]/L)**2)


for i in range(65,127):
    for j in range(0,64):
        v_init[i,j] =(g1/f1)*(-2*y1[64-j]/(L**2))*h_o*np.
            exp(-(x1[i-64]/L)**2-(y1[64-j]/L)**2)
for i in range(65,127):
    for j in range(65,129):
        v_init[i,j] =(g1/f1)*(-2*y1[j-64]/(L**2))*h_o*np.
            exp(-(x1[i-64]/L)**2-(y1[j-64]/L)**2)
for i in range(0,64):
    for j in range(65,129):
        v_init[i,j] =(g1/f1)*(-2*y1[j-64]/(L**2))*h_o*np.
```

```
                 exp(-(x1[64-i]/L)**2-(y1[j-64]/L)**2)
# Set the initial state of the model here by assigning to u[:],
    v[:] and h[:].
if experiment is '2d':

    h0 = h_init
    u0 = u_init
    v0 = v_init


# set the variable fields to the initial conditions
u[:] = u0
v[:] = v0
h[:] = h0


## PLOTTING
# Create several functions for displaying current state of the
    simulation
# Only one is used at a time - this is assigned to 'plot'
plt.ion()                           # allow realtime updates to plots
fig = plt.figure(figsize=(8*Lx/Ly, 8))   # create a figure with
    correct aspect ratio


# create a set of color levels with a slightly larger neutral
    zone about 0
nc = 12
colorlevels = np.concatenate([np.linspace(-1, -.05, nc),
    np.linspace(.05, 1, nc)])


def plot_all(u,v,h):
    hmax = np.max(np.abs(h))

    plt.plot()
    X, Y = np.meshgrid(hx, hy)
    plt.contourf(X/Lx, Y/Ly, h.T, cmap=plt.cm.RdBu,
        levels=colorlevels*absmax(h))
    #plt.colorbar()
    plt.title('h')
    plt.savefig('Shallow_water_equation_at_time'+np.str(t)+'dt'+
```

```
                        np.str(dt)+'.png',dpi=300)
        plt.pause(0.001)
        plt.draw()


im = None
def plot_fast(u,v,h):
    # only plots an imshow of h, much faster than contour maps
    global im
    if im is None:
        im = plt.imshow(h.T, aspect=Ly/Lx, cmap=plt.cm.RdBu,
            interpolation='bicubic')
        im.set_clim(-absmax(h), absmax(h))
    else:
        im.set_array(h.T)
        im.set_clim(-absmax(h), absmax(h))
    plt.pause(0.001)
    plt.draw()




plot = plot_all




## RUN
# Run the simulation and plot the state
c = time.clock()
nsteps = 1000
for i in range(nsteps):
    step()
    if i % plot_interval == 0:
        plot(*state)
        print('[t={:7.2f} u: [{:.3f}, {:.3f}], v: [{:.3f},
                      {:.3f}], h: [{:.3f}, {:.2f}]'.format(
            t/86400,
            u.min(), u.max(),
            v.min(), v.max(),
            h.min(), h.max()))
```

```
#print('fps: %r' % (tc / (time.clock()-c)))
```

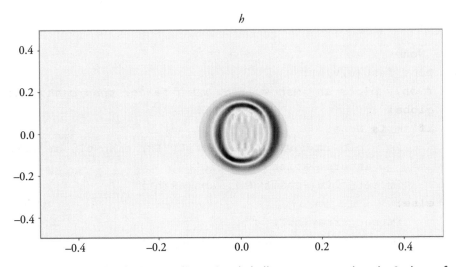

Figure 9.4 Geostrophic height in two-dimensional shallow water equations in β plane after 160 steps.

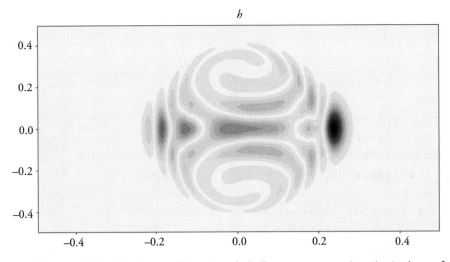

Figure 9.5 Geostrophic height in two-dimensional shallow water equations in β plane after 480 steps.

10

Numerical Methods for Solving Barotropic Equations

10.1 Introduction

Barotropic fluid is defined as a fluid whose density ρ is a function of pressure p only, i.e., $\rho = \rho(p)$. Examples of barotropic fluids are those that are homogeneous fluids (having constant and uniform density), an isothermal ideal gas (having constant temperature), or an isentropic ideal gas (having constant specific entropy). A fluid that is not barotropic is called a baroclinic fluid. For a baroclinic fluid, the fluid density depends on pressure and temperature as well as salinity and constituent concentration. For the barotropic fluid, as $\rho = \rho(p)$, lines of constant density (isopycnal) are parallel to lines of constant pressure (isobar) and, hence, there is no mechanism to generate relative vorticity (relative vorticity vector is defined as curl of relative velocity) in a barotropic fluid. Conversely, for a baroclinic fluid, the lines of constant density and constant pressure are no longer parallel. This gives rise to changes in relative vorticity and associated relative circulation as the net pressure force no longer passes through the center of mass of the fluid element in the case of baroclinic fluid.

As far as the atmosphere is concerned, it is relatively compressible. Hence, if the atmosphere is considered to be a barotropic fluid, it follows from the definition of barotropic fluid,that the only changes in density are brought about by changes in pressure. Moreover, a barotropic atmosphere would require that air temperature be horizontally constant, which leads to an absence of thermal wind and consequent vanishing of the vertical shear of the horizontal wind. As the tropics have relatively smaller horizontal air temperature gradients, the assumption of a barotropic atmosphere is a more reasonable assumption in the tropics than in the mid-latitudes.

As far as the oceans are concerned, water can be assumed to be pretty much incompressible. Although the oceans are stably stratified fluids (lighter water lies above heavier water, with water density increasing with depth), the vertical variation of density with depth over the oceans is very small ($\sim 4\,\text{kg/m}^3$) as compared to the density of the water ($1025\,\text{kg/m}^3$) itself so that the water density can be considered virtually to be a constant for a barotropic ocean. This means that there is no stratification (no variation of water density with depth) for a barotropic ocean. Conversely, if the ocean waters are baroclinic (water density depending on non-constant temperature, and also salinity), then there is stratification (variation of water density with depth).

10.2 Numerical Solution of a Non-divergent Barotropic Vorticity Equation on a β Plane – Linear Case

The barotropic equations describe "balanced" motion in a two-dimensional incompressible homogeneous fluid. The non-divergent linear barotropic vorticity equation on a beta plane is given by

$$\frac{\partial \zeta}{\partial t} + v\beta = 0, \tag{10.1}$$

where ζ is the vertical component of the relative vorticity vector, v is the meridional component of wind and β is the meridional gradient of the Coriolis parameter. Equation (10.1) states that the relative vorticity changes locally with time owing to advection of planetary vorticity. This equation is obtained from Equation (2.92) with the assumption that the nonlinear advection of the relative vorticity term is identically zero, i.e., $\vec{V} \cdot \nabla \zeta = 0$.

Equation (10.1) can be rewritten in terms of the stream function ψ with $\zeta = \nabla^2 \psi$; $v = \partial \psi / \partial x$ then becomes

$$\frac{\partial}{\partial t}(\nabla^2 \psi) = -\beta \frac{\partial \psi}{\partial x}. \tag{10.2}$$

The model domain may be either a doubly periodic domain, a channel or a bounded box with no normal flow on the sides, which means that the stream function is constant on the boundary

$$\psi = 0 \quad \text{on boundaries} \qquad \text{no-slip on solid boundaries} \tag{10.3}$$

$$\nabla \psi \cdot \hat{n} = 0 \quad \text{on boundaries} \qquad \text{no-penetration on solid boundaries} \tag{10.4}$$

where $\hat{n}$ is the outward directed unit normal.

Let $\chi = \partial \psi / \partial t$, then Equation (10.2) becomes

$$\nabla^2 \chi = -\beta \frac{\partial \psi}{\partial x}. \tag{10.5}$$

The following algorithm is utilized to solve Equation (10.5)

1. Solve the Poisson equation, Equation (10.5) to obtain χ at the initial time subject to the appropriate boundary conditions using the initial values of ψ over the region of interest using "successive over relaxation method (SOR)." Utilize the central difference scheme for the space derivative of ψ at the initial time in the RHS of Equation (10.5).

2. Utilize a suitable time differencing scheme, such as leapfrog time integration scheme to replace the time derivative of ψ to advance the ψ value to the next time over the region of interest. It is to be noted that for the first marching in time, one needs to utilize a forward in time difference scheme, which may be changed to the leapfrog time differencing scheme for later times (other than for the first time march).

3. Utilizing the ψ value at the next time over the region of interest, calculate the RHS of Equation (10.5) using central difference scheme for the space derivative of ψ and solve again the Poisson equation, Equation (10.5) using SOR method to obtain χ at the next time over the region of interest subject to the appropriate boundary conditions.

4. Increment the time index n and repeat steps (2) and (3) until the marching completes the entire simulation duration.

In effect, solving the linear non-divergent barotropic vorticity equation reduces to solving a two-dimensional Poisson equation using SOR method and updating the stream function ψ using a suitable time integration finite difference scheme.

An alternate way of solving Equation (10.1), which is the same as Equation (10.5) is to write the finite difference scheme for Equation (10.1) with the meridional component of wind v suitably replaced in terms of stream function ψ, as

$$\frac{\partial \zeta}{\partial t} = -\beta \frac{\partial \psi}{\partial x}. \tag{10.6}$$

Equation (10.6) can be rewritten in the finite difference form using leapfrog time difference scheme as

$$\zeta_i^{n+1} = \zeta_i^{n-1} - 2\Delta t (\beta \delta_x \overline{\psi}_j^x), \tag{10.7}$$

where the averaging and the derivative operators are defined as follows:

$$\overline{\psi}^x = \frac{\psi_{j+1/2} + \psi_{j-1/2}}{2},$$

(10.8)

$$\delta_x \psi = \frac{\psi_{j+1} - \psi_{j-1}}{2\Delta x}.$$

(10.9)

The following algorithm is utilized to solve the Equation (10.6):

1. Obtain the value of relative vorticity at the initial time over the region of interest from $\zeta = \nabla^2 \psi$, using appropriate space finite difference scheme, preferably the central finite difference scheme.

2. Calculate the RHS of Equation (10.6) at the initial time over the region of interest using the finite difference scheme as indicated in Equation (10.7).

3. Calculate the RHS of Equation (10.7) at the initial time over the region of interest

4. Advance the value of the relative vorticity to the next time using leapfrog time difference scheme; this will result in the knowledge of ζ_j^{n+1}.

5. Solve the Poisson equation $\nabla^2 \psi = \zeta$ at the next time using SOR method over the region of interest using appropriate boundary conditions.

6. Calculate the RHS of Equation (10.6) at the next time over the region of interest using ψ obtained at the next time using SOR method.

7. Calculate the RHS of Equation (10.7) at the next time over the region of interest.

8. Increment the time index n and repeat steps (4) to (7) until the marching completes the entire simulation duration.

10.3   Numerical Solution of a Non-divergent Barotropic Vorticity Equation on a f Plane – Nonlinear Case

The nonlinear non-divergent barotropic equations on a f plane is

$$\frac{\partial \zeta}{\partial t} + J(\psi, \zeta) = 0,$$

(10.10)

where ζ is the vertical component of the relative vorticity vector, ψ is the stream function and $J(\psi, \zeta)$ is the Jacobian operator, which is defined as

$$J(p,q) = \frac{\partial p}{\partial x}\frac{\partial q}{\partial y} - \frac{\partial p}{\partial y}\frac{\partial q}{\partial x}.$$

(10.11)

Equation (10.11) can also be written in the following flux forms

$$J(p,q) = \frac{\partial}{\partial x}\left(p\frac{\partial q}{\partial y}\right) - \frac{\partial}{\partial y}\left(p\frac{\partial q}{\partial x}\right) \quad \text{and} \tag{10.12}$$

$$J(p,q) = \frac{\partial}{\partial y}\left(q\frac{\partial p}{\partial x}\right) - \frac{\partial}{\partial x}\left(q\frac{\partial p}{\partial y}\right). \tag{10.13}$$

Because of these flux forms, if either p or q is either zero or constant on the boundary of a domain, then the domain integral of $\iint J(p,q)\,dxdy = 0$. Furthermore,

$$\langle p, J(p,q)\rangle = \left\langle J\left(\tfrac{1}{2}p^2, q\right)\right\rangle = 0. \tag{10.14}$$

$$\langle q, J(p,q)\rangle = \left\langle J\left(p, \tfrac{1}{2}q^2\right)\right\rangle = 0. \tag{10.15}$$

In the aforementioned case, p is ψ (a stream function) and q is ζ (the vertical component of relative vorticity). It has been shown that multiplying the vorticity equation by a stream function yields conservation of mean kinetic energy, while multiplying the vorticity equation by vorticity yields conservation of mean enstrophy.

In Section 4.11, we explained that to prevent nonlinear computational instability, Arakawa constructed a finite difference scheme that could conserve the mean values of the vorticity, kinetic energy and enstrophy; these same properties that the continuous governing nonlinear barotropic equations had conserved. Arakawa rightly deduced that construction of such a finite difference scheme would ensure that a systematic transfer of energy towards the highest wavenumbers would not occur and the nonlinear computational instability discovered by Phillips would be prevented. The method proposed by Arakawa for doing this is to find a form of the Jacobian term in Equation (10.10) that had the appropriate conservation properties exhibited by the continuous governing equations.

There are three alternate ways in which to write the Jacobian in Equation (10.10).

$$J_o^{++}(\psi, \zeta) = \frac{\partial\psi}{\partial x}\frac{\partial\zeta}{\partial y} - \frac{\partial\psi}{\partial y}\frac{\partial\zeta}{\partial x}. \tag{10.16}$$

$$J_o^{+\times}(\psi, \zeta) = \frac{\partial}{\partial x}\left(\psi\frac{\partial\zeta}{\partial y}\right) - \frac{\partial}{\partial y}\left(\psi\frac{\partial\zeta}{\partial x}\right). \tag{10.17}$$

$$J_o^{\times+}(\psi, \zeta) = \frac{\partial}{\partial y}\left(\zeta\frac{\partial\psi}{\partial x}\right) - \frac{\partial}{\partial x}\left(\zeta\frac{\partial\psi}{\partial y}\right). \tag{10.18}$$

Based on the aforementioned various expressions for Jacobians, one can consider several possible ways to construct second-order accurate finite difference approximations to the Jacobian. With the simplest centered space finite differencing, one requires values of ψ and ζ from a box of nine adjacent grid points to evaluate the Jacobians expressed in Equations (10.16) to (10.18) (refer to Figure 4.3 of Section 4.11). Let the grid size be equal in both x and y directions and let it be equal to d. Let ψ_n and ζ_n be the values of ψ and ζ at the point n.

Figure 10.1 shows the labels associated with a stencil of grid points involved in the finite difference expressions for the Arakawa Jacobian. Using these labels, the corresponding finite difference approximations to Equations (10.16)–(10.18) are as follows:

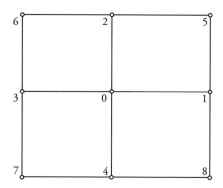

Figure 10.1   Stencil of grid points for evaluating the Arakawa Jacobian.

$$J_o^{++} = \frac{(\psi_1 - \psi_3)(\zeta_2 - \zeta_4) - (\psi_2 - \psi_4)(\zeta_1 - \zeta_3)}{4d^2}. \tag{10.19}$$

$$J_o^{+x} = \frac{\psi_1(\zeta_5 - \zeta_8) - \psi_3(\zeta_6 - \zeta_7) - \psi_2(\zeta_5 - \zeta_6) + \psi_4(\zeta_8 - \zeta_7)}{4d^2}. \tag{10.20}$$

$$J_o^{x+} = \frac{\zeta_2(\psi_5 - \psi_6) - \zeta_4(\psi_8 - \psi_7) - \zeta_1(\psi_5 - \psi_8) + \zeta_3(\psi_6 - \psi_7)}{4d^2}. \tag{10.21}$$

Arakawa (1960) showed that

1. J_o^{+x} conserves the integral of enstrophy, ζ^2,

2. $(J_o^{++} + J_o^{x+})/2$ conserves the integral of enstrophy, ζ^2,

3. J_o^{x+} conserves the integral of kinetic energy $\nabla\psi \cdot \nabla\psi$,

4. $(J_o^{++} + J_o^{+x})/2$ conserves the integral of kinetic energy $\nabla \psi \cdot \nabla \psi$,

5. $(J_o^{++} + J_o^{+x} + J_o^{x+})/3$ conserves the integral of enstrophy as well as integral of kinetic energy.

Equation (10.10) can be rewritten as follows:

$$\frac{\partial \zeta}{\partial t} + \left(\frac{\partial \psi}{\partial x} \frac{\partial \zeta}{\partial y} - \frac{\partial \psi}{\partial y} \frac{\partial \zeta}{\partial x} \right) = 0, \tag{10.22}$$

where the second and third terms in the LHS of Equation (10.22) correspond to the Jacobian operator $J(\psi, \zeta)$.

Equation (10.22) can be rewritten in the finite difference form using leapfrog time difference scheme as

$$\zeta_j^{n+1} = \zeta_j^{n-1} - 2\Delta t \left(\frac{J_o^{++} + J_o^{+x} + J_o^{x+}}{3} \right)_j^n, \tag{10.23}$$

a form that conserves the integral of enstrophy as well as integral of kinetic energy where $J_j^n = J(\psi_j^n, \zeta_j^n)$.

The following algorithm is utilized to solve Equation (10.23):

1. Obtain the value of relative vorticity at the initial time over the region of interest from $\zeta = \nabla^2 \psi$, using appropriate space finite difference scheme, preferably the central finite difference scheme.

2. Calculate the sum of the three Jacobian forms as seen in the RHS of Equation (10.23) at the initial time over the region of interest using the finite difference scheme as indicated in Equations (10.19)–(10.21).

3. Calculate the RHS of Equation (10.23) at the initial time over the region of interest.

4. Advance the value of the relative vorticity to the next time using leapfrog time difference scheme; this will result in the knowledge of ζ_j^{n+1}.

5. Solve the Poisson equation $\nabla^2 \psi = \zeta$ at the next time using the SOR method over the region of interest using appropriate boundary conditions.

6. Calculate the RHS of Equation (10.23) at the next time over the region of interest using the finite difference scheme as indicated in Equations (10.19)–(10.21).

7. Increment the time index n and repeat steps (4) to (6) until the marching completes the entire simulation duration.

10.4 Numerical Solution of a Non-divergent Barotropic Vorticity Equation on a β Plane – Nonlinear Case

The nonlinear non-divergent barotropic equations on a f plane is

$$\frac{\partial \zeta}{\partial t} + J(\psi, \zeta + f) = 0, \tag{10.24}$$

where ζ is the vertical component of the relative vorticity vector, ψ is the stream function, and $J(\psi, \zeta + f)$ is the Jacobian operator, which is defined as before. The only difference between Equations (10.10) and (10.24) is that in the latter there is an additional linear term βv in the vorticity equation due to advection of planetary vorticity. The solution of Equation (10.24) is very similar to solving Equation (10.10), except that the absolute vorticity $\zeta + f$ replaces ζ in Equations (10.16)–(10.21). In the β plane, these equations assume the following forms:

$$J_o^{++}(\psi, \zeta + f) = \frac{\partial \psi}{\partial x}\frac{\partial(\zeta + f)}{\partial y} - \frac{\partial \psi}{\partial y}\frac{\partial(\zeta + f)}{\partial x}. \tag{10.25}$$

$$J_o^{+x}(\psi, \zeta + f) = \frac{\partial}{\partial x}\left[\psi\frac{\partial(\zeta + f)}{\partial y}\right] - \frac{\partial}{\partial y}\left[\psi\frac{\partial(\zeta + f)}{\partial x}\right]. \tag{10.26}$$

$$J_o^{x+}(\psi, \zeta + f) = \frac{\partial}{\partial y}\left[(\zeta + f)\frac{\partial \psi}{\partial x}\right] - \frac{\partial}{\partial x}\left[(\zeta + f)\frac{\partial \psi}{\partial y}\right]. \tag{10.27}$$

Using these labels, (refer to Figure 10.1), the corresponding finite-difference approximations to Equations (10.25)–(10.27) are as follows:

$$J_o^{++} = \frac{(\psi_1 - \psi_3)[(\zeta + f)_2 - (\zeta + f)_4] - (\psi_2 - \psi_4)[(\zeta + f)_1 - (\zeta + f)_3]}{4d^2}. \tag{10.28}$$

$$J_o^{+x} = \frac{\psi_1[(\zeta + f)_5 - (\zeta + f)_8] - \psi_3[(\zeta + f)_6 - (\zeta + f)_7]}{4d^2} \\ - \frac{\psi_2[(\zeta + f)_5 - (\zeta + f)_6] - \psi_4[(\zeta + f)_8 - (\zeta + f)_7]}{4d^2}. \tag{10.29}$$

$$J_o^{x+} = \frac{(\zeta + f)_2(\psi_5 - \psi_6) - (\zeta + f)_4(\psi_8 - \psi_7) - (\zeta + f)_1(\psi_5 - \psi_8) + (\zeta + f)_3(\psi_6 - \psi_7)}{4d^2} \tag{10.30}$$

As before, Arakawa (1966) showed that the expressions for the three forms of Jacobians have the following properties:

1. J_o^{+x} conserves the integral of absolute enstrophy, $(\zeta + f)^2$.

2. $(J_o^{++} + J_o^{x+})/2$ conserves the integral of absolute enstrophy, $(\zeta + f)^2$.

3. J_o^{x+} conserves the integral of kinetic energy $\nabla\psi \cdot \nabla\psi$.

4. $(J_o^{++} + J_o^{+x})/2$ conserves the integral of kinetic energy $\nabla\psi \cdot \nabla\psi$.

5. $(J_o^{++} + J_o^{+x} + J_o^{x+})/3$ conserves the integral of absolute enstrophy as well as integral of kinetic energy.

Equation (10.24) can be rewritten as follows:

$$\frac{\partial \zeta}{\partial t} + \left(\frac{\partial \psi}{\partial x}\frac{\partial \zeta}{\partial y} - \frac{\partial \psi}{\partial y}\frac{\partial \zeta}{\partial x} \right) + \beta \frac{\partial \psi}{\partial x} = 0, \tag{10.31}$$

where the second and third terms in the LHS of Equation (10.31) correspond to the Jacobian operator $J(\psi, \zeta)$ and the last term in the LHS of Equation (10.31) corresponds to the advection of planetary vorticity.

Equation (10.31) can be rewritten in the finite difference form using the leapfrog time difference scheme as

$$\zeta_j^{n+1} = \zeta_j^{n-1} - 2\Delta t \left[\left(\frac{J_o^{++} + J_o^{+x} + J_o^{x+}}{3} \right)_j^n + \beta \delta_x \overline{\psi}_j^x \right], \tag{10.32}$$

a form that conserves the integral of absolute enstrophy as well as integral of energy, where $J_j^n = J(\psi_j^n, \zeta_j^n + f_j)$.

The following algorithm is utilized to solve Equation (10.32):

1. Obtain the value of relative vorticity at the initial time over the region of interest from $\zeta = \nabla^2 \psi$, using appropriate space finite difference scheme, preferably the central finite difference scheme.

2. Calculate the sum of the three Jacobian forms as seen in the RHS of Equation (10.32) at the initial time over the region of interest using the finite difference scheme as indicated in Equations (10.28)–(10.30).

3. Calculate the RHS of Equation (10.32) at the initial time over the region of interest.

4. Advance the value of the relative vorticity to the next time using leapfrog time difference scheme; this will result in the knowledge of ζ_j^{n+1}.

5. Solve the Poisson equation $\nabla^2 \psi = \zeta$ at the next time using the SOR method over the region of interest using appropriate boundary conditions.

6. Calculate the RHS of Equation (10.32) at the next time over the region of interest using the finite difference scheme as indicated in Equations (10.28)–(10.30).

7. Increment the time index n and repeat steps (4) to (6) until the marching completes the entire simulation duration.

10.5   Solving One-dimensional Linear Shallow Water Equations without Rotation

The one-dimensional linear shallow water equations without rotation is given by

$$\frac{\partial u}{\partial t} = -g\frac{\partial h}{\partial x}. \tag{10.33a}$$

$$\frac{\partial h}{\partial t} = -H\frac{\partial u}{\partial x}. \tag{10.33b}$$

Solution of Equations (10.33a) and (10.33b) can be performed easily using an explicit finite difference scheme and using leapfrog time integration scheme. The x-axis is discretized into a mesh having equal grid size Δx, i.e., $x_i = i\Delta x$, $i = 0, 1, 2, \ldots, M$. Moreover, time is discretized as follows: $t^n = n\Delta t$, $n = 0, 1, 2, \ldots, N$. The finite difference scheme of Equation (10.33b) is given by

$$h_i^{n+1} = h_i^{n-1} - 2\Delta t H\left(\frac{u_{i+1}^n - u_{i-1}^n}{2\Delta x}\right). \tag{10.34}$$

The finite difference expression of Equation (10.33a) is given by

$$u_i^{n+1} = u_i^{n-1} - 2\Delta t g\left(\frac{h_{i+1}^n - h_{i-1}^n}{2\Delta x}\right). \tag{10.35}$$

This scheme is conditionally stable with the requirement that the Courant number must be ≤ 1. As the leapfrog scheme is employed, the computational mode needs to be controlled using $2\Delta t$ and $2\Delta x$ filters.

10.6   Solving One-dimensional Linear Shallow Water Equations with Rotation

The one-dimensional linear shallow water equations with rotation is given by

$$\frac{\partial u}{\partial t} - fv = -g\frac{\partial h}{\partial x}, \tag{10.36}$$

$$\frac{\partial v}{\partial t} + fu = 0, \tag{10.37}$$

$$\frac{\partial h}{\partial t} = -H\frac{\partial u}{\partial x}, \tag{10.38}$$

Solution of Equations (10.36)–(10.38) can be performed easily using an explicit finite difference scheme and using the leapfrog time integration scheme. The x-axis is discretized into mesh having equal grid size Δx, i.e., $x_i = i\Delta x$, $i = 0, 1, 2, \ldots, M$. Moreover, time is discretized as follows: $t^n = n\Delta t$, $n = 0, 1, 2, \ldots, N$. The finite difference scheme of Equation (10.38) is given by

$$h_i^{n+1} = h_i^{n-1} - 2\Delta t H\left(\frac{u_{i+1}^n - u_{i-1}^n}{2\Delta x}\right). \tag{10.39}$$

The finite difference expression of the Equation (10.36) is given by

$$u_i^{n+1} = u_i^{n-1} - 2\Delta t\left[g\left(\frac{h_{i+1}^n - h_{i-1}^n}{2\Delta x} - fv_i^n\right)\right]. \tag{10.40}$$

The finite difference expression of Equation (10.37) is given by

$$v_i^{n+1} = v_i^{n-1} + 2\Delta t(-fu_i^n). \tag{10.41}$$

10.7 Solving One-dimensional Nonlinear Shallow Water Equations without Rotation

The one-dimensional nonlinear shallow water equations without rotation is given by

$$\frac{\partial u}{\partial t} + u\frac{\partial u}{\partial x} + g\frac{\partial h}{\partial x} = 0, \tag{10.42a}$$

$$\frac{\partial h}{\partial t} + u\frac{\partial h}{\partial x} + h\frac{\partial u}{\partial x} = 0. \tag{10.42b}$$

Equations (10.42) can be rewritten in flux form as follows:

$$\frac{\partial(hu)}{\partial t} + \frac{\partial}{\partial x}\left(hu^2 + \frac{1}{2}gh^2\right) = 0. \tag{10.43a}$$

$$\frac{\partial h}{\partial t} + \frac{\partial(uh)}{\partial x} = 0. \tag{10.43b}$$

Solution of Equations (10.43a) and (10.43b) can be performed using the following finite difference scheme. The x-axis is discretized into a mesh having equal grid size Δx, i.e., $x_i = i\Delta x$, $i = 0, 1, 2, \ldots, M$. Moreover, time is discretized as follows: $t^n = n\Delta t$, $n = 0, 1, 2, \ldots, N$. The finite difference scheme utilized for time is forward while that for space is central.

$$\left(\frac{\partial h}{\partial t}\right)_i^n = \frac{h_i^n - h_i^{n-1}}{\Delta t}. \tag{10.44}$$

$$\left[\frac{\partial (hu)}{\partial t}\right]_i^n = \left[h\frac{\partial u}{\partial t} + u\frac{\partial h}{\partial t}\right]_i^n = h_i^{n-1}\left(\frac{u_i^n - u_i^{n-1}}{\Delta t}\right) + u_i^{n-1}\left(\frac{h_i^n - h_i^{n-1}}{\Delta t}\right). \tag{10.45}$$

$$\left[\frac{\partial (hu)}{\partial x}\right]_i^n = \left[h\frac{\partial u}{\partial x} + u\frac{\partial h}{\partial x}\right]_i^n = h_i^{n-1}\left(\frac{u_{i+1}^n - u_{i-1}^n}{2\Delta x}\right) + u_i^{n-1}\left(\frac{h_{i+1}^n - h_{i-1}^n}{2\Delta x}\right). \tag{10.46}$$

$$\left[\frac{\partial (hu^2)}{\partial x}\right]_i^n = \left[2hu\frac{\partial u}{\partial x} + u^2\frac{\partial h}{\partial x}\right]_i^n = 2h_i^{n-1}u_i^{n-1}\left(\frac{u_{i+1}^n - u_{i-1}^n}{2\Delta x}\right) + (u^2)_i^{n-1}\left(\frac{h_{i+1}^n - h_{i-1}^n}{2\Delta x}\right). \tag{10.47}$$

$$\left[\frac{\partial}{\partial x}\left(\frac{1}{2}gh^2\right)\right]_i^n = gh_i^{n-1}\left(\frac{h_{i+1}^n - h_{i-1}^n}{2\Delta x}\right). \tag{10.48}$$

Using Equations (10.44) and (10.46), the finite difference expression for Equation (10.43b) becomes

$$\frac{h_i^n - h_i^{n-1}}{\Delta t} + h_i^{n-1}\left(\frac{u_{i+1}^n - u_{i-1}^n}{2\Delta x}\right) + u_i^{n-1}\left(\frac{h_{i+1}^n - h_{i-1}^n}{2\Delta x}\right) = 0. \tag{10.49}$$

Using Equations (10.45), (10.47), and (10.48), the finite difference expression for Equation (10.43a) becomes

$$h_i^{n-1}\left(\frac{u_i^n - u_i^{n-1}}{\Delta t}\right) + u_i^{n-1}\left(\frac{h_i^n - h_i^{n-1}}{\Delta t}\right) + 2h_i^{n-1}u_i^{n-1}\left(\frac{u_{i+1}^n - u_{i-1}^n}{2\Delta x}\right) + (u^2)_i^{n-1}\left(\frac{h_{i+1}^n - h_{i-1}^n}{2\Delta x}\right)$$
$$+ gh_i^{n-1}\left(\frac{h_{i+1}^n - h_{i-1}^n}{2\Delta x}\right). \tag{10.50}$$

Multiplying Equation (10.49) by u_i^{n-1}

$$u_i^{n-1}\frac{h_i^n - h_i^{n-1}}{\Delta t} + u_i^{n-1}h_i^{n-1}\left(\frac{u_{i+1}^n - u_{i-1}^n}{2\Delta x}\right) + (u^2)_i^{n-1}\left(\frac{h_{i+1}^n - h_{i-1}^n}{2\Delta x}\right) = 0. \tag{10.51}$$

Subtracting Equation (10.51) from Equation (10.50) gives the following equation,

$$h_i^{n-1}\frac{u_i^n - u_i^{n-1}}{\Delta t} + h_i^{n-1}u_i^{n-1}\left(\frac{u_{i+1}^n - u_{i-1}^n}{2\Delta x}\right) + gh_i^{n-1}\left(\frac{h_{i+1}^n - h_{i-1}^n}{2\Delta x}\right) = 0 \tag{10.52}$$

and

$$\frac{h_i^n - h_i^{n-1}}{\Delta t} + h_i^{n-1} \left(\frac{u_{i+1}^n - u_{i-1}^n}{2\Delta x} \right) + u_i^{n-1} \left(\frac{h_{i+1}^n - h_{i-1}^n}{2\Delta x} \right) = 0. \tag{10.49}$$

Equation (10.49) and (10.52) can be written as an algebraic system of linear equations of the form $Ax = b$ by keeping the known terms (u_i^{n-1}, h_i^{n-1}) on the RHS and unknown terms such as (u_i^n, h_i^n) on the LHS of the equation. Applying the initial and boundary conditions, the aforementioned linear system of algebraic equations can be solved.

10.8 Solving Two-dimensional Linear Shallow Water Equations without Rotation

The two-dimensional linear shallow water equations without rotation is given by

$$\frac{\partial u}{\partial t} = -g \frac{\partial h}{\partial x}, \tag{10.53a}$$

$$\frac{\partial v}{\partial t} = -g \frac{\partial h}{\partial y}, \tag{10.53b}$$

$$\frac{\partial h}{\partial t} = -H \left(\frac{\partial u}{\partial x} + \frac{\partial v}{\partial y} \right). \tag{10.53c}$$

Solution of Equations (10.53a)–(10.53c) can be performed easily using an explicit finite difference scheme and using the leapfrog time integration scheme. The x-axis is discretized into a mesh having equal grid size Δx and Δy, i.e., $x_i = i\Delta x$, $i = 0, 1, 2, \ldots, M_X$, $y_j = j\Delta y$, $j = 0, 1, 2, \ldots, M_Y$. Moreover, time is discretized as follows: $t^n = n\Delta t$, $n = 0, 1, 2, \ldots, N$. The finite difference scheme of Equation (10.53c) is given by

$$h_{i,j}^{n+1} = h_{i,j}^{n-1} - 2\Delta t H \left[\left(\frac{u_{i+1,j}^n - u_{i-1,j}^n}{2\Delta x} \right) + \left(\frac{v_{i,j+1}^n - v_{i,j-1}^n}{2\Delta y} \right) \right]. \tag{10.54}$$

The finite difference expression of Equations (10.53a) and (10.53b) are given by

$$u_{i,j}^{n+1} = u_{i,j}^{n-1} - 2\Delta t g \left(\frac{h_{i+1,j}^n - h_{i-1,j}^n}{2\Delta x} \right). \tag{10.55}$$

$$v_{i,j}^{n+1} = v_{i,j}^{n-1} - 2\Delta t g \left(\frac{h_{i,j+1}^n - h_{i,j-1}^n}{2\Delta y} \right). \tag{10.56}$$

The aforementioned scheme is conditionally stable with the requirement that the Courant number must be $\leq 1/\sqrt{2}$. As the leapfrog scheme is employed, the computational mode needs to be controlled using $2\Delta t$, $2\Delta x$, and $2\Delta y$ filters.

10.9 Solving Two-dimensional Linear Shallow Water Equations with Rotation on a β Plane

The two-dimensional linear shallow water equations with rotation on a β plane is given by

$$\frac{\partial u}{\partial t} - \beta y v = -g\frac{\partial h}{\partial x}, \tag{10.57a}$$

$$\frac{\partial v}{\partial t} + \beta y u = -g\frac{\partial h}{\partial y}, \tag{10.57b}$$

$$\frac{\partial h}{\partial t} = -H\left(\frac{\partial u}{\partial x} + \frac{\partial v}{\partial y}\right). \tag{10.57c}$$

Solution of Equations (10.57a)–(10.57c) can be performed easily using an explicit finite difference scheme and using the leapfrog time integration scheme. The x-axis can be discretized into a mesh having equal grid size Δx and Δy, i.e., $x_i = i\Delta x$, $i = 0, 1, 2, \ldots, M_X$, $y_j = j\Delta y$, $j = 0, 1, 2, \ldots, M_Y$. Moreover, time is discretized as follows: $t^n = n\Delta t$, $n = 0, 1, 2, \ldots, N$. The finite difference scheme of Equation (10.53c) is given by

$$h_{i,j}^{n+1} = h_{i,j}^{n-1} - 2\Delta t H\left[\left(\frac{u_{i+1,j}^n - u_{i-1,j}^n}{2\Delta x}\right) + \left(\frac{v_{i,j+1}^n - v_{i,j-1}^n}{2\Delta y}\right)\right]. \tag{10.58}$$

The finite difference expression of Equations (10.57a) and (10.57b) are given by

$$u_{i,j}^{n+1} = u_{i,j}^{n-1} + 2\Delta t\left[\beta y_j v_{i,j}^n - g\left(\frac{h_{i+1,j}^n - h_{i-1,j}^n}{2\Delta x}\right)\right]. \tag{10.59}$$

$$v_{i,j}^{n+1} = v_{i,j}^{n-1} - 2\Delta t\left[\beta y_j u_{i,j}^n + g\left(\frac{h_{i,j+1}^n - h_{i,j-1}^n}{2\Delta y}\right)\right]. \tag{10.60}$$

The aforementioned scheme is conditionally stable with the requirement that the Courant number must be $\leq 1/\sqrt{2}$. As the leapfrog scheme is employed, the computational mode needs to be controlled using $2\Delta t$, $2\Delta x$, and $2\Delta y$ filters.

10.10 Solving Two-dimensional Nonlinear Shallow Water Equations without Rotation

The two-dimensional nonlinear shallow water equations without rotation is given by

$$\frac{\partial u}{\partial t} + u\frac{\partial u}{\partial x} + v\frac{\partial u}{\partial y} = -g\frac{\partial h}{\partial x}. \tag{10.61a}$$

$$\frac{\partial v}{\partial t} + u\frac{\partial v}{\partial x} + v\frac{\partial v}{\partial y} = -g\frac{\partial h}{\partial y}. \tag{10.61b}$$

$$\frac{\partial h}{\partial t} + \frac{\partial (hu)}{\partial x} + \frac{\partial (vh)}{\partial y} = 0. \tag{10.61c}$$

Equations (10.61a)–(10.61c) can be rewritten in flux form as follows:

$$\frac{\partial (hu)}{\partial t} + \frac{\partial}{\partial x}\left(hu^2 + \frac{1}{2}gh^2\right) + \frac{\partial (uvh)}{\partial y} = 0, \tag{10.62a}$$

$$\frac{\partial (hv)}{\partial t} + \frac{\partial (uvh)}{\partial x} + \frac{\partial}{\partial y}\left(hv^2 + \frac{1}{2}gh^2\right) = 0, \tag{10.62b}$$

$$\frac{\partial h}{\partial t} + \frac{\partial (uh)}{\partial x} + \frac{\partial (vh)}{\partial y} = 0. \tag{10.62c}$$

Equations (10.62a)–(10.62c) are solved by discretization of the x-axis into a mesh having equal grid size Δx and Δy, i.e., $x_i = i\Delta x$, $i = 0,1,2,\ldots,M_X$, $y_j = j\Delta y$, $j = 0,1,2,\ldots,M_Y$. Moreover, time is discretized as follows: $t^n = n\Delta t$, $n = 0,1,2,\ldots,N$. The spatial derivatives are approximated using the second-order centered differences while the time derivatives were approximated using the second-order centered differences following the Crank–Nicolson method.

Equation (10.62c) can be approximated as follows:

$$h_{i,j}^{n+1} + \frac{\Delta t}{2}\left[\frac{(uh)_{i+1,j}^{n+1} - (uh)_{i-1,j}^{n+1}}{2\Delta x} + \frac{(vh)_{i,j+1}^{n+1} - (vh)_{i,j-1}^{n+1}}{2\Delta y}\right]^{*}$$

$$= h_{i,j}^{n} - \frac{\Delta t}{2}\left[\frac{(uh)_{i+1,j}^{n} - (uh)_{i-1,j}^{n}}{2\Delta x} + \frac{(vh)_{i,j+1}^{n} - (vh)_{i,j-1}^{n}}{2\Delta y}\right] \tag{10.63}$$

$$(uh)_{i,j}^{n+1} = (uh)_{i,j}^n - \Delta t \left[\frac{(hu^2)_{i+1,j}^n - (hu^2)_{i-1,j}^n}{2\Delta x} + \frac{g}{2} \frac{(h^2)_{i+1,j}^n - (h^2)_{i-1,j}^n}{2\Delta x} \right.$$

$$\left. + \frac{(uvh)_{i,j+1}^n - (uvh)_{i,j-1}^n}{2\Delta y} \right] \qquad (10.64)$$

$$(vh)_{i,j}^{n+1} = (vh)_{i,j}^n - \Delta t \left[\frac{(uvh)_{i+1,j}^n - (uvh)_{i-1,j}^n}{2\Delta x} + \frac{g}{2} \frac{(h^2)_{i,j+1}^n - (h^2)_{i,j-1}^n}{2\Delta y} \right.$$

$$\left. + \frac{(hv^2)_{i,j+1}^n - (hv^2)_{i,j-1}^n}{2\Delta y} \right]. \qquad (10.65)$$

The $*$ in Equation (10.63) indicates that the values of (hu) and (hv) at time step $(n+1)$ are to be obtained from Equations (10.64) and (10.65).

10.11   Solving Two-dimensional Nonlinear Shallow Water Equations with Rotation on a β Plane

The two-dimensional nonlinear shallow water equations with rotation on a β plane is given by

$$\frac{\partial u}{\partial t} + u\frac{\partial u}{\partial x} + v\frac{\partial u}{\partial y} - \beta y v = -g\frac{\partial h}{\partial x}, \qquad (10.66a)$$

$$\frac{\partial v}{\partial t} + u\frac{\partial v}{\partial x} + v\frac{\partial v}{\partial y} + \beta y u = -g\frac{\partial h}{\partial y}, \qquad (10.66b)$$

$$\frac{\partial h}{\partial t} + u\frac{\partial h}{\partial x} + v\frac{\partial h}{\partial y} = 0. \qquad (10.66c)$$

Equations (10.66a)–(10.66c) can be rewritten in flux form as follows:

$$\frac{\partial (hu)}{\partial t} + \frac{\partial}{\partial x}\left(hu^2 + \frac{1}{2}gh^2\right) + \frac{\partial (uvh)}{\partial y} - \beta yhv = 0, \qquad (10.67a)$$

$$\frac{\partial (hv)}{\partial t} + \frac{\partial (uvh)}{\partial x} + \frac{\partial}{\partial y}\left(hv^2 + \frac{1}{2}gh^2\right) + \beta yhu = 0, \qquad (10.67b)$$

$$\frac{\partial h}{\partial t} + \frac{\partial (uh)}{\partial x} + \frac{\partial (vh)}{\partial y} = 0. \qquad (10.67c)$$

Equations (10.67a)–(10.67c) are solved by discretization of the x-axis into a mesh having equal grid size Δx and Δy, i.e., $x_i = i\Delta x$, $i = 0, 1, 2, \ldots, M_X$, $y_j =$

$j\Delta y$, $j = 0, 1, 2, \ldots, M_Y$. Moreover, time is discretized as follows: $t^n = n\Delta t$, $n = 0, 1, 2, \ldots, N$. The spatial derivatives are approximated using the second-order centered differences while the time derivatives were approximated using the second-order centered differences following the Crank–Nicolson method Equation (10.67a) can be approximated as follows:

$$
h_{i,j}^{n+1} + \frac{\Delta t}{2} \left[\frac{(uh)_{i+1,j}^{n+1} - (uh)_{i-1,j}^{n+1}}{2\Delta x} + \frac{(vh)_{i,j+1}^{n+1} - (vh)_{i,j-1}^{n+1}}{2\Delta y} \right]^*
$$
$$
= h_{i,j}^{n} - \frac{\Delta t}{2} \left[\frac{(uh)_{i+1,j}^{n} - (uh)_{i-1,j}^{n}}{2\Delta x} + \frac{(vh)_{i,j+1}^{n} - (vh)_{i,j-1}^{n}}{2\Delta y} \right] \quad (10.68)
$$

$$
(uh)_{i,j}^{n+1} = (uh)_{i,j}^{n} - \Delta t \left[\frac{(hu^2)_{i+1,j}^{n} - (hu^2)_{i-1,j}^{n}}{2\Delta x} + \frac{g}{2} \frac{(h^2)_{i+1,j}^{n} - (h^2)_{i-1,j}^{n}}{2\Delta x} \right.
$$
$$
\left. + \frac{(uvh)_{i,j+1}^{n} - (uvh)_{i,j-1}^{n}}{2\Delta y} \right] + \Delta t \beta y_j (hv)_{i,j}^{n} \quad (10.69)
$$

$$
(vh)_{i,j}^{n+1} = (vh)_{i,j}^{n} - \Delta t \left[\frac{(uvh)_{i+1,j}^{n} - (uvh)_{i-1,j}^{n}}{2\Delta x} + \frac{g}{2} \frac{(h^2)_{i,j+1}^{n} - (h^2)_{i,j-1}^{n}}{2\Delta y} \right.
$$
$$
\left. + \frac{(hv^2)_{i,j+1}^{n} - (hv^2)_{i,j-1}^{n}}{2\Delta y} \right] - \Delta t \beta y_j (hu)_{i,j}^{n}. \quad (10.70)
$$

The $*$ in Equation (10.68) indicates that the values of (hu) and (hv) at time step $(n+1)$ are to be obtained from Equations (10.69) and (10.70).

10.12 Equivalent Barotropic Model

An improvement to the non-divergent barotropic equation would involve introduction of "horizontal divergence," which is what is observed in the shallow water equations. An alternate improvement to the non-divergent barotropic equation would be to assume that the wind direction is assumed constant with height while allowing for vertical variation of wind speed. This in essence is the underlying philosophy of the equivalent barotropic model. To derive the equivalent barotropic equation, consider the approximate form of the vorticity equation with non-vanishing of the horizontal divergence term, with the latter term replaced by invoking the continuity equation in pressure coordinates.

$$\frac{\partial \zeta}{\partial t} + \vec{V}_H \cdot \nabla_p (\zeta + f) = \tilde{f} \frac{\partial \omega}{\partial p} \tag{10.71}$$

It is assumed that planetary vorticity dominates over relative vorticity in the horizontal divergence term in Equation (10.71).

In this equivalent barotropic model, the wind direction is assumed constant with height while allowing for vertical variation of wind speed. The wind speed is hence expressed in the form

$$\vec{V} = A(p)\overline{V}, \tag{10.72}$$

where $A(p)$ is an empirical function of pressure and the bar over the velocity in Equation (10.72) represents the integral mean with respect to pressure, i.e.,

$$(\overline{}) = \frac{1}{p_o} \int_0^{p_o} \{\cdots\} \, dp. \tag{10.73}$$

Applying the bar operator to Equation (10.72), one obtains $\overline{V} = \overline{A}\,\overline{V}$, which indicates that $\overline{A} = 1$. Based on climatological data, A is less than 1 from surface to $600\,\text{hPa}$, has the value of 1 at $600\,\text{hPa}$ and has values greater than 1 at pressure less than $600\,\text{hPa}$ in the troposphere. Defining Equation (10.72) in terms of the vertical component of relative vorticity ζ, one obtains

$$\zeta = A(p)\overline{\zeta}. \tag{10.74}$$

Substituting Equations (10.72) and (10.74) and applying the integral operator, Equation (10.73) to Equation (10.71), one obtains

$$\frac{\partial \overline{\zeta}}{\partial t} + \overline{A}^2 \vec{V}_H \cdot \nabla_p \overline{\zeta} + \vec{V}_H \cdot \nabla_p f = \tilde{f}\frac{\omega_o}{p_o}, \tag{10.75}$$

where ω_o represents the vertical velocity at the lower boundary p_o. Recall $\overline{A} = 1$, the top boundary condition is $\omega = 0$ at the upper boundary $(p = 0)$. Expanding the vertical velocity w from the expression in terms of local derivative and advection terms

$$w = \frac{1}{g}\frac{d\phi}{dt}, \tag{10.76}$$

$$wg = \frac{\partial \phi}{\partial t} + \vec{V}_H \cdot \nabla_p \phi + \omega \frac{\partial \phi}{\partial p}. \tag{10.77}$$

Using hydrostatic equations in pressure coordinates $\partial \phi / \partial p = -\alpha$ and substituting for $\partial \phi / \partial p$, Equation (10.77) becomes

$$\omega = \rho \left(\frac{\partial \phi}{\partial t} + \vec{V}_H \cdot \nabla_p \phi - gw \right). \tag{10.78}$$

If the lower boundary is level, the kinematic boundary condition requires that $w_o = 0$. If the winds are assumed to be geostrophic, then $V_g \cdot \nabla \phi = 0$. Substituting the aforementioned two conditions in Equation (10.78), one obtains

$$\omega_o = \rho_o \left(\frac{\partial \phi}{\partial t} \right)_{p=p_o} = \rho_o A(p_o) \frac{\partial \bar{\phi}}{\partial t}. \tag{10.79}$$

The form of Equation (10.75) can be further simplified by defining the following:

$$V^* = \bar{A}^2 \bar{V} \quad \text{and} \quad \zeta^* = \bar{A}^2 \bar{\zeta} \tag{10.80}$$

Multiplying Equation (10.75) by $\bar{A}^2$ and using Equations (10.79) and (10.80), one obtains

$$\frac{\partial \zeta^*}{\partial t} + \vec{V}_H^* \cdot \nabla_p (\zeta^* + f) = \tilde{f} \frac{\rho_o A_o}{p_o} \frac{\partial \phi^*}{\partial t} = \frac{\tilde{f} A_o}{RT} \frac{\partial \phi^*}{\partial t}. \tag{10.81}$$

Note that Equation (10.81) has only one dependent variable, ϕ^*, as both ζ^* and $\vec{V}_H^*$ are related to ϕ^*, through the geostrophy relationship.

$$\zeta^* = \frac{1}{f} \nabla^2 \phi^* \quad \text{and} \quad \vec{V}_H^* = \frac{1}{f} \hat{k} \times \nabla \phi^*. \tag{10.82}$$

Equation (10.81) needs to be applied at the level p^* for which $A(p*) = \bar{A}^2$ as estimated from climatological data. As $\bar{A}^2 = 1.25$, p^* is somewhat higher in altitude than $600 \, hPa$ and is near about $500 \, hPa$. Substituting Equation (10.82) in Equation (10.81) results in a Helmholtz equation in $\partial \phi^* / \partial t$, which can be solved by successive overrelaxation methods. Knowledge of $\partial \phi^* / \partial t$ and invoking an appropriate time difference scheme (say leapfrog scheme) to replace the time derivative, will advance the solution of ϕ^* to the next time level. From the knowledge of ϕ^*, one can deduce ζ^* and $\vec{V}_H^*$ from Equation (10.82). A repetition of the aforementioned steps would help in marching the solution of the equivalent barotropic model to the desired time duration.

Exercises 10 (Questions and answers)

1. What is the Boussinesq approximation?

   Answer: The Boussinesq approximation refers to situations where the density fluctuations of the fluid are smaller compared to the mean density of the fluid. More specifically, whenever the Boussinesq approximation is applied, the variation of fluid density is confined only to the "buoyancy term" in the momentum equations.

2. What is an anelastic approximation?

   Answer: The development of the Boussinesq approximation to a system such as the atmosphere having a larger vertical extent that is comparable to or greater than the density scale height is called an anelastic approximation. The anelastic approximation recognizes the fact that a situation exists where the density field may vary markedly in the vertical while still possessing negligible density fluctuations within individual fluid element. An anelastic approximation uses the following assumption $\nabla \cdot (\bar{\rho}\vec{V}) = 0$ where $\bar{\rho} = \bar{\rho}(z)$ is some reference density.

3. Consider a two-dimensional Poisson equation $\nabla^2 u = f$ with the Dirichlet conditions $u = g$ on all the boundaries. Give the expression for the optimum value of the "successive over relaxation parameter" to solve the aforementioned elliptic partial differential equation assuming that the solution domain is divided into a uniform mesh with $N+2$ grid points in each of the coordinate directions.

   Answer: The optimum value of the "successive over relaxation parameter" to solve the two-dimensional Poisson equation with the Dirichlet conditions is given by the expression

   $$\omega_{\text{opt}} = \frac{2}{1 + \sin(\pi h)}, \qquad \text{where} \qquad h = \frac{1}{N+1}$$

4. The relaxation methods are examples of iterative methods to solve a system of linear algebraic equations. Discuss the convergence of the iterative methods.

   Answer: Any stationary iterative method can be written in the following general form $\mathbf{x}^{k+1} = \mathbf{B}\mathbf{x}^k + \mathbf{c}$. By subtracting the equation $\mathbf{x} = \mathbf{B}\mathbf{x} + \mathbf{c}$, from this, it is possible to obtain a relation between the errors in successive approximation as follows:

   $$\mathbf{x}^{k+1} - \mathbf{x} = \mathbf{B}(\mathbf{x}^k - \mathbf{x}) = \ldots \mathbf{B}^{k+1}(\mathbf{x}^o - \mathbf{x}).$$

   Let $\mathbf{B}$ have the following n eigenvalues $\lambda_1, \lambda_2, \ldots, \lambda_n$ with the corresponding eigenvectors $\mathbf{u}_1, \mathbf{u}_1, \ldots, \mathbf{u}_n$ that are assumed to be linearly independent. Then one can expand the initial error

   $$\mathbf{x}^o - \mathbf{x} = \alpha_1 u_1 + \alpha_2 u_2 + \cdots + \alpha_n u_n.$$ From this, one can obtain the error at the kth iteration as

   $$\mathbf{x}^k - \mathbf{x} = \alpha_1 \lambda_1^k u_1 + \alpha_2 \lambda_2^k u_2 + \cdots + \alpha_n \lambda_n^k u_n.$$

This means that the process converges from an arbitrary approximation if and only if $|\lambda_i| \leq 1$ for $i = 1, 2, \ldots, n$.

A necessary and sufficient condition for a stationary iterative method $\mathbf{x}^{k+1} = \mathbf{B}\mathbf{x}^k + \mathbf{c}$ to converge for an arbitrary initial approximation $\mathbf{x}^o$ is that the spectral radius $\rho(\mathbf{B})$ of $\mathbf{B}$ satisfies the following condition:

$$\rho(\mathbf{B}) = \max_{1 \leq i \leq n} |\lambda_i(\mathbf{B})| < 1.$$

5. For uniform mesh subject to Dirichlet boundary conditions, is there a relation between the spectral radius ρ and the optimum value of the "successive over relaxation parameter ω" for the solution of an elliptic PDE?

   Answer: For a uniform mesh subject to Dirichlet boundary conditions, the relation between the spectral radius ρ and the optimum value of the "successive over relaxation parameter ω" for solution of an elliptic PDE is given by

   $$\omega_{\text{opt}} = \frac{2}{1 + \sqrt{1 - \rho^2}}.$$

6. Show that the algebraic error obeys the residual equation while employing relaxation methods.

   Answer: The iterative method employed to solve the matrix equation $\mathbf{A}\mathbf{x} = \mathbf{f}$ proceeds in the following manner. A sequence of approximations $\mathbf{v}_0, \mathbf{v}_1, \ldots, \mathbf{v}_n$ of $\mathbf{x}$ is constructed that converges to $\mathbf{x}$. Let $\mathbf{v}_i$ be an approximation to $\mathbf{x}$ after the ith iteration. One may define the residual as $\mathbf{r}_i = \mathbf{f} - \mathbf{A}\mathbf{v}_i$ to be the computable measure of the deviation of $\mathbf{v}_i$ from $\mathbf{x}$. Then the algebraic error $\mathbf{e}_i$ of the approximation $\mathbf{v}_i$ is defined as $\mathbf{e}_i = \mathbf{x} - \mathbf{v}_i$. Thus, it is clear that the algebraic error $\mathbf{e}_i$ obeys the following residual equation $\mathbf{A}\mathbf{e}_i = \mathbf{r}_i$.

7. Mention the maximum and minimum properties of the Laplace equation.

   Answer: The "Maximum Principle" and "Minimum Principle" are embodied in the following statement. Let ϕ satisfies the Laplace equation $\nabla^2 \phi = 0$ in a volume V with surface S. Then the aforementioned "Maximum Principle" and "Minimum Principle" theorem states that both the minimum and maximum values of ϕ occur somewhere on S.

8. Write down the leapfrog time difference scheme with the Robert–Asselin filter to take care of the $2\Delta t$ wave while solving the equation $d\psi(t)/dt = f[\psi(t)]$.

   Answer: The leapfrog time difference scheme for the aforementioned equation is given by

   $$\psi(t + \Delta t) = \psi^*(t - \Delta t) + 2\Delta t\{f[\psi(t)]\}$$

   $$\psi^*(t) = \psi(t) + \gamma[\psi^*(t - \Delta t) - 2\psi(t) + \psi(t + \Delta t)]$$

   where $\gamma = 0.1$ is the filter coefficient and terms with superscript '*' refer to terms that has been subjected to the Robert–Asselin filter.

9. Write down the quasi-geostrophic shallow water equations in two dimensions in a β plane.

   Answer: The quasi-geostrophic shallow water equations in two dimensions in a β plane are given as:

   $$\frac{\partial u_g}{\partial t} + u_g \frac{\partial u_g}{\partial x} + v_g \frac{\partial u_g}{\partial y} - f_o v_a - \beta y v_g = -\frac{\partial (g h_a)}{\partial x},$$

   $$\frac{\partial v_g}{\partial t} + u_g \frac{\partial v_g}{\partial x} + v_g \frac{\partial v_g}{\partial y} + f_o u_a + \beta y u_g = -\frac{\partial (g h_a)}{\partial y},$$

   $$\frac{\partial (g h_g)}{\partial t} + u_g \frac{\partial (g h_g)}{\partial x} + v_g \frac{\partial (g h_g)}{\partial y} = -g h_o \left(\frac{\partial u_a}{\partial x} + \frac{\partial v_a}{\partial y} \right),$$

   where subscripts g and a refer to "geostrophic" and "ageostrophic" parts, and h_o is the mean depth of the fluid.

10. Write down the quasi-geostrophic vorticity equations in a β plane.

    Answer: The quasi-geostrophic vorticity equations in a β plane is given by

    $$\frac{\partial \zeta_g}{\partial t} + u_g \frac{\partial \zeta_g}{\partial x} + v_g \frac{\partial \zeta_g}{\partial y} - \beta v_g = -f_o \left(\frac{\partial u_a}{\partial x} + \frac{\partial v_a}{\partial y} \right)$$

11. What are the conditions that need to be satisfied for the successive over-relaxation (SOR) method to converge to the unique solution of the linear system $A\vec{x} = \vec{b}$?

    Answer: The conditions that need to be satisfied for the SOR method to converge to the unique solution of the linear system $A\vec{x} = \vec{b}$ is that, $a_{ii} \neq 0$, and that matrix A be positive definite and symmetric.

12. Indicate the number of operations per iteration for the SOR method.

    Answer: The number of operations per iteration for the SOR method is of the order $O(4n + 2cn)$, where n is the number of elements of the unknown vector $\vec{x}$, and c is the average number of elements per row.

Python examples

1. Solve the linear non-divergent vorticity equation in a beta plane $\frac{\partial \zeta}{\partial t} + v\beta = 0$, where ζ is the relative vorticity, β is the meridional gradient of Coriolis parameter, and v is the meridional component of velocity. Writing the aforementioned equation in terms of stream function ψ, one has the following equation $\frac{\partial \{\nabla^2 \psi\}}{\partial t} + \beta \frac{\partial \psi}{\partial x} = 0$; which can be written as $\nabla^2 \chi = -\beta \frac{\partial \psi}{\partial x}$, where $\chi = \frac{\partial \psi}{\partial t}$. The initial condition that is considered is the axisymmetric cyclonic

vortex with the following tangential wind profile (Chan and Williams, 1987, Analytical and Numerical Studies of the beta-effect in Tropical Cyclone Motion, Part I: Zero Mean Flow, *Journal of Atmospheric Sciences*, 44, 9, 1257–1265).

$V(r) = V_m \{r/r_m\} \exp[\frac{1}{b} \{1 - (\frac{r}{r_m})^b\}]$, where $V(r)$ is the tangential wind profile at any distance r from the center of the cyclonic vortex, V_m is the value of the maximum tangential wind, r_m is the radius of the maximum tangential wind, and b is a shape factor. The relative vorticity profile $\zeta(r)$ is obtained from the following expression $\zeta(r) = \frac{2V_m}{r_m}[1 - \frac{1}{r} \{\frac{r}{r_m}\}^b] \exp[\frac{1}{b} \{1 - (\frac{r}{r_m})^b\}]$. The python code for solving nonlinear nondivergent vorticity equation in a beta plane is modified from that developed by Prof G Vallis to solve the linear non-divergent vorticity equation in a beta plane with the initial condition corresponding to a cyclonic vortex. The code uses the following values for the various parameters, $V_m = 40$ m/s; $r_m = 100$ m; $\Delta x = 1/256$ m; $\Delta y = 1/256$ m; $b = 1$; $\beta = 12$;

```python
#python code

from __future__ import (division, print_function)
import matplotlib.pyplot as plt
import numpy as np

from numpy import pi, cos, sin
from numpy.fft import fftshift, fftfreq
from numpy.fft import rfft2, irfft2
####################
###Functions
def ft(phi):
    """Go from physical space to spectral space."""
    return rfft2(phi, axes=(-2, -1))


def ift(psi):
    """Go from spectral space to physical space."""
    return irfft2(psi, axes=(-2,-1))


def courant_number(psix, psiy, dx, dt):
    """Calculate the Courant Number given the velocity field and
        step size."""
    maxu = np.max(np.abs(psiy))
    maxv = np.max(np.abs(psix))
    maxvel = maxu + maxv
```

```
        return maxvel*dt/dx

def grad(phit):
    """Returns the spatial derivatives of a Fourier transformed
        variable.
    Returns (partial/partial x[F[phi]], partial/partial
        y[F[phi]]) i.e. (ik F[phi], il F[phi])"""
    global ik, il
    phixt = ik*phit        # d/dx F[phi] = ik F[phi]
    phiyt = il*phit        # d/dy F[phi] = il F[phi]
    return (phixt, phiyt)

def anti_alias(phit):
    """Set the coefficients of wavenumbers > k_mask to be
        zero."""
    k_mask = (8./9.)*(nk+1)**2.
    phit[(np.abs(ksq/(dk*dk)) >= k_mask)] = 0.0



def cart2pol(x, y):
    rho = np.sqrt(x**2 + y**2)
    phi = np.arctan2(y, x)
    return(rho, phi)

def pol2cart(rho, phi):
    x = rho * np.cos(phi)
    y = rho * np.sin(phi)
    return(x, y)

def high_wn_filter(phit):
    """Applies the high wavenumber filter of smith et al 2002"""
    filter_dec = -np.log(1.+2.*pi/nk)/((nk-kcut)**filter_exp)
    filter_idx = np.abs(ksq/(dk*dk)) >= kcut**2.
    phit[filter_idx] *= np.exp(filter_dec*(np.sqrt(ksq
        [filter_idx]/(dk*dk))-kcut)**filter_exp)

def forcing_spatial_mask(phi):
    """TODO: Make spatial mask for forcing"""
```

```
    phi[idx_sp_mask] *= 0.

_prhs, _pprhs  = 0.0, 0.0  # previous two right hand sides

def adams_bashforth(zt, rhs, dt):
    """Take a single step forward in time using Adams-Bashforth
       3."""
    global step, t, _prhs, _pprhs
    if step is 0:
        # forward euler
        dt1 = dt
        dt2 = 0.0
        dt3 = 0.0
    elif step is 1:
        # AB2 at step 2
        dt1 = 1.5*dt
        dt2 = -0.5*dt
        dt3 = 0.0
    else:
        # AB3 from step 3 on
        dt1 = 23./12.*dt
        dt2 = -16./12.*dt
        dt3 = 5./12.*dt

    newzt = zt + dt1*rhs + dt2*_prhs + dt3*_pprhs
    _pprhs = _prhs
    _prhs  = rhs
    return newzt
###############################
#Constants and variables

#########################
##########################
###Velocity and vorticity distribution from Chan and Wiiliams
    1987 JAS

#########################
rm=100 ### (rnormalized)rm=100 km
```

```
vm=40   ###m s-1
b=1
#######Using Chan and William 1987 eqn 2.10, 2.11
r=np.arange(0,1280*4,10)    ########## r=0 to r=1270


v_r=vm*(r/rm)*np.exp((1/b)*(1-(r/rm)**b))


z_r=(2*(vm/rm))*(1-((1/2)*(r/rm)**b))*np.exp((1/b)*(1-(r/rm)**b))


#z_r[0]=np.nan    ########


fig, (ax1,ax2) = plt.subplots(1,2,figsize=(10,5))
ax1.plot(r,v_r,color='r')
ax1.set_xlabel('Distance R from centre(km)')
ax1.set_ylabel(r'velocity($m$ $s^{-1}$)')
ax1.set_title('Velocity distribution')
ax2.plot(r,z_r,color='b')
ax1.set_xlabel('Distance R from centre(km)')
ax2.set_title('Vorticity distribution')
ax2.set_xlabel('Distance R from centre(km)')
ax2.set_ylabel(r'vorticity ($s^{-1}$)')
plt.savefig("Initial velocity and vorticity distribution.png",
    dpi=300)
##################################
#Convert polar to cartesian coordinates
theta=np.arange(0,np.pi/2,np.pi/(2*128*4))
vxq1=np.zeros((len(r),len(r)))
vyq1=np.zeros((len(r),len(r)))
zq1=np.zeros((len(r),len(r)))
zq1[:]=np.nan


for r1 in range(0,len(r)):
    for t in range(0,len(theta)):


        vxq1[np.int(r1*cos(theta[t])),np.int(r1*sin(theta[t]))]
            =v_r[r1]*cos(theta[t])
        vyq1[np.int(r1*cos(theta[t])),np.int(r1*sin(theta[t]))]
            =v_r[r1]*sin(theta[t])
```

```python
        if ~(np.isinf(z_r[r1])):
            zq1[np.int(r1*cos(theta[t])),np.int(r1*sin(theta
                [t]))]=z_r[r1]


####################
x=np.arange(0,len(theta),1)
y=np.arange(0,len(theta),1)
xx,yy=np.meshgrid(x,y)
vmq1=np.sqrt(vxq1**2+vyq1**2)
#######################
vxq2=np.copy(vxq1)
vxq3=np.copy(vxq1)
vxq4=np.copy(vxq1)

vyq2=np.copy(vyq1)
vyq3=np.copy(vyq1)
vyq4=np.copy(vyq1)


###############################
########Crosscheck_velocity ditribution for 2 qudrant(vmq1)
vmq2=np.sqrt(vxq2**2+vyq2**2)
##############Get velocity and vorticity distribution for all
    co-ordinates
vxup=np.hstack((np.fliplr(vxq2),vxq1))
vyup=np.hstack((np.fliplr(vyq2),vyq1))

vxdn=np.hstack((np.fliplr(vxq3),vxq4))
vydn=np.hstack((np.fliplr(vyq3),vyq4))


zrup=np.hstack((np.fliplr(zq1),zq1))
zrdn=np.hstack((np.fliplr(zq1),zq1))
zr1=np.vstack((np.flipud(zrup),zrdn))
#
vx1=np.vstack((np.flipud(vxup),vxdn))
vy1=np.vstack((np.flipud(vyup),vydn))
```

```
####
##Give cyclonic direction to wind:
###Quad1:
#vx=np.zeros((len(theta)*2,len(theta)*2))
#vy=np.zeros((len(theta)*2,len(theta)*2))
#vx=np.copy(vx1)
#vy=np.copy(vy1)
## ###Quad2:rleft: lower; v negative
#
#vx[256:,256:]=-1*vx1[256:,256:]
#
####Quad2:rleft: top; bothu and v negative
#vx[256:,:256]=-1*vx1[256:,:256]
#vy[256:,:256]=-1*vy1[256:,:256]
#
# ###Quad2:rleft: lower; v negative
#vy[:256,:256]=-1*vy1[:256,:256]
# ##Check velocity distribution for all coodrinates
#v_mag=np.sqrt((vx**2+vy**2))
#plt.figure()
#plt.imshow(v_mag,origin='lower')
#plt.title('Magnitude of Velocity magnitude')

zr1=np.copy(zr1[384:640,384:640])
plt.figure()
plt.imshow(zr1,origin='lower')
plt.title('Magnitude_of_vorticity_magnitude')
######################
######################
#zr1[np.isnan(zr1)]=0.0
###########################
######The code below this is based mainly on Prof. Vallis:

######
### Configuration
nx = 256
ny = 256                        # numerical resolution
Lx = 1.0
```

```
Ly = 1.0                          # domain size [m]
ubar = 0.00                       # background zonal velocity  [m/s]
beta = 12.0                        # beta-plane f = f0 + beta y
                                     [1/s 1/m]
n_diss = 2.0
tau = 0.1                         # coefficient of dissipation
                                  # smaller = more dissipation


#Poorly determined coefficients for forcing and dissipation
r_rayleigh = (1./50000.)/np.sqrt(10.)
forcing_amp_factor=100.0/np.sqrt(1.)
r_rayleigh = 0.
forcing_amp_factor=0.



t = 0.0
tmax = 1000
step = 0

ALLOW_SPEEDUP = True              # if True, allow the simulation to
                                          take a larger
SPEEDUP_AT_C  = 0.4             # timestep when the Courant number
                                          drops below
                                  # value of parameter SPEEDUP_AT_C
SLOWDN_AT_C = 0.6               # reduce the timestep when Courant
                                          number
                                  # is bigger than SLOWDN_AT_C
PLOT_EVERY_S = 100
PLOT_EVERY_S2 = 40


### Physical Domain
nl = ny
nk = nx/2 + 1
dx = Lx / nx
dy = Ly / ny
dt = 0.4 * 16.0 / nx              # choose an initial dt. This will
                                          change
                                  # as the simulation progresses to
```

```
                                      maintain
                               # numerical stability
dk = 2.0*pi/Lx
dl = 2.0*pi/Ly
y = np.linspace(0, Ly, num=ny)


y_arr = np.flipud(np.tile(y,(nx,1)).transpose())


xx = np.linspace(0, Lx, num=nx)
yy = 1. - np.linspace(0, Ly, num=ny)


k = dk*np.arange(0, nk, dtype=np.float64)[np.newaxis, :]
l = dl*fftfreq(nl, d=1.0/nl)[:, np.newaxis]


ksq = k**2 + l**2
ksq[ksq == 0] = 1.0
rksq = 1.0 / ksq                       # reciprocal 1/(k^2 + l^2)


ik = 1j*k                              # wavenumber mul. imaginary unit
                                           is useful
il = 1j*l                              # for calculating derivatives


##############################
###Calculate stream function
# initialise the transformed zeta
zt = ft(zr1)       ###zrnew if calculated from the dv_dx -du_dy
#anti_alias(zt)
z=ift(zt)
psit = -rksq * zt


#amp = forcing_amp_factor* np.max(np.abs(zr1))
#
## use the x-dimension for reference scale values
#nu = ((Lx/(np.floor(nx/3)*2.0*pi))**(2*n_diss))/tau


#filter_exp = 8.
#kcut = 30.
#
```

```
## Spectral Filter as per [Arbic and Flierl, 2003]
#wvx = np.sqrt((k*dx)**2 + (l*dy)**2)
#spectral_filter = np.exp(-23.6*(wvx-0.65*pi)**4)
#spectral_filter[wvx <= 0.65*pi] = 1.0
###########################
#plt.figure()
#plt.pcolormesh(xx,yy,psit)
###########################


#amp = forcing_amp_factor* np.max(np.abs(qi))


# initialise the storage arrays
#time_arr[0]=t


psit = -rksq * zt               # F[psi] = - F[zeta] / (k^2 + l^2)
psixt, psiyt = grad(psit)
psix = ift(psixt)
psiy = ift(psiyt)
###################################
### Diagnostic arrays
time_arr = np.zeros(1)
tot_energy_arr = np.zeros(1)
# initialise the storage arrays
time_arr[0]=t


psit = -rksq * zt               # F[psi] = - F[zeta] / (k^2 + l^2)
psixt, psiyt = grad(psit)
psix = ift(psixt)
psiy = ift(psiyt)


#urms=np.sqrt(np.mean(psix**2 + psiy**2))
#tot_energy=0.5*urms**2.
#tot_energy_arr[0]=tot_energy


###########Streamfunction#####
plt.rcParams['contour.negative_linestyle']= 'dashed'
plt.figure()
psi=ift(psit)
```

```python
cs1=plt.pcolormesh(xx,yy,psi*10**2)
cs=plt.contour(xx,yy,psi*10**2,colors='k')
plt.colorbar(cs1)


#################################
fig, (ax1,ax2) = plt.subplots(1,2,figsize=(10,5))
clevs=np.arange(-2,2.1,0.5)
cs=ax1.contourf(xx,yy,z,clevs,cmap='seismic',extend='both')

ax1.contour(xx,yy,z,clevs,colors='k')
#plt.colorbar(cs)
clevs1=np.arange(-2,2.1,0.5)
cbar=fig.colorbar(cs,ax=ax1,ticks=clevs1,orientation=
    'horizontal',extend='both')
cbar.ax.set_xlabel(r'Vorticity($s^{-1}$)')
#ax1.set_xlabel('Distance R from centre(km)')
#ax1.set_ylabel(r'velocity($m$ $s^{-1}$)')
ax1.set_title('Vorticity at{:.2f}s dt={:.2f}'.format(t, dt))
clevs=np.arange(-2,2.01,0.5)
cs2=ax2.contourf(xx,yy,psi*10**3,clevs,cmap='seismic',extend=
    'both')
ax2.contour(xx,yy,psi*10**3,clevs,colors='k')
#plt.colorbar(cs)
cbar2=fig.colorbar(cs2,ax=ax2,ticks=clevs1,orientation=
    'horizontal',extend='both')
cbar2.ax.set_xlabel(r'Streamfunction($10^{-3}$$ $m^{2}$ $s^{-1}$)')
ax2.set_title('streamfunction')
plt.savefig('Vorticity and streamfunction at time'+np.str(t)+
    'dt'+np.str(dt)+'.png',dpi=300)
print("t",t)
#plt.savefig("Initial velocity and Streamfunction.png",dpi=300)


#################################
#show_plot()
tplot = t + PLOT_EVERY_S
tplot2 = t + PLOT_EVERY_S2
```

```
#filter_testt=np.ones(zt.shape)

#filter_testt_shift = np.fft.fftshift(filter_testt, axes=(0,))


while t < tmax:
    # calculate derivatives in spectral space
    psit = -rksq * zt           # F[psi] = - F[zeta] / (k^2 + l^2)
    psixt, psiyt = grad(psit)
    zxt, zyt = grad(zt)

    # transform back to physical space for pseudospectral part
    z[:] = ift(zt)
    psix = ift(psixt)
    psiy = ift(psiyt)
    zx =   ift(zxt)
    zy =   ift(zyt)

    # Non-linear: calculate the Jacobian in real space
    # and then transform back to spectral space
    jac = psix * zy - psiy * zx + ubar * zx
    jact = ft(jac)
    jact=0.0 ###Linear domain

    # calculate the size of timestep that can be taken
    # (assumes a domain where dx and dy are of the same order)
    c = courant_number(psix, psiy, dx, dt)
    if c >= SLOWDN_AT_C:
        print('DEBUG:_Courant_No_>_0.8,_reducing_timestep')
        dt = 0.9*dt
    elif c < SPEEDUP_AT_C and ALLOW_SPEEDUP:
        dt = 1.1*dt

    rhs = -jact - beta*psixt    ####
    zt[:] = adams_bashforth(zt, rhs, dt)
    # if t > tplot:
```

```
psi=ift(psit)
fig, (ax1,ax2) = plt.subplots(1,2,figsize=(10,5))
clevs=np.arange(-2,2.1,0.5)
cs=ax1.contourf(xx,yy,z,clevs,cmap='seismic',extend='both')
ax1.contour(xx,yy,z,clevs,colors='k')
#plt.colorbar(cs)
clevs1=np.arange(-2,2.1,0.5)
cbar=fig.colorbar(cs,ax=ax1,ticks=clevs1,orientation=
    'horizontal',extend='both')
cbar.ax.set_xlabel(r'Vorticity($s^{-1}$)')
#ax1.set_xlabel('Distance R from centre(km)')
#ax1.set_ylabel(r'velocity($m$ $s^{-1}$)')
ax1.set_title('Vorticity_at{:.2f}s_dt={:.2f}'.format(t, dt))
clevs=np.arange(-2,2.01,0.5)
cs2=ax2.contourf(xx,yy,psi*10**3,clevs,cmap='seismic',extend
    ='both')
ax2.contour(xx,yy,psi*10**3,clevs,colors='k')
#plt.colorbar(cs)
cbar2=fig.colorbar(cs2,ax=ax2,ticks=clevs1,orientation=
    'horizontal',extend='both')
cbar2.ax.set_xlabel(r'Streamfunction($10^{-3}$_$m^{2}$
_____$s^{-1}$)')
ax2.set_title('streamfunction')
plt.savefig('Vorticity_and_streamfunction_at_time'+np.str(t)
    +'dt'+np.str(dt)+'.png',dpi=300)

t = t + dt
step = step + 1
plt.close()
###############################
```

2. Solve the nonlinear non-divergent vorticity equation in a beta plane $\frac{\partial \zeta}{\partial t} + u\frac{\partial \zeta}{\partial x} + v\frac{\partial \zeta}{\partial y} + v\beta = 0$, where ζ is the relative vorticity, β is the meridional gradient of Coriolis parameter, u and v are the zonal and meridional components of velocity. Writing the aforementioned equation in terms of stream function ψ, one has the following equation:

$\frac{\partial\{\nabla^2\psi\}}{\partial t} - \frac{\partial \psi}{\partial y}\frac{\partial \zeta}{\partial x} + \frac{\partial \psi}{\partial x}\frac{\partial \zeta}{\partial y} + \frac{\partial \psi}{\partial x}\beta = 0$, which can be written as $\nabla^2\chi - \frac{\partial \psi}{\partial y}\frac{\partial \zeta}{\partial x} + \frac{\partial \psi}{\partial x}\frac{\partial \zeta}{\partial y} + \frac{\partial \psi}{\partial x}\beta = 0$, where $\chi = \frac{\partial \psi}{\partial t}$. This equation can also be written as $\nabla^2\chi + J(\psi, \nabla^2\psi) + \frac{\partial \psi}{\partial x}\beta = 0$, where $J(\psi, \nabla^2\psi)$ is the Jacobian of stream function and relative vorticity. The initial condition that

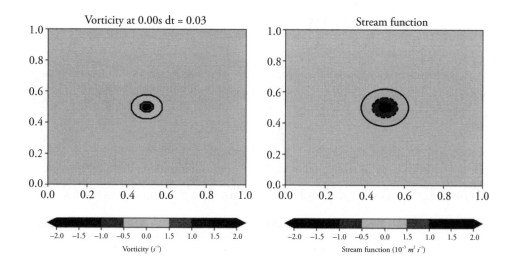

Figure 10.2 Vorticity and stream function at t = 0 for the linear non-divergent vorticity equation.

is considered is the axisymmetric cyclonic vortex with the following tangential wind profile (Chan and Williams, 1987, Analytical and Numerical Studies of the beta-effect in Tropical Cyclone Motion, Part I: Zero Mean Flow, *Journal of Atmospheric Sciences*, 44,9, 1257–1265).

The python code for solving the nonlinear nondivergent vorticity equation in a beta plane is modified from that developed by Prof G Vallis to solve the nonlinear non-divergent vorticity equation in a beta plane with the initial condition corresponding to a cyclonic vortex as given by Chan and Williams. The code uses the following values for the various parameters, $V_m = 40$ m/s; $r_m =100$ m; $\Delta x = 1/256$ m; $\Delta y = 1/256$ m; $b = 1$; $\beta = 12$;

```python
#python code

from __future__ import (division, print_function)
import matplotlib.pyplot as plt
import numpy as np

from numpy import pi, cos, sin
from numpy.fft import fftshift, fftfreq
from numpy.fft import rfft2, irfft2
#####################
###Functions
def ft(phi):
    """Go from physical space to spectral space."""
```

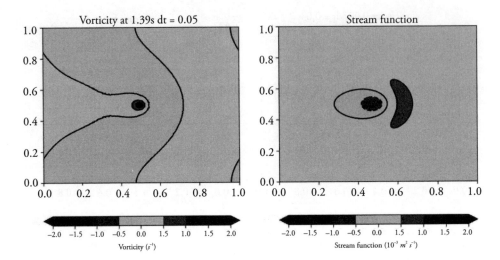

Figure 10.3 Vorticity and stream function at t = 1.39 s for the linear non-divergent vorticity equation.

```
    return rfft2(phi, axes=(-2, -1))

def ift(psi):
    """Go from spectral space to physical space."""
    return irfft2(psi, axes=(-2,-1))

def courant_number(psix, psiy, dx, dt):
    """Calculate the Courant Number given the velocity field and
        step size."""
    maxu = np.max(np.abs(psiy))
    maxv = np.max(np.abs(psix))
    maxvel = maxu + maxv
    return maxvel*dt/dx

def grad(phit):
    """Returns the spatial derivatives of a Fourier transformed
        variable.
    Returns (partial/partial x[F[phi]], partial/partial y
        [F[phi]]) i.e. (ik F[phi], il F[phi])"""
    global ik, il
    phixt = ik*phit          # d/dx F[phi] = ik F[phi]
```

```
    phiyt = il*phit          # d/dy F[phi] = il F[phi]
    return (phixt, phiyt)


def anti_alias(phit):
    """Set the coefficients of wavenumbers > k_mask to be
        zero."""
    k_mask = (8./9.)*(nk+1)**2.
    phit[(np.abs(ksq/(dk*dk)) >= k_mask)] = 0.0



def cart2pol(x, y):
    rho = np.sqrt(x**2 + y**2)
    phi = np.arctan2(y, x)
    return(rho, phi)


def pol2cart(rho, phi):
    x = rho * np.cos(phi)
    y = rho * np.sin(phi)
    return(x, y)


def high_wn_filter(phit):
    """Applies the high wavenumber filter of smith et al 2002"""
    filter_dec = -np.log(1.+2.*pi/nk)/((nk-kcut)**filter_exp)
    filter_idx = np.abs(ksq/(dk*dk)) >= kcut**2.
    phit[filter_idx] *= np.exp(filter_dec*(np.sqrt(ksq
        [filter_idx]/(dk*dk))-kcut)**filter_exp)


def forcing_spatial_mask(phi):
    """TODO: Make spatial mask for forcing"""
    phi[idx_sp_mask] *= 0.


_prhs, _pprhs  = 0.0, 0.0  # previous two right hand sides


def adams_bashforth(zt, rhs, dt):
    """Take a single step forward in time using Adams-Bashforth
        3."""
    global step, t, _prhs, _pprhs
    if step is 0:
```

```
        # forward euler
        dt1 = dt
        dt2 = 0.0
        dt3 = 0.0
    elif step is 1:
        # AB2 at step 2
        dt1 = 1.5*dt
        dt2 = -0.5*dt
        dt3 = 0.0
    else:
        # AB3 from step 3 on
        dt1 = 23./12.*dt
        dt2 = -16./12.*dt
        dt3 = 5./12.*dt

    newzt = zt + dt1*rhs + dt2*_prhs + dt3*_pprhs
    _pprhs = _prhs
    _prhs  = rhs
    return newzt
###################


#######################
########################
###Velocity and vorticity distribution from Chan and Wiiliams
    1987 JAS


#######################
rm=100 ### (rnormalized)rm=100 km
vm=40  ###m s-1
b=1
#######Using Chan and William 1987 eqn 2.10, 2.11
r=np.arange(0,1280*4,10)    ######### r=0 to r=1270


v_r=vm*(r/rm)*np.exp((1/b)*(1-(r/rm)**b))


z_r=(2*(vm/rm))*(1-((1/2)*(r/rm)**b))*np.exp((1/b)*(1-(r/rm)**b))


#z_r[0]=np.nan    ##########To avoid infinity value at centre
```

```
    when r=0

fig, (ax1,ax2) = plt.subplots(1,2,figsize=(10,5))
ax1.plot(r,v_r,color='r')
ax1.set_xlabel('Distance_R_from_centre(km)')
ax1.set_ylabel(r'velocity($m$_$s^{-1}$)')
ax1.set_title('Velocity_distribution')
ax2.plot(r,z_r,color='b')
ax1.set_xlabel('Distance_R_from_centre(km)')
ax2.set_title('Vorticity_distribution')
ax2.set_xlabel('Distance_R_from_centre(km)')
ax2.set_ylabel(r'vorticity_($s^{-1}$)')
plt.savefig("Initial_velocity_and_vorticity_distribution.png",
    dpi=300)
##################
#Convert polar to cartesian coordinates
theta=np.arange(0,np.pi/2,np.pi/(2*128*4))
vxq1=np.zeros((len(r),len(r)))
vyq1=np.zeros((len(r),len(r)))
zq1=np.zeros((len(r),len(r)))
zq1[:]=np.nan

for r1 in range(0,len(r)):
    for t in range(0,len(theta)):

        vxq1[np.int(r1*cos(theta[t])),np.int(r1*sin(theta[t]))]
            =v_r[r1]*cos(theta[t])
        vyq1[np.int(r1*cos(theta[t])),np.int(r1*sin(theta[t]))]
            =v_r[r1]*sin(theta[t])

        if ~(np.isinf(z_r[r1])):
            zq1[np.int(r1*cos(theta[t])),np.int(r1*sin(theta
                [t]))]=z_r[r1]


######################
x=np.arange(0,len(theta),1)
y=np.arange(0,len(theta),1)
xx,yy=np.meshgrid(x,y)
```

```
vmq1=np.sqrt(vxq1**2+vyq1**2)
#########################
vxq2=np.copy(vxq1)
vxq3=np.copy(vxq1)
vxq4=np.copy(vxq1)

vyq2=np.copy(vyq1)
vyq3=np.copy(vyq1)
vyq4=np.copy(vyq1)


###############################
########Crosscheck_velocity ditribution for 2 qudrant(vmq1)
vmq2=np.sqrt(vxq2**2+vyq2**2)
#############Get velocity and vorticity distribution for all
    co-ordinates
vxup=np.hstack((np.fliplr(vxq2),vxq1))
vyup=np.hstack((np.fliplr(vyq2),vyq1))

vxdn=np.hstack((np.fliplr(vxq3),vxq4))
vydn=np.hstack((np.fliplr(vyq3),vyq4))


zrup=np.hstack((np.fliplr(zq1),zq1))
zrdn=np.hstack((np.fliplr(zq1),zq1))
zr1=np.vstack((np.flipud(zrup),zrdn))
#
vx1=np.vstack((np.flipud(vxup),vxdn))
vy1=np.vstack((np.flipud(vyup),vydn))

####
##Give cyclonic direction to wind:
##Quad1:
vx=np.zeros((len(theta)*2,len(theta)*2))
vy=np.zeros((len(theta)*2,len(theta)*2))
vx=np.copy(vx1)
vy=np.copy(vy1)
# ###Quad2:rleft: lower; v negative
```

```python
vx[256:,256:]=-1*vx1[256:,256:]

###Quad2:rleft: top; bothu and v negative
vx[256:,:256]=-1*vx1[256:,:256]
vy[256:,:256]=-1*vy1[256:,:256]

 ###Quad2:rleft: lower; v negative
vy[:256,:256]=-1*vy1[:256,:256]
 ##Check velocity distribution for all coodrinates
v_mag=np.sqrt((vx**2+vy**2))
plt.figure()
plt.imshow(v_mag,origin='lower')
plt.title('Magnitude_of_Velocity_magnitude')

zr1=np.copy(zr1[384:640,384:640])
plt.figure()
plt.imshow(zr1,origin='lower')
plt.title('Magnitude_of_vorticity_magnitude')
######################
############################
#zr1[np.isnan(zr1)]=0.0    ####
############################
######The code below this is based mainly on Prof. Vallis:

##############
#Constants and variables

### Configuration
nx = 256
ny = 256                          # numerical resolution
Lx = 1.0
Ly = 1.0                          # domain size [m]
ubar = 0.00                       # background zonal velocity  [m/s]
beta = 12.0                       # beta-plane f = f0 + beta y
                                  #  [1/s 1/m]

n_diss = 2.0
tau = 0.1                         # coefficient of dissipation
                                  # smaller = more dissipation
```

```
#Poorly determined coefficients for forcing and dissipation
r_rayleigh = (1./50000.)/np.sqrt(10.)
forcing_amp_factor=100.0/np.sqrt(1.)
r_rayleigh = 0.
forcing_amp_factor=0.


t = 0.0
tmax = 1000
step = 0

ALLOW_SPEEDUP = True        # if True, allow the simulation to
                                        take a larger
SPEEDUP_AT_C  = 0.4         # timestep when the Courant number
                                        drops below
                            # value of parameter SPEEDUP_AT_C
SLOWDN_AT_C = 0.6           # reduce the timestep when Courant
                                        number
                            # is bigger than SLOWDN_AT_C
PLOT_EVERY_S = 100
PLOT_EVERY_S2 = 40

### Physical Domain
nl = ny
nk = nx/2 + 1
dx = Lx / nx
dy = Ly / ny
dt = 0.4 * 16.0 / nx        # choose an initial dt. This will
                                        change
                            # as the simulation progresses to
                               maintain
                            # numerical stability
dk = 2.0*pi/Lx
dl = 2.0*pi/Ly
y = np.linspace(0, Ly, num=ny)


y_arr = np.flipud(np.tile(y,(nx,1)).transpose())
```

```
xx = np.linspace(0, Lx, num=nx)
yy = 1. - np.linspace(0, Ly, num=ny)
# calculate the wavenumbers [1/m]
# The real FT has half the number of wavenumbers in one
    direction:
# FT_x[real] -> complex : 1/2 as many complex numbers needed as
    real signal
# FT_y[complex] -> complex : After the first transform has been
    done the signal
# is complex, therefore the transformed domain in second
    dimension is same size
# as it is in euclidean space.
# Therefore FT[(nx, ny)] -> (nx/2, ny)
# The 2D Inverse transform returns a real-only domain (nx, ny)
k = dk*np.arange(0, nk, dtype=np.float64)[np.newaxis, :]
l = dl*fftfreq(nl, d=1.0/nl)[:, np.newaxis]

ksq = k**2 + l**2
ksq[ksq == 0] = 1.0               # avoid divide by zero - set ksq
                                          = 1 at zero wavenum

rksq = 1.0 / ksq                  # reciprocal 1/(k^2 + l^2)


ik = 1j*k                         # wavenumber mul. imaginary unit
                                          is useful
il = 1j*l                         # for calculating derivatives


################################
###Calculate stream function
# initialise the transformed zeta
zt = ft(zr1)       ###zrnew if calculated from the dv_dx -du_dy
#anti_alias(zt)
z=ift(zt)
psit = -rksq * zt

amp = forcing_amp_factor* np.max(np.abs(zr1))
nu = ((Lx/(np.floor(nx/3)*2.0*pi))**(2*n_diss))/tau
#High wavenumber filter coefficients.
```

```python
filter_exp = 8.
kcut = 30.

# Spectral Filter as per [Arbic and Flierl, 2003]
wvx = np.sqrt((k*dx)**2 + (l*dy)**2)
spectral_filter = np.exp(-23.6*(wvx-0.65*pi)**4)
spectral_filter[wvx <= 0.65*pi] = 1.0
##########################
#plt.figure()
#plt.pcolormesh(xx,yy,psit)
####################################

#amp = forcing_amp_factor* np.max(np.abs(qi))
# initialise the storage arrays
#time_arr[0]=t

psit = -rksq * zt              # F[psi] = - F[zeta] / (k^2 + l^2)
psixt, psiyt = grad(psit)
psix = ift(psixt)
psiy = ift(psiyt)
##################################
### Diagnostic arrays
time_arr = np.zeros(1)
tot_energy_arr = np.zeros(1)
# initialise the storage arrays
time_arr[0]=t

psit = -rksq * zt              # F[psi] = - F[zeta] / (k^2 + l^2)
psixt, psiyt = grad(psit)
psix = ift(psixt)
psiy = ift(psiyt)

urms=np.sqrt(np.mean(psix**2 + psiy**2))
tot_energy=0.5*urms**2.
tot_energy_arr[0]=tot_energy

############Streamfunction##
plt.rcParams['contour.negative_linestyle']= 'dashed'
```

```
plt.figure()
psi=ift(psit)
cs1=plt.pcolormesh(xx,yy,psi*10**2)
cs=plt.contour(xx,yy,psi*10**2,colors='k')
plt.colorbar(cs1)


####################
fig, (ax1,ax2) = plt.subplots(1,2,figsize=(10,5))
clevs=np.arange(-2,2.1,0.5)
cs=ax1.contourf(xx,yy,z,clevs,cmap='seismic',extend='both')

ax1.contour(xx,yy,z,clevs,colors='k')
#plt.colorbar(cs)
clevs1=np.arange(-2,2.1,0.5)
cbar=fig.colorbar(cs,ax=ax1,ticks=clevs1,orientation=
    'horizontal',extend='both')
cbar.ax.set_xlabel(r'Vorticity($s^{-1}$)')
#ax1.set_xlabel('Distance R from centre(km)')
#ax1.set_ylabel(r'velocity($m$ $s^{-1}$)')
ax1.set_title('Initial vorticity at{:.2f}s dt={:.2f}'.format(t,
    dt))
clevs=np.arange(-2,2.01,0.5)
cs2=ax2.contourf(xx,yy,psi*10**3,clevs,cmap='seismic',extend
    ='both')
ax2.contour(xx,yy,psi*10**3,clevs,colors='k')
#plt.colorbar(cs)
cbar2=fig.colorbar(cs2,ax=ax2,ticks=clevs1,orientation=
    'horizontal',extend='both')
cbar2.ax.set_xlabel(r'Streamfunction($10^{-3}$ $m^{2}$ $s^{-1}$)')
ax2.set_title('Initial streamfunction')
plt.savefig("Initial velocity and Streamfunction.png",dpi=300)


##########################
#show_plot()
tplot = t + PLOT_EVERY_S
tplot2 = t + PLOT_EVERY_S2
```

```python
#filter_testt=np.ones(zt.shape)
#high_wn_filter(filter_testt) #enables the profile of the
    high_wn_filter to be compared with deln, the equivalent
    for hyperviscosity.
#filter_testt_shift = np.fft.fftshift(filter_testt, axes=(0,))


while t < tmax:
    # calculate derivatives in spectral space
    psit = -rksq * zt            # F[psi] = - F[zeta] / (k^2 + l^2)
    psixt, psiyt = grad(psit)
    zxt, zyt = grad(zt)

    # transform back to physical space for pseudospectral part
    z[:] = ift(zt)
    psix = ift(psixt)
    psiy = ift(psiyt)
    zx =   ift(zxt)
    zy =   ift(zyt)

    # Non-linear: calculate the Jacobian in real space
    # and then transform back to spectral space
    jac = psix * zy - psiy * zx + ubar * zx
    jact = ft(jac)
    #   jact=0.0 ###Linear domain

    # calculate the size of timestep that can be taken
    # (assumes a domain where dx and dy are of the same order)
    c = courant_number(psix, psiy, dx, dt)
    if c >= SLOWDN_AT_C:
        print('DEBUG:_Courant_No_>_0.8,_reducing_timestep')
        dt = 0.9*dt
    elif c < SPEEDUP_AT_C and ALLOW_SPEEDUP:
        dt = 1.1*dt

    rhs = -jact - beta*psixt
```

```
        zt[:] = adams_bashforth(zt, rhs, dt)
    #   deln = 1.0 / (1.0 + nu*ksq**n_diss*dt)

    # if t > tplot:
      psi=ift(psit)

      fig, (ax1,ax2) = plt.subplots(1,2,figsize=(10,5))
      clevs=np.arange(-2,2.1,0.5)
      cs=ax1.contourf(xx,yy,z,clevs,cmap='seismic',extend='both')

      ax1.contour(xx,yy,z,clevs,colors='k')
    #plt.colorbar(cs)
      clevs1=np.arange(-2,2.1,0.5)
      cbar=fig.colorbar(cs,ax=ax1,ticks=clevs1,orientation=
          'horizontal',extend='both')
      cbar.ax.set_xlabel(r'Vorticity($s^{-1}$)')
    #ax1.set_xlabel('Distance R from centre(km)')
    #ax1.set_ylabel(r'velocity($m$ $s^{-1}$)')
      ax1.set_title('Vorticity_at{:.2f}s_dt={:.2f}'.format(t, dt))
      clevs=np.arange(-2,2.01,0.5)
      cs2=ax2.contourf(xx,yy,psi*10**3,clevs,cmap='seismic',extend=
          'both')
      ax2.contour(xx,yy,psi*10**3,clevs,colors='k')
    #plt.colorbar(cs)
      cbar2=fig.colorbar(cs2,ax=ax2,ticks=clevs1,orientation=
          'horizontal',extend='both')
      cbar2.ax.set_xlabel(r'Streamfunction($10^{-3}$_$m^{2}$_$s^{-1}$)')
      ax2.set_title('streamfunction')
      plt.savefig('Vorticity_and_streamfunction_at_time'+np.str(t)
          +'dt'+np.str(dt)+'.png',dpi=300)
      t = t + dt
      step = step + 1
      plt.close()
##############################
```

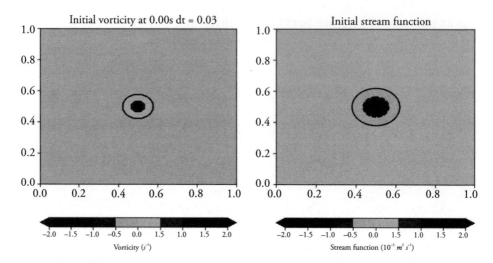

Figure 10.4　Vorticity and stream function at t = 0 for the nonlinear non-divergent vorticity equation.

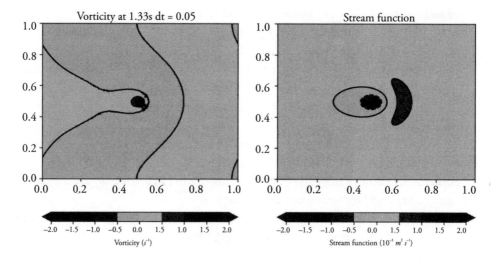

Figure 10.5　Vorticity and stream function at t = 1.33 s for the nonlinear non-divergent vorticity equation.

11

Numerical Methods for Solving Baroclinic Equations

11.1   Introduction

A baroclinic model of the atmosphere is one that does not invoke the assumptions of the barotropic model; i.e., a fluid whose density depends only upon pressure. Thus, this model is more general than a barotropic model; however, it is not fully general. Winds in the baroclinic model are still represented by geostrophic approximation. Geostrophic winds are essentially non-divergent. Because geostrophic winds are non-divergent, there exists a stream function ψ such that

$$\vec{V} = \hat{k} \times \nabla \psi \quad \text{and} \quad \zeta = \nabla^2 \psi, \tag{11.1}$$

where ψ is stream function, $\vec{V}$ is the horizontal wind and ζ is the vertical component of the relative vorticity. Moreover, as the geostrophic winds are non-divergent, they can be expressed as

$$\vec{V} = \frac{1}{f_o} \nabla \phi \quad \text{and} \quad \psi = \frac{1}{f_o} \phi. \tag{11.2}$$

However, most numerical weather prediction modelers employ baroclinic models that utilize primitive equations with hydrostatic approximation, but without quasi-geostrophic filtering. Quasi-geostrophic models were utilized for simpler problems where the main motivation was the understanding of atmospheric or oceanic dynamics. Historically, the height coordinate z was employed as the vertical coordinate. However, while utilizing primitive equations with hydrostatic approximation, it became apparent that employing pressure p as a vertical coordinate is more advantageous. The most commonly employed vertical coordinates are height z, pressure p, a normalized pressure coordinate σ, potential temperature θ, and some examples of hybrid coordinates (combination of the earlier mentioned coordinates). The most important requirement of the choice of vertical coordinates is that the vertical coordinate has to be a monotonic function of the height z.

11.2   Atmospheric Vertical Coordinates

Let any arbitrary variable $\zeta(x,y,z,t)$ be denoted as the vertical coordinate. It is of course assumed that ζ is a monotonic function of height z. In this section, the system of equations is derived for a generalised vertical coordinate $\zeta(x,y,z,t)$, where ζ is assumed to be related to the height z by a single-valued monotonic function. With transformation of the vertical coordinate z to ζ, a variable $u(x,y,z,t)$ becomes $u(x,y,\zeta(x,y,z,t),t)$. In the transformed coordinate, the horizontal coordinates remain the same. Let s represent x or y or t. Figure 11.1 shows a schematic diagram of the vertical coordinate transformation. From Figure 11.1, it can be seen that

$$\frac{C-A}{\Delta s} = \frac{B-A}{\Delta s} + \frac{C-B}{\Delta z}\frac{\Delta z}{\Delta s}, \tag{11.3}$$

$$\left(\frac{\partial u}{\partial s}\right)_\zeta = \left(\frac{\partial u}{\partial s}\right)_z + \left(\frac{\partial u}{\partial z}\right)_s \left(\frac{\partial z}{\partial s}\right)_\zeta, \tag{11.4}$$

where

$$\left(\frac{\partial u}{\partial \zeta}\right) = \left(\frac{\partial u}{\partial z}\right)\left(\frac{\partial z}{\partial \zeta}\right), \tag{11.5}$$

or

$$\left(\frac{\partial u}{\partial z}\right) = \left(\frac{\partial u}{\partial \zeta}\right)\left(\frac{\partial \zeta}{\partial z}\right). \tag{11.6}$$

Substituting Equation (11.6) into Equation (11.4), one obtains

$$\left(\frac{\partial u}{\partial s}\right)_\zeta = \left(\frac{\partial u}{\partial s}\right)_z + \left(\frac{\partial u}{\partial \zeta}\right)\left(\frac{\partial \zeta}{\partial z}\right)\left(\frac{\partial z}{\partial s}\right)_\zeta. \tag{11.7}$$

From Equation (11.7), assuming that $s=x$ and $s=y$, one can obtain an equation for a horizontal gradient of the scalar u in ζ coordinates, as

$$\nabla_\zeta u = \nabla_z u + \left(\frac{\partial u}{\partial \zeta}\right)\left(\frac{\partial \zeta}{\partial z}\right)\nabla_\zeta z \tag{11.8}$$

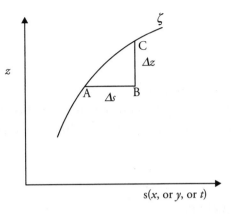

Figure 11.1 Schematic diagram showing the vertical coordinate transformation.

The horizontal divergence of a vector $\vec{V}$ is as follows:

$$\nabla_\zeta \cdot \vec{V} = \nabla_z \cdot \vec{V} + \left(\frac{\partial \vec{V}}{\partial \zeta}\right)\left(\frac{\partial \zeta}{\partial z}\right)\nabla_\zeta z. \tag{11.9}$$

The total derivative of $u(x, y, \zeta, t)$ becomes

$$\frac{du}{dt} = \left(\frac{\partial u}{\partial t}\right)_\zeta + \vec{V}\cdot\nabla_\zeta u + \dot{\zeta}\left(\frac{\partial u}{\partial \zeta}\right). \tag{11.10}$$

The horizontal pressure gradient force then becomes

$$\frac{1}{\rho}\nabla_z p = \frac{1}{\rho}\left[\nabla_\zeta p - \left(\frac{\partial p}{\partial \zeta}\right)\left(\frac{\partial \zeta}{\partial z}\right)\nabla_\zeta z\right]. \tag{11.11}$$

For an atmosphere under hydrostatic balance, $(\partial p/\partial \phi) = -\rho$ and, hence, Equation (11.11) becomes

$$\frac{1}{\rho}\nabla_z p = \frac{1}{\rho}\nabla_\zeta p + \nabla_\zeta \phi. \tag{11.12}$$

The horizontal momentum equations become

$$\frac{d\vec{V}}{dt} = -\alpha\nabla_\zeta p - \nabla_\zeta \phi - f\hat{k}\times\vec{V} + \vec{F}. \tag{11.13}$$

The hydrostatic equation, $(\partial p/\partial z) = -g\rho$, becomes

$$\left(\frac{\partial p}{\partial \zeta}\right)\left(\frac{\partial \zeta}{\partial z}\right) = -g\rho \tag{11.14}$$

or

$$\left(\frac{\partial p}{\partial \zeta}\right) = -\rho\left(\frac{\partial \phi}{\partial \zeta}\right). \tag{11.15}$$

The continuity equation can be derived from the conservation of mass principle for an infinitesimal air parcel. The hydrostatic balance equation shows that the mass of a parcel is proportional to the increase in pressure from the top to the bottom of the air parcel, i.e.,

$$g\,\Delta m = \Delta x \Delta y \Delta p. \tag{11.16}$$

From the hydrostatic equation,

$$\Delta p = \frac{\partial p}{\partial \zeta}\Delta \zeta. \tag{11.17}$$

Taking a logarithmic total derivative of Equation (11.16) and noting that

$$\frac{1}{\Delta x}\frac{d(\Delta x)}{dt} = \frac{\partial u}{\partial x} \tag{11.18}$$

and similarly for the other two derivatives with respect to y and p, one obtains

$$\frac{1}{\Delta y}\frac{d(\Delta y)}{dt} = \frac{\partial v}{\partial y} \tag{11.19}$$

and

$$\frac{1}{\Delta \zeta}\frac{d(\Delta \zeta)}{dt} = \frac{\partial \dot{\zeta}}{\partial \zeta}. \tag{11.20}$$

Utilizing Equations (11.17) to (11.20) in Equation (11.16), one obtains the continuity equation as

$$\frac{d}{dt}\left(\ln\frac{\partial p}{\partial \zeta}\right) + \nabla \cdot \vec{V}_H + \frac{\partial \dot{\zeta}}{\partial \zeta} = 0. \tag{11.21}$$

The thermodynamic equation is not changed except for the total derivative

$$c_p \frac{T}{\theta}\frac{d\theta}{dt} = c_p\frac{dT}{dt} - \alpha\frac{dp}{dt} = \dot{q}. \tag{11.22}$$

The appropriate kinematic lower boundary condition is that the surface of the earth is a material surface, i.e., the air flow can only be parallel to it, not normal. This means that once an air parcel touches the earth surface, it is "stuck" to it. The aforementioned lower boundary condition can be expressed as

$$\frac{d}{dt}(\zeta - \zeta_s) = 0 \quad \text{at} \quad \zeta = \zeta_s \tag{11.23}$$

or

$$\frac{d\zeta}{dt} = \frac{\partial \zeta_s}{\partial t} + \vec{V} \cdot \nabla \zeta_s \quad \text{at} \quad \zeta = \zeta_s. \tag{11.24}$$

The upper boundary condition at the top of the atmospheric model is not so well defined. While $z \to \infty$, $p \to 0$; however, in general there is no satisfactory manner of expressing the aforementioned upper condition for a finite vertical resolution model. Most baroclinic models assume a simple upper condition of a "rigid top," i.e., making the top surface, a material surface.

$$\frac{d\zeta}{dt} = 0 \quad \text{at} \quad \zeta = \zeta_T. \tag{11.25}$$

However, assuming the top to be a material surface is an artificial boundary condition that introduces spurious effects.

Earlier studies have shown that a rigid top introduces artificial "upsidedown" baroclinic instabilities in global models. However, if the top of the baroclinic model is sufficiently high, and there is adequate vertical resolution, the upward moving perturbations get damped in the model, and the spurious interaction with the artificial top may remain small. Alternatively, radiation conditions requiring that energy can only propagate upwards are generally used to minimize the spurious effects. However, these are not easy to implement.

11.3 Pressure as a Vertical Coordinate

As pressure p is monotonically decreasing with increase of height h, pressure meets the requirement for a vertical coordinate. The other advantages of pressure as the vertical coordinate is that (i) upper air observations are usually available at constant pressure surfaces, and hence, it would be convenient to directly utilize the upper air observations in numerical models as "initial conditions," (ii) the form of the continuity equation assumes the simplest non-divergent form, with the continuity equation becoming a *diagnostic equation*, and (iii) the horizontal pressure gradient force per unit mass terms assume simpler forms $-\nabla\phi$, resulting in the curl of these terms becoming identically zero when pressure is used as the vertical coordinate. Moreover, the geostrophic relationship assumes a simpler form $\vec{V}_H = (\frac{1}{f})\hat{k} \times \nabla\phi$, when pressure is used as the vertical coordinate. The total derivative operator for pressure as a vertical coordinate [refer to Equation (11.10)], becomes

$$\frac{d}{dt} \equiv \frac{\partial}{\partial t} + \vec{V}_H \cdot \nabla + \omega \frac{\partial}{\partial p}, \tag{11.26}$$

where $\omega = (dp/dt)$ is the vertical velocity in the pressure coordinate system. The governing equations become

$$\frac{d\vec{V}_H}{dt} = -\nabla_p \phi - f\hat{k} \times \vec{V}_H + \vec{F}, \tag{11.27}$$

with $\vec{F}$ being friction force per unit mass.

$$\frac{\partial \phi}{\partial p} = -\alpha, \tag{11.28}$$

$$\nabla_p \cdot \vec{V}_H + \frac{\partial \omega}{\partial p} = 0, \tag{11.29}$$

$$c_p \frac{T}{\theta} \frac{d\theta}{dt} = c_p \frac{dT}{dt} - \alpha \frac{dp}{dt} = \dot{q}. \tag{11.30}$$

The geostrophic and thermal wind relationships assume simple forms in the pressure coordinate system:

$$\vec{V}_g = \frac{1}{f}\hat{k} \times \nabla\phi, \tag{11.31}$$

$$\frac{\partial \vec{V}_g}{\partial p} = -\left(\frac{R}{fp}\right)\hat{k} \times \nabla T. \tag{11.32}$$

However, as the pressure surface intersects the surface, application of the bottom boundary condition is not simple in pressure coordinates.

$$\omega = \frac{\partial p_s}{\partial t} + \vec{V}_H \cdot \nabla p_s \quad \text{at} \quad p = p_s. \tag{11.33}$$

Equation (11.33) requires knowing the time rate of change of p_s. From Equation (11.29),

$$\frac{\partial p_s}{\partial t} + \vec{V}_H \cdot \nabla p_s = -\int_0^\infty (\nabla_p \cdot \vec{V}_H) dp. \tag{11.34}$$

Figure 11.2 shows a schematic diagram illustrating the height, pressure, and sigma vertical coordinates in the atmosphere.

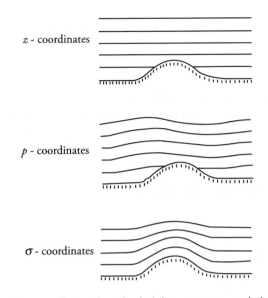

z - coordinates

p - coordinates

σ - coordinates

Figure 11.2 Schematic diagram illustrating the height, pressure, and sigma vertical coordinates in the atmosphere.

11.4 Sigma (σ) as a Vertical Coordinate

The sigma (σ) as a vertical coordinate was proposed mainly to alleviate the limitation of the pressure coordinate system caused especially due to the difficulty in prescribing the lower boundary condition. The sigma coordinate or the normalized pressure coordinate is defined as

$$\sigma = \frac{p}{p_s}, \tag{11.35}$$

where $p_s(x,y,t)$ is the surface pressure. The sigma coordinate is by far the most widely used vertical coordinate in atmospheric models, as at the earth surface, $\sigma = 1$, and at the top of the atmosphere ($p = 0$), $\sigma = 0$. This ensures that at both the top and the bottom, the boundary conditions are $\dot{\sigma} = 0$. More generally, allowing for a rigid top at the model top at a finite pressure $p_T = $ constant,

$$\sigma = \frac{p - p_T}{p_s - p_T} = \frac{p - p_T}{\pi} \quad \text{with} \quad \dot{\sigma} = 0 \quad \text{at} \quad \sigma = 0 \text{ and } 1. \tag{11.36}$$

The continuity equation in the sigma coordinate system becomes

$$\frac{d(lnp_s)}{dt} = -\nabla_\sigma \cdot (\vec{V}_H) - \frac{\partial}{\partial \sigma}(\dot{\sigma}). \tag{11.37}$$

The surface pressure tendency equation in the sigma coordinate system becomes

$$\frac{\partial p_s}{\partial t} = -\nabla_\sigma \cdot \int_0^1 (p_s \vec{V}_H)\, d\sigma. \tag{11.38}$$

after incorporating the boundary conditions.

Substituting back into the continuity equation, one can determine $\dot{\sigma}$ diagnostically from the horizontal wind field $\vec{V}_H$. The governing equations of atmospheric motion in sigma coordinates is as follows:

$$\frac{d\vec{V}_H}{dt} = -\nabla_\sigma \phi - RT\nabla_\sigma q - f\hat{k}\times \vec{V}_H + \vec{F}, \tag{11.39}$$

$$\frac{dT}{dt} = kT\left(\frac{\dot{\sigma}}{\sigma} - \frac{\partial \dot{\sigma}}{\partial \sigma} - \nabla_\sigma \cdot \vec{V}_H\right), \tag{11.40}$$

$$\frac{\partial q}{\partial t} + \vec{V}_H \cdot \nabla_\sigma q = -\left(\frac{\partial \dot{\sigma}}{\partial \sigma} + \nabla_\sigma \cdot \vec{V}_H\right), \tag{11.41}$$

$$\sigma \frac{\partial \phi}{\partial \sigma} = -RT, \tag{11.42}$$

with $q = \ln(ps)$, $k = R/c_p$, ϕ is the geopotential height and all other variables have the usual meaning. The chief advantage of the sigma coordinate system is the relative ease of implementing the lower boundary condition. Moreover, it is easy to utilize the sigma coordinate system to couple with boundary layer parameterizations because of the almost homogeneous vertical resolution near the surface associated with the sigma coordinate system. However, in the presence of steep topography, the small-scale structures in the coordinate surfaces lead to errors in the calculation of advection and the associated imbalances in the discretization of the horizontal pressure gradient may result in spurious motions over mountains. Furthermore, while resolving deep valleys in the model topography, a computationally expensive truly horizontal computation of horizontal diffusion becomes necessary. In Equation (11.38), the atmosphere is assumed to be adiabatic.

11.5   Eta (η) as a Vertical Coordinate

The eta (η) coordinate system was introduced to overcome the disadvantage of the sigma (σ) coordinate system, especially related to the computation of the pressure gradient force. The eta vertical coordinate is defined as follows:

$$\eta = \frac{p - p_T}{p_s - p_T}\eta_s, \quad \text{where} \quad \eta_s = \frac{p_{\text{ref}}(z_s) - p_T}{p_{\text{ref}}(0) - p_T}, \tag{11.43}$$

with p_T being the pressure at the top of the atmosphere (model top), p_s is the surface pressure at the model bottom boundary, z_s is the surface height at the model bottom boundary, p_{ref} is the reference pressure, which is a function of z. For example, the standard atmospheric pressure is an example of the reference pressure. Utilizing this definition of the eta vertical coordinate, mountains are modeled as three-dimensional grid boxes with the η-coordinate surfaces coinciding with the free tops of the mountain boxes. From the expression of Equation (11.43), it is clear that the eta coordinate is a generalized version of the sigma coordinate and that it is possible to utilize the Eta model with the sigma (σ) vertical coordinate by setting $\eta_s = 1$ in Equation (11.43). The governing equations of atmospheric motion in sigma coordinates is as follows:

$$\frac{d\vec{V}_H}{dt} = -\nabla_\eta \phi - \frac{RT}{p}\nabla_\eta p - f\hat{k} \times \vec{V}_H + \vec{F}, \tag{11.44}$$

$$\frac{dT}{dt} = -\frac{kT\omega}{p}, \tag{11.45}$$

$$\frac{\partial}{\partial \eta}\left(\frac{\partial p}{\partial t}\right) + \nabla \cdot \left(\vec{V}_H \frac{\partial p}{\partial \eta}\right) + \frac{\partial}{\partial \eta}\left(\dot{\eta}\frac{\partial p}{\partial \eta}\right) = 0, \tag{11.46}$$

$$\frac{\partial \phi}{\partial \eta} = -\frac{RT}{p}\frac{\partial p}{\partial \eta}, \tag{11.47}$$

$$\omega = \frac{dp}{dt} = -\int_0^\eta \nabla_\eta \cdot \left(\vec{V}_H \frac{\partial p}{\partial \eta}\right) d\eta + \vec{V}_H \cdot \nabla_\eta p, \tag{11.48}$$

$$\frac{\partial p_s}{\partial t} = -\int_0^{\eta_s} \nabla_\eta \cdot \left(\vec{V}_H \frac{\partial p}{\partial \eta}\right) d\eta, \tag{11.49}$$

$$\dot{\eta}\frac{\partial p}{\partial \eta} = -\frac{\eta}{\eta_s}\frac{\partial p}{\partial t} - \int_0^\eta \nabla_\eta \cdot \left(\vec{V}_H \frac{\partial p}{\partial \eta}\right) d\eta, \tag{11.50}$$

where the variables have their usual meaning. It is clear from Equation (11.43) that for the eta coordinate system, it follows that

$$\eta = 0 \quad \text{at} \quad p = p_T, \qquad \eta = 1 \quad \text{at} \quad z = z_S = 0, \qquad \eta = \eta_S \quad \text{at} \quad z = z_S. \qquad (11.51)$$

The top and bottom boundary conditions for the vertical velocity and the pressure are as follows:

$$\dot{\eta} = 0 \quad \text{at} \quad \eta = 0 \text{ and } \eta_S, \qquad p = p_T = \text{constant at } \eta = 0. \qquad (11.52)$$

11.6   Isentropic Vertical Coordinate

Potential temperature (θ) was proposed as a vertical coordinate as θ is a conserved variable for adiabatic atmospheric motion. The main advantage of using θ as a vertical coordinate is that, except for diabatic heating processes, "vertical" motion $\dot{\theta}$ is approximately zero in these coordinates. This feature reduces considerably the finite difference errors in regions involving fronts, where pressure or z-coordinates tend to have large errors associated with a poorly resolved vertical motion. It is pertinent to note that potential temperature is not conserved when (i) diabatic heating or cooling occurs or (ii) mixing of air parcels with different properties occurs. Examples of diabatic processes include condensation, evaporation, sensible heating from surface, radiative heating, and radiative cooling

It is important to note that (i) isentropic surfaces slope downward toward warm air, which is exactly opposite to the slope of pressure surfaces and (ii) isentropic surfaces slope much more steeply than pressure surfaces for the same thermal gradient.

Using the definition of potential temperature,

$$\theta = T \left(\frac{p_o}{p} \right)^{R/c_p}. \qquad (11.53)$$

From Equation (11.53), it is clear that on a constant θ surface, an isobar (constant pressure line) must also be an isotherm (line of constant temperature). From the equation of ideal gas (relation between pressure, temperature, and density), one can conclude that on a constant θ surface, an isobar (constant pressure line) must also be an isopycnal (line of constant density). Hence, on a constant θ surface, pressure advection is also equivalent to thermal advection, i.e., when wind blows from high pressure to low pressure on a constant θ surface, the same is also associated with warm air advection.

Taking the logarithm and differentiating Equation (11.53), one obtains

$$\frac{d\theta}{\theta} = \frac{dT}{T} - \frac{R}{c_p}\frac{dp}{p} = \frac{dT}{T} + \frac{1}{c_p}\frac{d\phi}{T}. \tag{11.54}$$

Defining the Exner function as

$$\pi = \frac{c_p T}{\theta} = c_p \left(\frac{p}{p_o}\right)^{R/c_p} \tag{11.55}$$

and the Montgomery potential as

$$M = c_p T + \phi, \tag{11.56}$$

Equation (11.54) in terms of the Exner function π and Montgomery potential M is of the form

$$\frac{\partial M}{\partial \theta} = \pi. \tag{11.57}$$

Equation (11.57) is hydrostatic equation.

The horizontal pressure gradient term becomes very simple in the isentropic coordinate system; hence, the horizontal momentum equation is given by

$$\frac{d\vec{V}_H}{dt} = -\nabla_\theta M - f\hat{k} \times \vec{V}_H. \tag{11.58}$$

The continuity equation in isentropic coordinate becomes

$$\frac{d}{dt}\left(\ln\frac{\partial p}{\partial\theta}\right) + \nabla_\theta \cdot \vec{V}_H + \frac{\partial\dot{\theta}}{\partial\theta} = 0. \tag{11.59}$$

According to *Ertel's theorem*, the potential vorticity (q) is a conserved quantity for adiabatic, frictionless flow. This general property can be expressed in its simplest form in the isentropic coordinate as follows:

$$\frac{dq}{dt} = 0, \quad \text{where} \quad q = -g\left(f + \hat{k}\cdot\nabla_\theta \times \vec{V}\right)\frac{\partial\theta}{\partial p} \tag{11.60}$$

Despite the obvious advantages of the isentropic coordinate system, they have also two important disadvantages. The first disadvantage is a familiar one, that the isentropic surfaces intersect the ground (which was also the case for the pressure coordinates), leading to a situation where it is difficult to ensure strict conservation of mass. The second disadvantage of using the isentropic coordinate system is that only statically stable solutions are allowed, as the vertical coordinate (θ) has to vary monotonically with height. There are obvious situations, for example over hot surfaces, where this assumption is not true. Moreover, over regions of low static stability, the vertical resolution of isentropic coordinates can be inadequate. To overcome the aforementioned disadvantages of the isentropic coordinate system and the other vertical coordinates that were discussed in earlier sections, hybrid sigma–theta coordinates have been proposed and are being utilized.

The following are some of the advantages of isentropic coordinates: (i) for synoptic scale motions, in the absence of diabatic processes, isentropic surfaces are material surfaces, i.e., air parcels are thermodynamically bound to the constant θ-surface. Horizontal flow along an isentropic surface contains the adiabatic component of vertical motion; (ii) moisture transport on an isentropic surface is three-dimensional; patterns are more spatially and temporally coherent than on pressure surface; (iii) isentropic surfaces tend to run parallel to frontal zones making the variation of basic meteorological quantities such as horizontal wind components, air temperature, and moisture (u, v, T, q) more gradual along them, (iv) the vertical spacing between isentropic surfaces is a measure of the dry static stability. Convergence (divergence) between two isentropic surfaces increases (decreases) the static stability in the layer; and (v) the slope of a constant θ surface is directly related to the thermal wind.

The following are some of the disadvantages of isentropic coordinates: (i) over regions of neutral or super adiabatic lapse rates, isentropic surfaces are not well-defined, i.e., they are multi-valued with respect to pressure; (ii) over areas having near-neutral lapse rates, there is poor vertical resolution of atmospheric features in isentropic coordinates. However, over stable frontal zones, there is very good vertical resolution; (iii) diabatic processes significantly disrupt the continuity of isentropic surfaces and hence, isentropic coordinates are not very useful where diabatic processes dominate; and (iv) isentropic surfaces tend to intersect the ground at steep angles and hence, require careful analysis near the ground.

11.7 Vertical Staggering

The two commonly employed vertical staggering approaches are (i) Charney and Phillips the Charney–Phillips grid, proposed by Charney and Phillips, and (ii) Lorenz grid, proposed by Lorenz. Figure 11.3 depicts the configuration of variables for the Charney–Phillips (right) grid and the Lorenz grid (left). In the Lorenz grid, most of the dependent prognostic variables (such as horizontal wind components, temperature, and geopotential) except for the vertical velocity are prescribed at the center of the layer with the vertical velocities prescribed at the boundary between the vertical layers. However, for the Charney–Phillips grid, the vertical staggering is such that the vertical velocity and the geopotential are prescribed at the boundary between the layers while the other prognostic dependent variables such as horizontal wind components and temperature are prescribed at the center of the vertical layers. Both grids have some advantages and some disadvantages. The Lorenz grid allows simple quadratic conservation and also ensures that the requirement of no flux is enforced easily at the top and the bottom boundary. However, the Lorenz permits the development of a spurious computational mode, as the geopotential in the hydrostatic equation (and, hence, the acceleration of the wind components) is insensitive to the temperature oscillations of $2\Delta\sigma$ wavelengths. However, in the Charney–Phillips grid, the vertical staggering is relatively more consistent with the hydrostatic equation as compared to the Lorenz grid and, hence, there is no additional computational mode developed in the Charney–Phillips mode. Furthermore, studies have indicated that utilizing the Charney–Phillips grid leads to improved accuracy in the representation of vertical wave propagation. If one were to utilize a non-staggered grid in the vertical direction that allows for the implementation of higher order differences in the vertical direction, it would invariably lead to the presence of additional computational modes in the solution.

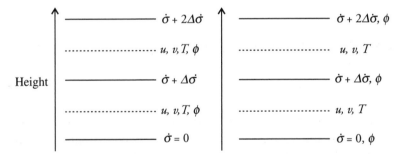

Figure 11.3 Staggered grid in the vertical following (i) Lorentz grid (left), and (ii) Charney–Phillips grid (right).

11.8   Two-layer Quasi-geostrophic Equation

The two-layer quasi-geostrophic model is one of the simplest and the most familiar of the baroclinic models that utilizes the quasi-geostrophic approximation. In quasi-geostrophic theory, the following assumptions are made: (i) the horizontal winds are approximately geostrophic; (ii) the atmosphere is approximately hydrostatic; (iii) advection is dominated by geostrophic winds; and (iv) small vertical temperature perturbations are present. These assumptions allow the simplification of the three-dimensional equations of motion while allowing the retention of the time derivative terms that ensures that the governing equations are prognostic. The quasi-geostrophic model has been applied to investigate the dynamics of the mid-latitude synoptic scale motion in isobaric coordinates. The quasi-geostrophic vorticity equation using pressure as a vertical coordinate is of the form

$$\nabla^2 \left(\frac{\partial \phi}{\partial t} \right) + J[\phi, \zeta + f] - f_o^2 \frac{\partial \omega}{\partial p} = 0. \tag{11.61}$$

The thermodynamic energy equation assuming no diabatic heating, continuity equation, and hydrostatic equation in isobaric coordinates are as follows:

$$\frac{\partial \theta}{\partial t} + \vec{V}_H \cdot \nabla \theta + \omega \frac{\partial \theta}{\partial p} = 0, \tag{11.62}$$

$$\frac{\partial \omega}{\partial p} + \nabla^2 \chi = 0, \tag{11.63}$$

$$\frac{\partial \phi}{\partial p} + \frac{RT}{p} = 0, \tag{11.64}$$

where $\vec{V}_H$ is the horizontal velocity, given by

$$\vec{V}_H = \frac{1}{f_o} \hat{k} \times \nabla \phi \quad \text{and} \quad \zeta = \frac{1}{f_o} \nabla^2 \phi, \tag{11.65}$$

where ϕ is the geopotential, ζ is the vertical component of the relative vorticity vector, $\omega = (dp/dt)$ is the vertical velocity in isobaric coordinates, θ is the potential temperature, T is the absolute temperature, f_o is the average (constant) value of Coriolis parameter, and χ is the velocity potential for the divergent wind component. The quasi-geostrophy theory assumes that $\vec{V}_H$ in Equation (11.62) is to be obtained from Equation (11.65) and the value of $(\partial \theta / \partial p)$ is assumed to be constant. The atmosphere is divided into four layers ($\Delta p = 250\,\text{hPa}$) of equal thickness (250 hPa thickness) between the earth surface ($p = 1000\,\text{hPa}$) and the top of the atmosphere ($p = 0$). Figure 11.4 shows the schematic vertical structure of the two-level model. One needs to utilize the central finite difference scheme to evaluate the pressure derivatives besides employing the following boundary conditions

$$\omega = \frac{dp}{dt} = 0 \quad \text{at} \quad p = p_o \text{ and } p = 0. \tag{11.66}$$

$p = 0$ 0 hPa ——————————————————————— 0

250 hPa ——————————————————————— 1

500 hPa ——————————————————————— 2

750 hPa ——————————————————————— 3

$p = p_o$ 1000 hPa ——————————————————————— 4

Figure 11.4   Vertical structure of the two-level model.

The system of equations then become

$$\nabla^2 \left(\frac{\partial \psi}{\partial t} \right) + J[\psi, \nabla^2 \psi + f] + J[\tau, \nabla^2 \tau] = 0, \tag{11.67}$$

$$\nabla^2 \left(\frac{\partial \tau}{\partial t} \right) + J[\tau, \nabla^2 \psi + f] + J[\psi, \nabla^2 \tau] - \frac{2 f_o}{p_o} \omega = 0, \tag{11.68}$$

$$\sigma_o \nabla^2 \omega - 2^k f_o^2 R^{-1} \omega = \frac{p_o}{4} \nabla^2 J(\psi, \theta) - \frac{2^{k-1} f_o p_o}{R} \left[J[\tau, \nabla^2 \psi + f] + J[\psi, \nabla^2 \tau] \right], \tag{11.69}$$

$$\theta = 2^{k+1} f_o R^{-1} \tau, \tag{11.70}$$

where the changed dependent variables are

$$\psi = \frac{\phi_1 + \phi_3}{2 f_o}, \tag{11.71}$$

$$\tau = \frac{\phi_1 - \phi_3}{2 f_o}, \tag{11.72}$$

$$\theta = \theta_2 \quad \text{and} \quad \omega = \omega_2, \tag{11.73}$$

where the subscripts refer to the levels as indicated in Figure 11.4, whereas $\sigma_o = -(\partial \theta / \partial p)$ is the constant static stability. Equations (11.67) and (11.68) are known as mean vorticity and thermal vorticity equations. Equation (11.69) is the omega equation applied at level 2, whereas Equation (11.70) is a form of thermal wind equation.

Alternatively, one can apply the vorticity equation at levels 1 and 3, that would lead to the following equations

$$\frac{\partial \zeta_1}{\partial t} + \vec{V}_1 \cdot \nabla(\zeta_1 + f) = \overline{f}\left(\frac{\partial \omega}{\partial p}\right)_1 = \overline{f}\left(\frac{\omega_2 - \omega_0}{2\Delta p}\right). \tag{11.74}$$

$$\frac{\partial \zeta_3}{\partial t} + \vec{V}_3 \cdot \nabla(\zeta_3 + f) = \overline{f}\left(\frac{\partial \omega}{\partial p}\right)_3 = \overline{f}\left(\frac{\omega_4 - \omega_2}{2\Delta p}\right). \tag{11.75}$$

As before, one assumes that the vertical velocity in the pressure coordinates at the bottom and at the top is zero, i.e., $\omega_4 = 0$; $\omega_0 = 0$. Moreover, one can utilize the following definitions:

$$\vec{V} = \frac{\vec{V}_1 + \vec{V}_3}{2}, \qquad \vec{V}_T = \frac{\vec{V}_1 - \vec{V}_3}{2}, \qquad \zeta = \frac{\zeta_1 + \zeta_3}{2}, \qquad \zeta_T = \frac{\zeta_1 - \zeta_3}{2}. \tag{11.76}$$

Equations (11.76) indicate that $\vec{V}$ and $\vec{V}_T$ refer to the mean wind and the thermal wind respectively. Moreover, ζ and ζ_T refer to the mean vorticity and the thermal vorticity of the atmospheric layer between levels 1 and 3 as indicated in Equation (11.76).

Substituting $\vec{V}_1 = \vec{V} + \vec{V}_T$, $\vec{V}_3 = \vec{V} - \vec{V}_T$, $\zeta_1 = \zeta + \zeta_T$, $\zeta_3 = \zeta - \zeta_T$ in Equations (11.74) and (11.75), and adding and subtracting the aforesaid equations after substitution one obtains

$$\frac{\partial \zeta}{\partial t} + \vec{V} \cdot \nabla(\zeta + f) + \vec{V}_T \cdot \nabla \zeta_T = 0. \tag{11.77}$$

$$\frac{\partial \zeta_T}{\partial t} + \vec{V} \cdot \nabla \zeta_T + \vec{V}_T \cdot \nabla(\zeta + f) = \frac{\overline{f}\omega_2}{2\Delta p}. \tag{11.78}$$

The thermodynamic energy equation in terms of potential temperature θ and in the absence of diabatic heating in pressure coordinates is

$$\frac{\partial \theta}{\partial t} + \vec{V} \cdot \nabla \theta + \omega \frac{\partial \theta}{\partial p} = 0. \tag{11.79}$$

Using the definition of potential temperature and the equation of state for ideal gas and hydrostatic equation in pressure coordinates

$$\theta = T\left(\frac{p_o}{p}\right)^{R/c_p}, \qquad p\alpha = RT, \qquad \frac{\partial \phi}{\partial p} = -\alpha. \tag{11.80}$$

Using Equation (11.80) in Equation (11.79), one obtains

$$\frac{\partial}{\partial t}\left(\frac{\partial \phi}{\partial p}\right) + \vec{V} \cdot \nabla \left(\frac{\partial \phi}{\partial p}\right) + \sigma \omega = 0, \tag{11.81}$$

where

$$\sigma = -\alpha \left[\frac{\partial (\ln \theta)}{\partial p}\right].$$

Applying Equation (11.81) in level 2, one obtains Equation (11.84) after substituting the following expressions for $\vec{V}$, $\vec{V}_T$, ζ, and ζ_T. Using geostrophic approximations,

$$\vec{V} = \frac{g}{f}\hat{k} \times \nabla z, \qquad \vec{V}_T = \frac{g}{f}\hat{k} \times \nabla h, \qquad \zeta = \frac{g}{f}\nabla^2 z, \qquad \zeta_T = \frac{g}{f}\nabla^2 h \tag{11.82}$$

where

$$z = \frac{z_1 + z_3}{2}, \qquad h = \frac{z_1 - z_3}{2}, \tag{11.83}$$

$$\frac{\partial h}{\partial t} + \vec{V}_2 \cdot \nabla h + (\Delta p \sigma \omega_2)/g = 0. \tag{11.84}$$

Equations (11.77), (11.78), and (11.84) can be written in terms of z, h, and ω_2 as can be seen from Equations (11.82) and (11.83). This forms a close set of three equations in three unknowns.

A further simplification is obtained by eliminating ω_2 from Equations (11.78) and (11.84) to obtain a modified equation, Equation (11.78):

$$\frac{\partial \zeta_T}{\partial t} + \vec{V} \cdot \nabla \zeta_T + \vec{V}_T \cdot \nabla(\zeta + f) - \frac{\overline{f}g}{4\sigma(\Delta p)^2}\left(\frac{\partial h}{\partial t} + \vec{V} \cdot \nabla h\right) = 0. \tag{11.85}$$

Now Equations (11.77) and (11.85) can be written in terms of z and h forming a set of two equations in two unknowns. An important characteristic of this model is the fact that the vertical velocity field ω_2 is directly available from Equation (11.84) with the knowledge of $(\partial h/\partial t)$ obtained by solving Equation (11.85). A negative value of ω_2 indicates upward vertical motion and, hence, the possibility of cloud formation and rainfall. Furthermore, a *diagnostic equation* for ω_2 can be obtained as indicated in the following by applying the Laplacian operator to Equation (11.84) and subtracting the aforementioned equation from Equation (11.78):

$$\sigma \nabla^2 \omega_2 + \frac{\overline{f}^2 \omega_2}{2(\Delta p)^2} = -\left(\frac{g}{\Delta p}\right)\nabla^2(\vec{V} \cdot \nabla h) + \frac{\overline{f}}{\Delta p}\left[\vec{V} \cdot \nabla \zeta_T + \vec{V} \cdot \nabla(\zeta + f)\right]. \tag{11.86}$$

Equation (11.86) is of the Helmholtz type, which can be directly solved using the successive over relaxation method.

11.9   Multi-level Models

The two-layer quasi-geostrophic model discussed in Section 11.6 is easily extendable to more vertical levels, the latter providing for better vertical resolution. Figure 11.5 provides a schematic figure that shows the vertical indexing and specifications of various dependent variables for a multi-level baroclinic atmospheric model. Figure 11.5 shows that the stream function ψ (rotational part) and the velocity potential χ (divergent part) are prescribed at odd levels (mid-point of the different layers), whereas the vertical velocity ω is prescribed at the boundary between the different layers (at the even numbered levels). The above prescription of rotational, divergent and vertical motion may be carried out by solving the vorticity equation for the height tendency $(\partial \omega / \partial p)$ at the odd levels, $1, 3, \ldots, n-1$ and solving a thermal vorticity equation at the even levels $2, 4, \ldots, n-2$ as shown in Figure 11.5. Moreover, the diagnostic ω equation may be derived and solved over the three-dimensional grid to obtain ω at the even levels $2, 4, \ldots, n-2$. Solving the diagnostic ω equation and obtaining ω fields at the even vertical levels (employing suitable boundary conditions at the top and the bottom of the model levels), the term in the vorticity equation can be calculated using a central finite difference scheme and the equation can be solved for the height tendency by successive overrelaxation method. The solution may be advanced in time by employing a suitable time difference scheme.

The vorticity equation as expressed using quasi-geostrophic approximation in the pressure coordinate is given as

$$\frac{g}{f} \nabla^2 \left(\frac{\partial z}{\partial t} \right) + \vec{V} \cdot \nabla (\zeta + f) = \overline{f} \frac{\partial \omega}{\partial p}, \tag{11.87}$$

where

$$\zeta = \frac{g}{f} \nabla^2 z,$$

The thermodynamic energy equation is written in the pressure coordinate as

$$\frac{\partial}{\partial t} \left(\frac{\partial z}{\partial p} \right) + \vec{V} \cdot \nabla \left(\frac{\partial z}{\partial p} \right) + \sigma \omega = -Q, \tag{11.88}$$

where σ is a function of pressure p only and Q is the diabatic heating rate. For the multi-level baroclinic model, the vorticity equation, Equation (11.87) is prescribed at

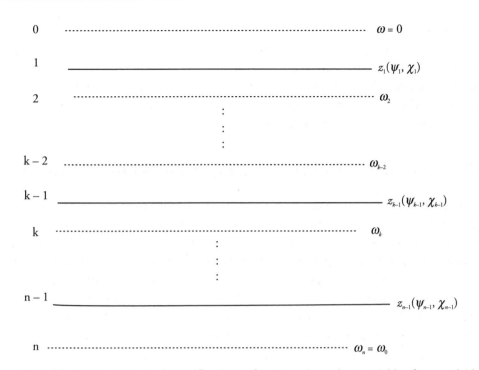

Figure 11.5 Vertical indexing and specifications of various dependent variables for a multi-level baroclinic atmospheric model.

the odd vertical levels, $1,3,5,\ldots,n-1$ and the diagnostic ω equation is prescribed at the even vertical levels $2,4,6,\ldots,n-2$. The diagnostic ω equation can be derived by eliminating the time derivatives in Equations (11.87) and (11.88), by differentiating Equation (11.87) with respect to p and applying the Laplacian operator to Equation (11.88), and then obtaining the difference between the aforementioned two equations. This derivation of ω equation results in the following equation:

$$\sigma \nabla^2 \omega + \frac{\overline{f}^2}{2} \frac{\partial^2 \omega}{\partial p^2} = \overline{f} \frac{\partial}{\partial p} \left[\vec{V} \cdot \nabla \zeta_T + \vec{V} \cdot \nabla(\zeta + f) \right] - \nabla^2 \left(\vec{V} \cdot \nabla \frac{g \partial z}{\partial p} \right) - \nabla^2 Q$$

$$= \overline{f} \frac{\partial}{\partial p} \left[\vec{V} \cdot \nabla \zeta_T + \vec{V} \cdot \nabla(\zeta + f) \right] + \frac{R}{p} \nabla^2 \left(\vec{V} \cdot \nabla T \right) - \nabla^2 Q. \quad (11.89)$$

The RHS of Equation (11.89) contains terms that involve (i) the vertical variation of vorticity advection, (ii) the Laplacian of horizontal temperature advection, and (iii) the Laplacian of the diabatic heating. Equation (11.89) may be expressed in finite difference form by utilizing the central finite difference of the Laplacian operator and the second-order derivative of ω with respect to pressure p. The nonlinear vertical variation of vorticity advection term and the nonlinear term involving the Laplacian of the horizontal temperature advection are to be treated in such a manner that nonlinear computational instability is avoided. The lateral and vertical boundary conditions for ω needs to be prescribed; these are usually taken as zero except for the lower boundary condition at the earth surface. The diagnostic ω Equation (11.89) can then be solved using the successive overrelaxation method.

11.10   Limited Area Primitive Equation Atmospheric Model

The limited area primitive equation atmospheric model over the Australian region (McGregor et al. 1978) is a multi-level semi implicit model in which the governing equations of atmospheric motion are written in the flux form and include an option to represent topography together with parameterization schemes for dry and moist convection, vertical and horizontal diffusion and surface turbulent transfer of heat, momentum, and moisture. The vertical distribution of the dependent variables is based on the Lorentz grid having six vertical layers (refer to Figure 11.6 with air temperature, horizontal wind components, water vapour mixing ratio and geopotential prescribed at the middle of the vertical layers whereas the vertical velocity is prescribed at the boundary between the vertical layers) with the horizontal grid size of 250 km requiring a time step of 36 minutes while employing the semi-implicit method of integration.

The governing equations in flux form are as follows.

Horizontal momentum equations:

$$\frac{\partial(p_*u)}{\partial t} = -m^2\left[\frac{\partial}{\partial x}\left(\frac{up_*u}{m}\right) + \frac{\partial}{\partial y}\left(\frac{vp_*u}{m}\right)\right] - \frac{\partial(\dot{\sigma}p_*u)}{\partial\sigma} + fp_*v - m\left(p_*\frac{\partial\phi}{\partial x} + RT\frac{\partial p_*}{\partial x}\right)$$
$$+ F_x + D_u. \quad (11.90)$$

$$\frac{\partial(p_*v)}{\partial t} = -m^2\left[\frac{\partial}{\partial x}\left(\frac{up_*v}{m}\right) + \frac{\partial}{\partial y}\left(\frac{vp_*v}{m}\right)\right] - \frac{\partial(\dot{\sigma}p_*v)}{\partial\sigma} - fp_*u - m\left(p_*\frac{\partial\phi}{\partial y} + RT\frac{\partial p_*}{\partial y}\right)$$
$$+ F_y + D_v. \quad (11.91)$$

Thermodynamic energy equation:

$$\frac{\partial(p_*T)}{\partial t} = -m^2\left[\frac{\partial}{\partial x}\left(\frac{up_*T}{m}\right) + \frac{\partial}{\partial y}\left(\frac{vp_*T}{m}\right)\right] - \frac{\partial(\dot{\sigma}p_*T)}{\partial\sigma}$$
$$+ \frac{RT}{\sigma c_p}\left\{p_*\dot{\sigma} + \sigma\left[\frac{\partial p_*}{\partial t} + m\left(u\frac{\partial p_*}{\partial x} + v\frac{\partial p_*}{\partial y}\right)\right]\right\} + p_*H + D_T. \quad (11.92)$$

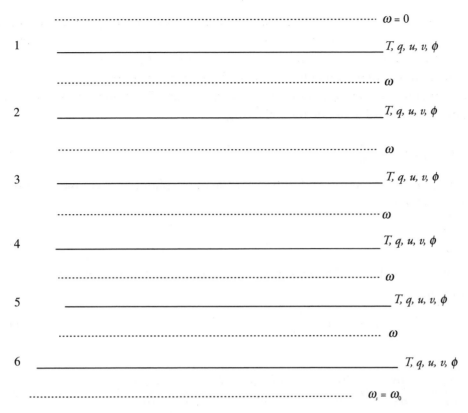

Figure 11.6 Vertical indexing and specifications of various dependent variables for a six layer baroclinic atmospheric model.

Moisture equation

$$\frac{\partial(p_*q)}{\partial t} = -m^2\left[\frac{\partial}{\partial x}\left(\frac{up_*q}{m}\right) + \frac{\partial}{\partial y}\left(\frac{vp_*q}{m}\right)\right] - \frac{\partial(\dot\sigma p_*q)}{\partial\sigma} + p_*Q + D_q. \tag{11.93}$$

Continuity equation

$$\frac{\partial p_*}{\partial t} = -m^2\left[\frac{\partial}{\partial x}\left(\frac{p_*u}{m}\right) + \frac{\partial}{\partial y}\left(\frac{p_*v}{m}\right)\right] - p_*\frac{\partial\dot\sigma}{\partial\sigma}. \tag{11.94}$$

Hydrostatic equation

$$\frac{\partial\phi}{\partial\sigma} = -\frac{RT}{\sigma}. \tag{11.95}$$

In these equations, u and v are the horizontal zonal (east–west) and meridional (north–south) wind components, m is the map factor for a *Lambert conformal projection*, p_*, the surface pressure, f is the Coriolis parameter, T the air temperature, ϕ the geopotential, q the water vapour mixing ratio, H the diabatic heating, Q the moisture source sink term, F_x and F_y are the skin friction coefficients, and D_u, D_v, D_T, and D_q are the lateral diffusion of horizontal momentum, heat, and moisture respectively.

Introducing a climatological mean air temperature $T_o(\sigma)$ and a reference surface pressure $p_o = 1000\,\text{hPa}$, define the following:

$$W = p_* \dot\sigma + \sigma \frac{\partial p_*}{\partial t}. \tag{11.96}$$

Assuming $\dot\sigma = 0$ at the top and bottom of the atmospheric model, one obtains from Equation (11.96),

$$\frac{\partial p_*}{\partial t} = W_*, \tag{11.97}$$

where W_* is the value of W at the surface. Equations (11.90) to (11.95) may be rewritten in terms of

$$\hat{u} = \frac{p_* u}{m} \quad \text{and} \quad \hat{v} = \frac{p_* v}{m} \tag{11.98}$$

as

$$\frac{\partial \hat{u}}{\partial t} + p_o \frac{\partial \phi}{\partial x} + RT_o \frac{\partial p_*}{\partial x} = -m \left[\frac{\partial(u\hat{u})}{\partial x} + \frac{\partial(v\hat{u})}{\partial y} \right] - \frac{\partial(\dot\sigma \hat{u})}{\partial \sigma} + f\hat{v} + (p_o - p_*)\frac{\partial \phi}{\partial x}$$
$$+ R(T_o - T)\frac{\partial p_*}{\partial x} + \frac{F_x}{m} + \frac{D_u}{m}, \tag{11.99}$$

$$\frac{\partial \hat{v}}{\partial t} + p_o \frac{\partial \phi}{\partial y} + RT_o \frac{\partial p_*}{\partial y} = -m \left[\frac{\partial(u\hat{v})}{\partial x} + \frac{\partial(v\hat{v})}{\partial y} \right] - \frac{\partial(\dot\sigma \hat{v})}{\partial \sigma} - f\hat{u} + (p_o - p_*)\frac{\partial \phi}{\partial y}$$
$$+ R(T_o - T)\frac{\partial p_*}{\partial y} + \frac{F_y}{m} + \frac{D_v}{m}, \tag{11.100}$$

$$\frac{\partial T}{\partial t} - \frac{W}{p_*}\left(\frac{\alpha T}{\sigma} - \frac{\partial T}{\partial \sigma} \right) - \frac{\sigma W_*}{p_*}\frac{\partial T}{\partial \sigma} = -\frac{m^2}{p_*}\left(\hat{u}\frac{\partial T}{\partial x} + \hat{v}\frac{\partial T}{\partial y} \right) + \frac{\alpha m^2 T}{p_*^2}\left(\hat{u}\frac{\partial p_*}{\partial x} + \hat{v}\frac{\partial p_*}{\partial y} \right)$$
$$+ H + \frac{D_T}{p_*}, \tag{11.101}$$

where $\alpha = R/c_p$.

$$\frac{\partial(p_* q)}{\partial t} = -m^2 \left[\frac{\partial(\hat{u} q)}{\partial x} + \frac{\partial(\hat{v} q)}{\partial y} \right] - \frac{\partial(\dot\sigma p_* q)}{\partial \sigma} + p_* Q + D_q. \tag{11.102}$$

$$\frac{\partial p_*}{\partial t} = -m^2 \left(\frac{\partial \hat{u}}{\partial x} + \frac{\partial \hat{v}}{\partial y} \right) - p_* \frac{\partial \dot{\sigma}}{\partial \sigma}. \tag{11.103}$$

Equation (11.103) can be rewritten using Equation (11.96) as follows:

$$\frac{\partial W}{\partial \sigma} - m^2 \left(\frac{\partial \hat{u}}{\partial x} + \frac{\partial \hat{v}}{\partial y} \right) = 0. \tag{11.104}$$

Finally, the hydrostatic equation remains the same:

$$\frac{\partial \phi}{\partial \sigma} = -\frac{RT}{\sigma}. \tag{11.105}$$

11.10.1 Finite difference equations for the limited area primitive equation atmospheric model

Define the averaging and derivative operators as

$$\overline{A}^t = \frac{A(t+\Delta t/2) + A(t-\Delta t/2)}{2} \quad \text{and} \quad A_t = \frac{A(t+\Delta t/2) - A(t-\Delta t/2)}{\Delta t}. \tag{11.106}$$

Equations (11.99) to (11.102) and Equations (11.104) and (11.105) can be rewritten using leapfrog time differencing and time averaging appropriately the gravity wave terms as in the semi-implicit method:

$$\overline{p}_*^{2t} - p_*(t - \Delta t) = \Delta t \overline{W}_*^{2t}. \tag{11.107}$$

Horizontal momentum equations:

$$\overline{\hat{u}}^{2t} + \Delta t p_o \frac{\partial \overline{\phi}^{2t}}{\partial x} + \Delta t R T_o \frac{\partial \overline{p}_*^{2t}}{\partial x} = a + \Delta t \frac{\partial}{\partial x} [p_o \phi_* + R T_o p_*(t - \Delta t)], \tag{11.108}$$

where

$$a = \hat{u}(t - \Delta t) - m\Delta t \left[\frac{\partial(u\hat{u})}{\partial x} + \frac{\partial(v\hat{u})}{\partial y} \right] + \Delta t \left[(p_o - p_*) \frac{\partial \phi}{\partial x} + R(T_o - T) \frac{\partial p_*}{\partial x} - \frac{\partial(\dot{\sigma}\hat{u})}{\partial \sigma} + f\hat{v} + \frac{F_x}{m} + \frac{D_u}{m} \right]$$
$$- \Delta t \frac{\partial}{\partial x} [p_o \phi_* + R T_o p_*(t - \Delta t)],$$

$$\overline{\hat{v}}^{2t} + \Delta t p_o \frac{\partial \overline{\phi}^{2t}}{\partial y} + \Delta t R T_o \frac{\partial \overline{p}_*^{2t}}{\partial y} = b + \Delta t \frac{\partial}{\partial y} [p_o \phi_* + R T_o p_*(t - \Delta t)], \tag{11.109}$$

where

$$b = \hat{v}(t-\Delta t) - m\Delta t \left[\frac{\partial(u\hat{v})}{\partial x} + \frac{\partial(v\hat{v})}{\partial y}\right] + \Delta t \left[(p_o - p_*)\frac{\partial \phi}{\partial y} + R(T_o - T)\frac{\partial p_*}{\partial y} - \frac{\partial(\dot{\sigma}\hat{v})}{\partial \sigma} - f\hat{u} + \frac{F_y}{m} + \frac{D_v}{m}\right]$$

$$- \Delta t \frac{\partial}{\partial y}\left[p_o\phi_* + RT_o p_*(t-\Delta t)\right].$$

Thermodynamic energy equation:

$$\overline{T}^{2t} - \frac{\Delta t}{p_*}\left(\frac{\alpha T_o}{\sigma} - \frac{\partial T_o}{\partial \sigma}\right)\overline{W}^{2t} - \frac{\sigma\Delta t}{p_o}\frac{\partial T_o}{\partial \sigma}\overline{W}_*^{2t} = T(t-\Delta t) - \frac{m^2\Delta t}{p_*}\left(\hat{u}\frac{\partial T}{\partial x} + \hat{v}\frac{\partial T}{\partial y}\right)$$

$$+ \frac{\Delta t\,\alpha m^2 T}{p_*^2}\left(\hat{u}\frac{\partial p_*}{\partial x} + \hat{v}\frac{\partial p_*}{\partial y}\right) + \Delta t\left[\frac{1}{p_*}\left(\frac{\alpha T}{\sigma} - \frac{\partial T}{\partial \sigma}\right)W - \frac{1}{p_o}\left(\frac{\alpha T_o}{\sigma} - \frac{\partial T_o}{\partial \sigma}\right)W\right]$$

$$+ \Delta t\left[\frac{\sigma}{p_*}\frac{\partial T}{\partial \sigma}W_* - \frac{\sigma}{p_o}\frac{\partial T_o}{\partial \sigma}W_*\right] + \Delta t\left(H + \frac{D_T}{p_*}\right) = c. \quad (11.110)$$

Moisture equation:

$$(\overline{p}_*\overline{q})^t = \Delta t \left\{-m^2\left[\frac{\partial(\hat{u}q)}{\partial x} + \frac{\partial(\hat{v}q)}{\partial y}\right] - \frac{\partial(\dot{\sigma}p_*q)}{\partial \sigma} + p_*Q + D_q\right\}. \quad (11.111)$$

Continuity equation:

$$\frac{\partial \overline{W}^{2t}}{\partial \sigma} + m^2\left(\frac{\partial \overline{\hat{u}}^{2t}}{\partial x} + \frac{\partial \overline{\hat{v}}^{2t}}{\partial y}\right) = 0. \quad (11.112)$$

Hydrostatic equation:

$$\frac{\partial \phi}{\partial \sigma} = -\frac{RT}{\sigma}. \quad (11.113)$$

11.10.2 Solution procedure

Let k be the vertical differencing index. The thermodynamic energy equation, Equation (11.110) may then be written as

$$\overline{T}_k^{2t} - \frac{\Delta t}{2p_o}\left(\frac{\alpha T_o}{\sigma} - \frac{\partial T_o}{\partial \sigma}\right)\left(\overline{W}_{k-1}^{2t} + \overline{W}_k^{2t}\right) - \sigma\frac{\Delta t}{p_o}\frac{\partial T_o}{\partial \sigma}\overline{W}_*^{2t} = c_k. \quad (11.114)$$

Define

$$T_k = \begin{bmatrix} T_1 \\ \vdots \\ T_6 \end{bmatrix} \quad (11.115)$$

as the unknown vector. Noting that in the Lorentz grid, W values are located at mid-levels with $W_* = W_6$, one obtains

$$\left(\overline{T}_k^{2t}\right) - A\left(\overline{W}_k^{2t}\right) = (c_k),$$
(11.116)

where

$$A = \begin{bmatrix} S_1 & 0 & 0 & 0 & 0 & G_1 \\ S_2 & S_2 & 0 & 0 & 0 & G_2 \\ 0 & S_3 & S_3 & 0 & 0 & G_3 \\ 0 & 0 & S_4 & S_4 & 0 & G_4 \\ 0 & 0 & 0 & S_5 & S_5 & G_5 \\ 0 & 0 & 0 & 0 & S_6 & S_6 + G_6 \end{bmatrix}$$
(11.117)

and

$$S_k = \frac{\Delta t}{2p_o}\left[\frac{\alpha T_o}{\sigma_k} - \left(\frac{\partial T_o}{\partial \sigma}\right)_k\right] \quad \text{and} \quad G_k = \frac{\Delta t}{p_o}\sigma_k\left(\frac{\partial T_o}{\partial \sigma}\right)_k.$$
(11.118)

The hydrostatic equation is utilized to relate temperature T to geopotential ϕ. Assuming that the temperature varies between the levels as a linear function of $\ln \alpha$, the hydrostatic equation, Equation (11.113) may be integrated as

$$\phi_k = \phi_{k+1} + \beta_k(T_k + T_{k+1}), \quad k = 1,2,3,4,5$$
(11.119)

$$\phi_6 = \beta_{65}T_5 + \beta_6 T_6 + \phi_*,$$
(11.120)

where β_k and β_{65} are constants depending only on the choice of σ levels and ϕ_* is the surface geopotential value. In matrix-vector notation, this becomes

$$(\phi_k - \phi_*) = B(T_k),$$
(11.121)

where

$$B = \begin{bmatrix} \beta_1 & \beta_1 + \beta_2 & \beta_2 + \beta_3 & \beta_3 + \beta_4 & \beta_4 + \beta_5 + \beta_{65} & \beta_6 + \beta_5 \\ 0 & \beta_2 & \beta_2 + \beta_3 & \beta_3 + \beta_4 & \beta_4 + \beta_5 + \beta_{65} & \beta_6 + \beta_5 \\ 0 & 0 & \beta_3 & \beta_3 + \beta_4 & \beta_4 + \beta_5 + \beta_{65} & \beta_6 + \beta_5 \\ 0 & 0 & 0 & \beta_4 & \beta_4 + \beta_5 + \beta_{65} & \beta_6 + \beta_5 \\ 0 & 0 & 0 & 0 & \beta_5 + \beta_{65} & \beta_6 + \beta_5 \\ 0 & 0 & 0 & 0 & \beta_{65} & \beta_6. \end{bmatrix}$$
(11.122)

Substituting Equations (11.108) and (11.109) in Equation (11.112), one obtains

$$-\frac{1}{\Delta t}\frac{\partial \overline{W}^{2t}}{\partial \sigma} + m^2 \nabla^2 \left(p_o \overline{\phi}^{2t} + RT_o \overline{p}_*^{2t} \right) = \frac{m^2}{\Delta t}\left(\frac{\partial a}{\partial x} + \frac{\partial b}{\partial y} \right) + m^2 \nabla^2 \left[p_o \phi_* + RT_o p_* (t - \Delta t) \right].$$

(11.123)

Substituting the terms from Equation (11.107) to the aforementioned equation, one obtains

$$-\frac{1}{\Delta t}\frac{\partial \overline{W}^{2t}}{\partial \sigma} + m^2 \nabla^2 \left(p_o \overline{\phi}^{2t} - p_o \phi_* + RT_o \overline{W}_*^{2t} \right) = \frac{m^2}{\Delta t}\left(\frac{\partial a}{\partial x} + \frac{\partial b}{\partial y} \right),$$

(11.124)

We note that

$$\frac{\partial \overline{W}^{2t}}{\partial \sigma} \approx \frac{\overline{W}_k^{2t} - \overline{W}_{k-1}^{2t}}{\Delta \sigma_k}.$$

(11.125)

Using Equation (11.125) in Equation (11.124), one obtains

$$C\left(\overline{W}_k^{2t} \right) + m^2 p_o \nabla^2 \left(\overline{\phi}_k^{2t} - \phi_* \right) + m^2 E \left(\nabla^2 \overline{W}_k^{2t} \right) = m^2 (d_k),$$

(11.126)

where

$$C = \begin{bmatrix} -\gamma_1 & 0 & 0 & 0 & 0 & 0 \\ -\gamma_2 & -\gamma_2 & 0 & 0 & 0 & 0 \\ 0 & -\gamma_3 & -\gamma_3 & 0 & 0 & 0 \\ 0 & 0 & -\gamma_4 & -\gamma_4 & 0 & 0 \\ 0 & 0 & 0 & -\gamma_5 & -\gamma_5 & 0 \\ 0 & 0 & 0 & 0 & -\gamma_6 & -\gamma_6 \end{bmatrix}$$

(11.127)

with

$$\gamma_k = \frac{1}{\Delta t \Delta \sigma_k} \quad \text{and} \quad d_k = \frac{1}{\Delta t}\left(\frac{\partial a_k}{\partial x} + \frac{\partial b_k}{\partial y} \right).$$

(11.128)

and

$$E = R\Delta t \begin{bmatrix} 0 & 0 & 0 & 0 & 0 & (T_o)_1 \\ 0 & 0 & 0 & 0 & 0 & (T_o)_2 \\ 0 & 0 & 0 & 0 & 0 & (T_o)_3 \\ 0 & 0 & 0 & 0 & 0 & (T_o)_4 \\ 0 & 0 & 0 & 0 & 0 & (T_o)_5 \\ 0 & 0 & 0 & 0 & 0 & (T_o)_6 \end{bmatrix}$$

(11.129)

Using Equations (11.116) and (11.121), Equation (11.126) becomes

$$CA^{-1}\left(\overline{T}_k^{2t}\right) + m^2\left(p_o B + EA^{-1}\right)\nabla^2\overline{T}_k^{2t} = m^2(d_k) + CA^{-1}(c_k) + m^2 EA^{-1}\left(\nabla^2 c_k\right). \quad (11.130)$$

Equation (11.130) is of the Helmholtz type for $\overline{T}_k^{2t}$. The other dependent variables may be obtained by substitution.

Equation (11.116) gives

$$\overline{W}_k^{2t} = A^{-1}\left(\overline{T}_k^{2t}\right) - A^{-1}(c_k). \quad (11.131)$$

Equation (11.107) becomes

$$\overline{p}_*^{2t} = p_*(t - \Delta t) + \Delta t \overline{W}_*^{2t}. \quad (11.132)$$

Equation (11.121) becomes

$$\left(\overline{\phi}_k^{2t} - \phi_*\right) = B\left(\overline{T}_k^{2t}\right). \quad (11.133)$$

Finally, Equations (11.108) and (11.109) give the horizontal wind components

$$\overline{u}^{2t} = a - \Delta t \frac{\partial}{\partial x}\left\{p_o\left(\overline{\phi}^{2t} - \phi_*\right) + RT_o\left[\overline{p}_*^{2t} - p_*(t - \Delta t)\right]\right\}. \quad (11.134)$$

$$\overline{v}^{2t} = b - \Delta t \frac{\partial}{\partial y}\left\{p_o\left(\overline{\phi}^{2t} - \phi_*\right) + RT_o\left[\overline{p}_*^{2t} - p_*(t - \Delta t)\right]\right\}. \quad (11.135)$$

Exercises 11 (Questions and answers)

1. For the vertical staggering of the hydrostatic equation using the Exner function, mention the advantages of the vertical staggering of the Charney–Phillips over the Lorenz type of vertical staggering.

   Answer: The hydrostatic equation in terms of the Exner function Π is given by

   $$\frac{\partial \phi}{\partial \Pi} = -\theta \quad \text{and} \quad \Pi = c_p\left(\frac{p}{p_o}\right)^{R/c_p},$$

   where ϕ, is the geopotential and θ is the potential temperature. In Lorentz vertical staggering, where winds and temperature are at the same vertical level, one obtains

   $$\phi_{i-1} - \phi_i = -\tfrac{1}{2}\left(\theta_{i-1} + \theta_i\right)\left(\Pi_{i-1} + \Pi_i\right).$$

   The same vertical staggering of the hydrostatic equation using the Charney–Phillips scheme, where the winds and the temperature are at different vertical levels, is given as

$$\phi_{i-1} - \phi_i = -\theta_{i-1/2}\left(\Pi_{i-1} + \Pi_i\right).$$

A zigzag perturbation of the potential temperature could get completely decoupled from the dynamics of the discrete system when applied to the Lorenz vertical staggering scheme.

2. Why does the Lorenz grid have the extra degree of freedom and, hence, the associated computational mode?

   Answer: When the atmospheric equations are discretized on the Lorenz vertically staggered grid, there exists one more degree of freedom than necessary for the number of physical modes, resulting in the associated computational mode. The extra degree of freedom for the Lorentz vertically staggered grid is due to the placement of one of the thermodynamic variables relative to the boundaries. The additional averaging that results in the vertical momentum equation leads to the computational mode manifesting itself as a 'two-grid' wave in one or more of the thermodynamic prognostic variables. The aforementioned associated mode for the Lorentz vertically staggered grid is nonphysical, and does not properly propagate. It can interact nonlinearly with the other modes, creating overall inaccuracy.

3. Mention the important differences between the ability of the two types of grid that are commonly employed for vertical staggering (Lorenz grid and Charney–Phillips grid).

   Answer: The Lorenz and Charney–Phillips vertical staggered grids have been compared for their ability to capture the steady state of a set of atmospheric equations that simultaneously represent the large-scale dynamics of the atmosphere and the planetary boundary layer. The most important differences between the two grids are as follows:

   (a) Lorenz grid is preferred for modeling the steady-state boundary layer.

   (b) For the coupled (atmosphere–planetary boundary layer) steady state, the Charney–Phillips vertically staggered grids are preferred as they use the averaging of the potential temperature gradient and either averaging of eddy diffusivity or the Richardson number in boundary layer terms.

   (c) The Charney–Phillips vertically staggered grids that use averaging of shear behave poorly owing to a suppression of a negative feedback, which leads to discontinuities in the vertical structure of the predicted fields.

4. Mention the disadvantages of the Lorentz computational mode for the unsteady transient problem.

   Answer: The structure of the Lorentz computational mode manifests in the distinct two-grid wave pattern in the potential temperature field. The phase speed of this wave is approximately given by the reference wind above the boundary layer. For the dynamics-only case (without the coupling of the atmosphere–planetary boundary layer), hydrostatic balance results in the aforementioned two-grid wave contaminating both thermodynamic fields. Moreover, the assumption of geostrophic balance along with the non-resting reference state above the boundary layer results in the two-grid wave contaminating all fields.

5. Mention the disadvantage of the Lorentz vertical staggering due to baroclinic instability.

   Answer: The Lorentz vertical staggering causes spurious amplification of short waves due to baroclinic instability.

6. Mention the advantage of the Charney–Phillips vertical staggered scheme.

   Answer: The Charney–Phillips vertical staggered scheme has in general better wave dispersion properties.

7. Mention the advantage of the Lorentz vertical staggering scheme.

   Answer: Conservation properties are more easily satisfied with the Lorentz vertical staggering scheme.

8. Indicate the problem of the pressure gradient force in the sigma coordinate system.

   Answer: The pressure gradient force term in the sigma coordinate system has the following expression

   $$-\nabla_\sigma \phi - \frac{RT}{p} \nabla_\sigma p$$

   In the presence of steep topography, while employing the sigma coordinate system, both the aforementioned pressure gradient force terms may assume large magnitudes and contribute to a large error.

9. How is the problem highlighted in question Exercise 11 Q 8 circumvented?

   Answer: The problem highlighted in Exercise 11 Q 8 is circumvented by adopting η coordinate system that uses quasi-horizontal surfaces only rather than σ coordinate system.

10. What are the disadvantages of using a step-wise terrain in the non-hydrostatic η model?

    Answer: The disadvantages of using a stepwise terrain in the non-hydrostatic η model is that this leads to an artificial separation of flow in the lee of the mountain; this is attributed to a spurious generation of vorticity on the steps.

11. What is the "blocking effect" of the η coordinate system?

    Answer: As the η coordinate system utilizes quasi-horizontal surfaces, air prefers to move around rather than over the terrain. This is called the blocking effect.

12

Boundary Conditions

12.1 Introduction

In the earlier chapters, the reader has been introduced to various equations of motion of the atmosphere, such as non-divergent vorticity equation, shallow water barotropic equations, quasi-geostrophic equations, and baroclinic equations. The aforementioned model equations for the atmosphere represents both an initial value problem and a boundary-value problem. For global atmospheric models, the required boundary conditions would correspond to both upper and lower boundaries of the atmosphere. There are no lateral boundaries for a global model as the model computational domain is naturally periodic. However, for regional atmospheric models that have a limited area of computational domain, the governing equations cannot be solved without specifying the nature of the lateral-boundary conditions. These lateral-boundary conditions for limited area models provide a means of obtaining the values of the dependent variables at these boundary points that correspond to lateral boundaries. For operational meteorological forecasts that employ a limited area regional atmospheric model, the lateral boundary values are obtained by interpolation from values of dependent variables at grid points of a previously run global atmospheric forecast model. For non-operational researchers working with a limited area regional atmospheric model, the lateral boundary values are obtained from archived and gridded regional or global analysis, the latter obtained by combining the optimal atmospheric model output with all possible atmospheric observations.

Both global and regional models require the upper and lower boundary values to be assigned. Real atmosphere does not have a definite upper value. However, unlike the real atmosphere, the model atmosphere does not extend to infinity; hence, it is necessary to define an artificial upper boundary for the model atmosphere and provide upper boundary values for the dependent variables at these artificial upper boundaries. The choice of the upper artificial boundary or lid impacts the computational costs. Upward-propagating internal-gravity waves that are generated by mountains or by deeply convective and organized systems can extend to great heights in the atmosphere. The most commonly employed upper boundary conditions in atmospheric models such as the rigid lid condition or free surface condition can reflect these vertically propagating internal gravity waves and distort the model solution. It is to be noted that in the real atmosphere such erroneous downward propagating waves arising due to reflection at the artificial upper model boundary do not manifest. Hence, it is important to devise an approach for prescribing the upper boundary conditions in a manner that would minimize or prevent such reflection from the artificial upper boundary.

For the prescription of lateral boundary conditions for a limited area regional atmospheric model, it is indicated that the lateral boundary values be obtained from coarser resolution gridded model output obtained from either global or regional models. To accomplish this, one has to transfer model values of dependent variables from the coarser model to the finer model through their common boundary. The conditions that are needed at these common boundaries are known as lateral-boundary conditions. As fluid is allowed to pass freely through these boundaries, they are referred to as open boundaries. An open boundary is defined as a computational boundary at which disturbances originating in the interior of the computational domain are allowed to leave it without affecting or worsening the interior solution. Another important requirement for the open boundary is that disturbances originating in the exterior domain are free to enter our computational model domain without distortions. The conditions that are prescribed at open boundaries are also known as open boundary conditions.

12.2 Upper Boundary Conditions

As mentioned in the previous section, it is necessary to devise an approach for prescribing the upper boundary conditions in a manner that would minimize or prevent reflection of vertically propagating gravity waves from the artificial upper boundary. A workaround solution for overcoming this limitation in the model involves employing a gravity-wave absorbing layer, or sponge layer, immediately below the model top, to prevent the wave from reaching the top and reflecting. This sponge layer that absorbs waves can be provided by assigning a greatly enhanced, artificial horizontal and/or vertical diffusion (viscosity), where the viscosity increases from the standard value at the bottom of the layer to a maximum at the top of the boundary. One disadvantage of introducing a sponge layer is that the absorbing layer may need to be thick, spanning a large number of model layers and, hence, contributing to higher computational cost. Moreover, the overall effectiveness of the absorption sponge layer depends on the wavelength of the gravity wave, the thickness of the absorbing layer, and the distribution of viscosity in the layer. It is to be noted that employing a shallow absorbing layer with a very large, but computationally stable, viscosity may not lead to an effective solution to the reflection problem as large gradients in viscosity also produces wave reflections.

In the sponge layer approach, as mentioned earlier, an absorbing layer in the upper portion of the model is defined. The eddy viscosity coefficients K_h (horizontal) and K_v (vertical) have default values at the bottom of the absorbing sponge layer that increase to a specified maximum value at the top of the absorbing sponge layer. Eddy viscosity coefficients act to reduce horizontal and/or vertical gradients of meteorological fields where they exist within the model atmosphere. The form of horizontal and vertical eddy viscosity coefficients (K_{hd} and K_{vd}) that are applicable within the damping sponge layer is given as follows:

$$K_{hd} = \frac{(\Delta x)^2}{\Delta t}\, \gamma_g\, \cos\left[\frac{\pi}{2}\left\{\frac{z_T - z}{z_D}\right\}\right], \tag{12.1}$$

$$K_{vd} = \frac{(\Delta z)^2}{\Delta t}\, \gamma_g\, \cos\left[\frac{\pi}{2}\left\{\frac{z_T - z}{z_D}\right\}\right], \tag{12.2}$$

where K_{hd} and K_{vd} are the eddy viscosity coefficients applicable in the sponge layer; γ_g is the specified non-dimensional damping coefficient assuming values that range between 0.01 and 0.1, z_T is the height of the model top, and z_D is the depth of the damping sponge layer. The default value for z_D is taken as 5 km. At the top of the model, where $z = z_T$, the cosine functions of Equations 12.1 and 12.2 equal the value of 1 and the eddy viscosity coefficients K_{hd} and K_{vd} assume their highest values, whereas at the bottom of the damping layer, where $z_T - z = z_D$, the cosine functions equal the value of 0 and the eddy viscosity coefficients K_{hd} and K_{vd} assume their lowest zero values.

An alternative approach for damping the gravity waves before they reach the upper boundary is to introduce a Rayleigh damping layer below the model top, where the model dependent variables relax toward a predetermined reference state. For example, the Rayleigh damping term in a prognostic equation will appear as

$$\frac{\partial \beta}{\partial t} = -\tau(z)[\beta - \bar{\beta}], \tag{12.3}$$

where β is any dependent variable, $\bar{\beta}$ is the reference state value of that variable, and $\tau(z)$ is a Rayleigh damping function that extends over a vertical extent z_D and also increases upward from the base $z_T - z_D$ of the Rayleigh damping layer to its top (z_T). It is to be noted that reference state fields are a function of height z only. The reference state vertical velocity is assumed to be zero. While employing a pressure-based terrain following vertical coordinates, the height of the vertical coordinate surfaces changes with time. Thus, the reference state values on model coordinate surfaces must be adjusted accordingly at each time step. The Rayleigh damping function takes the following form:

$$\tau(z) = \gamma_d \, \sin^2 \left[\frac{\pi}{2} \left\{ 1 - \frac{z_T - z}{z_D} \right\} \right], \tag{12.4}$$

where the Rayleigh damping is applied only over the damping layer depth specified by z_D. The γ_d is a specific damping coefficient having a value of $0.003\,\text{s}^{-1}$. From Equation 12.4, it is clear that at the top of the atmospheric model, where $z = z_T$, the square of the sine function is equal to 1 with the damping function assuming its maximum value of $-\gamma_d(\beta - \bar{\beta})$, whereas at the bottom of the Rayleigh damping layer, where $z_T - z = z_D$, the square of the sine function is equal to 0 and Rayleigh damping is not applied. There is another alternate Rayleigh damping formulation in which an implicit form of Rayleigh damping is applied to the vertical velocity only. In this formulation, Rayleigh damping is applied to the perturbation vertical velocity within the acoustic time step loop of the model time integration and influences both the coupled vertical velocity and geopotential.

Besides the aforementioned methods, there are many other ways of handling the upper boundary condition in the atmospheric model. In the "free surface formulation," the atmospheric model top is represented by a free surface, where the model atmosphere is treated as a fluid that lies beneath a second fluid above the top of the model atmosphere. It is assumed that although the height of the free surface itself may change, there exists no flow across the free surface. However, even free surface boundaries reflect vertically propagating waves, necessitating the use of an absorbing or damping layer in the upper portions of the simulation domain. Lastly, the atmospheric model top may be represented by a radiative boundary condition. In the radiation upper boundary condition formulation, the energy associated with vertically propagating waves is allowed to radiate upward and out of the simulation domain. In atmospheric models that utilize a radiative upper boundary condition, no absorbing or damping layer is utilized; the top of the model is fixed at some user-specified altitude. When the radiation upper boundary condition is applied, the values of the dependent variables at the upper boundary are modified during the integration to minimize wave reflection. It is to be noted that the term "radiation upper boundary condition" refers to the fact that waves are meant to be radiated through the upper model boundary, and not reflect from it.

12.3   Lower Boundary Conditions

In limited area atmospheric models, only the lower boundary is physical; the other boundaries are arbitrary to a certain extent. The lower boundary is identified as the interface between the atmospheric model's lowest level with the model topography. The accuracy of the lower boundary condition depends on the formulations of surface physics and parameterization used in the model as well as the correctness of the observational information available for snow cover, soil temperature, soil moisture, soil type, and vegetation cover. Vertical velocity at the ground is usually set to zero, except for an upslope or downslope component due to flow along the model topography. However, horizontal winds are predicted as average for the lowest layer rather than at the ground or anemometer level. Near-surface winds are then empirically determined. As most atmospheric models predict near-surface conditions using energy balance principles, model errors are inevitable considering the inadequate handling of terrain, albedo, the amount of rainwater available for evaporation from the surface, lake and sea temperature, vegetation cover, as well as the accuracy of the method for simulating soil–vegetation–atmosphere interaction, together with inadequacies related to the model representation of various physical processes.

12.4   Lateral Boundary Conditions

A closed boundary is one in which there exists rigid walls that exert a physical barrier on the flow against exiting the computational model domain. For rigid walls, no-slip (i.e., both the tangential and normal components of fluid flow are assumed to vanish) conditions are usually applicable. An open boundary is one in which the model computational domain ends; however, the fluid (atmosphere) extends beyond the model computational domain. Most of the limited area atmospheric models assume open lateral boundaries and require prescription of open (lateral) boundary conditions (OBC). The aforementioned open lateral boundary conditions serve the following two main purposes, (i) they should permit waves to propagate out from the computational model domain without reflecting back and affecting the interior solution, and (ii) it should be possible to obtain the interior solution using external fields, the latter to be obtained from observations or models that encompass a larger domain.

Radiation condition

Ideally, lateral boundary conditions should be based on observed data. However, the best possible scenario for obtaining lateral boundary conditions (OBCs) in atmospheric models is to use boundary conditions based on another forecast model, the latter preferably of a coarser resolution. Many of the processes in the atmosphere and oceans are processes that involve propagation of waves. Early efforts of developing OBCs were formulations based on the simple wave equation. In its simplest form, the one-dimensional linear wave equation is as follows:

$$\frac{\partial \phi}{\partial t} + c\frac{\partial \phi}{\partial n} = 0, \tag{12.5}$$

where ϕ represents the dependent variable, c is the component of the phase velocity in the direction normal to the boundary, whereas $\frac{\partial \phi}{\partial n}$ denotes the derivative of the dependent variable normal to the open boundary. Imposing Equation 12.5 as an OBC, is referred to as the "radiation condition." When one utilizes Equation 12.5 as an OBC, it is assumed that the disturbances passing through the open boundary consists of waves. In the case of Equation 12.5, only one Fourier component is considered. One of the drawbacks in using Equation 12.5 as an OBC is that the value of the phase velocity c is unknown.. While implementing Equation 12.5 as an OBC, let it be assumed that $0 \leq x \leq L$ and $0 \leq t \leq T$ with x = 0 being the natural boundary and x = L being the open boundary. Discretizing the x and t domains using $x_j = (j-1)\Delta x$ and $t_n = n\Delta t$, employing the upwind scheme for $c \geq 0$, and applying for x = L, i.e., for j = J, one obtains:

$$\frac{\phi_{J,n+1} - \phi_{J,n}}{\Delta t} + c\frac{\phi_{J,n} - \phi_{J-1,n}}{\Delta x} = 0 \tag{12.6}$$

$$\phi_{J,n+1} = (1-r)\phi_{J,n} + r\phi_{J-1,n}, \tag{12.7}$$

where $r = \frac{c\Delta T}{\Delta x}$. Equation 12.7 shows that the radiation condition is essentially an interpolation of values from the interior and at previous times. The problem is that the value of the phase velocity 'c' is unknown and, hence, r is not known. The following expression is used to obtain 'r':

$$r' = -\frac{\phi_{J-1,n} - \phi_{J-1,n-1}}{\phi_{J-1,n-1} - \phi_{J-2,n-1}}, \tag{12.8}$$

$$r = r' \text{ for } 0 \leq r' \text{ and } r = 0 \text{ for } r' \leq 0. \tag{12.9}$$

Consequently, one uses Equation 12.9 to substitute for 'r' in Equation 12.7 when determining the new boundary value $\phi_{J,n+1}$ at time level n + 1. In case the other boundary at x = 0 also has an open boundary, the inequality sign in Equation 12.9 must be reversed to ensure that the phase velocity has a negative value.

Sponge condition

Alternative options for obtaining the OBC are (i) inclusion of a sponge layer and (ii) the flux relaxation scheme (FRS). For implementing the sponge condition, one needs to extend the computational model domain outside the interior region, i.e., the region of interest, to include an area where the energy contained in that part of the solution leaving the interior region is gradually decreased. This objective is achieved by gradually increasing the relative importance of those terms that are associated with diffusive or frictional processes as the interior solution is advected or propagated into the exterior or extended domain.

Flux relaxation scheme

The flux relaxation scheme (FRS) is to some extent similar to the sponge OBC, as (i) the FRS, like the sponge OBC scheme, requires one to extend the computational model domain to include an exterior or extended domain; (ii) moreover, the scheme acts basically like a sponge OBC scheme, in which the solution in the exterior or extended domain gets progressively suppressed as the interior solution moves into the exterior or extended domain.

One of the advantages of the FRS compared to the sponge scheme is that it allows one to specify an exterior solution. The FRS can therefore also be used as a one-way nesting condition, in which an exterior solution is specified using, say a coarser grid model that encompasses a much larger area. The FRS can also be employed as a nesting method to transfer information from global and regional models to limited area models, the latter having a finer horizontal resolution.

The FRS method successfully modifies the numerical solution in an overlap zone, called the FRS zone, for each time step, based on the interior solution and a specified exterior solution. The FRS zone usually has a limited number of grid points, about 5 to 20. It is to be noted that the FRS zone is an extension of the interior domain and thus, extends the computational model domain. Within the FRS zone, the solution relaxes towards an a-priori specified exterior solution, called the "outer solution." The relaxation is accomplished by specifying a weighing function that for each grid point in the FRS zone computes a weighted mean between the specified outer solution and the interior solution computed from the governing equations. The FRS method has a distinct disadvantage due to the increase in the computational model domain and the associated costs in computation. This disadvantage is ameliorated to a certain extent as the FRS method allows one to specify an outer solution that provides an effective means to minimize the error. Another disadvantage that is associated with the FRS method when employed as an OBC is that the solution does not necessarily conserve fundamental properties such as volume (or mass).

12.5   One-way and Two-way Interactive Nesting

Open, or free, lateral boundary conditions (LBC) provide for values of the dependent variables to be specified externally based on predictions available from either a global model output or from gridded analyzes of data. It is to be noted that there are fundamentally two approaches, known as (i) one-way nesting approach and (ii) two-way nesting approach for providing the LBCs from a gridded coarser-resolution model output. The "two-way interactive nesting approach" involves the simultaneous integration of both the limited area model (LAM) and a gridded coarser-resolution model within which the LAM is embedded. As the name indicates, it provides for the information to flow between the two domains (LAM and the coarser grid model that has the LAM within it) in both directions. Conversely, in the one-way interactive nesting approach, the gridded coarser-resolution model is integrated first over the time duration of interest and the LBC are prescribed using the model output from the gridded coarser-resolution model. Hence, the transfer of information in the one-way interactive nesting approach is in one direction only, i.e., from the coarser gridded model to the LAM. Alternatively, one can employ an analysis of observations instead of the gridded coarser-resolution model output to transfer information in the one-way interactive nesting approach. It is clear that in both approaches, the meteorological information from the gridded coarser-resolution domain must be able to enter the LAM or fine-mesh domain, while the inertia-gravity and other waves must be able to freely exit the fine mesh domain. Moreover, in the two-way interacting nesting approach, the information from the LAM or fine mesh domain can also affect the solution on the gridded coarser-resolution domain, which can feed back to the LAM or the fine mesh domain. For a two-way interactive nesting approach, both model domains (coarser and the finer domains) are run together at the same time and both the domains interact with each other. While the information from the coarse domain to the finer domain is introduced through its boundary, the feedback from the finer grid to the coarse grid happens all over the finer domain interior, as the coarser domain dependent variables values are replaced by a combination of finer domain dependent variable values. It is fairly obvious that the two-way interactive nesting approach is a more desirable approach as compared to the one-way interactive nesting approach, for providing improved model solutions as by employing the former, a convective–precipitation system (fine scale) can influence its large-scale environment (large-scale), which can then feed back to the mesoscale (fine scale). It is to be noted that LAMs that utilize a two-way interacting nested grid system must generally obtain LBC for their coarsest-resolution domain from a previously run global model or from analysis of observations. Hence, in both the aforementioned nesting strategies, the use of a one-way interacting nesting interface condition is almost always necessary.

Consider an example of a two-way interactive nesting approach that has a ratio of 3:1 between the grid size of the coarse domain grid (CDG) and the fine domain grid (FDG). It is usual practice to choose the corresponding time steps for time marching of both the coarse domain grid and the fine domain grid to have the same ratio of 3:1 so as to ensure the same Courant number for both the domain integrations, i.e., the time step on the fine domain grid (Δt_{FDG}) is one-third of the time step of the coarse domain grid (Δt_{CDG}). The following sequential steps elucidate the integration chronology for a time refinement factor of three. Assuming all model variables at time 't' are known, the coarse grid model (CGM) is integrated from time 't' to '$t + \Delta t_{CDG}$' The coarse grid model variables are linearly interpolated in time at intervals separated by '$[\Delta t_{CDG}]/3$' onto the corresponding fine grid model time steps (Δt_{FDG}). Boundary values for the fine domain grid are interpolated from the coarse domain grid solution at time '$t + [\Delta t_{CDG}]/3$', and are provided at the overlapping interface boundary points at '$t = t + [\Delta t_{CDG}]/3$'. Then, the fine grid model is advanced three time steps of '$[\Delta t_{CDG}]/3$', and each time, the fine grid model uses the coarse domain grid solution at the overlapping interface boundary points until the fine grid solution reaches the same time level '$t' + \Delta t_{CDG}$' as the coarse grid model. The computed fine model grid dependent variable values all over the finer domain interior may then be used to update the coarse grid model values at the coarse grid model grid points, the latter lying entirely within the fine domain grid. Now all the model variables at time '$t + \Delta t_{CDG}$' are known and one can repeat the aforementioned procedure to march the integration from '$t + \Delta t_{CDG}$' to '$t + 2\Delta t_{CDG}$' and so on.

While employing the one-way interacting nesting approach, techniques are employed that commonly filter or damp small scales in the fine-mesh solution near the boundary, by utilizing a wave-absorbing or sponge region near the lateral boundary. This sponge region prevents internal reflection of outward-propagating waves through an enhanced diffusion as well as truncation of the time derivatives. Mostly, the fine grid is forced with large-scale conditions through a relaxation or diffusion term in the one-way interacting nesting approach. While employing a two-way interacting nesting approach, a variety of methods are employed for the interface condition between the two domains for interpolating the coarser-grid solution to the finer grid, and for filtering the finer-grid solution that is fed back to the coarser grid. However, it is well recognized that in some practical situations, the one-way interactive nesting approach is the more desirable option available. In situations involving operational nested modeling systems, one-way nesting approach provides for the coarse-grid model forecast to be completed first, and the model output products from the forcasts be made available readily to forecasters who will perform the more computationally expensive calculations that are associated with the finer model domain. Moreover, in situations where there exist significant computer-memory limitations, it makes sense to limit the model integrations to one model grid at a time.

Exercises 12 (Questions and answers)

1. What is "no-slip boundary condition"?

   Answer: The "no-slip boundary condition" that is normally invoked while dealing with flows of viscous fluids assumes that at a solid boundary, the fluid will have zero velocity relative to the boundary. If the solid boundary is at rest and the viscous fluid flows over the solid boundary, both the tangential and normal components of fluid velocity will vanish at the solid boundary for the "no-slip boundary condition."

2. Is "no-slip boundary condition" applicable for flows of ideal fluids?

   Answer: "No-slip boundary condition" is normally applicable only for viscous fluid flows. For an ideal fluid, being incompressible and non-viscous, the fluid at the solid boundary may be allowed to slip, i.e., the tangential component of the velocity need not vanish at the solid boundary. However, the normal components of fluid velocity will vanish at the solid boundary for the ideal fluid flows.

3. What are the typical dynamic surface and bottom boundary conditions applied in an ocean model?

   Answer: The ocean model has the following bottom boundary $z = -H$ (x,y), where the bottom topography $H(x,y)$ is a function of horizontal coordinates. The ocean model is assumed to extend in the vertical direction with $-H(x, y) \leq z \leq \eta(x, y, t)$, where $\eta(x, y, t)$ is the surface elevation. The dynamic boundary conditions at the free surface are then given by $\nabla \{z - \eta\} \cdot T \mid_{z=\eta} = \tau^s$,

   where T is the viscous stress tensor, τ^s is the surface stress vector whose x and y components τ_x^s and τ_y^s are calculated from meteorological data using bulk formulae. At the bottom boundary, the dynamic boundary conditions are the no-slip conditions $\nabla\{z+H\}\cdot u\,|_{z=-H}=0$

4. What are the typical kinematic surface and bottom boundary conditions applied in an ocean model?

Answer: The ocean model has the following bottom boundary $z=-H(x,y)$, where the bottom topography $H(x,y)$ is a function of horizontal coordinates. The ocean model is assumed to extend in the vertical direction with $-H(x,y)\leq z\leq \eta(x,y,t)$, where $\eta(x,y,t)$ is the surface elevation. The typical kinematic surface and bottom boundary conditions applied in an ocean model are as follows:

$$w\,|_{z=\eta}=\frac{\partial\eta}{\partial t}+u\,|_{z=\eta}\frac{\partial\eta}{\partial x}+v\,|_{z=\eta}\frac{\partial\eta}{\partial y}+\{E-P\}$$

and

$$w\,|_{z=-H}=-u\,|_{z=-H}\frac{\partial\eta}{\partial x}+v\,|_{z=-H}\frac{\partial H}{\partial y}.$$

5. Which is the simplest lateral boundary condition that can be applied to a limted area model? How well does the aforementioned lateral boundary condition perform?

Answer: The simplest lateral boundary condition is called the "interpolation" boundary condition. It merely interpolates the coarse grid model data to the boundary points on the nested grid for all the dependent variables on each nested grid time step. The aforementioned "open" boundary condition for nested grids allows both inflow and outflow. However, its performance is considered the worst among nested-grid BCs used in limited area models.

6. What is meant by 'filtered sponge boundary condition'?

Answer: For poorly resolved solutions, there is a marked difference in the amplitudes and wavelengths of the solutions on both the coarse grid as well as on the fine grid. This causes a mismatch between the coarse grid and the fine grid solutions that results in larger reflections even in two-way nesting. A work-around solution to reduce the amplitude of the reflections is to filter the coarse-grid data to remove the problematic high-frequency modes before using it in the sponge boundary condition. For implementing this, a filtered sponge boundary condition is introduced in which the dependent variables of the coarse grid are smoothed with the fourth-order filter in the sponge zone that does not alter the actual coarse-grid solution. The fourth-order filter is generally utilized as it is a simple scale-selective filter that effectively damps short wavelengths while having little effect on better-resolved wavelengths and disturbances.

7. What are the situations where utilizing one-way nesting approach has its limitations?

Answer: Utilizing one-way nesting approach has its limitations while investigating situations involving baroclinic instability or gravity wave breaking. Furthermore, another limitation associated with one-way nesting approach is that troughs, ridges, and disturbances may move with different speeds in each of the two meshes (coarse and fine), contributing to solutions that "diverge" from one another. This divergence of solutions can contribute to discontinuities and distortions along the fine mesh boundaries.

8. What is the difference between the standard one-way nesting approach and the continuous one-way nesting approach? Which among the two approaches provides better results?

Answer: While employing the standard one-way nesting approach, the coarse and the fine grid meshes are run sequentially; first the coarse grid mesh run is completed and then the fine grid mesh run is initiated. The coarse grid values are stored at fixed time steps, say at every one-hour duration. These hourly values then provide the lateral boundary conditions for the fine grid mesh. However, as the fine grid mesh requires the lateral boundary conditions at each fine grid mesh time step, the latter being typically much smaller than one hour, it is necessary to interpolate with respect to time from the hourly coarse grid mesh values. In the continuous one-way nesting approach, both the coarse and fine meshes are run simultaneously and the coarse grid mesh values provide the lateral boundary conditions for the fine grid mesh at each coarse grid time step, rather than at one-hour duration as indicated in the standard one-way nesting approach. This allows the fine grid mesh to account for the coarse grid variations on a time scale that is finer than one hour. Thus, the continuous one-way nesting approach produces improved and better results as compared to the standard one-way nesting approach.

9. What is the difference between the continuous one-way nesting approach and the two-way nesting approach? Which among the two approaches provides better results?

In both the two approach, both the coarse and fine meshes are run simultaneously. Moreover, in both the approaches, the coarse grid mesh values provide the lateral boundary conditions for the fine grid mesh at each coarse grid time step. The only difference between continuous one-way nesting approach and the two-way nesting approach is that while in the former case, there are no feedbacks from fine grid mesh to the coarse grid mesh, in the two way nesting approach, there is a feedback from the fine grid mesh to the coarse grid mesh. This feedback involves the entire area of the coarse grid mesh that encompasses the fine grid mesh. The coarse grid mesh fields in this area are replaced by a combination of fine grid scale values. As the fine grid scale solution may introduce smaller scale waves into the coarse grid mesh that might lead to noisy solutions at the coarse grid mesh, some amount of smoothing is applied to the fine grid scale solution.

10. Are the advantages of the two-way nesting approach more apparent at the fine grid mesh or at the coarse grid mesh? What are the disadvantages of the two-way nesting approach?

The advantage of the two-way nesting approach is more apparent at the fine grid mesh as compared to the coarse grid mesh for the following reason. In the two-way nesting approach, during the feedback from the fine grid mesh to the coarse grid mesh, coarse grid values are overwritten and, hence, the coarse grid mesh fields obtained from the fine grid mesh solution in the overlapping area are no longer a solution to the coarse grid equations. This can induce perturbations that can generate gravity waves in the coarse grid mesh solution. The disadvantage of the two-way nesting approach is that this essentially requires additional computational resources as one needs to perform double interpolation at each of the coarse grid mesh time step.

13

Lagrangian and Semi-Lagrangian Schemes

13.1 Introduction

While discussing the various methods that are employed numerically to solve the governing equations of the atmosphere, the view point considered was mainly "Eulerian." In the Eulerian description of fluid motion, the motion is described with respect to a coordinate system that is fixed in space. These invariably result in the description of fluid properties as either a "scalar field" or a "vector field" and the evolution of fluid properties as the evolution of the resultant scalar field/vector field in space and time. In essence, in the Eulerian view point, the evolution of fluid properties were defined with respect to the time evolution of the fluid properties at fixed points in space through the use of the partial derivative with respect to time. This resulted in fluid properties being expressed as a function in space and time. In the Eulerian description of fluid motion, one is not bothered about the location and/or velocity of any particular "fluid particle," but rather about the velocity, acceleration, etc. of whatever fluid particle happens to be at a particular location of interest at a particular time. However, as fluid is treated as a "continuum," the Eulerian description of fluid motion is usually preferred in fluid mechanics. It is also known that the Eulerian description of fluid flow results in the appearance of nonlinear advection terms. Chapter 4.9 outlined the issues that arise because of the presence of nonlinear advection terms leading to nonlinear computational instability and the ways that one has to take recourse to overcome nonlinear computational instability.

Hence, it is not surprising that atmospheric scientists started taking serious note of the "Lagrangian" view point in terms of the Lagrangian description of fluid motion during the late 1950s and later years. In the Lagrangian description of fluid flow, individual fluid particles are "marked" and, hence, their positions, velocities, etc.

can be "tracked" as a function of time and their initial position. If there were only a few fluid particles to be considered in the flow domain, the Lagrangian description would be desirable. However, fluid is treated as a continuum and, hence, it is not possible to track each and every fluid particle in a complex flow field. Hence, the Lagrangian description of fluid motion is rarely employed in fluid mechanics, even though it has the very distinct advantage of completely avoiding the nonlinear terms in the governing equations and, hence, sidestepping the difficulties of avoiding or preventing nonlinear computational instability.

Figure 13.1 shows the difference between the Eulerian and the Lagrangian numerical models. In the Eulerian numerical model, one has fixed computational regions, or regions conveniently called "cells." Each and every fluid property, such as pressure, density, or velocity, is approximated in some way within each of the cells in the Eulerian viewpoint. For example, the Eulerian method usually employs cell-averaged properties (quantities), where each and every fluid variable is represented by a single value within each cell. In order to obtain a time-dependent simulation of fluid flow in the Eulerian framework, the movement of fluid particles between cells is modeled by considering "fluxes" across the cell-interfaces. Hence, in the Eulerian numerical model, the two basic constructions are cells and fluxes. However, the two basic constructions within the Lagrangian numerical model are individual fluid particles that are "tagged" and their trajectories; the latter are defined as the paths of tagged fluid particles with respect to time (refer to Figure 13.1). Each fluid particle in the Lagrangian numerical model carries values for all the fluid properties, and moves with the velocity of the fluid flow.

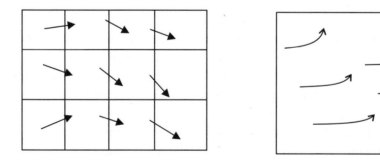

Figure 13.1   Differences between the Eulerian (left panel) and the Lagrangian (right panel) numerical models.

13.2 Fully Lagrangian Scheme

The Lagrangian frames of reference employed to describe the Lagrangian description of fluid motion are typically both space and time dependent. The fully Lagrangian numerical method is one in which the fluid particles are "tracked" in time; i.e., these fluid particles are suitably tagged with physical properties of the fluid dynamical system. Such fully Lagrangian numerical methods are often named as "gridless" particle methods. These methods adopt a discretization of partial differential operators on a Lagrangian mesh of particles, thereby simulating the time evolution of advection terms in a fully Lagrangian framework such that no interpolation of advected fields is required.

Lagrangian schemes have a far less severe restriction than Eulerian schemes on the time step, especially when the flow is strongly advection dominated, as is the case for most scenarios in atmospheric applications. For the aforementioned flows, the fully Lagrangian form of the evolution equations are more tractable numerically than are the corresponding governing equations in the Eulerian form. Hence, the time-step length for a Lagrangian scheme may be substantially higher and much beyond what would be an acceptable time step for an Eulerian scheme. Furthermore, the Lagrangian scheme, also provides for accurate numerical solutions without the need to consider the stability of the numerical schemes.

All said and done, Lagrangian numerical schemes are still nothing more than methods of approximation and as such subject to limitations of some kind or other. For any typical realistic fluid flow involving the atmosphere, there are so many manifested scales that are present in the atmospheric flows that the basic particle–trajectory model (fully Lagrangian numerical model) may not provide effective numerical solutions. For example, in the fully Lagrangian numerical model, an initially evenly distributed collection of fluid particles may, with time, become broken up into clusters and voids. A resultant consequence of this would be a loss of model representation for the fluid as a whole.

13.3 Semi-Lagrangian Scheme

Semi-Lagrangian schemes circumvent the aforementioned limitation of the fully Lagrangian schemes, by retaining the same fixed computational grid (like the Eulerian method) throughout the duration of a simulation. This ensures that the semi-Lagrangian schemes succeed in retaining the desirable properties of both the fully Lagrangian as well as the Eulerian formalisms while overcoming the limitations of both the formalisms. Moreover, like the fully Lagrangian schemes, the semi-

Lagrangian schemes allow larger time step length ensuring integrations by tracing trajectories of fluid particles. Furthermore, like the Eulerian scheme, the semi-Lagrangian schemes preserve the computational grid throughout the simulation duration. This "win–win" situation is achieved by obtaining the numerical solution on the fixed computational grid at the end of each time step, rather than tracing the long-term histories of individual particles.

13.3.1   Linear one-dimensional advection equation with constant velocity

In this section, the linear one-dimensional advection equation with constant velocity is solved using the semi-Lagrangian scheme. The linear one-dimensional advection equation with constant velocity a is as follows:

$$\frac{\partial u}{\partial t} + a\frac{\partial u}{\partial x} = 0, \tag{13.1}$$

subject to the initial condition

$$u(x,0) = u_0(x). \tag{13.2}$$

In the Lagrangian formulation, Equation (13.1) is given by

$$\frac{du}{dt} = 0. \tag{13.3}$$

The analytical solution to Equation (13.1) is obtained using the method of characteristics, where in the aforementioned equation is utilized to obtain a system of two first-order ordinary differential equations as follows:

$$\frac{dx}{dt} = a \tag{13.4}$$

$$\frac{du}{dt} = 0, \tag{13.5}$$

where the two functions, $x(t)$ and $u[x(t),t]$, have to be determined. Equation (13.4) describes the trajectory of the fluid particle moving with the advection velocity a. If the fluid particle is at the point $x = x_0$ when $t = 0$, then the solution of Equation (13.4) is

$$x(t) = x_0 + at, \quad t > 0. \tag{13.6}$$

Equation (13.5) for the advected quantity u has the following solution:

$$u[x(t),t] = \text{constant}. \tag{13.7}$$

Equation (13.7) states that u remains constant along a characteristic. Combining the two equations Equations (13.6) and (13.7), one obtains

$$u(x_0 + at, t) = u(x_0, 0), \quad t > 0. \tag{13.8}$$

From Equation (13.8), one obtains the solution of the linear one-dimension advection equation with constant velocity as

$$u(x,t) = u_0(x - at) \tag{13.9}$$

The overall basic outline of the method described here can be utilized to obtain a numerical scheme to solve the linear one-dimensional advection equation with constant velocity.

13.3.2 Semi-Lagrangian scheme to solve the linear one-dimensional advection equation with constant velocity

The numerical semi-Lagrangian scheme described in this section employs a finite difference (grid point) representation of the solution similar to the Eulerian grid. Envisage a grid having the grid-size in space and time as Δx and Δt respectively. Let u_j^n be the approximate to the solution $u(x_j, t^n)$, where j and n are the space and time indices, defined as $x_j = j\Delta x$ and $t^n = n\Delta t$. Towards designing a semi-Lagrangian scheme, it is assumed that the solution is known at time $t = t^n$. Then, one employs a forward marching scheme in time using the method of characteristics. That is, given the function $u(x, t^n)$ as the initial condition, one wishes to obtain the finite difference solution, u_j^{n+1}, at time t^{n+1} at each of the grid points x_j.

Figure 13.2 shows the particular *characteristic* that passes through the identified grid point $x = x_J$ at time t^{n+1}. The characteristic for Equation (13.1) is a straight line with slope equal to $1/a$, as can be identified from Equation (13.4). Extending the characteristic shown in Figure 13.2 at grid point $x = x_J$ at time t^{n+1}, back to the previous time step, i.e., to time, $t = t^n$, the characteristic passes through a point $x = x_d$ at $t = t^n$. It is clear from Figure 13.2 that the point $x = x_d$ is not a grid point and actually lies between $x = x_j$ and $x = x_{j+1}$. The point $x = x_d$ is called the departure point for the fluid particle that starting at the previous time $t = t^n$, had arrived at the identified grid-point $x = x_J$ at the next later time $t = t^{n+1}$

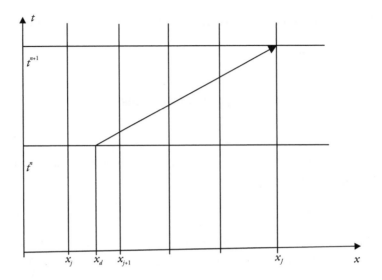

Figure 13.2   Schematic of the semi-Lagrangian scheme for constructing a numerical solution.

Continuing to design the semi-Lagrangian scheme, the assumption that the analytical solution u remains constant on any characteristic is utilized, i.e., the analytical solution u at $x = x_J$ at time $t = t^{n+1}$ is the same as u at $x = x_d$ and $t = t^n$. Hence,

$$u_J^{n+1} = u_d^n. \tag{13.10}$$

The departure point of the characteristic will not in general coincide with a grid point, rather it would ideally lie between two grid points, as shown in Figure 13.2. In order to obtain the value of solution u at $x = x_d$ and $t = t^n$, (the RHS of Equation (13.10), there is a need to interpolate and obtain u at $x = x_d$ and $t = t^n$, from the known values of u at $x = x_j$ and $t = t^n$, and u at $x = x_{j+1}$ and $t = t^n$. For convenience, one can utilize linear interpolation of the finite difference values at $x = x_j$ and $x = x_{j+1}$ for $t = t^n$. The linear interpolation formula is given by

$$u_d^n = \left(\frac{x_d - x_j}{x_{j+1} - x_j} \right) u_{j+1}^n + \left(\frac{x_{j+1} - x_d}{x_{j+1} - x_j} \right) u_j^n. \tag{13.11}$$

It is to be noted that u_d^n is an approximation of the analytical solution $u(x = x_d, t = t^n)$. From Equations (13.10) and (13.11), one obtains the following numerical scheme for obtaining the finite difference solution at time $t = t^{n+1}$ from the finite difference data at $t = t^n$.

$$u_j^{n+1} = \left(\frac{x_d - x_j}{x_{j+1} - x_j}\right) u_{j+1}^n + \left(\frac{x_{j+1} - x_d}{x_{j+1} - x_j}\right) u_j^n. \qquad (13.12)$$

Equation (13.12) relates the solution at $x = x_J$ at time t^{n+1}, i.e., $u(x = x_J, t = t^{n+1})$, to the solution at the previous time $(t = t^n)$ at the grid points $x = x_j$ and $x = x_{j+1}$ and is equal to the solution at the previous time $(t = t^n)$ at the location of the departure point $(x = x_d)$ for the characteristic.

The aforementioned semi-Lagrangian scheme essentially provides for a numerical scheme using the method of characteristics for solving the linear one-dimensional advection equation with constant velocity. In essence, one assumes that the solution is known at the initial time $t = t^n$ at all grid points. For each and every grid point $(x = x_J)$ at the next time $t = t^{n+1}$, the location of the departure point at the previous time $t = t^n$, of the characteristic is determined. Once the location of the departure point $(x = x_d)$ at the previous time $t = t^n$, is determined (from the knowledge of the slope of the characteristic) for all the grid points at the time $t = t^{n+1}$, the solution u at $x = x_d$ (location of the departure point) at the time $t = t^n$, is obtained by interpolation of the solution available at the nearby grid points $(x = x_j$ and $x = x_{j+1})$ at the time $t = t^n$. As the solution u is same along any characteristic, the solution u at $t = t^{n+1}$ is the same as the solution obtained by interpolation at the location of the departure point $(x = x_d)$ at the previous time $t = t^n$. A repetition of the aforementioned procedure provides the marching of the solution at the next time $t = t^{n+2}$.

13.3.3 Semi-Lagrangian scheme to solve the linear one-dimensional advection equation with non-constant velocity – three time level scheme

We now apply the basic idea of the semi-Lagrangian scheme to solve the linear one-dimensional advection equation with non-constant velocity by employing a three time level scheme. Consider solving the following equation:

$$\frac{du}{dt} = \frac{\partial u}{\partial t} + \frac{dx}{dt}\frac{\partial u}{\partial x} = 0. \qquad (13.13)$$

Figure 13.3 depicts the schematic diagram of the three time level technique for the semi-Lagrangian scheme for solving the linear one-dimensional advection equation with non-constant velocity. To implement this, one has to solve the "trajectory equation" to determine $x(t^n - \Delta t)$, the location of the departure point x_d at twice the previous time step before the present time $(t^n + \Delta t)$ for the fluid particle at the grid point x_j.

$$\frac{dx}{dt} = a(x,t), \tag{13.14}$$

subject to the condition $x(t^{n+1}) = x_j$, in order to find $x(t^n - \Delta t)$.

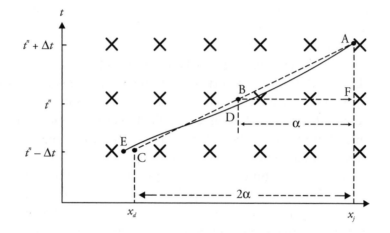

Figure 13.3   Schematic diagram of the three time level technique for the semi-Lagrangian scheme to solve linear one-dimensional advection equation with non-constant velocity and no source term.

Equation (13.14) is discretized by employing central finite difference scheme in both space and time; this ensures second-order accuracy in both space and time. This procedure is called the implicit mid-point rule. The mid-point rule is obtained by assuming that the velocity remains constant at its mid-step value during each time step. This assumption ensures that each trajectory is linear, and the mid-point of such a trajectory is the average of the positions of its end points. Employing the aforementioned approximations together with a three-time level scheme, one obtains the following discretization of Equation (13.14):

$$\frac{x_j - x_d}{2\Delta t} = a(x_m, t^n), \tag{13.15}$$

where

$$x_m = \frac{x_j + x_d}{2}. \tag{13.16}$$

It may be noted that one has to solve Equation (13.15) to find the location of the departure point x_d at the previous two time steps back for a fluid particle present at x_j at time t^{n+1}. From Equations (13.15) and (13.16), it is clear that departure point x_d cannot be explicitly found in terms of the other quantities such as x_j, t^n, and Δt

as x_d appears in both sides of Equation (13.15) implicitly. Defining the displacement along the trajectory as 2α, where $2\alpha = x_j - x_d$, one can write Equation (13.15) in terms of the displacement 2α and solve the resulting equation implicitly as follows (where superscript r is iteration number):

$$\alpha^{(r+1)} = \Delta t a(x_j - \alpha^{(r)}, t^n). \tag{13.17}$$

In Figure 13.3, the fluid particle is present at point A (location x_j at time $(t^n + \Delta t)$, i.e., at time level t^{n+1}), whereas the exact trajectory is shown by the curve ADE. The straight line ABC is a linear approximation of the curve ADE and, hence, is an approximate linear trajectory. The mid-point rule (Equation (13.16)) requires that the velocity data is to be used at the mid-point of the linear approximation to the trajectory ABC, i.e., velocity data at point B. The location of the point B (x_m) would itself be a function of the displacement (distance from x_j and x_d) and, hence, depend on the location of the departure point x_d. This clearly shows the implicit nature of Equation (13.17). Note that the velocity data at point B would be equal to the reciprocal of the value of the slope of the straight line CB in Figure 13.3. Discretizing the Lagrangian equation, Equation (13.13), using central time differencing scheme along the approximate fluid trajectory CBA, one gets

$$\frac{u(x_j, t^n + \Delta t) - u(x_j - 2\alpha, t^n - \Delta t)}{2\Delta t} = 0, \tag{13.18}$$

where α is the distance BF the fluid particle travels in the x direction in time Δt. Thus, Equation (13.18) indicates that if one knows the value of α, then the value of u at the arrival point A (location x_j at time $t^n + \Delta t$) is the value of u at its upstream point C (location $x_j - 2\alpha$ at the previous two time steps back $t^n - \Delta t$).

The superscript $(r+1)$ is the estimate of α in the $(r+1)$th iteration in Equation (13.17). Usually a few iterations are adequate to obtain an accurate solution of Equation (13.17) to within the expected tolerance. Once the displacement α has been obtained from solving Equation (13.17), the location of the departure point can be easily determined from $x_d = x_j - 2\alpha$. In the aforementioned mid-point rule, the velocity $a(x,t)$ has to be obtained at the mid-point of the trajectory; the mid-point of the trajectory may not coincide with the grid point. Hence, obtaining the velocity $a(x,t)$ at the mid-point of the trajectory would require interpolation of the spatially varying data. Moreover, in a flow situation in the atmosphere, the velocity field is also a variable that needs to be forecasted or predicted from the governing equations of motion of the atmosphere. However, like the other predicted variables, the velocity will also be available over the defined grid at fixed intervals of time, separated by Δt, i.e., at times $0, t^1, t^2, t^3, \ldots, t^n$. However, Equation (13.17) requires that the velocity

$a(x,t)$ be available at time t^n. To evaluate u and a between mesh points (refer to Equations (13.17) and (13.18) where the location of the departure point is in general not a mesh point), there is a need to employ spatial interpolation. The following sequence of steps defines the semi-Lagrangian scheme for solving the linear one-dimensional advection equation with non-constant velocity using the three time level scheme in the absence of a source term

1. Solve Equation (13.17) iteratively for the half-displacement α for all grid points x_j at time $t^n + \Delta t$ using some initial guess (the value of α at the previous time step can be utilized as a first guess value) and employing an appropriate interpolation formula in space for the velocity.

2. Evaluate u at upstream location $x_j - 2\alpha$ and at time $t^n - \Delta t$ using an appropriate interpolation formula.

3. Evaluate u at arrival location x_j and at time $t^n - \Delta t$ using Equation (13.18).

13.3.4   Semi-Lagrangian scheme to solve the linear one-dimensional advection equation with non-constant velocity – two time level scheme

The existing semi-Lagrangian schemes are known to employ either a two time level scheme (Section 9.2.4) or a three time level scheme (Section 13.3.3). The fundamental advantage of a two time level scheme over a three time level scheme is that the former is potentially twice as fast as the latter. This is true as the three time level schemes require time steps that are typically half the required size of the time steps of the two time level scheme for the same level of time truncation error. It is, however, important to guarantee that the accuracy of the time-integration scheme be not sacrificed and the scheme remains second-order accurate in time even if it utilizes two time level schemes. This ensures that the provision of a larger time step (and the associated stability) for the two time level scheme also does not compromise on the second-order accuracy, as a stable scheme with a larger time step without being sufficiently accurate would not be an ideal choice. The important task is then to address the question of efficiently determining the trajectories that are at least second-order accurate in time.

Referring to the algorithm of the three time level scheme (Section 13.3.3) with the assumption that $a(x,t)$ is known at time t^n, independently of $u(x,t)$ at the same time, it is clear from Equation (13.17) that it is possible to determine the trajectory as a is known at time t^n. Knowledge of the trajectory (solution of Equation (13.17) and, hence, α) will allow one to obtain the value of u from time $t^n - \Delta t$ to $t^n + \Delta t$ using Equation (13.18) without needing to know the value of u at time t^n. Extending this argument, it should be possible to obtain the value of u at time $t^n + 3\Delta t$ using the values of u at time $t^n + \Delta t$ and values of a at time $t^n + 2\Delta t$. This clearly shows that one has two decoupled independent integrations, the first one utilizing values of u at odd time steps (for example $t^n + \Delta t$) and a at even time steps (for example $t^n + 2\Delta t$) to obtain u at $t^n + 3\Delta t$, whereas the second one utilizies values of u at even time steps (for example $t^n + 2\Delta t$) and a at odd time steps (for example $t^n + 3\Delta t$) to obtain u at $t^n + 4\Delta t$. It is apparent that either one of the aforementioned indicated possibilities is adequate, as one will leapfrog two time steps always, say, from $t^n - \Delta t$ to $t^n + \Delta t$. This ensures that it is possible in the aforementioned situation to envisage a two time level scheme for the advected quantity u and save half the computational costs by simply reassigning and relabeling the time levels $t^n - \Delta t$, t^n, and $t^n + \Delta t$ (refer to Section 13.3.3 for the three time level scheme) to t^n, $t^n + \Delta t/2$ and $t^n + \Delta t$ (refer to Figure 13.4 for the two time level scheme). This is the essence of the two time level semi-Lagrangian scheme that has a second-order accurate trajectory, which ensures larger time step with second-order accuracy in time through a two time level scheme.

The semi-Lagrangian scheme for solving linear one-dimensional advection equation with non-constant velocity has to resolve a velocity a that is not a constant, i.e., in general $a = a(x,t)$. In such cases of non-constant velocity, one takes recourse to "implicit mid-point" rule. To implement this rule, one has to solve the trajectory equation to determine $x(t^n)$, the location of the departure point x_d at the previous time for the fluid particle at the grid point x_j.

$$\frac{dx}{dt} = a(x,t), \tag{13.19}$$

subject to the condition $x(t^{n+1}) = x_j$, in order to find $x(t^n)$. Equation (13.19) is discretized by employing central finite difference scheme in both space and time ensuring second-order accuracy in both space and time; this procedure is called the implicit mid-point rule. The mid-point rule is obtained by assuming that the velocity remains constant at its mid-step value during each time step. This assumption ensures that each trajectory is linear, and the mid-point of such a trajectory is just the average of the positions of its end points. Employing these approximations together with a two time level scheme, one obtains the following discretization of Equation (13.19):

$$\frac{x_j - x_d}{\Delta t} = a(x_m, t^n + \Delta t/2), \tag{13.20}$$

where

$$x_m = \frac{x_j + x_d}{2}. \tag{13.21}$$

Recall, that one has to solve Equation (13.19) to find the location of the departure point x_d at the previous time for a fluid particle present at x_j at time t^{n+1}. From Equations (13.20) and (13.21), it is clear that departure point x_d cannot be explicitly found in terms of the other quantities such as x_j, t^n, and Δt as x_d appears in both sides of Equation (13.20) implicitly. Defining the displacement along the trajectory as α where $\alpha = x_j - x_d$, one can write Equation (13.20) in terms of the displacement α and solve the resulting equation implicitly as follows (where superscript r is iteration number):

$$\alpha^{(r+1)} = \Delta t a(x_j - \alpha^{(r)}/2, t^n + \Delta t/2). \tag{13.22}$$

Figure 13.4 depicts the schematic diagram of the two time level technique for the semi-Lagrangian scheme for solving the linear one-dimensional advection equation with non-constant velocity and no source term. In Figure 13.4, the fluid particle is present at point A (location x_j at time $t^n + \Delta t$, i.e., at time level t^{n+1}), whereas the exact trajectory is shown by the curve ADE. The straight line ABC is a linear approximation of the curve ADE and, hence, is an approximate linear trajectory. The mid-point rule (Equation (13.21)) requires that the velocity data is to be used at the mid-point of the linear approximation to the trajectory ABC, i.e., velocity data at point B. The location of the point B (x_m) would itself be a function of the displacement (distance from x_j and x_d) and, hence, depend on the location of the departure point x_d. This clearly shows the implicit nature of Equation (13.22). Note that the velocity data at point B would be equal to the reciprocal of the value of the slope of the straight line CB in Figure 13.4.

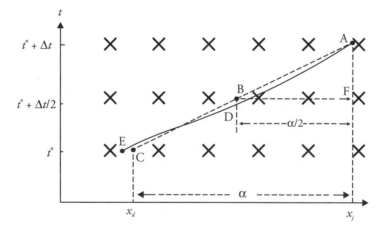

Figure 13.4 Schematic diagram of the two time level technique for the semi-Lagrangian scheme for solving the linear one-dimensional advection equation with non-constant velocity and no source term.

Discretizing the Lagrangian equation, Equation (13.13) along the approximate fluid trajectory CBA, one obtains

$$\frac{u(x_j, t^n + \Delta t) - u(x_j - \alpha, t^n)}{\Delta t} = 0, \tag{13.23}$$

where $\alpha/2$ is the distance BF the fluid particle travels in the x direction in time $\Delta t/2$. Thus, Equation (13.23) indicates that if one knows the value of $\alpha/2$, then the value of u at the arrival point A (location x_j at time $t^n + \Delta t$) is the value of u at its upstream point C (location $x_j - \alpha$ at the previous time t^n).

The superscript $(r+1)$ is the estimate of α in the $(r+1)$th iteration. Usually, a few iterations are adequate to obtain an accurate solution of Equation (13.20) to within the expected tolerance. Once the displacement α has been obtained from solving Equation (13.20), the location of the departure point can be easily determined from $x_d = x_j - \alpha$. In the aforementioned mid-point rule, the velocity $a(x,t)$ has to be obtained at the mid-point of the trajectory; the mid-point of the trajectory may not coincide with the grid point. Hence, obtaining the velocity $a(x,t)$ at the mid-point of the trajectory would require interpolation of the spatially varying data. Moreover, in a flow situation in the atmosphere, the velocity field is also a variable that needs to be forecasted or predicted from the governing equations of motion of the atmosphere. However, like the other predicted variables, the velocity will also be available over the defined grid at fixed intervals of time, separated by Δt, i.e., at times $0, t^1, t^2, t^3, \ldots, t^n$. However, Equation (13.20) requires that the velocity $a(x,t)$ be available at time $t^n + \Delta t/2$. The only way to make the velocity $a(x,t)$ be available at later times $t > t^n$ is by using appropriate extrapolation formulas (of second-order accurate or third-order accurate) of the existing velocity data such as

$$a^{n+1/2} = \frac{1}{2}\left(3a^n - a^{n-1}\right) + O(\Delta t^2) \tag{13.24}$$

$$a^{n+1/2} = \frac{1}{8}\left(15a^n - 10a^{n-1} + 3a^{n-2}\right) + O(\Delta t^3) \tag{13.25}$$

Hence, for the case of the linear one-dimensional advection equation with non-constant velocity, one can determine the location of the departure point x_d at the previous time t^n by interpolating velocity data in space and extrapolating velocity data in time and solving Equation (13.22) implicitly. Once the location of the departure point is determined at the previous time for a fluid particle located at x_j in the present time t^{n+1}, the value of the solution u at the departure point location x_j at the previous time t^n can be found by interpolating the solution u from the gridded known values of the solution u at the previous time t^n. Once the value of the solution u is obtained at the departure point location x_d at the previous time t^n, the solution u at the grid point

x_j at the present time t^{n+1} will be the same as that u at x_d at the previous time t^n. A repetition of this procedure provides for the marching of the solution at the next time $t = t^{n+2}$.

To sum up, the following sequence of steps defines the semi-Lagrangian scheme for solving the linear one-dimensional advection equation with non-constant velocity:

1. Solve Equation (13.22) iteratively for the half-displacement $\alpha/2$ for all grid points x_j at time $t^n + \Delta t$ using some initial guess (the value of $\alpha/2$ at the previous time step can be utilized as a first guess value) and appropriate interpolation formula in space for the velocity and extrapolation formula in time for the velocity.

2. Evaluate u at an upstream location $x_j - \alpha$, and at time t^n using an appropriate interpolation formula.

3. Evaluate u at arrival location x_j, and at time $t^n + \Delta t$ using Equation (13.23).

13.3.5 Semi-Lagrangian scheme to solve the linear one-dimensional advection equation with non-constant velocity in the presence of a source term using two time level scheme

The linear one-dimensional advection equation with non-constant velocity in the presence of a source term is given by

$$\frac{\partial u}{\partial t} + a\frac{\partial u}{\partial x} = g(x,t), \tag{13.26}$$

subject to the initial condition

$$u(x,0) = u_0(x). \tag{13.27}$$

In the Lagrangian description of fluid motion, Equation (13.13) is re-written as the evolution equation along any trajectory as

$$\frac{du}{dt} = g[x(t),t]. \tag{13.28}$$

For Equation (13.28) with a source term, the solution u is not conserved along any characteristic, unlike the cases that were discussed in Sections 13.3.2, 13.3.3, and 13.3.4. Assuming the trajectory equation, Equation (13.20) to remain the same, one can define the semi-Lagrangian scheme for Equations (13.20) and (13.28) using the following two-time level time integration scheme as

$$\frac{u(x_j, t^n + \Delta t) - u(x_j - \alpha, t^n)}{\Delta t} = g[x_j - \alpha/2, t^n + \Delta t/2]. \tag{13.29}$$

$$\alpha^{(r+1)} = \Delta t a(x_j - \alpha^{(r)}/2, t^n + \Delta t/2). \tag{13.30}$$

The following sequence of steps defines the semi-Lagrangian scheme for solving the linear one-dimensional advection equation with non-constant velocity in the presence of a source term using the two time level scheme:

1. Solve Equation (13.30) iteratively for the half-displacement $\alpha/2$ for all grid points x_j at time $t^n + \Delta t$ using some initial guess (the value of $\alpha/2$ at the previous time step can be utilized as a first guess value) and an appropriate interpolation formula in space for the velocity and extrapolation formula in time for the velocity.

2. Evaluate u at an upstream location $x_j - \alpha$, and at time t^n using an appropriate interpolation formula.

3. Evaluate the source term g at the mid-point of the approximate straight line trajectory, i.e., at the location $[x_j - (\alpha/2), t^n + (\Delta t/2)]$ using an appropriate interpolation formula in space and an extrapolation formula in time.

4. Evaluate u at the arrival location x_j and at time $t^n + \Delta t$ using Equation (13.29), the latter incorporating the source term value at the mid-point of the approximate straight line trajectory.

13.3.6 Stability of the semi-Lagrangian scheme to solve the linear one-dimensional advection equation

Consider the linear one-dimensional advection equation

$$\frac{\partial u}{\partial t} + a \frac{\partial u}{\partial x} = 0. \tag{13.31}$$

The distance traveled during the time interval Δt by a fluid parcel arriving at point x_j at time $t^n + \Delta t$ is $a\Delta t$. Hence, this fluid parcel originates from a point x_d at the previous time t^n. Hence,

$$x_d = x_j - a\Delta t \tag{13.32}$$

Typically, x_d will not coincide with a grid point and is most likely to lie between two grid points, say between x_{j-p} and x_{j-p-1}, where p is an integer. Let q be the fraction of the grid length from x_d to x_{j-p}. Then,

$$a\Delta t = x_j - x_d = x_j - [x_{j-p} + x_d - x_{j-p}] = x_j - [x_{j-p} - (x_{j-p} - x_d)] = (x_j - x_{j-p})$$
$$+ (x_{j-p} - x_d)$$

$$= p\Delta x + q\Delta x \tag{13.33}$$

Using linear interpolation to obtain u_d^n, one gets

$$u_j^{n+1} = u_d^n = (1-q)u_{j-p}^n + qu_{j-p-1}^n \tag{13.34}$$

For the case, $p = 0$, Equation (13.34) becomes identical to the upstream finite difference scheme with q given by $c\Delta t/\Delta x$.

Investigating the stability of the semi-Lagrangian scheme for the linear one-dimensional linear equation utilizing the von Neumann stability method, a solution of the following form is assumed

$$u_j^n = u_0 \lambda^n e^{ikx_j}. \tag{13.35}$$

Substituting Equation (13.35) in Equation (13.34), one obtains the amplification factor λ to be given by

$$\lambda = (1-q)e^{-ikp\Delta x} + qe^{-ik(p+1)\Delta x} \tag{13.36}$$

$$\lambda = \left[1 - q\left(1 - e^{-ik\Delta x}\right)\right]e^{-ikp\Delta x}. \tag{13.37}$$

From Equation (13.37), one obtains

$$|\lambda|^2 = \left|e^{-ikp\Delta x}\right|^2 \left|(1-q) + qe^{-ik\Delta x}\right|^2$$

$$= \left|[(1-q) + q\cos(k\Delta x) - iq\sin(k\Delta x)]^2\right|$$

$$= [(1-q) + q\cos(k\Delta x)]^2 + q^2\sin^2(k\Delta x)$$

$$= (1-q)^2 + 2(1-q)q\cos(k\Delta x) + q^2\cos^2(k\Delta x) + q^2\sin^2(k\Delta x)$$

$$= (1 - 2q + q^2) + 2q(1-q)\cos(k\Delta x) + q^2$$

$$= 1 - 2q(1-q)[1 - \cos(k\Delta x)] \tag{13.38}$$

As $0 \le 1 - \cos(k\Delta x) \le 2$, the RHS of Equation (13.38) will assume the following value for $1 - \cos(k\Delta x) = 2$

$$|\lambda|^2 = 1 - 4q(1-q) = 1 - 4q + 4q^2 = (1-2q)^2 \le 1, \tag{13.39}$$

where q is the fraction of the grid length from x_d to x_{j-p}. Moreover, for $1 - \cos(k\Delta x) = 0$, Equation (13.38) will assume the following value:

$$|\lambda|^2 = 1. \tag{13.40}$$

Equations (13.39) and (13.40) show conclusively that the amplification factor is less than or equal to unity and, hence, the semi-Lagrangian scheme is unconditionally stable. It is clear that the semi-Lagrangian scheme is stable always as $0 \leq q \leq 1$. This condition clearly shows that the semi-Lagrangian scheme is stable if the interpolating grid points are the two nearest grid points to the departure point. From Equation (13.38), it is clear that the amplification factor is unity (neutral case) for $q = 0$ and $q = 1$; this neutral case is associated with no requirement for interpolation. From Equation (13.38), it is clear that heavy damping occurs for the shortest wavelength (wavelength $= 2\Delta x$ and $q = 0.5$) and that damping decreases with increase in wavelength. A peculiarity of the semi-Lagrangian scheme (for a constant wind case) is that for a given value of q, the phase errors and damping decreases with increase of p, i.e., the departure point can be located precisely using only the wind at the arrival point. It is possible to utilize quadratic interpolation rather than linear interpolation in Equation (13.34). A semi-Lagrangian scheme that utilizes quadratic interpolation can be shown to be absolutely stable if the u_d^n is computed by interpolating from the three nearest grid points. It is found that the semi-Lagrangian scheme that uses quadratic interpolation has lower damping as compared to the scheme that uses linear interpolation; however, the phase representation of the scheme is not improved with the use of quadratic interpolation.

13.3.7 Semi-Lagrangian scheme to solve the forced advection equation with non-constant velocity – three time level scheme

We will apply the basic idea of the semi-Lagrangian scheme to solve the forced advection equation with non-constant velocity by employing a three time level scheme. Consider solving the following equation:

$$\frac{du}{dt} + G(X,t) = R(X,t), \tag{13.41}$$

where

$$\frac{du}{dt} = \frac{\partial u}{\partial t} + V(X,t) \cdot \nabla u \tag{13.42}$$

and

$$\frac{dX}{dt} = V(X,t). \tag{13.43}$$

In Equations (13.41) to (13.43), X is the position vector in the one-dimensional, two-dimensional, and three-dimensional grid, G and R are forcing terms, u is the advected quantity, and t is time.

A semi-Lagrangian approximation of Equations (13.41) to (13.43) using a three time level scheme is given by

$$\frac{u^+ - u^-}{2\Delta t} + \frac{1}{2}(G^+ + G^-) = R^0 \tag{13.44}$$

and

$$\alpha = \Delta t V(X - \alpha, t), \tag{13.45}$$

where the superscripts '+', '0' and '−' refer to the evaluation of the arrival point $(X, t + \Delta t)$, evaluation of the mid-point of the trajectory $(X - \alpha, t)$, and the evaluation of the departure point $(X - 2\alpha, t - \Delta t)$, respectively. Both the forcing functions G and R are supposed to be known. In Equations (13.44) and (13.45), X refers to a point in a regular one-dimensional, two-dimensional, and three-dimensional grid whereas α denotes the vector displacements. Equations (13.44) and (13.45) provide a second-order accurate approximation to Equations (13.41) to (13.43), where G is evaluated as the time average of the values of G at the arrival and the departure points of the trajectory while R is evaluated at the mid-point of the trajectory. As in the one-dimensional case, the sequence of steps to be followed are as follows:

1. Calculate the trajectories by iteratively solving Equation (13.45) for the vector displacements α for all the grid points X, using some initial guess value of α (α value at the previous time step).
2. Evaluate $u - \Delta t G$ at the upstream points $X - 2\alpha$ at time $t - \Delta t$ using an interpolation formula.
3. Evaluate $2\Delta t R$ at the location of the mid-point $X - \alpha$ of the trajectories at time t using an interpolation formula.
4. Extrapolate G at arrival location X and at time $t + \Delta t$ by using extraplotion formula in time.
5. Evaluate u at arrival points at time $t + \Delta t$ using

$$u(X, t + \Delta t) = (u - \Delta t G)_{(X - 2\alpha, t - \Delta t)} + 2\Delta t R_{X - \alpha, t} - \Delta t G_{X, t + \Delta t}$$

$$= (u - \Delta t G)^- + 2\Delta t R^0 - \Delta t G^+ \tag{13.46}$$

13.4 Numerical Domain of Dependence

It is clear from the earlier discussions that each computed value of the solution u_j^n depends on the previously computed values as well as the initial conditions. The numerical domain of dependence of the solution at a grid point is defined as the set of grid points that would influence the value of the solution u_j^n. It is clear that the numerical scheme cannot provide for an accurate solution if the fluid (air) parcel arriving at a grid point $j\Delta x$ at time $n\Delta t$ were to originate outside the numerical domain of dependence corresponding to the grid point $j\Delta x$. This follows from the definition of the numerical domain of dependence itself as any fluid parcel originating outside the numerical domain of dependence would not carry with itself the necessary information. Furthermore, it may so happen that when the fluid parcel originates outside the numerical domain of dependence, the associated errors are very large resulting in the numerical solution deviating considerably from the expected physical solution. One way to avoid this situation is to ensure that the numerical domain of dependence does include the physical trajectory of the fluid (air) parcel. This requirement that the physical trajectory of the fluid (air) parcel is always included within the domain of dependence is automatically taken care of in the semi-Lagrangian scheme.

Figure 13.5 depicts schematically the domain of dependence of the solution at grid point $(j\Delta x, n\Delta t)$ indicated by continuous slant lines. The figure also depicts the fluid (air) parcel trajectory as the line of stars that originate from left and move towards right. The fluid (air) parcel trajectory as depicted in Figure 13.5 originates outside the numerical domain of dependence; hence, any numerical scheme that is Eulerian in nature would have serious errors associated with the Eulerian solution. As the line of stars represents the fluid (air) parcel trajectory, all points along the physical trajectory would have the same value of the solution u_j^n. From Figure 13.5, it is clear that the fluid (air) parcel is located between grid points $(j-1)\Delta x$ and $(j-2)\Delta x$ at time $(n-1)\Delta t$. The solution value at the departure point at time $(n-1)\Delta t$ can be obtained by interpolation from neighboring grid points. As these neighboring grid points are found within the domain of dependence and the solution values at these neighboring grid points are utilized in obtaining the solution value at the departure point, the semi-Lagrangian scheme ensures that the solution values at all points of the physical trajectory carries the necessary information to propagate the solution forward in time.

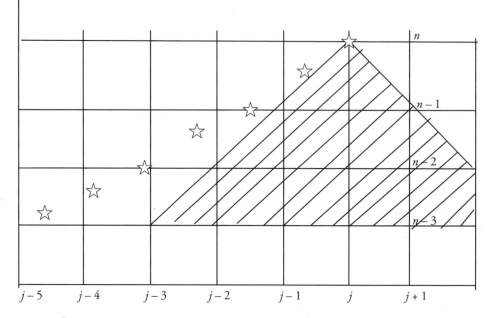

Figure 13.5 Schematic diagram of the domain of dependence of the solution at grid point $(j\Delta x, n\Delta t)$ indicated by continuous slant lines. The line of stars that originate from left and move towards right represent a fluid (air) parcel trajectory.

13.5 Semi-Lagrangian Scheme to Solve Shallow Water Equations

Let G discussed in Section 13.3.7 be not known at time $t + \Delta t$ as G depends on another dependent variable. This situation will naturally arise in a coupled set of partial differential equations (PDEs). As the governing equations of atmosphere involve a coupled set of partial differential equations, it would be interesting to investigate how the semi-Lagrangian scheme can be fruitfully applied to such a coupled set of PDEs. A very important set of coupled PDEs relevant to the atmosphere manifests in the shallow water equations. In this section, the semi-Lagrangian scheme is applied to the shallow water equations. The shallow water equations with rotation is as follows:

$$\frac{du}{dt} - fv = -\frac{\partial \phi}{\partial x} \tag{13.47}$$

$$\frac{dv}{dt} + fu = -\frac{\partial \phi}{\partial y} \tag{13.48}$$

$$\frac{d(\ln\phi)}{dt} = -\left(\frac{\partial u}{\partial x} + \frac{\partial v}{\partial y}\right) \tag{13.49}$$

where u and v are the zonal and meridional velocity components, f is the Coriolis parameter and ϕ is the geopotential of the free surface of the fluid above a flat bottom. Equations (13.50) to (13.52) provide for a two-time level integration scheme with semi-implicit semi-Lagrangian scheme implementation:

$$\frac{u^+ - u^0}{\Delta t} - \frac{1}{2}[(fv)^+ + (fv)^o] = -\frac{1}{2}(\phi_x^+ + \phi_x^o) \tag{13.50}$$

$$\frac{v^+ - v^0}{\Delta t} + \frac{1}{2}[(fu)^+ + (fu)^o] = -\frac{1}{2}(\phi_y^+ + \phi_y^o) \tag{13.51}$$

$$\frac{\ln\phi^+ - \ln\phi^0}{\Delta t} = -\frac{1}{2}\left[\left(\frac{\partial u}{\partial x} + \frac{\partial v}{\partial y}\right)^+ + \left(\frac{\partial u}{\partial x} + \frac{\partial v}{\partial y}\right)^o\right] \tag{13.52}$$

In Equations (13.50) to (13.52), advection terms are treated as time differences along the trajectories and all other terms are treated as time averages along the trajectories leading to a second-order accurate time difference scheme. Quantities with superscripts + and 0 refer to quantities that correspond to the arrival point and departure point, respectively. Equations (13.50) and (13.51) are rewritten as Equations (13.53) and (13.54):

$$u^+ = -\frac{\Delta t}{2}(a\phi_x^+ + b\phi_y^+) + \text{known terms} \tag{13.53}$$

$$v^+ = -\frac{\Delta t}{2}(a\phi_y^+ - b\phi_x^+) + \text{known terms} \tag{13.54}$$

where

$$a = \left[1 + \left(\frac{f\Delta t}{2}\right)^2\right]^{-1} \quad \text{and} \quad b = \left(\frac{f\Delta t}{2}\right)a \tag{13.55}$$

Taking the horizontal divergence of the LHS of Equations (13.53) and (13.54) and eliminating the horizontal divergence obtained above in the first term of the square bracket of the RHS in Equation (13.52), one gets the following elliptic boundary value problem

$$\left[(a\phi_x)_x + (a\phi_y)_y + (b\phi_y)_x - (b\phi_x)_y - 4\frac{\ln\phi}{(\Delta t)^2}\right]_{x,t+\Delta t} = \text{known terms} \tag{13.56}$$

The solution algorithm for solving the shallow water equations with rotation using a two-time level integration scheme with semi-implicit semi-Lagrangian scheme implementation is as follows:

1. Extrapolate the velocity a using Equation (13.24) and solve Equation (13.22) iteratively to obtain the displacements α_j for all grid points x_j using values at the previous time step as the first guess values and utilizing an appropriate interpolation formula. It is pertinent to note that this step is required to be performed just once every time step as the same physical trajectory is used for advecting all the three quantities (u, v, and ϕ).

2. Calculate upstream values (superscript $'o'$) in Equations (13.50) to (13.52) by first computing derivative terms such as u_x and then evaluating quantities upstream.

3. Solve the elliptic boundary value problem, Equation (13.56), to obtain $\phi(x, t + \Delta t)$.

4. Back substitute $\phi(x, t + \Delta t)$ in Equations (13.53) and (13.54) to obtain $u(x, t + \Delta t)$ and $v(x, t + \Delta t)$.

13.5.1 Advantages of semi-Lagrangian scheme as compared to Eulerian scheme

It is obvious that the semi-Lagrangian scheme is more computationally expensive for one time-step as compared with the Eulerian scheme, as in the former, for each grid point, it is necessary to determine the positions of the departure point as well as the medium point along the physical trajectory and to interpolate some of the quantities at these points. However, unlike the Eulerian scheme, the semi-Lagrangian scheme allows one to employ larger time steps. Moreover, the semi-Lagrangian scheme is unconditionally stable with the stability condition only requiring that any two physical trajectories do not cross each other. This stability condition is much less restrictive and much less severe than the stability condition (CFL condition) that a Eulerian scheme has to satisfy. In practice, for applying the semi-Lagrangian scheme in the atmosphere, the stability requirement of trajectories not crossing each other is generally satisfied and two iterations are adequate in most situations to achieve second-order accuracy.

13.6 Interpolation

Linear interpolation formula as utilized in Equation (13.34) are easy to implement and also have very interesting properties. It is known that the integration using the semi-Lagrangian scheme with time ensures that the initially positive data remains positive with time. Moreover, no new local extrema (maxima or minima) are introduced in time while integrating the semi-Lagrangian scheme with time. Finally, the area beneath the initial profile remains same at all times while using the semi-Lagrangian scheme for time integration. The last property mentioned here corresponds to the conservation of mass in the advection process and is a very desirable property. If one were to define the numerically advected mass after n time steps to be

$$M_n = \sum_j u_j^n \Delta x. \tag{13.57}$$

Calculating M_{n+1} by utilizing Equation (13.34), one obtains for $p = 0$ and for periodic boundary conditions:

$$M_{n+1} = \sum_j u_j^{n+1} \Delta x = (1-q) \sum_j u_j^n \Delta x + q \sum_j u_{j-1}^n \Delta x = \sum_j u_j^n \Delta x = M_n. \tag{13.58}$$

Equation (13.58) shows that the numerically advected mass after n time steps is the same as the numerically advected mass after $n+1$ time steps, providing proof that the semi-Lagrangian scheme provides for conservation of mass in the advection process.

The quadratic interpolation formula is as follows:

$$u_d^{n+1} = \frac{1}{2}(1-q)(2-q)u_{j-p-1}^n + q(2-q)u_{j-p}^n + \frac{1}{2}q(q-1)u_{j-p+1}^n \tag{13.59}$$

while the cubic interpolation formula is given by

$$u_d^{n+1} = -\frac{1}{6}(1-q)q(1+q)u_{j-p-1}^n + \frac{1}{2}q(1+q)(2-q)u_{j-p-1}^n$$
$$+ \frac{1}{2}(2-q)(1-q)(1+q)u_{j-p}^n - \frac{1}{6}(2-q)(1-q)qu_{j-p+1}^n \tag{13.60}$$

Comparing the amplitude and phase errors for the cubic interpolation with linear interpolation shows that the amplitude for the cubic interpolation scheme remains closer to the analytic value (one) over a wider range of wave-numbers as compared to the linear interpolation scheme. However, both cubic and linear interpolation schemes show severe attenuation in amplitude for the Courant number of one-half. As far as phase error is concerned, cubic interpolation performs better as compared to the linear interpolation at all Courant numbers for wave numbers having phase $\phi = k\Delta x$ that are within $\pi/2$ radians of zero, i.e., $\phi \leq \pi/2$. For $\phi = \pi/2$, the corresponding wavelength equals $4\Delta x$. Both linear and cubic interpolation formulas when used in the semi-Lagrangian scheme have similar phase errors for wavelengths less than $4\Delta x$, i.e., for phase $|\phi| > \pi/2$, the two schemes have practically identical phase errors. However, for low Courant number values, the cubic interpolation performs better as compared to the linear interpolation schemes.

Exercises 13 (Questions and answers)

1. Regarding the stability of the semi-Lagrangian scheme to solve the linear one-dimensional advection equation, show that neutral solutions are obtained when no interpolation is required, i.e., for the special case where the location of the departure point at the earlier time is a grid point itself.

   Answer: The expression for the amplification factor in the aforementioned case is

   $$|\lambda|^2 = 1 - 2q(1-q)[1 - \cos(k\Delta x)]$$

   where q is the fraction of the grid length from x_d to x_{j-p}. The location of the departure point is given as

   $$u_j^{n+1} = u_d^n = (1-q)u_{j-p}^n + qu_{j-p-1}^n$$

   For the case, $q = 0$, it is clear that the location of the departure point at the previous time is a grid point $(j-p)$, whereas for the case, $q = 1$, the location of the departure point at the previous time is a grid point $(j-p-1)$. In each of the aforementioned special cases, the departure point at the previous time is itself a grid point and, hence, there is no requirement to obtain the u value at the departure location using interpolation. From the expression of amplification factor, for these two special cases ($q = 0$ and $q = 1$), the amplification factor λ becomes unity, i.e., the semi-Lagrangian method provides for neutral solutions when $q = 0$ and for $q = 1$.

2. For question Exercise 13.1 Q 1, show that complete extinction of solution occurs for the smallest wave and for $q = 1/2$.

   Answer: For the smallest wave, $k\Delta x = \pi, \cos(k\Delta x) = -1$. Hence, from the expression for the amplification factor, one obtains for the value $q = 1/2$,

   $$|\lambda|^2 = 1 - 2q(1-q)[1 - \cos(k\Delta x)] = 1 - 2\tfrac{1}{2}(1 - \tfrac{1}{2})[1 - (-1)] = 0$$

3. Outline the quasi-cubic interpolation method that is found to provide equivalent results to cubic interpolation while employing the semi-Lagrangian scheme for atmospheric models.

   Answer: The quasi-cubic interpolation technique can be illustrated by two-dimensional interpolation on a regular grid. Let the target point be $(x_i + \alpha, y_j + \beta)$. In the first step, four interpolations are performed in the x-direction: linear (rather than the usual cubic) interpolations to the points $(x_i + \alpha, y_{j-1})$ and $(x_i + \alpha, y_{j+2})$, and cubic interpolations to the points $(x_i + \alpha, y_j)$ and $(x_i + \alpha, y_{j+1})$. In the second step, one cubic interpolation is performed in the y-direction, to evaluate the field at the target point.

4. For applying the semi-Lagrangian scheme to obtain an approximate trajectory using an explicit three time level scheme, as applied to an atmospheric model, indicate how the RHS terms are considered for easier and effective interpolation.

Answer: Let the general form of the model equations be given by

$$\frac{dX}{dt} = \frac{\partial X}{\partial t} + A(x) = R = L + N,$$

where A is the three-dimensional advection operator, L is the linear part of R while N is the remaining (nonlinear) terms of R. Approximating this equation as

$$\frac{X^+ - X^-}{2\Delta t} = R^0,$$

where superscripts '$+$', '0', and '$-$', respectively denote evaluation at the arrival point $(\tilde{x}, t + \Delta t)$, the mid-point of the trajectory $(\tilde{x} - \alpha, t)$, and at the location of the departure point $(\tilde{x} - 2\alpha, t - \Delta t)$. As the mid-point and the departure point will not in general coincide with model grid points, X^- and R^0 must be determined by interpolation. It is more economical (the same also provides better results) to evaluate the RHS as

$$R^0 = \tfrac{1}{2}[R(\tilde{x} - 2\alpha, t) + R(\tilde{x}, t)]$$

as only a single interpolation [of the combined field $X(t - \Delta t) + \Delta t R(t)$ at the point $(\tilde{x} - 2\alpha)$] is then required to determine X^+.

5. Are there any disadvantages with the semi-Lagrangian scheme?

Answer: The semi-Lagrangian scheme is an unconditionally stable scheme that does not require the CFL condition to be satisfied. Hence, one can obtain stable solutions employing a much larger time step when applying the semi-Lagrangian scheme. However, the semi-Lagrangian scheme does not necessarily always provide for solutions that are conservative.

14

Spectral Methods

14.1 Introduction

For solving the governing equations of atmospheric motion that give rise to a coupled set of nonlinear partial differential equations, the emphasis for a solution procedure was restricted to the method of finite differences. The method of finite differences is quite popular and is still widely employed. However, as mentioned in the earlier chapters, despite its popularity, it has a number of problems associated with truncation error, linear and nonlinear instability, not to mention, the associated amplitude and phase errors of finite difference schemes. Hence, continued efforts have been initiated to seek and find a more accurate and much improved alternate solution procedure that does not have the limitations of the method of finite differences. The Galerkin method (or series expansion method) is one such solution procedure that provides for a more accurate solution to the governing equations of atmospheric motions as manifested in a coupled system of nonlinear partial differential equations.

14.2 Series Expansion Method

Series expansion method (or the Galerkin method) is a technique that provides for a more accurate solution to the governing equations of atmospheric motions. The method forms part of the general class of methods that include the two well-known methods: (i) spectral methods and (ii) finite element method. Broadly speaking, in the Galerkin method, one approximates functions as a linear combination of prescribed expansion functions, the latter known as "basis functions." The basis functions are known to possess nice properties. For a continuous function $u(x)$, one can write

$$u(x) = \sum_{j=1}^{N} u_j \phi_j(x), \tag{14.1}$$

where $\phi_j(x)$, $j = 1, 2, \ldots, N$, are the basis functions, which satisfy any boundary conditions on $u(x)$. The coefficients u_j are the unknown coefficients that form a vector of N numbers.

Consider a partial differential equation of the following form with $\mathscr{L}$ being an ordinary differential operator, operating on u:

$$\mathscr{L}(u) = f(x) \qquad \text{for} \qquad a \le x \le b. \tag{14.2}$$

Substituting Equation (14.1) in Equation (14.2) leads to a mismatch between the LHS and RHS of Equation (14.2) that provides for a definition of the residual $\varepsilon(x)$ given by

$$\varepsilon(x) = \mathscr{L}(u) - f(x) = \mathscr{L}\left[\sum_{j=1}^{N} u_j \phi_j(x)\right] - f(x). \tag{14.3}$$

If the ordinary differential operator $\mathscr{L}$ is linear and the basis functions, $\phi_j(x)$ are the eigenfunctions of $\mathscr{L}$, then the residual can be set to zero for the whole domain and the resulting N algebraic equations can be solved for the unknown coefficients u_j. More generally, if $\mathscr{L}$ is a nonlinear ordinary differential operator; the subsequent section would describe the general approaches for solving for the unknown coefficients u_j.

There are three broad strategies to obtain the unknown coefficient in the series expansion method

1. minimization of the of l_2-norm of the residual,

$$\{\|\varepsilon(x)\|_2\}^2 = \int [\varepsilon(x)]^2 dx. \tag{14.4}$$

2. the collocation method, where one sets the residual to zero at a discrete set of positions x_j (for example x_j may refer to a regular grid $x_j = j\Delta x$), i.e., $\varepsilon(x_j) = 0$ for all $j = 1, 2, \ldots, N$

3. the Galerkin method, which requires that the residual to be orthogonal to each of the basis functions, i.e.,

$$\int \phi_j \varepsilon(x)\, dx = 0, \qquad \text{for all} \quad j = 1, 2, \ldots, N. \tag{14.5}$$

It is to be noted that the collocation method is employed in the pseudo-spectral methods whereas the Galerkin method is used extensively in the finite element method. The spectral methods is a special case of the series expansion method where the l_2-norm and the Galerkin method become equivalent.

There are many variants in the aforementioned methods; for example, in the Petrov–Galerkin method, the residual is made orthogonal to a set of test functions, $\theta_j(x)$, which may be different from the basis function set, $\phi_j(x)$, i.e.,

$$\int \theta_j \varepsilon(x)\,dx = 0, \qquad \text{for all} \quad j = 1, 2, \ldots, N. \tag{14.6}$$

For the aforementioned reason, the Petrov–Galerkin method is considered more general than the Galerkin method, as for the latter, one chooses $\theta_j(x) = \phi_j(x)$.

14.3   Spectral Methods and Finite Difference Method

Spectral methods are examples of the series expansion method, for which the basis functions form an orthogonal set.

$$\int \phi_i \phi_j\,dx = 0, \quad \text{for} \quad i \neq j. \tag{14.7}$$

The choice of deciding on a particular orthogonal basis set is largely dictated by the domain geometry and boundary conditions. It is obvious that while employing spectral methods to solve global atmospheric motions, one would choose "spherical harmonics" as the orthogonal basis functions.

The error in satisfying the ordinary differential equation, Equation (14.2), with the N terms of the sum given in Equation (14.1) is as follows:

$$\varepsilon_N = \mathscr{L}\left[\sum_{j=1}^{N} u_j \phi_j(x)\right] - f(x). \tag{14.8}$$

The Galerkin method requires that the error ε_N be orthogonal to each basis function $\phi_j(x)$ in the following sense:

$$\int_a^b \varepsilon_N \phi_i\,dx = 0, \qquad i = 1, 2, \ldots, N. \tag{14.9}$$

Substituting Equation (14.8) in Equation (14.9), one obtains

$$\int_a^b \phi_i \mathscr{L}\left[\sum_{j=1}^{N} u_j \phi_j(x)\,dx\right] - \int_a^b \phi_i f(x)\,dx = 0, \qquad i = 1, 2, \ldots, N. \tag{14.10}$$

Equation (14.10) reduces the solution of ordinary differential equation, Equation (14.2), to a system of N algebraic equations involving the unknown coefficients u_j to the "transforms" of the forcing function. Once the unknown coefficients u_j are determined by solving the system of N algebraic equations, the same can be substituted in Equation (14.1) to obtain the unknown $u(x)$. This procedure is quite general and can be applied to more complicated ordinary and partial differential equations involving more dependent and independent variables. The spectral method takes care of nonlinear computational instability. Furthermore, in the spectral methods, as space derivatives are evaluated analytically, there is no further need to approximate space derivatives with finite differences as is seen in the method of finite difference.

14.3.1 Spectral methods as applied to a linear one-dimension advection equation

The one-dimensional linear advection equation with constant velocity c is given by

$$\frac{\partial \rho}{\partial t} + c\frac{\partial \rho}{\partial x} = 0, \tag{14.11}$$

for $-\pi \leq x \leq \pi$. Employing a Fourier series expansion for $\rho(x,t)$, one gets

$$\rho(x,t) = \sum_{k=-N}^{N} a_k(t)\, e^{ikx}, \tag{14.12}$$

where periodic domain is assumed. As $\rho(x,t)$ is real, the unknown coefficients in Equation (14.12) satisfy the conditions, $a_k = a_{-k}^*$, where a_k^* is the complex conjugate of a_k.

Substituting Equation (14.12) in Equation (14.11), one obtains

$$\frac{\partial a_k}{\partial t} + icka_k = 0 \qquad \text{for all } k. \tag{14.13}$$

Solving Equation (14.13), one gets $a_k(t) = e^{-ickt}$. Substituting $a_k(t)$ in Equation (14.12), one obtains the solution that all waves propagate with the correct phase speed.

Note that according to the definition of "orthogonality," two complex functions, $g(x)$ and $h(x)$, are orthogonal over the domain S, if the integral of the product of one function with the complex conjugate of the other function is zero over the domain, S, i.e.,

$$\int g(x)h^*(x)\,dx = 0. \tag{14.14}$$

Applying the spectral methods, and using Equations (14.11) and (14.12) in Equation (14.5), one obtains

$$\int_{-\pi}^{\pi} \left\{ \frac{\partial}{\partial t} \left[\sum_{k=-N}^{N} a_k(t) e^{ikx} \right] + c \frac{\partial}{\partial x} \left[\sum_{k=-N}^{N} a_k(t) e^{ikx} \right] \right\} \left(e^{ijx} \right) dx = 0 \qquad \text{for } j = -N, \ldots, N$$

(14.15)

Calculating the LHS of Equation (14.15) requires calculating the following integral:

$$\int_{-\pi}^{\pi} e^{-ijx} e^{ikx} dx = \begin{cases} 2\pi & \text{if } j = k \\ 0 & \text{if } j \neq k \end{cases}$$

(14.16)

Using Equation (14.16) in Equation (14.15), one obtains

$$2\pi \frac{\partial a_k}{\partial t} + 2\pi i k c a_k = 0.$$

(14.17)

Equation (14.17) is identical to Equation (14.13), the latter obtained by direct substitution. It is clear from Equation (14.17), that spectral methods provides the same result as the direct substitution. Moreover, except possibly for time discretization error, the spectral methods do not introduce any amplitude error or any phase speed error, unlike the method of finite differences.

14.3.2 Spectral methods as applied to a linear second-order ordinary differential equation

The second-order ordinary differential equation that is considered is as follows:

$$\frac{d^2 u}{dx^2} = f(x); \quad 0 \leq x \leq \pi$$

(14.18)

with boundary conditions

$$u(0) = 0 \quad \text{and} \quad u(\pi) = 0$$

(14.19)

The following basis functions are chosen for solving the second-order ordinary differential equation (14.18).

$$\phi_j = \sin(jx), \qquad j = 1, 2, \ldots, N$$

(14.20)

Choice of the basis functions in Equation (14.20) follow from the fact that they are orthogonal and also identically satisfy both the boundary conditions (14.19). Using the aforementioned basis functions, one obtains

$$\mathscr{L}(u) = \frac{d^2 u}{dx^2} = \mathscr{L}\left[\sum_{j=1}^{N} u_j \phi_j\right] = \sum_{j=1}^{N} (-j^2) u_j \phi_j \qquad (14.21)$$

The residual (or error) ε_N equals

$$\varepsilon_N = \sum_{j=1}^{N} (-j^2) u_j \phi_j - f(x) \qquad (14.22)$$

The requirement that the error be orthogonal for each basis function assumes the following form:

$$-\sum_{j=1}^{N} j^2 u_j \int_0^\pi \phi_i \phi_j \, dx = \int_0^\pi \phi_i f(x) \, dx, \qquad i = 1, 2, \ldots, N \qquad (14.23)$$

Equation (14.23) has to be solved as a system of N algebraic equations to obtain values of the unknown coefficients u_i. However, one can use the following standard results to obtain an expression for the coefficients u_i:

$$\int_0^\pi \sin(ix) \sin(jx) \, dx = \frac{1}{2} \int_0^\pi [\cos(i-j)x - \cos(i+j)x] \, dx = \frac{\pi}{2} \delta_{ij}, \qquad (14.24)$$

where δ_{ij} is the *Kronecker delta* with

$$\delta_{ij} = \begin{cases} 1 & \text{if } i = j \\ 0 & \text{if } i \neq j \end{cases}$$

Utilizing Equation (14.24), which is nothing but the orthogonality condition in Equation (14.23), one obtains the values of the unknown coefficients u_i:

$$u_i = -\frac{2}{\pi i^2} \int_0^\pi \phi_i f \, dx \qquad (14.25)$$

Substituting the unknown coefficients u_i in Equation (14.1) and utilizing the form of the basis function from Equation (14.20), one can obtain the solution $u(x)$.

14.3.3 Spectral methods as applied to a partial differential equation involving time

As in the atmosphere, time variation is important, spectral methods are applied to a partial differential equation involving time as follows:

$$\frac{\partial u}{\partial t} + \mathscr{L}(u) = 0, \qquad \text{for } a \leq x \leq b \text{ and } t > 0 \qquad (14.26)$$

where the differential operator $\mathscr{L}$ may be nonlinear. As before, we approximate $u(x,t)$ with a finite series of N terms as

$$u(x,t) = \sum_{j=1}^{N} u_j(t)\phi_j(x), \tag{14.27}$$

where the coefficients u_j are functions of time t and $\phi_j(x)$; the basis functions are functions of x. In the Galerkin method, it is always more convenient to treat the time differencing using the method of finite difference.

Substituting Equation (14.27) to Equation (14.26), calculating the error involved, multiplying with the basis function $\phi_j(x)$, and finally integrating over the domain from a to b, we obtain

$$\sum_{j=1}^{N} \frac{du_j}{dt} \int_a^b \phi_i\phi_j \, dx + \int_a^b \phi_i \mathscr{L}\left[\sum_{j=1}^{N} u_j\phi_j\right] dx = 0, \qquad i = 1,2,\dots,N \tag{14.28}$$

Equation (14.28) represents a system of N coupled ordinary differential equations in the unknown coefficients $u_j(t)$. Utilizing a suitable finite difference scheme in time for the time derivatives in Equation (14.28), the aforementioned set of N ordinary differential equations can in principle be solved.

14.3.4   Spectral methods and energy conservation

Energy conserving finite difference schemes are extremely important and well known. The spectral methods naturally lead to energy conservation in equations with quadratic energy variants. To show this, multiply Equation (14.26) by u and integrate with respect to x from a to b, to get

$$\int_a^b \frac{\partial}{\partial t}\left(\frac{u^2}{2}\right) dx = -\int_a^b u\mathscr{L}(u) \, dx \tag{14.29}$$

For an energy conserving system, the operator $\mathscr{L}(u)$ must satisfy the following condition

$$\int_a^b u\mathscr{L}(u) \, dx = 0 \tag{14.30}$$

where u is any function that satisfies the boundary conditions. In this case, Equation (14.29) becomes

$$\frac{d}{dt}\int_a^b \frac{u^2}{2} \, dx = 0 \tag{14.31}$$

Equation (14.31) shows energy conservation for the exact equation. To demonstrate that the same result holds for the finite sum in Equation (14.27), multiply the ith equation of Equation (14.28) by u_i and sum from $i = 1$ to $i = N$, to get

$$\int_a^b \left[\sum_{i=1}^N u_i \phi_i \frac{\partial}{\partial t} \left(\sum_{j=1}^N u_j \phi_j \right) \right] dx = - \int_a^b \left(\sum_{i=1}^N u_i \phi_i \right) \mathscr{L} \left(\sum_{j=1}^N u_j \phi_j \right) dx \qquad (14.32)$$

The integral on the RHS of Equation (14.32) vanishes due to Equation (14.30) as the function given by Equation (14.27) satisfies the boundary condition. Hence, Equation (14.32) can be written as

$$\int_a^b \frac{\partial}{\partial t} \left\{ \sum_{i=1}^N \left[\frac{(u_i \phi_i)^2}{2} \right] \right\} dx = 0 \qquad (14.33)$$

Equation (14.33) expresses the energy conservation for the spectral approximation to the spatial variation. The actual degree of energy conservation would, however, depend on the time differencing scheme that one would employ.

14.3.5 Spectral methods applied to nonlinear one-dimensional advection equation

Consider the nonlinear one-dimensional advection equation (also called Burgers equation)

$$\frac{\partial u}{\partial t} + u \frac{\partial u}{\partial x} = 0, \qquad 0 \le x \le 2\pi \qquad (14.34)$$

with periodic boundary conditions. The natural choice of basis function are $\phi_j(x) = e^{ijx}$. We then define the solution as

$$u(x,t) = \sum_{j=-K}^K \hat{u}_j(t) \phi_j(x) = \sum_{j=-K}^K \hat{u}_j(t) e^{ijx} \qquad (14.35)$$

Substituting Equation (14.35) in Equation (14.34), one obtains

$$\sum_{j=-K}^K \frac{d\hat{u}_j}{dt} \phi_j(x) + \sum_{j=-K}^K \hat{u}_j \phi_j(x) \sum_{m=-K}^K \hat{u}_m \frac{d\phi_m(x)}{dx} = 0 \qquad (14.36)$$

As indicated in the previous section, the next step is to multiply Equation (14.36) with the $N+1$ test functions ϕ_l together with integrating over the domain to obtain a system of $N+1$ ordinary differential equations with $l = 0, 1, 2, \ldots, N$.

$$\int_0^{2\pi} \phi_l(x) \sum_{j=-K}^{K} \frac{d\hat{u}_j}{dt} \phi_j(x)\,dx + \int_0^{2\pi} \phi_l(x) \sum_{j=-K}^{K} \hat{u}_j \phi_j(x) \sum_{m=-K}^{K} \hat{u}_m \frac{d\phi_m(x)}{dx}\,dx = 0 \quad (14.37)$$

Equation (14.37) can be rewritten as follows using the product and derivative properties of ϕ_l:

$$\sum_{j=-K}^{K} \frac{d\hat{u}_j}{dt} \int_0^{2\pi} \phi_j(x)\phi_l(x)\,dx + \sum_{j=-K}^{K}\sum_{m=-K}^{K} im\hat{u}_j\hat{u}_m \int_0^{2\pi} \phi_l(x)\phi_{j+m}(x)\,dx = 0 \quad (14.38)$$

Employing the orthogonality relations

$$\int_0^{2\pi} \phi_j(x)\phi_{-l}(x)\,dx = 2\pi\delta_{jl} \quad (14.39)$$

where δ_{jl} is Kronecker delta. This gives

$$\sum_{j=-K}^{K} \frac{d\hat{u}_j}{dt} \delta_{-lj} + \sum_{j=-K}^{K}\sum_{m=-K}^{K} im\hat{u}_j\hat{u}_m\delta_{-l,j+m} = 0 \quad (14.40)$$

which eliminates one summation in each term yielding

$$\frac{d\hat{u}_{-l}}{dt} + \sum_{j=-K,-l=j+m,|m|\leq K}^{K} im\hat{u}_j\hat{u}_m = 0 \quad (14.41)$$

After replacing $-l$ by l in Equation (14.41), one finally obtains a system of nonlinear equations for the unknown coefficients u_l

$$\frac{d\hat{u}_l}{dt} + \sum_{j=-K,l=j+m,|m|\leq K}^{K} im\hat{u}_j\hat{u}_m = 0 \quad (14.42)$$

The appearance of the nonlinear advection term in the original equation, Equation (14.34), results in the appearance of the complex summation term in Equation (14.42). The number of operations of the complex summation in the second term of Equation (14.42) is $O(K^2)$, which makes the aforementioned summation computationally expensive.

14.3.6 Spectral methods applied to nonlinear one-dimensional advection equation – handling the nonlinear term

It is clear from the previous section that the appearance of a nonlinear term gives rise to a term having a very complex summation leading to a very expensive computational exercise. In this section, a work around is suggested to handle the nonlinear term that effectively reduces considerable computational time. To illustrate this, consider the product $w_j = w(x_j), j = 1, 2, \ldots, N$ of two grid functions $u_j = u(x_j)$ and $v_j = v(x_j)$ in physical space, i.e.,

$$w_j = u_j v_j \tag{14.43}$$

Transform $u(x_j)$ and $v(x_j)$ from the physical space to spectral space

$$u_j = \sum_{|k| \le K} \hat{u}_k e^{ikx_j}, \qquad \hat{u}_k = \frac{1}{N} \sum_{j=1}^{N} u_j e^{-ikx_j} \tag{14.44}$$

$$v_j = \sum_{|m| \le K} \hat{v}_m e^{imx_j}, \qquad \hat{v}_m = \frac{1}{N} \sum_{j=1}^{N} v_j e^{-imx_j} \tag{14.45}$$

Utilize the discrete form of the orthogonality relation

$$\frac{1}{N} \sum_{j=1}^{N} e^{ikx_j} e^{imx_j} = \delta_{k,-m+nN} \tag{14.46}$$

The n in Equation (14.46) corresponds to an arbitrary multiple of N. Utilizing Equations (14.44) to (14.46) in Equation (14.43), one obtains

$$\hat{w}_l = \frac{1}{N} \sum_{|k| \le K} \sum_{|m| \le K} \hat{u}_k \hat{v}_m \sum_{j=1}^{N} e^{ikx_j} e^{i(m-l)x_j} = \sum_{k=-K, l=k+m, |m| \le K}^{K} \hat{u}_k \hat{v}_m \tag{14.47}$$

Equation (14.47) brings out the well-known result that a multiplication in physical space corresponds to a convolution in spectral space. This result can be put to good use to obtain an efficient evaluation of a nonlinear product of two variables given in spectral space $\hat{u}_k, \hat{v}_m = im\hat{u}_m$ (as in Equation (14.47) using the following procedure).

1. Transform $\hat{u}_m$, $\hat{v}_m$ to physical space using Fast Fourier transform (FFT); $u_j = F^{-1}(\hat{u}_m)$; $v_j = F^{-1}(\hat{v}_m)$.

2. Multiplication is performed in the physical space $w_j = u_j v_j$

3. Transform back to spectral space: $\hat{w}_l = F(w_j)$ so that

$$\hat{w}_l = \sum_{k=-K, l=k+m, |m| \le K}^{K} im\hat{u}_k \hat{u}_m,$$

which helps us to obtain the complex summation in Equation (14.42).

14.3.7 Spectral methods applied to barotropic vorticity equation on a β plane

Spectral methods can be applied to the barotropic vorticity equation on a β plane. The appropriate basis functions for a β plane are the Fourier basis functions where the fields are periodic in x and y. The barotropic vorticity equation in spherical coordinates may be written as

$$\frac{\partial \zeta}{\partial t} = -V \cdot \nabla(\zeta + f), \tag{14.48}$$

where the horizontal gradient operator and horizontal wind velocity are given in spherical coordinates as

$$\nabla \equiv e_\theta \frac{1}{a} \frac{\partial}{\partial \theta} + e_\lambda \frac{1}{a \sin \theta} \frac{\partial}{\partial \lambda} \qquad \text{and} \qquad V = v_\theta e_\theta + v_\lambda e_\lambda,$$

where a is the radius of Earth, θ is the co-latitude, λ is longitude, v_θ and v_λ are the velocity components in the meridional (increasing θ) and zonal (increasing λ) directions, f is the Coriolis parameter, Ω is the angular velocity of Earth's rotation and ζ is the relative vorticity. Equation (14.48) then becomes

$$\frac{\partial \zeta}{\partial t} = -\frac{1}{a} \left(v_\theta \frac{\partial}{\partial \theta} + \frac{v_\lambda}{a \sin \theta} \frac{\partial}{\partial \lambda} \right) (\zeta + 2\Omega \cos \theta) \tag{14.49}$$

and

$$\zeta = \frac{1}{a \sin \theta} \left[\frac{\partial}{\partial \theta} (v_\lambda \sin \theta) - \frac{\partial v_\theta}{\partial \lambda} \right] \tag{14.50}$$

In terms of stream function ψ, the horizontal velocity vector and its components are given by

$$V = \hat{k} \times \nabla \psi, \qquad v_\lambda = \frac{1}{a} \frac{\partial \psi}{\partial \theta}, \qquad v_\theta = -\frac{1}{a \sin \theta} \frac{\partial \psi}{\partial \lambda}. \tag{14.51}$$

With the above relations, Equation (14.49) becomes

$$\nabla_s^2 \left(\frac{\partial \psi}{\partial t} \right) = \frac{1}{a^2 \sin \theta} \left(\frac{\partial \psi}{\partial \lambda} \frac{\partial}{\partial \theta} - \frac{\partial \psi}{\partial \theta} \frac{\partial}{\partial \lambda} \right) (\nabla_s^2 \psi + 2\Omega \cos \theta), \tag{14.52}$$

where ∇_s^2 is the Laplacian in spherical coordinates given by

$$\nabla_s^2 \equiv \frac{1}{a^2 \sin \theta} \left[\frac{\partial}{\partial \theta} \left(\sin \theta \frac{\partial}{\partial \theta} \right) + \frac{1}{\sin \theta} \frac{\partial^2}{\partial \lambda^2} \right]$$

The stream function ψ is represented in terms of spherical harmonics, which are the solutions Y_n^m of the equation

$$a^2 \nabla_s^2 Y_n^m + n(n+1) Y_n^m = 0, \tag{14.53}$$

where the eigenvalues are given by $n(n+1)/a^2$, and $|m|$ is the planetary wavenumber and $n - |m|$ is the number of zeros between the poles. Moreover, n must be greater than or equal to $|m|$.

The functions Y_n^m are expressible in terms of *Legendre functions* P_n^m

$$Y_n^m(\theta, \lambda) = e^{im\lambda} P_n^m(\theta) \tag{14.54}$$

Y_n^m are normalized in the sense

$$\frac{1}{4\pi} \int |Y_n^m|^2 \, \delta S = 1. \tag{14.55}$$

Substituting Equation (14.54) in Equation (14.53), one gets the following ordinary differential equation in P_n^m

$$\frac{d^2 P_n^m}{d\theta^2} + \cot\theta \frac{dP_n^m}{d\theta} + \left[n(n+1) - \frac{m^2}{\sin\theta} \right] P_n^m = 0 \tag{14.56}$$

Equation (14.56) is called the *Legendre equation* and the solution of Equation (14.56), P_n^m are known as *Legendre functions of order m* and degree n. Whereas n is a positive integer, m can be either positive integer or negative integer with $|m| \leq n$. The characteristic solutions of Equation (14.56) are expressible as a set orthonormal functions such that

$$\int_0^\pi P_n^m P_s^m \sin\theta \, d\theta = \delta_{ns}, \tag{14.57}$$

where δ_{ns} is the Kroneckar delta. The order m may take on negative integer values for which the Legendre function P_n^m is given by

$$P_n^{-m} = (-1)^m P_n^m \tag{14.58}$$

For integer values of n and m, the Legendre functions P_n^m are simply polynomials, the orders of which increase with n. The stream function $\psi(\theta, \lambda, t)$ may be expressed as the finite sum as follows:

$$\psi(\theta, \lambda, t) = a^2 \Omega \sum_{n=|m|}^{n'} \sum_{m=-m'}^{m'} \psi_n^m(t) Y_n^m(\theta, \lambda) \tag{14.59}$$

As the series indicated in Equation (14.59) is finite, disturbances of sufficiently small-scale are not represented, which constitutes to a truncation error; however, ignoring sufficiently small-scale motion is not necessarily undesirable. The harmonic coefficients ψ_n^{-m} are in general complex, and the condition necessary for ψ to be real is that

$$\psi_n^{-m} = (-1)^m \psi_n^{-m*},$$

where $*$ represents complex conjugate.

The stream function tendency can be obtained from Equation (14.59) and is as follows:

$$\frac{\partial \psi(\theta,\lambda,t)}{\partial t} = a^2 \Omega \sum_{n=|m|}^{n''} \sum_{m=-m''}^{m''} \frac{d}{dt}(\psi_n^m) Y_n^m \tag{14.60}$$

Substituting the expression for the RHS of Equation (14.60) to the LHS of Equation (14.52), multiplying the resulting equation by $Y_n^{-m} \sin\theta$ and integrating from 0 to 2π with respect to λ and integrating from 0 to π with respect to θ, one gets

$$\frac{d\psi_n^m(t)}{dt} = \frac{2i\Omega \psi_n^m(t)}{n(n+1)} + \frac{i\Omega}{2} \sum_{s=|r|}^{n'} \sum_{r=-m'}^{m'} \sum_{k=|j|}^{n'} \sum_{j=-m'}^{m'} \psi_k^j(t) \psi_s^r(t) H_{kns}^{jmr} \tag{14.61}$$

where

$$H_{kns}^{jmr} = \begin{cases} \frac{s(s+1)-k(k+1)}{n(n+1)} \int_0^\pi P_n^m \left[jP_k^j \frac{dP_s^r}{d\theta} - r\frac{dP_k^j}{d\theta} P_s^r \right] d\theta & \text{if } j+r = m \\ 0 & \text{otherwise} \end{cases} \tag{14.62}$$

The quantities H_{kns}^{jmr} are called *interaction coefficients*, which are zero unless $k+n+s$ equals an odd integer and $(k-s) < n < (k+s)$.

After the RHS of Equation (14.61) is determined, the future values of the expansion coefficients ψ_n^m are obtained by extrapolating them in time (as in the time differencing in time) from the knowledge of LHS of Equation (14.61), say using the leapfrog time integration scheme.

It has been pointed out (Robert 1966) that the components u and v of the wind field constitute pseudo-scalar fields globally, and as such are not well suited to be represented in terms of scalar spectral expansions; instead it is suggested that one employ the following variables

$$U = u\cos\phi \quad \text{and} \quad V = v\sin\phi \tag{14.63}$$

where ϕ is the latitude. It is suggested that variables as defined in Equation (14.63) would be more appropriate for global spectral representation.

The non-divergent velocity flow field may be represented as before in terms of a scalar stream function ψ as given by

$$V = \hat{k} \times \nabla \psi \tag{14.64}$$

Moreover, the vertical component of relative vorticity ζ is defined as before as

$$\zeta = \hat{k} \cdot \nabla \times V = \nabla^2 \psi \tag{14.65}$$

The modified components of the wind variables defined in Equation (14.63) can be then defined in terms of stream function ψ as follows:

$$U = -\frac{\cos\phi}{a}\frac{\partial\psi}{\partial\phi} \quad \text{and} \quad V = \frac{1}{a}\frac{\partial\psi}{\partial\lambda} \tag{14.66}$$

The conservation of absolute vorticity [Equation (14.48)] can be written, with substitution of Equation (14.66) and an expansion into spherical polar coordinates as

$$\frac{\partial}{\partial t}(\nabla^2\psi) = -\frac{1}{a\cos^2\phi}\left[\frac{\partial}{\partial\lambda}(U\nabla^2\psi) + \cos\phi\frac{\partial}{\partial\phi}(V\nabla^2\psi)\right] - 2\frac{\Omega V}{a}, \tag{14.67}$$

where ϕ, λ, a and Ω are the latitude, longitude, radius of Earth and Earth's rotation rate, respectively. Substituting Equation (14.66) in Equation (14.67), one obtains

$$\frac{\partial}{\partial t}(\nabla^2\psi) = \frac{1}{a^2\cos\phi}\left[\frac{\partial\psi}{\partial\phi}\frac{\partial}{\partial\lambda}(\nabla^2\psi) - \frac{\partial\psi}{\partial\lambda}\frac{\partial}{\partial\phi}(\nabla^2\psi)\right] - 2\frac{\Omega}{a^2}\frac{\partial\psi}{\partial\lambda} \tag{14.68}$$

The calculation of the interaction coefficients H_{kns}^{jmr} is a lengthy and complicated task. However, the spectral methods has the following advantage, i.e., the nonlinear computational instability is completely avoided as all the nonlinear interactions are computed analytically and, hence, all the contributions to the wave numbers outside the truncated series are automatically not taken into consideration.

14.3.8   Types of truncation

The spectral methods as applied to global spherical atmospheric models utilizes the truncated series of spherical harmonics, as the latter are employed naturally as the basis functions. In general, there are two types of truncation. Consider an arbitrary variable Φ, which is a function of latitude ϕ and longitude λ that can be expanded in the following spherical harmonics truncated series:

$$\Phi(\lambda,\phi) = \sum_{m=-M}^{M} \sum_{n=|m|}^{N} \Phi_n^m P_n^m(\sin\phi)e^{im\lambda}, \tag{14.69}$$

where

$$P_n^m(\mu) = \sqrt{\frac{(2n+1)(n-m)!}{(n+m)!}} \frac{(1-\mu^2)^{m/2}}{2^n n!} \frac{d^{n+m}}{d\mu^{n+m}}(\mu^2-1)^n \tag{14.70}$$

is the normalized associated Legendre function of the first kind of order m and degree n. Φ_n^m are complex expansion coefficients, m is the zonal wavenumber and n is the total wavenumber. M defines the zonal (east–west) truncation and N the meridional (north–south) truncation, and $\mu = \sin\phi$.

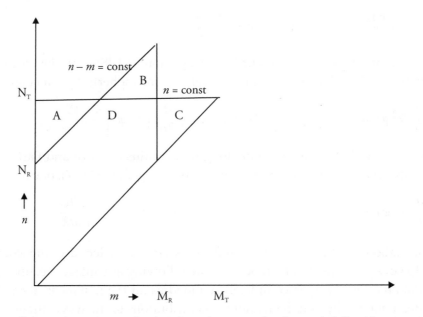

Figure 14.1 Schematic representation by triangular truncation (M_T, N_T) defined by areas $A+D+C$ and by rhomboidal truncation (M_R, N_R) defined by the areas $D+B$. The x-axis is the zonal wavenumber m whereas the y-axis is the total wavenumber n. Only positive values of m are shown in the figure.

For the triangular truncation, all the components with $n > N_T$ are set to zero and the truncated series for the triangular truncation assumes the following form:

$$\Phi(\lambda,\phi,t) = \sum_{m=-N_T}^{N_T} \sum_{n=|m|}^{N_T} \Phi_n^m(t)P_n^m(\sin\phi)e^{im\lambda} \tag{14.71}$$

For the *parallelogrammic truncation*, all the components with $|m| > M_R$ or $n > |m| + J$ are set to zero. Here, M_R, M_T, N_R, N_T and J are positive integer constants. The truncated series for the parallelogrammic truncation assumes the following form:

$$\Phi(\lambda,\phi,t) = \sum_{m=-M_R}^{M_R} \sum_{n=|m|}^{|m|+J} \Phi_n^m(t) P_n^m(\sin\phi) e^{im\lambda} \tag{14.72}$$

Usually M_R is chosen as equal to J in which case, the parallelogrammic truncation is called the *rhomboidal truncation*. Figure 14.1 shows the schematic representation by triangular truncation (M_T,N_T) defined by areas $A + D + C$ and by rhomboidal truncation (M_R,N_R) defined by the areas $D + B$. As indicated in Figure 14.1, the appropriate zonal and meridional truncations are (M_R,N_R) and (M_T,N_T) for the rhomboidal and triangular truncation.

For the triangular truncation, $N_T = M_T$ and for the rhomboidal truncation, $N_R = |m| + M_R$. In Figure 14.1, both rhomboidal and triangular truncations have the same number of degrees of freedom. The rhomboidal truncation is defined by the areas $D+B$, whereas the triangular truncation is defined by the areas $A+D+C$. For example, rhomboidal truncation, indicated as R_{20}, implies that $M_R = 20$, $N_R = |m| + 20$. Triangular truncation, indicated as T_{20}, implies that $M_T = 20$, $N_T = 20$.

The triangular truncation has some inherent advantages as compared to the rhomboidal truncation. A field expanded in a triangular truncated series using spherical harmonics as a basis function is invariant under rotation of the polar axis. Furthermore, the triangular truncation preserves the two-dimensional isotropy of scale resolution. Moreover, observational studies suggest that the triangular truncation maximises the variance explained for a given number of spectral degrees of freedom. In the representation of initial data, observational studies suggest that the triangular truncation is indeed superior. However, it does not necessarily imply that a model that is superior in the representation of the initial data will indeed provide for a more accurate forecast.

14.3.9   Advantages of spectral methods over the method of finite differences

It was earlier noted in Section 14.3 that the spectral methods take care of nonlinear computational instability. Moreover, it was mentioned in Section 14.3 that one distinct advantage of the spectral methods, is that unlike the finite difference method, the space derivatives are evaluated analytically in the spectral methods. From this, it is clear that the spectral methods are much more accurate than most finite difference methods for the same number of degrees of freedom. For example, the linear advection is treated exactly by the spectral methods provided that the initial field in resolved. Finite difference methods, however, provide for false dispersion as the short waves move too slowly. Moreover, the spectral methods produce no aliasing and, hence, do not give rise to nonlinear computational instability as interactions involving shorter waves outside of the truncated set are excluded. Conversely, the finite difference methods lead to falsely reflected interactions with shorter waves back onto longer waves. Utilizing the Arakawa Jacobian forms in the finite difference expressions, provides relief from aliasing for the finite difference methods as far as production of spurious energy is concerned. However, the finite difference methods give rise to phase errors in the interacting waves. The most important error in the spectral methods involves the neglect of interactions with wave components that are outside of the original set; this causes an error in the waves that are represented by the basis functions. Hence, the error in the spectral solution will occur in the scales described by the basis functions.

When the method of finite difference is applied to solve the barotropic vorticity equation, Equation (14.48), the Poisson equation for $(\partial \psi / \partial t)$ has to be solved for each time step to obtain the stream function tendency value. For example, envisage solving Equation (14.68) using the method of finite differences. Assuming $\chi = (\partial \psi / \partial t)$, Equation (14.68) becomes a Poisson equation in $\nabla^2 \chi$, which is solved by treating the nonlinear advection terms using Arakawa Jacobian finite difference form. Once χ is determined by solving Equation (14.68) as a Poisson equation using the method of finite difference, ψ for the next time step is obtained by writing χ using a time finite difference expression. Knowledge of ψ for the next time step, leads to the known RHS of Equation (14.68), which again can be solved as a Poisson equation in χ using the method of finite differences and the aforementioned procedure is continued till the integration period is completed. Hence, it is clear that to solve the barotropic vorticity equation, Equation (14.48), using the method of finite difference, one has to solve a Poisson equation for χ at each and every time step if one utilizes the method of finite differences to solve the barotropic vorticity equation. However, employing the spectral methods to solve the barotropic vorticity equation, does not entail the solution of Poisson equation for $(\partial \psi / \partial t)$.

Consider solving the barotropic vorticity equation in a beta plane. The barotropic vorticity equation in a beta plane is given in terms of stream function ψ as

$$\frac{\partial}{\partial t}(\nabla^2 \psi) + \hat{k} \times \nabla \psi \cdot \nabla(\nabla^2 \psi) + \beta \frac{\partial \psi}{\partial x} = 0 \qquad (14.73)$$

Assuming that the stream function is periodic in both x and y, i.e., ψ satisfies

$$\psi(x+2\pi/k, y+2\pi/l, t) = \psi(x,y,t) \tag{14.74}$$

The appropriate orthogonal basis functions for the aforementioned case of the β-plane geometry having the periodicity conditions, are of the form

$$\phi_{mn}(x,y) = e^{i(mkx+nly)} \tag{14.75}$$

The orthogonal basis functions Φ_m^n also turn out to be eigen-solutions of the following Poisson equation:

$$\nabla^2\phi + b\phi = 0 \tag{14.76}$$

where the eigenvalues b are given by

$$b = m^2k^2 + n^2l^2 \tag{14.77}$$

When the spectral methods is invoked to solve the barotropic vorticity equation, the aforementioned equation does not entail the solution of Poisson equation for $(\partial\psi/\partial t)$, as the orthogonal basis functions are themselves eigensolutions of Poisson equation (14.76).

The greatest disadvantage of the spectral methods is in calculating the nonlinear term that appears as the sum in Equation (14.61). The interaction coefficients H_{kns}^{jmr} is usually computed just once and stored for use during the integration of the equation. The disadvantage with spectral methods is that if there are N degrees of freedom, the number of operations that are needed to compute the nonlinear term goes as N^2 for the spectral model whereas the number of operations that are required to compute the nonlinear term is N for most finite difference methods. Hence, for high resolution (large N), the form of the spectral methods that utilizes calculating the nonlinear term using interaction coefficients requires much larger computer time than finite difference methods. However, in later years, a method that circumvents the aforementioned problem was proposed that addressed the calculation of nonlinear terms in spectral methods. Section 14.3.6 provided details of this method that successfully addressed the issues arising out of the appearance of a nonlinear term in spectral methods.

14.3.10   Transform method

Section 14.3.6 outlined a method that successfully addressed the issues arising out of the appearance of a nonlinear term in spectral methods. This method, called the "transform method," that avoided the use of the interaction coefficient method for handling the nonlinear term in spectral methods was formulated independently by Orszag (1970) and Eliasen, Machenhauer, and Rasmussen (1970). The inherent difficulty with the interaction coefficient method for computing nonlinear terms is that it requires multiplication of two series (together), which is very time consuming. In essence, the transform method sums the series at certain spatial grid points (in physical space) and these fields are multiplied together (in physical space) at each point to calculate the nonlinear terms. The next step in the transform method requires that the nonlinear terms so calculated in physical space be transformed back to spectral space. Hence, in effect the aforementioned method completely avoids the need to multiply two series (in spectral space) whenever a nonlinear term appears in the governing equation that needs to be solved by the spectral methods. By ensuring that the multiplication of the two fields (nonlinear term) is performed in the physical space and the same (product) is then transformed to the spectral space ensures that the multiplication of two series is completely avoided in the transform method. Furthermore, the effectiveness of the aforementioned transform method is enhanced by the existence of efficient transform methods. For example, in spherical coordinates, the fast Fourier transform used in longitude and the Legendre integrals in latitude are evaluated by Gaussian quadrature. The transform method is far superior to the interaction coefficient method for high resolution solutions over a spherical domain.

14.4   Spectral Methods for Shallow Water Equations

We will now apply spectral methods to the shallow water equations in spherical coordinates. The nonlinear shallow water equations in spherical coordinates with rotation is as follows:

$$\frac{\partial \vec{V}}{\partial t} = -(\zeta + f)\hat{k} \times \vec{V} - \nabla \left(\phi' + \frac{\vec{V} \cdot \vec{V}}{2} \right) \tag{14.78}$$

$$\frac{\partial \phi'}{\partial t} = -\nabla \cdot (\phi' \vec{V}) - \overline{\phi} \delta \tag{14.79}$$

where δ is the horizontal divergence, ζ is the vertical component of the relative vorticity. Geopotential Φ has been split up into the mean part $\overline{\phi}$ and a perturbation part ϕ'. f is the Coriolis parameter and $\vec{V}$ is the horizontal velocity. The velocity is characterized by the rotational and divergent parts as follows:

$$\vec{V} = \hat{k} \times \nabla\psi + \nabla\chi = \frac{U}{\cos\phi}\hat{i} + \frac{V}{\cos\phi}\hat{j} \tag{14.80}$$

The vorticity and divergence equations are obtained from Equation (14.78) by taking $\hat{k} \cdot \nabla \times$ and $\nabla\cdot$, respectively

$$\frac{\partial\zeta}{\partial t} = -\nabla \cdot (\zeta + f)\vec{V} \tag{14.81}$$

$$\frac{\partial\delta}{\partial t} = \hat{k} \cdot \nabla \times [(\zeta + f)\vec{V}] - \nabla^2\left(\phi' + \frac{\vec{V} \cdot \vec{V}}{2}\right) \tag{14.82}$$

where vorticity ζ and divergence δ become

$$\zeta = \nabla^2\psi \quad \text{and} \quad \delta = \nabla^2\chi \tag{14.83}$$

In spectral models, it is convenient to replace the equation of motion (horizontal momentum equations) by the vorticity and divergence equations because Equation (14.83) assumes a simple form when spherical harmonics are used as basis functions. This form of the equation is also more convenient for application of semi-implicit method. The vorticity equation, Equation (14.81), and the divergence equation, Equation (14.82), can now be expanded with the use of Equations (14.80) and (14.83) to give

$$\frac{\partial}{\partial t}(\nabla^2\psi) = -\frac{1}{a\cos^2\phi}\left[\frac{\partial}{\partial\lambda}(U\nabla^2\psi) + \cos\phi\frac{\partial}{\partial\phi}(V\nabla^2\psi)\right] - 2\Omega\left(\sin\phi\nabla^2\chi + \frac{V}{a}\right) \tag{14.84}$$

$$\frac{\partial}{\partial t}(\nabla^2\chi) = \frac{1}{a\cos^2\phi}\left[\frac{\partial}{\partial\lambda}(V\nabla^2\psi) - \cos\phi\frac{\partial}{\partial\phi}(U\nabla^2\psi)\right] + 2\Omega\left(\sin\phi\nabla^2\psi - \frac{U}{a}\right)$$
$$- \nabla^2\left(\frac{U^2 + V^2}{2\cos^2\phi} + \phi'\right) \tag{14.85}$$

The continuity equation, Equation (14.79), becomes

$$\frac{\partial\phi'}{\partial t} = -\frac{1}{a\cos^2\phi}\left[\frac{\partial}{\partial\lambda}(U\phi') + \cos\phi\frac{\partial}{\partial\phi}(V\phi')\right] - \bar{\phi}\nabla^2\chi \tag{14.86}$$

The two components of the velocity field in Equation (14.80) may be expressed as

$$U = -\frac{\cos\phi}{a}\frac{\partial\psi}{\partial\phi} + \frac{1}{a}\frac{\partial\chi}{\partial\lambda} \quad \text{and} \quad V = \frac{\cos\phi}{a}\frac{\partial\chi}{\partial\phi} + \frac{1}{a}\frac{\partial\psi}{\partial\lambda} \tag{14.87}$$

Equations (14.84) to (14.86) are the predictive equations for χ, ψ, and ϕ' whereas Equation (14.87) represents the *diagnostic expressions* for U and V. The nonlinear terms in these equations are in a convenient form for the application of the "transform method," which was presented in the earlier section.

Each of the dependent variables χ, ψ, and ϕ' are expanded in terms of the spherical harmonic basis functions as indicated:

$$\psi = a^2 \sum_{m=-M}^{M} \sum_{n=|m|}^{|m|+M} \psi_{m,n} Y_n^m, \qquad \chi = a^2 \sum_{m=-M}^{M} \sum_{n=|m|}^{|m|+M} \chi_{m,n} Y_n^m, \qquad \phi' = a^2 \sum_{m=-M}^{M} \sum_{n=|m|}^{|m|+M} \phi_{m,n} Y_n^m$$

(14.88)

$$U = a \sum_{m=-M}^{M} \sum_{n=|m|}^{|m|+M+1} U_{m,n} Y_n^m, \qquad V = a \sum_{m=-M}^{M} \sum_{n=|m|}^{|m|+M+1} V_{m,n} Y_n^m, \qquad (14.89)$$

These expansions are for the rhomboidal wave number truncation. Equation (14.87) are transformed and the result is

$$U_{m,n} = (n-1)D_{m,n}\psi_{m,n-1} - (n+2)D_{m,n+1}\psi_{m,n+1} + im\chi_{m,n} \qquad (14.90)$$

$$V_{m,n} = -(n-1)D_{m,n}\chi_{m,n-1} + (n+2)D_{m,n+1}\chi_{m,n+1} + im\psi_{m,n} \qquad (14.91)$$

It is to be noted that the expansions for U and V must extend one degree above the expansions for χ and ψ. The quantities needed for the nonlinear terms are obtained by evaluating the sums in Equations (14.88) and (14.89) at equally spaced points in longitude and at Gaussian latitudes. The required products are computed at each grid point and the products are then Fourier transformed in longitude as follows:

$$U\nabla^2\psi = a \sum_{m=-M}^{M} A_m e^{im\lambda}, \qquad V\nabla^2\psi = a \sum_{m=-M}^{M} B_m e^{im\lambda} \qquad (14.92)$$

$$U\phi' = a^3 \sum_{m=-M}^{M} C_m e^{im\lambda}, \qquad V\phi' = a^3 \sum_{m=-M}^{M} D_m e^{im\lambda}, \qquad \frac{U^2+V^2}{2} = a^2 \sum_{m=-M}^{M} E_m e^{im\lambda}$$

(14.93)

The spectral equations are formed by substituting Equations (14.88), (14.89), (14.92) and (14.93) into Equations (14.84) to (14.86) and multiplying each equation by Y_n^{m*} and integrating over the domain. With the use of the orthogonality condition, the equations finally reduce to the following:

$$-n(n+1)\frac{\partial}{\partial t}\psi_{m,n} = \frac{1}{2}\int_{-1}^{1}\frac{1}{1-\mu^2}\left(imA_mP_n^m - B_m\frac{dP_n^m}{d\mu}\right)d\mu + 2\Omega\left[n(n-1)D_{m,n}\chi_{m,n-1}\right.$$
$$\left. +(n+1)(n+2)D_{m,n+1}\chi_{m,n+1} - V_{m,n}\right] \quad (14.94)$$

$$-n(n+1)\frac{\partial}{\partial t}\chi_{m,n} = \frac{1}{2}\int_{-1}^{1}\frac{1}{1-\mu^2}\left(imB_mP_n^m + A_m\frac{dP_n^m}{d\mu}\right)d\mu - 2\Omega\left[n(n-1)D_{m,n}\psi_{m,n-1}\right.$$
$$\left. +(n+1)(n+2)D_{m,n+1}\psi_{m,n+1} + U_{m,n}\right] + n(n+1)(E_{m,n} + \phi_{m,n}) \quad (14.95)$$

$$\frac{\partial}{\partial t}\phi_{m,n} = -\frac{1}{2}\int_{-1}^{1}\frac{1}{1-\mu^2}\left(imC_mP_n^m - D_m\frac{dP_n^m}{d\mu}\right)d\mu + \overline{\phi}n(n+1)\chi_{m,n}, \quad (14.96)$$

where

$$E_{m,n} = \frac{1}{2}\int_{-1}^{1}\frac{E_m}{1-\mu^2}P_n^m d\mu \text{ and } \mu = \sin\phi \quad (14.97)$$

The integrals in Equations (14.94) to (14.96) are evaluated by the Gaussian quadrature formula as before, but this time, $(5M+1)/2$ Gaussian latitudes are required. Moreover, the required number of longitudinal grid points is $3M+1$.

Equations (14.94) to (14.96) are in a form that is very convenient for the application of the semi-implicit method. For this method, all the terms are evaluated explicitly except for the term involving $\phi_{m,n}$ that is treated implicitly in Equation (14.95) and all the terms are evaluated explicitly in Equation (14.95) except for the terms involving $\chi_{m,n}$ that is treated implicitly. These two equations (14.95) and (14.96) can be easily solved for $\phi_{m,n}(t+\Delta t)$, from which Equations (14.94) and (14.95) can be solved explicitly. It is to be noted that to solve the same shallow water equations, say for $\phi(t+\Delta t)$, using the semi-implicit finite difference method, one will have to solve the Helmholtz equation for every time step. Thus, in solving the shallow water equations using the spectral methods, one can utilize a much larger time step with almost the same computational effort per time step.

The use of the transform method together with the semi-implicit method have made the spectral primitive equation models very competitive as far as comparison with finite difference models for global prediction is concerned. Studies have also indicated that the spectral models are very useful while solving baroclinic models. All operational numerical weather forecasting centers worldwide are presently employing the spectral models for their prediction as they provide for as good or better forecasts when compared with finite difference models while using the same amount of computer time.

It should be noted that energy is not exactly conserved in the spectral model as applied to the shallow water equations. This is because the kinetic energy for the shallow water equations is proportional to $\phi \vec{V} \cdot \vec{V}$, which is of the cubic energy form. However, the nonlinear terms are computed very accurately in spectral models and various spectral model studies with shallow water equations shows that the energy is in fact very nearly conserved.

14.5   Pseudo-spectral Methods

In the pseudo-spectral methods, the partial differential equations are solved point wise in physical space in a finite-difference like manner. However, the space derivatives that appear in the PDEs are calculated using orthogonal functions such as Fourier integrals or Chebyshev polynomials. The term "pseudo-spectral" refers to the spatial part of the PDE. Consider solving a differential equation

$$\mathbf{L}\{\mathbf{u}(\mathbf{X})\} = \mathbf{s}(\mathbf{X}), \tag{14.98}$$

where $\mathbf{L}$ is a linear spatial differential operator such that $\mathbf{L} = \frac{\partial^2}{\partial z^2}$ in one dimension. One seeks a numerical solution $\mathbf{u}^N(\mathbf{X})$ such that the residual $\mathbf{R}(\mathbf{X})$, defined as

$$\mathbf{R}(\mathbf{X}) = \mathbf{L}\{\mathbf{u}^N(\mathbf{X})\} - \mathbf{s}(\mathbf{X}) \text{is small} \tag{14.99}$$

The general procedure for applying the pseudo-spectral methods is as follows:

(i) choose a finite set of trial functions (also known as expansion functions) ϕ_j, j = 0,1,2,..,N-1, and expand u^N in these functions, as follows:

$$u^N(X) = \sum_{j=0}^{N-1} \hat{u}_j \phi_j(X) \tag{14.100}$$

(ii) choose a set of test functions χ_n , n = 0,1,2,..., N-1 and demand that the inner product of trial functions with the residual is zero, i.e.,

$$(\chi_n, \mathbf{R}) = 0; \; for \, n = 0, 1, 2, ..., N - 1 \tag{14.101}$$

For the pseudo-spectral methods, one requires

$$(\chi_n, \mathbf{R}) = [\delta(\mathbf{X} - \mathbf{X_n}), \mathbf{R}] = \mathbf{R}(\mathbf{X_n}) = \mathbf{L}\{\mathbf{u}^N(\mathbf{X_N})\} - \mathbf{s}(\mathbf{X_n}), \tag{14.102}$$

where δ is the Dirac delta function and the solution at the special points (also known as collocation points) are supposed to be exact. Substituting Equation

(14.100) in Equation (14.102), one obtains

$$\sum_{j=0}^{N-1} \hat{u}_j L\phi_j(X_n) - s(X_n) = 0; \ n = 0, 1, 2, \ldots, N-1 \tag{14.103}$$

Equation (14.103) provides for N equations to determine the unknown N coefficients, $\hat{u}_j$. For periodic boundary conditions, one employs trigonometric functions (Fourier series) as trial functions, whereas for problems involving non-periodic boundary conditions, one utilizes orthogonal polynomials such as Chebyschev polynomials. Assuming periodic boundary conditions and with $L = \frac{\partial^2}{\partial z^2}$, the differential equation becomes

$$\frac{d^2}{dz^2} u = s \tag{14.104}$$

One can choose the trial functions as

$$\phi_j = exp[ik_j X] \tag{14.105}$$

(iii) In the pseudo-spectral methods, the objective is to find the expansion coefficients $\hat{u}_j$ such that the residual

$$\sum_{j=0}^{N-1} \hat{u}_j L\phi_j(X_n) - s(X_n) = 0; \ n = 0, 1, 2, \ldots, N-1 \ vanishes \ at \ colocation \ points \tag{14.106}$$

Equation (14.106) is rewritten in terms of trial functions ϕ_j from Equation (14.105)

$$\sum_{j=0}^{N-1} \hat{u}_j L e^{[-ik_j X_n]} - s(X_n) = 0; \ vanishes \tag{14.107}$$

If the spatial operator L is also a linear operator, then $Lexp\{-ik_k X_n\} = h(k_k) Lexp\{-ik_k X_n\}$ and Equation (14.107) then becomes

$$\sum_{j=0}^{N-1} \hat{u}_j h(k_j) e^{[-ik_j X_n]} - s(X_n) = 0; \ n = 0, 1, 2, \ldots, N-1 \tag{14.108}$$

For the aforementioned one-dimensional case, choose 'z' and 'k_j' as equally spaced as follows: $X_n = n\Delta$; $k_j = (2\pi j)/(N\Delta)$; where Δ is the spatial resolution. The condition on the residuals, then becomes

$$\sum_{j=0}^{N-1} \hat{u}_j h(k_j) e^{[-2\pi i j n / N]} - s(X_n) = 0; \ n = 0, 1, 2, \ldots, N-1 \tag{14.109}$$

Using the definition of the discrete Fourier transform (DFT) of $u_n = u(x_n)$ as follows:

$$\hat{u}_j = \frac{1}{\sqrt{N}} \sum_{n=o}^{N-1} u_n exp[2\pi i n j/N] \tag{14.110}$$

and the inverse DFT as

$$u_n = \frac{1}{\sqrt{N}} \sum_{j=0}^{N-1} \hat{u}_j exp[-2\pi i n j/N] \tag{14.111}$$

it can be shown that with a specific choice of k_j and X_n, one gets

$$\frac{1}{\sqrt{N}} \sum_{n=0}^{N-1} exp[2\pi i n j'/N] exp[-2\pi i n j/N] = \delta(j - j') \tag{14.112}$$

The condition on the residual (Equation 14.109) can be written using the DFT as follows:

$$DFT^{-1} \left\{ \hat{u}_j h(k_j) \right\} (X_n) - s(X_n) = 0 \tag{14.113}$$

Applying DFT to Equation (14.113), one gets

$$\left\{ \hat{u}_j h(k_j) \right\} - DFT \left\{ s(X_n) \right\} (k_j) = 0 \tag{14.114}$$

It is possible to obtain the unknown N coefficients, $\hat{u}_j$ from Equation (14.114) and substituting the same in Equation (14.100), one can obtain the numerical solution $u^N(X)$.

Exercises 14 (Questions and answers)

1. The spherical harmonics are eigenfunctions of the two-dimensional Laplacian on a sphere, i.e., over spherical coordinates (λ, θ, r). However, in most atmospheric models, the vertical coordinate is the σ coordinate. Are the spherical harmonics still eigenfunctions of the two-dimensional Laplacian in such a σ coordinate?

   Answer: Using σ as the vertical coordinate, one finds

   $$\nabla_\sigma^2 Y_n^m \neq \frac{-n(n+1)}{r^2} Y_n^m$$

   as r is not a constant over the horizontal surface (σ = constant). However, the aforementioned equation remains a very good approximation to invoke and utilize.

2. The spherical harmonics are also eigenfunctions of the zonal differential operator.

   Answer: This as manifested in $(\partial Y_n^m / \partial \lambda) = im(Y_n^m)$ follows directly from the definition of the spherical harmonics $Y_n^m(\lambda, \theta) = e^{im\lambda} P_n^m(\sin \theta)$.

3. Write down the recurrence formula for the meridional derivative of the associated Legendre function of the first kind.

   Answer: The recurrence formula for the meridional derivative of the associated Legendre function of the first kind is given as

   $$(\mu - 1)\frac{dP_n^m}{d\mu} = n\varepsilon_{n+1}^m P_{n+1}^m - (n+1)\varepsilon_n^m P_{n-1}^m \quad \text{where} \quad \varepsilon_n^m = \sqrt{\frac{n^2 - m^2}{4n^2 - 1}}$$

4. What are the advantages of a triangular truncation?

   Answer: The main advantage of a triangular truncation is that it is isotropic. If $\overline{F}_1$ is an expansion of F triangularly truncated at wavenumber M in one spherical coordinate system (λ_1, θ_1) and F_2 in another system (λ_2, θ_2), then $\overline{F}_1 = \overline{F}_2$. This results in uniform resolution in a triangularly truncated system on a sphere.

5. What is the chief disadvantage of a rhomboidal truncation?

   Answer: The main disadvantage of a rhomboidal truncation is its non-uniform resolution on a sphere. The rhomboidal truncation corresponds to a uniform resolution in the east–west direction only. Moreover, rhomboidal truncation is invariant only under rotation around the Earth's axis.

6. How is the "pole problem" taken care of in spectral models?

   Answer: While employing the method of finite differences near the polar regions, one notices convergence of meridians towards the poles. Because of this, there is a severe restriction of the allowed time step that would avoid numerical instability as one approaches the poles. This problem is known as the "pole problem." An immediate work-around solution for this would be a situation that would provide for a uniform resolution. Working with spectral models that employ spherical harmonics together with triangular truncation would not give rise to a pole problem, as there is isotropic resolution over the sphere. The pole problem would remain if one were to employ the spectral methods using bi-Fourier expansion.

7. Do the spectral models provide solution for aliasing errors?

   Answer: Spectral models when employed do not suffer aliasing errors that are related to the computation of the quadratic terms and, hence, do not grow without bound from nonlinear computational instability. However, in instances where there are triple product terms or higher product terms, spectral models are not completely free from aliasing errors.

8. What are the disadvantages of the explicit finite difference methods if the grid points are distributed over a sphere in a uniform latitude-longitude grid? How can they be circumvented?

Answer: Application of explicit finite difference methods as applied to a grid point model would require very small time steps to maintain stability as there is convergence of meridians near the pole and the need to satisfy the "CFL condition." This disadvantage can be circumvented by employing spherical harmonics in grid point models.

9. What are the advantages and disadvantages of using the spectral methods?

   Answer: One chief advantage of the spectral methods, is that for a particular expansion of a dependent variable with respect to the (truncated spherical harmonics expansion), i.e., with respect to the expansion in the basis function, the horizontal derivatives of the dependent variables are exact. Furthermore, the advantage of employing spherical harmonics in spectral methods enables one to easily compute derivatives such as Laplacian based on the known properties of the expansion functions. Additionally, both energy and enstrophy are exactly conserved in the spectral methods. Moreover, one can obtain uniform resolution throughout the sphere in the spectral methods. The disadvantage of the spectral methods is the appearance of the unphysical Gibbs oscillation. Another disadvantage with the spectral methods is the lack of a highly efficient algorithm to transform the spectral formulation to grid point space, the latter typically on a latitude–longitude grid.

10. What are the minimum number of meridional latitudinal circles necessary to avoid aliasing errors while employing the "triangular truncation" in the spectral methods?

    Answer: Errors due to aliasing manifest whenever a product of two or more truncated spherical harmonic expansions in the spectral methods contains higher order Fourier modes in longitude λ and higher order functions in μ that are not present in the original truncation. While employing the "triangular truncation," at truncation to wave number M, the computation of the coefficients in the spherical harmonic expansion requires the evaluation of polynomials of degree $3M$. Hence, the exact evaluation by Gaussian quadrature of the integral of this polynomial to calculate the spherical harmonic coefficients require a minimum number of $(3M+1)/2$ meridional latitudinal circles for a triangular truncation to avoid aliasing errors.

11. What are the minimum number of meridional latitudinal circles necessary to avoid aliasing errors while employing the "rhomboidal truncation" in the spectral methods?

    Answer: While employing the "rhomboidal truncation," at truncation to wave number M, the computation of the coefficients in the spherical harmonic expansion requires the evaluation of polynomials of degree $5M$. Hence, the exact evaluation by Gaussian quadrature of the integral of this polynomial to calculate the spherical harmonic coefficients require a minimum number of $(5M+1)/2$ meridional latitudinal circles for a rhomboidal truncation to avoid aliasing errors.

12. How is the representation of the vector velocity field handled in terms of spherical harmonics expansion in the spectral methods?

Answer: The velocity components (zonal and meridional components) u and v are not conveniently approximated by spherical harmonics as artificial discontinuities in u and v are present at the poles, unless the horizontal wind speed vanishes at the pole. This difficulty is circumvented by replacing the prognostic equations in u and v by the prognostic equations in the relative vorticity and horizontal divergence. Moreover, in all expressions involving u and v, the variables are replaced by the transformed velocities, U and V, where $U = u\cos\phi$, $V = v\cos\phi$, where ϕ is the latitude. It is to be noted at the poles, both U and V vanish identically and are hence free from artificial discontinuities.

Python examples

1. The following python code solves the two-dimensional Poisson equation $\frac{\partial^2 \psi}{\partial x^2} + \frac{\partial^2 \psi}{\partial y^2} = \zeta(x,y)$, where $\psi(x,y)$ is the stream function and $\zeta(x,y)$ is the relative vorticity. It is assumed that $\zeta(x,y)$ is known over the region of interest and that the Dirichlet boundary conditions on $\psi(x,y)$ is prescribed at all the four boundaries. The relative vorticity $\zeta(x,y)$ corresponds to a cyclonic vortex in the northern hemisphere with the center of the vortex over the center of domain (x_o, y_o).

   The initial condition that is considered is the axisymmetric cyclonic vortex with the following tangential wind profile (Chan and Williams, 1987, Analytical and Numerical Studies of the beta-effect in Tropical Cyclone Motion, Part I: Zero Mean Flow, *Journal of Atmospheric Sciences*, 44, 9, 1257–1265).

   $V(r) = V_m \left\{ r/r_m \right\} exp\left[\frac{1}{b} \left\{ 1 - (\frac{r}{r_m})^b \right\} \right]$, where $V(r)$ is the tangential wind profile at any distance r from the center of the cyclonic vortex, V_m is the value of the maximum tangential wind, r_m is the radius of the maximum tangential wind and b is a shape factor. The relative vorticity profile $\zeta(r)$ is obtained from the following expression $\zeta(r) = \frac{2V_m}{r_m} [1 - \frac{1}{r} \left\{ \frac{r}{r_m} \right\}^b] exp[\frac{1}{b} \left\{ 1 - (\frac{r}{r_m})^b \right\}]$. The python code for solving nonlinear non-divergent vorticity equation in a beta plane is taken from Prof G Vallis and the python code was modified to solve the linear non-divergent vorticity equation in a beta plane with the initial condition corresponding to a cyclonic vortex as given earlier. The code uses the following values for the various parameters, $V_m = 40$ m/s; $r_m = 100$ km; $\Delta x = 1/256$ km; $\Delta y = 1/256$ km; and $b = 1$.

   As illustrated in Chan and Williams (1987), from Poisson equation, $\frac{\partial^2 \psi}{\partial x^2} + \frac{\partial^2 \psi}{\partial y^2} = \zeta(x,y$ expanding the stream function $\psi(x,y)$ as Fourier transform, one gets

$A(k,l) = \int_0^{2\pi} \int_0^{2\pi} \psi(x,y) e^{i(kx+ly)} dxdy.$ Hence $\int_0^{2\pi} \int_0^{2\pi} \nabla^2 \psi(x,y) e^{i(kx+ly)} dxdy = -(k^2 + l^2) \int_0^{2\pi} \int_0^{2\pi} \psi(x,y) e^{i(kx+ly)} dxdy.$ Hence, $A(k,l) = -\frac{1}{(k^2+l^2)} \int_0^{2\pi} \int_0^{2\pi} \zeta(x,y) e^{i(kx+ly)} dxdy.$ The stream function $\psi(x,y)$ is then obtained from

$$\psi(x,y) = \frac{1}{2\pi} \sum_k \sum_l A(k,l) e^{-(kx+ly)}$$

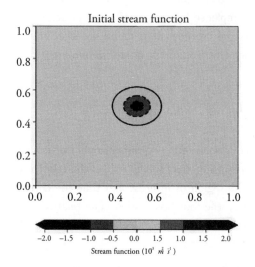

Figure 14.2   Initial stream function for the linear non-divergent vorticity equation.

The python code for solving this Poisson equation is available in the python examples section of Chapter 10.

15

Finite Volume and Finite Element Methods

15.1 Introduction

Except for Chapter 14 that discussed spectral methods, all earlier chapters have discussed the method of finite difference that still remains a popular method for solving partial differential equations. While employing the finite difference method, the differential form of the conservation law of fluid flow was utilized and the partial derivatives present in the differential form of the conservation equations were approximated using the appropriate finite difference expressions. In this chapter, the finite volume method will be introduced and applied to solve fluid flow problems. The basis of the finite volume method is the utilization of the conservation law of fluid flow in the integral form.

15.2 Integral Form of Conservation Law

The physical ("fluid" in this case) system is governed by a set of conservation laws that include conservation of mass, conservation of momentum, and conservation of energy. These conservation laws are mostly written in differential form. The method of finite difference requires the differential form of the conservation laws, whereas the finite volume method entails the integral form of the conservation laws for a fixed physical domain. Assume the existence of a physical domain, Ω, with the boundary of the domain, indicated by $\partial\Omega$. Then, the canonical conservation equation assuming that the physical domain is fixed is of the following form:

$$\frac{d}{dt}\int_{\Omega} U\,dV + \int_{\partial\Omega} \vec{F}(U)\cdot\hat{n}\,dS = \int_{\Omega} S(U,t)\,dV \tag{15.1}$$

where U is the conserved state, $\vec{F}$ is the flux of the conserved state, $\hat{n}$ is the outward directed unit normal on the boundary of the domain, and S denotes a source term. By applying the Gauss divergence theorem, the aforementioned conservation law can be written as a partial differential equation. The Gauss divergence theorem states that

$$\int_\Omega \nabla \cdot \vec{F} \, dV = \int_{\partial\Omega} \vec{F} \cdot \hat{n} \, dS \tag{15.2}$$

Equation (15.1) then becomes

$$\frac{d}{dt} \int_\Omega U \, dV + \int_\Omega \nabla \cdot \vec{F} \, dV = \int_\Omega S(U,t) \, dV \tag{15.3}$$

Applying the Leibnitz rule for differentiation under the integral sign and combining the volume integrals, one obtains

$$\int_\Omega \left(\frac{\partial U}{\partial t} + \nabla \cdot \vec{F} - S \right) dV = 0 \tag{15.4}$$

As Equation (15.4) must be valid for any arbitrary domain, Ω, the integrand that appears in Equation (15.4) must be zero everywhere, or equivalently,

$$\frac{\partial U}{\partial t} + \nabla \cdot \vec{F} = S \tag{15.5}$$

Equation (15.5) specifies the conservation law in the differential form as a partial differential equation.

15.2.1 Integral form of conservation law of mass

Assuming that the conserved state is the density of the fluid, (i.e., $U = \rho$), flux, $\vec{F}$ being equal to the mass flux $\rho\vec{V}$ and the absence of source term ($S = 0$), Equation (15.5) becomes the statement of conservation of mass in differential form.

$$\frac{\partial \rho}{\partial t} + \nabla \cdot (\rho\vec{V}) = 0 \tag{15.6}$$

The integral form of the conservation of mass can be obtained suitably from Equation (15.1) by substituting $U = \rho$, $\vec{F}$ being equal to the mass flux $\rho\vec{V}$ and $S = 0$.

$$\frac{d}{dt} \int_\Omega \rho \, dV + \int_{\partial\Omega} (\rho\vec{V}) \cdot \hat{n} \, dS = 0 \tag{15.7}$$

15.2.2   Convection equation

Convection is the most dominant physical transport mechanism for many applications in fluid mechanics. Diffusion effects are confined to limited regions in many applications. In this section, the convection equation is derived using the conservation law as given in Equation (15.1) by specifying U to be the "conserved" scalar quantity, and the flux $\vec{F}$ being equal to $U\vec{V}$, and for simplicity, the source term is assumed to be zero, i.e., $S = 0$. These specifications result in the conserved scalar equation

$$\frac{\partial U}{\partial t} + \nabla \cdot (U\vec{V}) = 0 \tag{15.8}$$

Expanding Equation (15.8) and assuming that the velocity field is non-divergent, $\nabla \cdot \vec{V} = 0$, one obtains the convection equation

$$\frac{\partial U}{\partial t} + \vec{V} \cdot \nabla U = 0 \tag{15.9}$$

Equation (15.9) states that along the streamwise direction (i.e., convecting with the velocity $\vec{V}$), the quantity U does not change its value, or in other words, following its motion, the quantity U is conserved. For a constant velocity field, say, $\vec{V}(\boldsymbol{x},t) = \boldsymbol{v}$, the solution of the convection equation is given by

$$U(\boldsymbol{x},t) = U_o(\xi), \tag{15.10}$$

where $\xi = \boldsymbol{x} - \boldsymbol{v}t$, with $U_o(\boldsymbol{x})$ as the distribution of U at time $t = 0$.

15.2.3   Integral form of linear momentum conservation equation

Consider the conservation of linear momentum over a control volume Ω. Then,

$$\frac{d}{dt}\int_{\Omega}\rho\vec{V}\,dV = -\int_{\partial\Omega}(\rho\vec{V})\vec{V}\cdot\hat{n}\,dA + \int_{\Omega}\rho\vec{b}\,dV + \int_{\partial\Omega}\overline{\overline{T}}\cdot\hat{n}\,dA, \tag{15.11}$$

where the term on the LHS represents the time rate of change of the total momentum in Ω; the first term in the RHS represents the net momentum flux across boundary of Ω; the second and third terms on the RHS of Equation (15.11) represent body forces acting on Ω and the external contact forces acting on $\partial\Omega$. In Equation (15.11), $\vec{b}$ is the body force per unit mass acting on the fluid while $\overline{\overline{T}}$ is the *Cauchy stress tensor*.

Applying Gauss's theorem to Equation (15.11) and rearranging the terms, one obtains

$$\frac{d}{dt}\int_{\Omega}\rho\vec{V}\,dV + \int_{\Omega}\nabla \cdot (\rho\vec{V}\vec{V})\,dV - \int_{\Omega}\rho\vec{b}\,dV - \int_{\Omega}\nabla \cdot \overline{\overline{T}}\,dV = 0 \tag{15.12}$$

Assuming $\rho\vec{V}$ is smooth, one applies the Leibnitz integral rule to get

$$\int_\Omega \left[\frac{\partial(\rho\vec{V})}{\partial t} + \nabla \cdot (\rho\vec{V}\vec{V}) - \rho\vec{b} - \nabla \cdot \overline{\overline{T}} \right] dV = 0 \tag{15.13}$$

As Ω is arbitrary, the integrand of Equation (15.13) must identically vanish providing the differential form of conservation of linear momentum.

15.3   Finite Volume Method

The finite volume method applied to solving one-dimensional fluid problems is based on dividing the spatial domain into intervals (called finite volumes or cells) and making in each of the cells/finite volumes an approximation of the integral of the conservative variables. At each time step, these values are updated using approximations of the flux at the ends of the intervals, as discussed in the following sub-sections.

15.3.1   Finite volume method applied to one-dimensional scalar conservation equation

The one-dimensional scalar conservation equation is given as follows:

$$\frac{\partial u}{\partial t} + \frac{\partial f}{\partial x} = 0, \quad u = u(x,t), \quad f = f(u), \tag{15.14}$$

where u is the conserved scalar quantity and f is the flux function. Typically, the flux function f depends on the scalar variable u. Equation (15.14) reduces to the well-known one-dimensional linear advection equation if $f = au$, where a is a constant.

Figure 15.1 shows a schematic diagram of examples of structured or conformal mesh. A structured or conformal mesh is one in which every interior vertex in the domain is connected to the same number of neighboring vertices. The left panel of Figure 15.1 shows a typical cell, a typical face and a typical node (vertex). The structured mesh may be a regular mesh or a body fitted mesh. Figure 15.2 shows a schematic diagram of examples of unstructured or non-conformal mesh. It is clear from Figure 15.2 that a non-conformal mesh is characterized by the fact that the vertices of a cell may fall on the faces of neighboring cells.

An alternate way to discretize Equation (15.14), is to divide the spatial domain x into finite volumes/cells and integrate the associated equation in each cell, thereby transforming it into an integral form. For simplicity, assume that the envisaged finite volumes/cells have the same equal length (Δx) and the same constant time step (Δt).

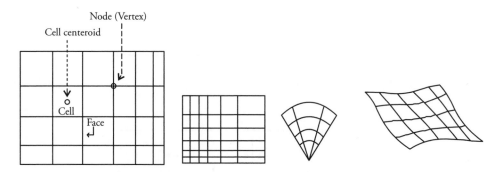

Figure 15.1 Schematic diagram of examples of structured or conformal mesh.

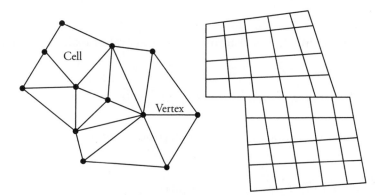

Figure 15.2 Schematic diagram of examples of unstructured or non-conformal mesh.

Thus, the spatial and temporal domains will be

$$I_i = \left[x_{i-1/2}, x_{i+1/2}\right] = \left[x_i - \Delta x/2, x_i + \Delta x/2\right] \tag{15.15}$$

$$I_n = \left[t^n, t^{n+1}\right] = \left[n\Delta t, (n+1)\Delta t\right] \tag{15.16}$$

The integral in the cell (refer Figure 15.3) of Equation (15.14) becomes

$$\int_{x_{i-1/2}}^{x_{i+1/2}} \left(\frac{\partial u}{\partial t} + \frac{\partial f}{\partial x}\right) dx = 0 \tag{15.17}$$

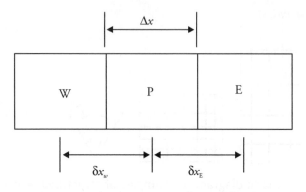

Figure 15.3   Arrangement of control volume.

The integral involving $(\partial f/\partial x)$ term can be integrated exactly, so that Equation (15.17) becomes

$$\int_{x_{i-1/2}}^{x_{i+1/2}} \frac{\partial u}{\partial t}\, dx + f\left[u(x_{i+1/2},t)\right] - f\left[u(x_{i-1/2},t)\right] = 0 \tag{15.18}$$

As the interval ends $x_{i-1/2}$ and $x_{i+1/2}$ do not depend on time, using Leibnitz rule, one obtains

$$\frac{\partial}{\partial t}\int_{x_{i-1/2}}^{x_{i+1/2}} u(x,t)\, dx + f\left[u(x_{i+1/2},t)\right] - f\left[u(x_{i-1/2},t)\right] = 0 \tag{15.19}$$

Defining u_i^n as the spatial average of the function $u(x,t)$ over the interval (length) I_i at time $t^n = n\Delta t$, as

$$u_i^n = \frac{1}{\Delta x}\int_{x_{i-1/2}}^{x_{i+1/2}} u(x,t)\, dx \tag{15.20}$$

Integrating Equation (15.19) between the times t^n and t^{n+1}, the time derivative disappears from the first term of Equation (15.20), resulting in

$$\int_{x_{i-1/2}}^{x_{i+1/2}} \left[u(x,t^{n+1}) - u(x,t^n)\right] dx + \int_{t^n}^{t^{n+1}} \left\{f\left[u(x_{i+1/2},t)\right] - f\left[u(x_{i-1/2},t)\right]\right\} dt = 0 \tag{15.21}$$

Equation (15.21) shows that the value of u in I_i only changes along time Δt because of the value of the flux f at the ends of I_i. Using Equation (15.20), one obtains

$$\left(u_i^{n+1} - u_i^n\right)\Delta x + \int_{t^n}^{t^{n+1}} \left\{f\left[u(x_{i+1/2},t)\right] - f\left[u(x_{i-1/2},t)\right]\right\} dt = 0 \tag{15.22}$$

In this equation, the values of the integral of f at points $x_{i-1/2}$ and $x_{i+1/2}$ will not be generally known; hence, we define the time-averaged cell face fluxes as

$$\hat{f}_{i+1/2} = \frac{1}{\Delta t} \int_{t^n}^{t^{n+1}} f\left[u(x_{i+1/2}, t)\right] dt \tag{15.23a}$$

$$\hat{f}_{i-1/2} = \frac{1}{\Delta t} \int_{t^n}^{t^{n+1}} f\left[u(x_{i-1/2}, t)\right] dt \tag{15.23b}$$

Using Equations (15.23a) and (15.23b) in Equation (15.22), one obtains

$$u_i^{n+1} = u_i^n - \frac{\Delta t}{\Delta x} \left[\hat{f}_{i+1/2} - \hat{f}_{i-1/2}\right]. \tag{15.24}$$

The explicit expression is obtained from Equation (15.24) for the solution u at each cell at the next time t^{n+1}, from its value in the previous time t^n at the same cell together with knowledge of the numerical fluxes $\hat{f}_{i-1/2}$ and $\hat{f}_{i+1/2}$ at the ends of the cell. These numerical fluxes represent approximations of the time average of the physical flux at the edges of the cell and, depending on the way these are calculated, one gets different schemes. To calculate them, the variables in cells adjacent to I_i are used

$$\hat{f}_{i-1/2} = \phi\left(u_{i-m}^n, u_{i-m+1}^n, \ldots, u_{i+l}^n\right), \tag{15.25}$$

where m and l are two non negative integers and ϕ is a certain function.

In hyperbolic partial differential equations, the solution (information) propagates at a finite speed. Hence, it appears reasonable to assume that one can obtain the value of the flux $\hat{f}_{i-1/2}$ from u_{i-1}^n and u_i^n (the average values of the scalar variable on both sides of the boundary $x_{i-1/2}$), whereas $\hat{f}_{i+1/2}$ is obtained from u_i^n and u_{i+1}^n. Hence, the general expression of Equation (15.25) for hyperbolic problems assumes the following form:

$$\hat{f}_{i-1/2} = \phi\left(u_{i-1}^n, u_i^n\right) \qquad \text{and} \qquad \hat{f}_{i+1/2} = \phi\left(u_i^n, u_{i+1}^n\right) \tag{15.26}$$

Equation (15.24) allows one to obtain the solution (variable) values at the next time level. To provide a good approximation of the law of conservation, the aforementioned method of solution should be convergent. The requirement of "convergence" entails that the numerical solution converges to the solution of the differential equation when $\Delta x \to 0$, $\Delta t \to 0$. According to the *Lax equivalence theorem*, "For a consistent finite difference method as applicable to a well-posed linear initial value problem, the method is convergent if and only if it is stable." Hence, it is clear from the Lax equivalence theorem that consistency and stability are the two requirements for ensuring convergence.

A finite difference scheme is said to be "consistent" if it represents faithfully the differential equation when $\Delta x \to 0$, $\Delta t \to 0$. As one is obtaining the numerical flux from the values of the variable u in the neighboring cells, and if the variable u has the same value v in all of the cells, the result must be the same in each one. Hence, there is a requirement for a consistency condition to be met by the function ϕ, which is given by

$$\phi(v, v, \dots, v) = \hat{f}(v) \tag{15.27}$$

Moreover, continuity for the variable u is also required, i.e.,

$$\phi\left(u_{i-1}^n, u_i^n\right) \to \hat{f}(v) \quad \text{as} \quad u_{i-1}^n \to v, \quad u_i^n \to v \tag{15.28}$$

A finite difference scheme is said to be "stable" if a small error introduced at anytime step, is not amplified indefinitely in time but remains bounded along the integration process.

A conservative scheme for the scalar conservation law (15.14) is a numerical method of the form [Equation (15.24)] that fulfils the condition given in Equation (15.25). It is noted that by applying a conservative scheme to a set of contiguous cells $M, M+1, \dots, N$, the result verifies the same property [Equation (15.21)] of the exact solution (the value of the variable u in the cell I_i only changes in time Δt due to the value of the flux f at the ends of the cell I_i). Indeed, adding the values of u_i^{n+1} obtained from Equation (15.24) for any set of consecutive cells, and multiplying by Δx and rearranging, one obtains

$$\left[\sum_{i=M}^N u_i^{n+1} - \sum_{i=M}^N u_i^n\right] \Delta x + \left[\hat{f}_{N+1/2} - \hat{f}_{M-1/2}\right] \Delta t = 0 \tag{15.29}$$

since fluxes at the cell boundaries cancel each other, except for the fluxes at the ends, $x = x_{M-1/2}$ and $x = x_{N+1/2}$.

15.3.2   Godunov scheme for a scalar equation

The importance of conservation schemes arises from the *Lax–Wendroff theorem* that states, "if a consistent conservative scheme converges, the result is a weak solution of the equation." A "weak solution" to an ordinary or partial differential equation is a function for which the derivatives may not all exist but which is nonetheless deemed to satisfy the equation in some precisely defined sense. Starting with a differential equation, one can rewrite it in a form such that no derivatives of the solution of the equation show up in that form. Such a form is called the weak formulation, and the solutions to the weak formulation are known as weak solutions. A differential equation may have solutions that are not differentiable; such solutions can be obtained using the weak formulation. Weak solutions are important to investigate as many differential equations that naturally arise while modeling real-world phenomena do not admit sufficiently smooth solutions, and the only way of solving such equations is using the weak formulation.

Godunov demonstrated the first conservative extension of the "first-order upwind scheme" to nonlinear systems of conservation laws. The Godunov first-order upwind method is a conservative scheme having the form as given in Equation (15.24), where the numerical fluxes at the boundaries of the cells, $f_{i-1/2}$ and $f_{i+1/2}$ are calculated using solutions of local "Riemann problems." A Riemann problem, is a specific initial value problem comprising a conservation equation together with piecewise constant initial data that has a single discontinuity in the domain of interest. Riemann problems appear naturally in finite volume methods for the solution of conservation law equations owing to the discreteness of the grid. Godunov indicated that this requires solving a Riemann problem in every time step at every boundary between two cells, taking as initial values at each side of the boundary, the average values of the variable in the previous time step.

It is assumed that in each time step t^n, variable u is piecewise constant, taking on each cell I_i the value given by Equation (15.20). This would result in a pair of steady states at each boundary of I_i : (u_{i-1}, u_i) on the left and (u_i, u_{i+1}) on the right, both of which can be considered as a local Riemann problem, originating at $x = 0, t = 0$.

Hence, in the left side, $x = x_{i-1/2}$, one obtains

$$\frac{\partial u}{\partial t} + \frac{\partial f}{\partial x} = 0 \tag{15.30}$$

$$u(x,0) = u_o(x) = u_{i-1}^n \quad \text{for} \quad x < 0 \quad \text{and} \quad u(x,0) = u_o(x) = u_i^n \quad \text{for} \quad x > 0 \tag{15.31}$$

and in the right side, $x = x_{i+1/2}$, one obtains

$$\frac{\partial u}{\partial t} + \frac{\partial f}{\partial x} = 0 \tag{15.32}$$

$$u(x,0) = u_o(x) = u_i^n \quad \text{for} \quad x < 0 \quad \text{and} \quad u(x,0) = u_o(x) = u_{i+1}^n \quad \text{for} \quad x > 0 \tag{15.33}$$

Let $\tilde{u}(x,t)$ be the combined solution of the Riemann problem (u_{i-1}^n, u_i^n) and of the Riemann problem (u_i^n, u_{i+1}^n) in I_i. As $\tilde{u}(x,t)$ is the exact solution of the conservation law, Equation (15.30), one introduces it in the integral form [Equation (15.21)], with spatial and temporal domains, as follows:

$$I_i = \left[x_{i-1/2}, x_{i+1/2} \right] = [x_i - \Delta x/2, x_i + \Delta x/2] \quad \text{and} \quad I_n = [0, \Delta t] \tag{15.34}$$

$$\int_{x_i-1/2}^{x_i+1/2} \tilde{u}(x,\Delta t)\,dx = \int_{x_i-1/2}^{x_i+1/2} \tilde{u}(x,0)\,dx - \int_0^{\Delta t} f[\tilde{u}(x_{i+1/2},t)]\,dt + \int_0^{\Delta t} f[\tilde{u}(x_{i-1/2},t)]\,dt = 0$$

$$(15.35)$$

Defining

$$u_i^n = \frac{1}{\Delta x}\int_{x_i-1/2}^{x_i+1/2} \tilde{u}(x,0)\,dx \qquad (15.36a)$$

$$u_i^{n+1} = \frac{1}{\Delta x}\int_{x_i-1/2}^{x_i+1/2} \tilde{u}(x,\Delta t)\,dx \qquad (15.36b)$$

Equation (15.24) is written as

$$u_i^{n+1} = u_i^n + \frac{\Delta t}{\Delta x}\left[\hat{f}_{i+1/2} - \hat{f}_{i-1/2}\right] \qquad (15.37)$$

with

$$\hat{f}_{i+1/2} = \frac{1}{\Delta t}\int_0^{\Delta t} f\left[\tilde{u}(x_{i+1/2},t)\right]dt \qquad (15.38a)$$

$$\hat{f}_{i-1/2} = \frac{1}{\Delta t}\int_0^{\Delta t} f\left[\tilde{u}(x_{i-1/2},t)\right]dt \qquad (15.38b)$$

The integrands in Equation (15.38) depends on the exact solution to the Riemann problem; at each end of the cell, along the time axis at $x = 0$, Equation (15.38) is represented as

$$\tilde{u}(x_{i-1/2},t) = u_{i-1/2}(0) \qquad \text{and} \qquad \tilde{u}(x_{i+1/2},t) = u_{i+1/2}(0) \qquad (15.39)$$

whereby

$$f_{i-1/2} = f[u_{i-1/2}(0)] \qquad \text{and} \qquad f_{i+1/2} = f[u_{i+1/2}(0)] \qquad (15.40)$$

Thus, Godunov's scheme considers the problem to be solved as a succession of states, constant in each finite volume. At each time step, a Riemann problem at the boundary of each cell is solved, taking the exact solutions of each local problem as the fluxes in these boundaries. These exact solutions must be calculated according to the equation in question. Finally, the spatial averaging of the dependent variables in each cell is performed.

15.3.3   Rankine–Hugoniot jump condition

The preceding section dealt with the Godunov's scheme, in which the integral form is discretized, while seeking for a weak solution to the differential equation, Equation

(15.30), when the initial condition is discontinuous (Riemann problem). It is true that any function u that is a solution to the partial differential equation is differentiable and will be a weak solution. Assuming that there exists a "discontinuity" in a weak solution $u(x,t)$, the function will take values u_L and u_R at both sides of the discontinuity. Then the following expression, known as the Rankine–Hugoniot jump condition, is to be satisfied:

$$(u_R - u_L)S = f(u_R) - f(u_L),$$ (15.41)

where S is the speed at which the jump is transmitted. For the transport equation [Equation (15.30)], the flux f is assumed to be linearly related to the scalar variable u through $f = au$, where a is a constant. Substituting in Equation (15.41), one obtains

$$S = \frac{[f(u)]}{[u]} = \frac{f(u_R) - f(u_L)}{u_R - u_L} = \frac{au_R - au_L}{u_R - u_L} = a$$ (15.42)

where $[\cdot]$ indicates the *jump* across the discontinuity. The speed at which the jump is transmitted in the linear one-dimensional transport equation is the same as the speed of the wave.

15.3.4 Finite volume method for one-dimensional linear heat equation

Consider the general linear one-dimensional heat conduction equation as applied to a rod of unit length.

$$\frac{\partial u}{\partial t} - \frac{\partial}{\partial x}\left[k(x)\frac{\partial u}{\partial x}\right] = S(x,t)$$ (15.43)

Equation (15.43) differs from the earlier mentioned linear one-dimensional heat conduction equation in terms of having a source term $S(x,t)$ and assuming that the coefficient of thermal conductivity of the material is not a constant. It is assumed that $u = u(x,t)$ is the temperature of the rod and $k(x)$ is strictly positive. Let Equation (15.43) be subjected to the following initial condition:

$$u(x,0) = f(x)$$ (15.44)

and the Neumann (heat flux) boundary conditions:

$$\left.\frac{\partial u}{\partial x}\right|_{x=0} = 0 \quad \text{and} \quad \left.\frac{\partial u}{\partial x}\right|_{x=1} = 0$$ (15.45)

The given spatial dimension x is discretized into N equal size grid cells that have size $h = 1/N$. Define $x_j = jh + h/2$ so that x_j is the value of x at the center of the cell j (refer

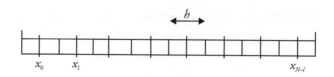

Figure 15.4   Discretization of space with grid size h.

Figure 15.4) whereas the x values at the edges of the cell j are $x_{j-1/2}$ and $x_{j+1/2}$. In the finite volume method, the unknowns approximate the average of the solution over the grid cell. Hence, assuming that $q_j(t)$ is the approximation as defined by

$$q_j(t) \approx u_j(t) = \frac{1}{h} \int_{x_{i-1/2}}^{x_{i+1/2}} u(x,t)\,dx \tag{15.46}$$

integrating Equation (15.43) over cell j and dividing by h, one obtains

$$\frac{1}{h} \int_{x_{j-1/2}}^{x_{j+1/2}} \frac{\partial u(x,t)}{\partial t}\,dx = \frac{1}{h} \int_{x_{j-1/2}}^{x_{j+1/2}} \frac{\partial}{\partial x}\left[k(x)\frac{\partial u(x,t)}{\partial x}\right]dx + \frac{1}{h} \int_{x_{j-1/2}}^{x_{j+1/2}} S(x,t)\,dx$$

$$= \frac{k(x_{j+1/2})\dfrac{\partial u(x_{j+1/2},t)}{\partial x} - k(x_{j-1/2})\dfrac{\partial u(x_{j-1/2},t)}{\partial x}}{h} + \frac{1}{h} \int_{x_{j-1/2}}^{x_{j+1/2}} S(x,t)\,dx \tag{15.47}$$

On defining the flux F as follows:

$$F_j(t) = F(x_j,t) = -k(x_j)\frac{\partial u(x_j,t)}{\partial x} \tag{15.48}$$

and the local average of the source,

$$S_j(t) = \frac{1}{h} \int_{x_{i-1/2}}^{x_{i+1/2}} S(x,t)\,dx \tag{15.49}$$

one obtains the exact update formula,

$$\frac{du_j(t)}{dt} = -\frac{F_{j+1/2}(t) - F_{j-1/2}(t)}{h} + S_j(t) \tag{15.50}$$

Figure 15.5 represents the fluxes $F_{j-1/2}$ (to the left of the grid cell j) and $F_{j+1/2}$ (to the right of the grid cell j) that provide a measure of the heat flowing out through the left and right boundary of the cell. It is to be noted that Equation (15.50) is an example of the conservation law in integral form as follows:

$$\frac{d}{dt}\int_V u\,dV + \int_S \vec{F}\cdot\hat{n}\,dS = \int_V S\,dV, \tag{15.51}$$

where V is the interval $[x_{j-1/2}, x_{j+1/2}]$ where $|V| = h$.

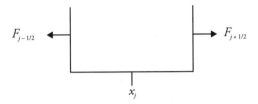

Figure 15.5   Representation of flux to the left and to the right of the grid cell j.

To utilize the update formula as indicated in Equation (15.50), one needs to approximate the fluxes $F_{j-1/2}$, and $F_{j+1/2}$ as follows where second-order central difference approximation is employed for the spatial derivative:

$$F_{j-1/2}(t) = -k(x_{j-1/2})\frac{\partial u(x_{j-1/2},t)}{\partial x} = -k(x_{j-1/2})\frac{u(x_j,t)-u(x_{j-1},t)}{h} = -k_{j-1/2}\frac{u_j(t)-u_{j-1}(t)}{h} \tag{15.52}$$

Similarly,

$$F_{j+1/2}(t) = -k(x_{j+1/2})\frac{\partial u(x_{j+1/2},t)}{\partial x} = -k(x_{j+1/2})\frac{u(x_{j+1},t)-u(x_j,t)}{h} = -k_{j+1/2}\frac{u_{j+1}(t)-u_j(t)}{h}, \tag{15.53}$$

where we use the notation $k_j = k(x_j)$. Utilizing Equation (15.46), one can approximate the fluxes as

$$F_{j-1/2}(t) \approx \tilde{F}_{j-1/2}(t) = -k_{j-1/2}\frac{q_j(t)-q_{j-1}(t)}{h} \tag{15.54}$$

$$F_{j+1/2}(t) \approx \tilde{F}_{j+1/2}(t) = -k_{j+1/2}\frac{q_{j+1}(t)-q_j(t)}{h} \tag{15.55}$$

When these are utilized in the numerical scheme for the interior points $1 \le j \le N-2$,

$$\begin{aligned}
\frac{dq_j(t)}{dt} &= -\frac{\tilde{F}_{j+1/2}(t)-\tilde{F}_{j-1/2}(t)}{h} + S_j(t) \\
&= \frac{k_{j+1/2}\left[q_{j+1}(t)-q_j(t)\right] - k_{j-1/2}\left[q_j(t)-q_{j-1}(t)\right]}{h^2} + S_j(t)
\end{aligned} \tag{15.56}$$

which may be expressed as

$$\frac{dq_j}{dt} = \frac{k_{j+1/2}q_{j+1} - \left(k_{j+1/2}+k_{j-1/2}\right)q_j + k_{j-1/2}q_{j-1}}{h^2} + S_j \tag{15.57}$$

for $j = 1, 2, \ldots, N-2$.

To obtain the solution of Equation (15.57), one needs to update the formulae for the boundary points $j = 0$ and $j = N-1$. This solution must be obtained by incorporating the boundary conditions, introducing ghost cells $j = -1$ and $j = N$ that are located outside the domain. Given the Neumann boundary conditions $(\partial u/\partial x) = 0$ at $x = 0$, it is assumed that it is possible to extend the solution domain for $x < 0$, i.e., outside the domain and approximate as follows:

$$0 = \left.\frac{\partial u}{\partial x}\right|_{x=0} = \frac{u(x_o,t) - u(x_{-1},t)}{h} = \frac{u_o(t) - u_{-1}(t)}{h} \tag{15.58}$$

Replacing u_j by the approximation q_j, one obtains an expression for q_{-1} in terms of q_o as the boundary value as given

$$q_{-1}(t) = q_0(t) \tag{15.59}$$

Equation (15.59) can be substituted into Equation (15.57) for $j = 0$, giving

$$\frac{dq_o}{dt} = \frac{k_{1/2}q_1 - \left(k_{1/2}+k_{-1/2}\right)q_o + k_{-1/2}q_{-1}}{h^2} + S_o = k_{1/2}\frac{q_1 - q_o}{h^2} + S_o \tag{15.60}$$

Similarly, one can obtain for $j = N-1$, such that $q_N(t) = q_{N-1}(t)$ and applying it to Equation (15.57) for $j = N-1$,

$$\frac{dq_{N-1}}{dt} = k_{N-3/2}\frac{q_{N-2} - q_{N-1}}{h^2} + S_{N-1} \tag{15.61}$$

Writing in matrix form Equations (15.57), (15.60), and (15.61), one obtains

$$\frac{d}{dt}\begin{bmatrix} q_o \\ q_1 \\ \vdots \\ q_{N-2} \\ q_{N-1} \end{bmatrix} = \frac{1}{h^2}\begin{bmatrix} -k_{1/2} & k_{1/2} & & & \\ k_{1/2} & -(k_{1/2}+k_{3/2}) & k_{3/2} & & \\ & & \ddots & & \\ & & k_{N-5/2} & -(k_{N-5/2}+k_{N-3/2}) & k_{N-3/2} \\ & & & k_{N-3/2} & -k_{N-3/2} \end{bmatrix}\begin{bmatrix} q_o \\ q_1 \\ \vdots \\ q_{N-2} \\ q_{N-1} \end{bmatrix} + \begin{bmatrix} S_o \\ S_1 \\ \vdots \\ S_{N-2} \\ S_{N-1} \end{bmatrix}$$

$$\tag{15.62}$$

Equation (15.62) is a system of linear ordinary differential equation (ODE) of the form

$$\frac{dq(t)}{dt} = Aq(t) + S(t),$$ (15.63)

where the original partial differential equation is now approximated to a system of ODEs, where the matrix A is a discrete approximation of the second-order differential operator $\frac{\partial}{\partial x}\left[k(x)\frac{\partial}{\partial x}\right]$ including its boundary conditions.

15.4 Finite Element Method

Presently, most operational numerical weather prediction (NWP) centers worldwide employ methods that are either based on the finite difference method (considered ideal for the regional domain), or, the spectral transform method (considered ideal for the global domain). It is difficult to find models using the aforementioned methods that scale optimally on massively parallel computers. This is also applicable for higher order finite volume methods.

Regional atmospheric models (or limited area models) consider atmospheric flows over a subsection of the earth's surface. Examples of regional models include mesoscale models, which typically extend to hundreds of kilometers in the horizontal, as well as cloud resolving models, which encompass approximately up to tens of kilometers in the horizontal. The finite difference method is the better choice among various other methods for regional models for the following reasons, (i) it is relatively easy to implement finite difference methods on a Cartesian grid, (ii) grid generation for finite difference method is straightforward, (iii) the finite difference method is considerably efficient on a single processor, or on a small number of processors within a shared memory architecture (e.g., vector machines), and (iv) constructing upwind and higher order discretization schemes using finite difference method is simple.

Global atmospheric models solve the governing equations of the evolution of the atmospheric system over the entire Earth surface and, hence, over a spherical surface. Spherical harmonics are the most obvious choice for the basis functions while employing the spectral transform method. Most operational NWP centers employ spectral transform method while utilizing global domains. The spectral transform method is known to provide for greater accuracy as compared to the finite difference method. Spherical harmonics are a combination of Fourier and Legendre transforms.

Let n be the number of grid points along the zonal and meridional directions. Applying the spectral transform method, Fourier transforms are evaluated along the zonal direction using a fast Fourier transform (FFT) approach requiring $O[n\ln(n)]$ number of operations. The spectral transform method, utilizes Legendre transform in the meridional direction that requires $O(n^3)$ number of operations. Hence, the total number of operations while employing the spectral transform method is $O[n \ln(n) + n^3]$, which scales adversely as n increases, i.e., essentially with increase of the horizontal resolution of the model. Spectral transform methods were developed during the age of smaller, shared-memory computing machines, which did not require communicating data across processors. As the architectures evolved from shared to distributed memory, the communication overhead became more important. For the spectral transform method, despite its distinct advantages of having least dispersion error, the all-to-all communication required by both the FFT and Legendre transform poses a barrier to scalability. Therefore, the spectral transform method, despite being more accurate and efficient with smaller and shared memory machines, cannot compete in the present computer age that has millions of processors.

Chapter 14 introduced spectral methods as an example of the *Galerkin method*. In the Galerkin method the dependent variables (unknowns) are represented as a summation of functions, called the "basis function" that have a prescribed spatial structure. Usually, the coefficients associated with each basis function are a function of time. The aforementioned Galerkin representation ensures that the partial differential equations with two independent variables (x and t) are transformed into a set of ordinary differential equations for the coefficients. The resulting set of ordinary differential equations are solved using finite difference in time. The two most popular examples of the Galerkin method are (i) *spectral methods*, and (ii) *finite element method*. The spectral methods uses the orthogonal (global) functions as basis functions, whereas, the finite element method uses functions that are zero except in a limited region where they are low order polynomials.

Finite element methods as applied to atmospheric models provide an excellent alternative to both the finite difference and the spectral transform methods in the current age of massively parallel computers, since the finite element method has demonstrated high parallel efficiency, on large to very large machines owing to its small parallel communication footprint. In the finite element method, the solution is sought on an element-wise basis and each element communicates information to the others only through its shared boundaries. When the finite element grid is partitioned into smaller portions of the global domain, the only information that needs to be exchanged among the subdomains of the partition is that on the boundary that each subdomain shares with its neighbors. However, for a finite difference method, the finite difference stencil is such that differentiation on each node in the domain requires information from a set of adjacent nodes that change with the order of differentiation. Hence, for the finite difference method, while partitioning the domain, some nodes will belong to two overlapping subdomains and, hence, additional communication is essential. By contrast, the finite element method requires low communication by its very design.

15.4.1 Finite element method as applied to an ordinary differential equation

The finite element method approximates the dependent variable as a finite series expansion in terms of linearly independent analytical functions, as in the spectral methods. In the finite element method (like the spectral methods), the dependent variable is defined over the whole domain of interest. The two elementary steps in the finite-element method are: (i) expand the dependent variables in terms of a set of low order polynomials (which are the "basis functions"), which are only locally non-zero, and (ii) substitute these expansions to the governing partial differential equations and orthogonalize the error with respect to some test functions.

The finite element method is employed to solve the second-order ordinary differential equation as follows:

$$\frac{d^2u}{dx^2} = f(x), \qquad 0 \leq x \leq \pi \tag{15.64}$$

subject to the boundary conditions

$$u(0) = u(\pi) = 0 \tag{15.65}$$

For the finite element method, the interval $0 \leq x \leq \pi$, is divided into $N+1$ elements such that $(N+1)\Delta x = \pi$. Here, the basis functions are chosen as "tent-shaped" piecewise linear function as shown in Figure 15.6.

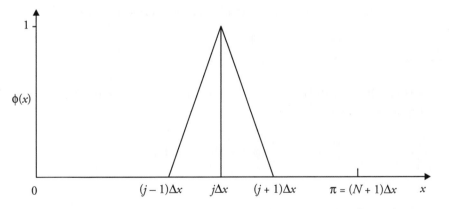

Figure 15.6 Schematic of a tent-shaped linear piecewise function used in finite element method.

Here, it is assumed that the dependent variable u is expanded as a finite series with $\phi_j(x)$ as the basis function and u_j as the unknown coefficients, as indicated here:

$$u(x) = \sum_{j=1}^{N} u_j \phi_j(x) \tag{15.66}$$

The point $x = j\Delta x$ is the "model point" and the basis function ϕ_j has a value 1 at this model point. The basis function is defined such that it deceases linearly to zero at $x = x_{j-1} = (j-1)\Delta x$ and also at $x = x_{j+1} = (j+1)\Delta x$ (refer Figure 15.6). The basis function is also assumed to be zero everywhere else. Figure 15.6 shows that the form of the basis function $\phi_j(x)$ can be defined as follows:

$$\phi_j(x) = \begin{cases} 0 & \text{if} \quad x < x_{j-1} \text{ or } x > x_{j+1} \\ \dfrac{x - x_{j-1}}{\Delta x} & \text{if} \quad x_{j-1} \leq x \leq x_j \\ \dfrac{x_{j+1} - x}{\Delta x} & \text{if} \quad x_j \leq x \leq x_{j+1} \end{cases} \tag{15.67}$$

The boundary conditions as indicated in Equation (15.65) are automatically satisfied based on this definition of the basis function. It is to be noted that the unknown coefficient u_j is actually the value of the function $u(x)$ at $x = j\Delta x$ as $\phi(j\Delta x) = 1$ and $\phi(i\Delta x) = 0$ for $i \neq j$. Substituting the form of the basis function [Equation (15.67)] using Equation (15.66) in the differential equation (15.64), and orthogonalizing the error with respect to the basis functions, one obtains

$$\sum_{j=1}^{N} u_j \int_0^\pi \phi_i \frac{d^2 \phi_j}{dx^2} dx - \int_0^\pi \phi_i f(x) dx = 0, i = 1, 2, \ldots .N \tag{15.68}$$

The first term in the LHS of Equation (15.68) is rewritten as follows:

$$\sum_{j=1}^{N} u_j \int_0^\pi \left[\frac{d}{dx} \left(\phi_i \frac{d\phi_j}{dx} \right) - \frac{d\phi_i}{dx} \frac{d\phi_j}{dx} \right] dx - \int_0^\pi \phi_i f(x) dx = 0 \tag{15.69}$$

The first term in the LHS of Equation (15.69) vanishes as all the ϕs are zero at $x = 0$ and at $x = \pi$. Equation (15.69) then becomes

$$-\sum_{j=1}^{N} u_j \int_0^\pi \left(\frac{d\phi_i}{dx} \frac{d\phi_j}{dx} \right) dx = \int_0^\pi \phi_j f(x) dx, \quad i = 1, 2, \ldots, N \tag{15.70}$$

Differentiating Equation (15.67) with respect to x yields the following:

$$\frac{d\phi_j}{dx} = \begin{cases} 0 & \text{if} \quad x < x_{j-1} \text{ or } x > x_{j+1} \\ \dfrac{1}{\Delta x} & \text{if} \quad x_{j-1} \le x \le x_j \\ -\dfrac{1}{\Delta x} & \text{if} \quad x_j \le x \le x_{j+1} \end{cases} \tag{15.71}$$

Substituting Equation (15.71) in Equation (15.70), the LHS of Equation (15.70) becomes equal to three non-vanishing terms as follows:

$$-\sum_{j=1}^{N} u_j \int_0^\pi \left(\frac{d\phi_i}{dx} \frac{d\phi_j}{dx} \right) dx = \frac{u_{i-1}\Delta x - 2u_i \Delta x + u_{i+1}\Delta x}{(\Delta x)^2} \tag{15.72}$$

For evaluating the RHS of Equation (15.70), one can approximate $f(x)$ in terms of the basis functions as follows:

$$f(x) = \sum_{j=1}^{N} f_j \phi_j \tag{15.73}$$

RHS of Equation (15.70) then becomes using Equation (15.73)

$$\sum_{j=1}^{N} f_j \int_0^\pi \phi_i \phi_j \, dx = \sum_{j=1}^{N} f_j \int_{x_{i-1}}^{x_{i+1}} \phi_i \phi_j \, dx \tag{15.74}$$

Using $\xi = x - x_i = x - i\Delta x$, this integral can be expanded into three integrals of the following form:

$$\int_0^\pi \phi_i f(x) \, dx = -f_{i-1} \int_{-\Delta x}^0 \frac{\xi(\xi + \Delta x)}{(\Delta x)^2} dx + 2f_i \int_{-\Delta x}^0 \frac{(\xi + \Delta x)^2}{(\Delta x)^2} dx + f_{i+1} \int_0^{\Delta x} \frac{\xi(\Delta x - \xi)}{(\Delta x)^2} \tag{15.75}$$

After evaluating the terms in the RHS of Equation (15.75) and using Equation (15.75) along with Equation (15.72) in Equation (15.70), one obtains

$$\frac{u_{i-1} - 2u_i + u_{i+1}}{(\Delta x)^2} = \frac{f_{i+1} + 4f_i + f_{i-1}}{6} \tag{15.76}$$

Equation (15.76) applies for $2 \le i \le N-1$ and the equations for $i = 1$ and $i = N$ are obtained by removing any terms in $i = 0$ or $i = N + l$. The equation may be solved by Gaussian elimination method. Comparing the final form of the finite element form [Equation (15.76)] with the finite difference form (where the LHS is the same as in Equation (15.76) and the RHS is f_i where $u_i = u(i\Delta x)$), they are exactly the same in the

LHS with the forcing term (RHS) in the finite element method appearing as a weighted average. Once the forcing function form $f(x)$ is known, the finite element method can be used to solve the system of linear algebraic equations (15.76). If the forcing function is sinusoidal in form, it turns out that the finite element solution is considerably more accurate as compared to the finite difference method for the shorter wavelengths.

As an example, consider how one can represent a field ϕ in finite element notation when one is given the values of ϕ at equally spaced points along the x-direction. Let the points be given by x_j (the nodes) and the values of the dependent variable by ϕ_j (the nodal value)(refer Figure 15.7). Assume that ϕ varies linearly between the nodes; known as a "piecewise linear fit" and, hence, the behaviour of ϕ within an element (the region between the nodes) is determined by the nodal values. If we define a set of basis functions $e_j(x)$ given by the hat (chapeau) function (refer to Figure 15.7), the field ϕ can be represented by

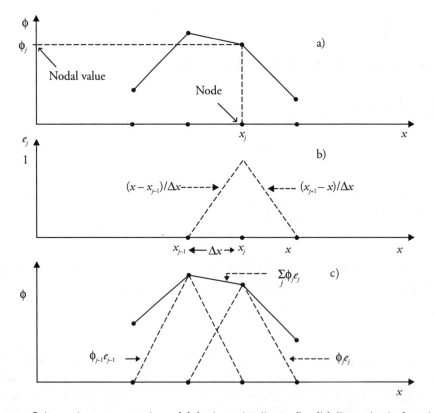

Figure 15.7 Schematic representation of (a) piecewise linear fit, (b) linear basis function, and (c) how a piecewise linear fit is a linear combination of basis functions.

$$\phi = \sum_j \phi_j e_j(x) \tag{15.77}$$

Figure 15.7 shows the schematic representation of (a) piecewise linear fit, (b) linear basis function, and (c) how a piecewise linear fit is a linear combination of basis functions.

15.4.2 Finite element method as applied to one-dimensional advection equation

In this section, the finite element method with linear elements will be applied to the one-dimensional linear advection equation given by

$$\frac{\partial u}{\partial t} + c\frac{\partial u}{\partial x} = 0 \tag{15.78}$$

subject to periodic boundary conditions, $u(x+L,t) = u(x,t)$ and initial condition $u(x,0) = f(x)$. Define a mesh of points $x_j = (j-1)\Delta x$, with $j = 1,2,\ldots,N+1$ and $\Delta x = L/N$. It is assumed that the finite element approximation to the exact solution has a piecewise linear representation using the x_j as the nodes

$$u(x,t) = \sum_{j=1}^{N+1} u_j(t)\,e_j(x) \tag{15.79}$$

Substituting Equation (15.79) in Equation (15.78) and obtaining the expression for the residual (error) as

$$R = \sum_j \frac{du_j}{dt}e_j + c\sum_j u_j \frac{de_j}{dx} \tag{15.80}$$

we employ the Galerkin method by assuming that the test function is the basis function to obtain

$$\int R e_i \, dx = 0, \qquad i = 1,2,\ldots,N+1 \tag{15.81}$$

Substituting Equation (15.80) in Equation (15.81), one obtains

$$\sum_j \frac{du_j}{dt} \int_0^L e_i e_j \, dx + c\sum_j u_j \int_0^L \frac{de_j}{dx}e_i \, dx = 0, \qquad i = 1,2,\ldots,N+1 \tag{15.82}$$

As the basis functions are hat functions, there are several combinations of i and j for which the integral is going to be zero. For a given j, there will be non-zero contributions only for $i = j+1, i = j$, and $i = j-1$ (i.e., for $x_{j-1} \leq x_j \leq x_{j+1}$). It is possible to show that

$$\int e_{j+1}e_j \, dx = \frac{1}{6}\Delta x, \quad \int e_j^2 \, dx = \frac{2}{3}\Delta x, \quad \int \frac{de_{j\pm 1}}{dx}e_j \, dx = \pm\frac{1}{2}, \quad \int \frac{de_j}{dx}e_j \, dx = 0,$$

(15.83)

$$\int e_{j\pm p}e_j \, dx = 0, \quad \int \frac{de_{j\pm p}}{dx}e_j \, dx = 0$$

(15.84)

where $p > 1$. Using Equations (15.83) and (15.84) in Equation (15.82), one obtains

$$\frac{1}{6}\left(\frac{du_{j+1}}{dt} + 4\frac{du_j}{dt} + \frac{du_{j-1}}{dt}\right) + c\left(\frac{u_{j+1} - u_{j-1}}{2\Delta x}\right) = 0$$

(15.85)

This implicit scheme has a smaller truncation error for the space derivatives as compared to the fourth-order finite difference scheme. To solve Equation (15.85), in practice, it is assumed that F_j^n represents the time derivative of u at node j and at time level n. That is,

$$\frac{du_j^n}{dt} = F_j^n$$

(15.86)

Substituting Equation (15.86) in Equation (15.85), one obtains

$$\frac{1}{6}\left(F_{j+1}^n + 4F_j^n + F_{j-1}^n\right) = -c\left(\frac{u_{j+1} - u_{j-1}}{2\Delta x}\right) \qquad \text{for all } j$$

(15.87)

As the RHS of Equation (15.87) is known, the set of simultaneous linear equations can be solved in principle for all the F_j^n. Once all the F_j^n are determined for all j, one can then employ a time marching scheme (say the leap frog scheme) to obtain

$$u_j^{n+1} = u_j^{n-1} + 2\Delta t F_j^n$$

(15.88)

To investigate the stability of this scheme, combining Equations (15.87) and (15.88) result in

$$u_{j+1}^{n+1} + 4u_j^{n+1} + u_{j-1}^{n+1} = (1+6\alpha)u_{j-1}^{n-1} + 4u_j^{n-1} + (1-6\alpha)u_{j+1}^{n-1},$$

(15.89)

where $\alpha = c\Delta t/(\Delta x)$ is the Courant number. Applying the von Neumann method of stability analysis, the scheme as given in Equation (15.89) is conditionally stable with the requirement that the Courant number $\alpha \leq \sqrt{3}$. This scheme appears to be less restrictive in the choice of the time step as compared with the CTCS (centered in time and centered in space) finite difference scheme. Further analysis shows that the scheme is neutral and that the results of the scheme are similar to the results of the fourth-order leapfrog finite difference scheme. The main disadvantage of the scheme is that it is an implicit scheme. Moreover, this is a three-time level scheme (with time levels, $n+1$, n, and $n-1$). It is possible to obtain a two-time level scheme by replacing the leapfrog scheme as applied to Equation (15.88) by the Crank–Nicolson scheme with a weighted mean of the advection terms at time levels n and $n+1$ with weights β^n and β^{n+1}, respectively. The corresponding expressions for Equations (15.87) and (15.88) are then as follows:

$$\frac{1}{6}\left(F_{j+1}^n + 4F_j^n + F_{j-1}^n\right) = \beta^{n+1}A_j^{n+1} + \beta^n A_j^n, \tag{15.90}$$

where A refers to the advection terms.

$$u_j^{n+1} = u_j^n + \Delta t F_j^n \tag{15.91}$$

Equations (15.90) and (15.91) can be combined to obtain

$$(1-3\alpha\beta^{n+1})u_{j-1}^{n+1} + 4u_j^{n+1} + (1+3\alpha\beta^{n+1})u_{j+1}^{n+1} = (1+3\alpha\beta^n)u_{j-1}^n + 4u_j^n + (1-3\alpha\beta^n)u_{j+1}^n \tag{15.92}$$

The scheme as expressed in Equation (15.92) is an implicit scheme but has only two time levels. The stability analysis using von Neumann method indicates that for $\beta^n > 1/2$, the scheme is unstable and for $\beta^n \le 1/2$, it is absolutely stable. Moreover, for $\beta^n = \beta^{n+1} = 1/2$, the scheme is neutral with no damping.

15.4.3 Finite element method as applied to one-dimensional linear Helmholtz equation

The solution of the one-dimensional linear Helmholtz equation using the finite element method is derived in this section. The one-dimensional linear Helmholtz equation is as follows:

$$\frac{d^2\psi}{dx^2} - \alpha^2\psi = 0 \tag{15.93}$$

We expand the dependent variable $\psi(x)$ in terms of basis function $e_j(x)$ and constant coefficients ψ as follows:

$$\psi(x) = \sum_j \psi_j e_j(x) \tag{15.94}$$

We then substitute Equation (15.94) into Equation (15.93) and calculate the residual R as

$$R = \sum_j \psi_j \frac{d^2 e_j}{dx^2} - \alpha^2 \sum_j \psi_j e_j \tag{15.95}$$

Employing the Galerkin method by assuming that the test function is the basis function to obtain

$$\int R e_i \, dx = 0, \qquad i = 1, 2, \ldots, N+1 \tag{15.96}$$

and substituting Equation (15.95) in Equation (15.96), one obtains

$$\sum_j \psi_j \left(\frac{d^2 e_j}{dx^2}, e_i \right) - \alpha^2 \sum_j \psi_j (e_j, e_i) = 0, \tag{15.97}$$

where the inner product of two functions f and g are denoted by

$$(f, g) = \int_0^L f g \, dx \tag{15.98}$$

Equation (15.97) using the notation of Equation (15.98) is expressed as

$$\sum_j \psi_j \int \frac{d^2 e_j}{dx^2} e_i \, dx - \alpha^2 \sum_j \psi_j \int e_j e_i \, dx = 0 \tag{15.99}$$

The first term in the LHS of Equation (15.99) can be integrated by parts, to obtain

$$\int \frac{d^2 e_j}{dx^2} e_i \, dx = \left[\frac{d e_j}{dx} e_i \right]_0^L - \int \frac{d e_i}{dx} \frac{d e_j}{dx} \, dx \tag{15.100}$$

The first term in the RHS of Equation (15.100) is zero for all $i \neq 1$ and $i \neq N+1$. Substituting the expression in Equation (15.100) in Equation (15.99), we note that all derivatives are of the first-order and, hence, linear elements can be utilized. The matrix of the resulting set of equations can in principle be solved.

Exercises 15 (Questions and answers)

1. What is a Reimann solver with reference to finite volume method?

   Answer: Consider solving the following partial differential equation that represents mass conservation equation on an arbitrary unstructured grid

   $$\frac{\partial \rho}{\partial t} + \nabla \cdot (\rho \vec{V}) = 0$$

   In the finite volume method, one first takes integral of the equation over the area S of a grid cell, which results in

   $$\frac{\partial \overline{\rho}}{\partial t} + \int_S \nabla \cdot (\rho \vec{V}) \, dS = 0$$

   By applying Gauss theorem, one can convert the surface integral in this equation into a line integral along boundary of the grid-box, resulting in

   $$\frac{\partial \overline{\rho}}{\partial t} + \oint_L (\rho \vec{V}) \cdot \hat{n} \, dl = 0$$

   It is convenient to denote fluxes through boundaries with $\overline{F}$. The aforementioned equation then becomes

   $$\frac{\partial \overline{\rho}}{\partial t} + \oint_L \overline{F} \cdot \hat{n} \, dl = 0$$

   In order to update the average value in the grid cell, one takes integral in time, yielding

   $$\overline{\rho}(t + \Delta t) = \overline{\rho}(t) - \int_t^{t + \Delta t} \left(\oint_L \overline{F} \cdot \hat{n} \, dl \right) dt$$

   Hence, the time change of averaged value in the cell is attributed to the time integral of fluxes across boundaries. An estimation of the time integral of fluxes is referred to as the Riemann solver. Knowledge of the Riemann solver helps one to obtain the time change of averaged values in the cell.

2. What are the advantages and disadvantages of the finite volume method?

   Answer: The advantages of the finite-volume methods are that they are easy to implement when applied to quasi-uniform and unstructured grids. Moreover, finite volume methods can satisfy most mimetic constraints. Moreover, finite volume methods use only local operators and, hence, they are good for parallelization. The disadvantage of the finite volume method is that they usually have low (usually third) order of accuracy.

3. What is a staggered finite volume scheme?

   Answer: In the staggered mesh arrangement employed in the staggered finite volume scheme, the velocity components are distributed on the cell faces, whereas the pressure and other scalar quantities are defined at the cell centers. This ensures fully coupled velocity and pressure fields. In a Cartesian mesh, the pressure and all other scalar variables are located in the center of each control volume. The velocity components u, v, and w are shifted in the x, y, and z directions, respectively. For each velocity component, a shifted control volume has to be introduced.

   The equations are then formally integrated over the respective control volumes. Thus, conservation equations for scalar quantities, such as, density and energy, are discretized on the volume centered on the point (x_i, y_j, z_k), whereas the x-momentum conservation equation is integrated over the volume centered on the location of u, i.e., with center $(x_{i+1/2}, y_j, z_k)$. In a similar manner, y-momentum and z-momentum conservation equations are integrated over the volumes centered on the location of v and w, respectively.

4. What are discontinuous Galerkin methods?

   Answer: The method where discontinuous basis functions are adopted as a Galerkin approximation of the continuous problem in conservative form, is called a discontinuous Galerkin (DG) method.

5. What are the similarities and differences between the finite volume method and the discontinuous Galerkin method?

   Answer: Both the finite volume methods, and the discontinuous Galerkin methods provide discrete conservation laws that reproduce on each control volume the fundamental physical balances that characterize the continuous problem. Hence, they both represent a good choice for the approximation of problems whose solution have discontinuities, and where classical solutions are not properly defined. However, unlike finite volume methods, the discontinuous Galerkin methods provide for higher order accuracy without extending the computational stencil, thus allowing for a good scalability on parallel architectures.

6. What is the difference between the Rayleigh–Ritz method and the finite element method?

   Answer: The Rayleigh–Ritz method uses global functions that are formulated over the entire domain and, hence, has no limitations to differentiability of these functions in most cases. However, the finite element method divides the entire domain into many small elements and uses functions that are defined only for the localised domains. For finite element methods, while the basis functions are element-related functions, the same for the Rayleigh–Ritz method are valid for the whole domain and have to satisfy the boundary conditions.

7. What is the difference between finite element method and boundary element method?

   Answer: The finite element method can be envisaged as a method of finding approximate solutions for partial differential equations or as a method to transform partial differential equations into algebraic equations; the latter are easier to solve. As the solution for the field variable satisfying both the boundary conditions and the differential equation is not known a priori, one begins with an assumed trial solution. The assumed trial solution, in general, does not satisfy the differential equation exactly and leaves a domain residual defined as the error in satisfying the differential equation. In general, the domain residual varies from point to point within the domain and cannot be exactly reduced to zero everywhere. One can choose to make it vanish at select points within the domain, however, it is better to render the residual very small, over the entire domain. Thus, the weighted sum of the domain residual computed over the entire domain is rendered zero. The trial solutions satisfy the applicable boundary conditions for the finite element method, and, hence, result in nonzero domain residual. For the boundary element method, as trial solutions so chosen implicitly satisfies the differential equation but not the boundary conditions, there is a resulting nonzero boundary residual.

8. What are the advantages of the finite element method?

   Answer: The finite element method has several advantages. It is possible to solve problems that have complicated and complex geometries using finite element methods through the use of unstructured meshes. Moreover, the variational form available in finite element method supports the straightforward and stable implementations of many of the standard boundary conditions. Moreover, for the finite element method, it is often possible to rigorously analyze properties such as stability and accuracy.

9. What are the disadvantages of the finite element method?

   Answer: The finite element methods in general find applications on problems with built-in variational principles. However, for some fully nonlinear and non-divergence form problems, the use of finite element methods is somewhat limited. The finite element methods are heavily dependent on numerical integration, where quadrature rules can sometimes contribute to severe difficulty. Finite element methods are also computationally very expensive. Moreover, programme writing while applying finite element method is inherently harder.

16
Ocean Models

16.1 Introduction

Although the governing equations of the evolution of the oceanic system are not very different from the atmospheric system, there are a few examples of ocean models that do not have a similar or equivalent atmospheric model. Furthermore, there are distinct differences between ocean and atmospheric models, for example, ocean models require larger computer resources in terms of the CPU time, core memory, as well as disk storage. As the eddies in the ocean are much smaller than the atmospheric weather systems, the grid resolution needed in ocean modeling is therefore much finer as compared to atmospheric models. Owing to considerations of stability in terms of CFL condition, fine grid resolution in ocean models also translates into smaller time steps while using explicit finite difference schemes. The time step is determined by the speed of the fast moving surface gravity waves while solving an ocean free surface model using explicit finite difference scheme. The smaller time step in the free surface ocean models while employing explicit finite difference scheme necessitates larger number of time marchings and, hence, longer CPU time to integrate over the specified integration or the forecast duration. Utilizing an implicit/semi-implicit method of finite difference will take care of the stability requirements as these schemes are unconditionally stable; however, the CPU time required for such schemes would depend on the rate of convergence of the iterative method used to solve the matrix equation (for fully implicit schemes) and the rate of convergence of the iterative method used to solve the resulting elliptic partial differential equation (for semi-implicit schemes).

It is well known that unlike the atmospheric system, the ocean system is poorly observed. The only information that satellites provide are about ocean bathymetry, sea surface temperature, sea surface salinity, ocean color, coral reefs, and sea and lake ice. Data assimilation is a technique whereby observational data are combined with output from a numerical model to produce an optimal estimate of the evolving state of the atmospheric or the oceanic system. Owing to the lack of ocean observations, data assimilation methods are inherently more difficult to implement for ocean models. Data-assimilative ocean models require even much larger computational resources than the ones that are running in the free-forecast mode. The enhanced computational resources that go into a data-assimilative ocean model include additional memory requirements and enhanced CPU time requirements and these depend on the complexity of the data assimilation scheme. Experience with data-assimilative ocean models provide sufficient evidence that data assimilation schemes take much longer computational time to produce an optimal estimate of the ocean system as compared to the computational time for time marchings for the ocean forecast, even while employing a simple "Optimal Interpolation" type data assimilation scheme. Hence, it goes without saying that the data-assimilative ocean models are prohibitively expensive in terms of the computational costs when they employ sophisticated data assimilation schemes such as Kalman filters and adjoint methods.

16.2   Sverdrup Model for Ocean Circulation

Sverdrup model is an example of an ocean model that has no equivalence in the atmosphere. Sverdrup's theory (model) of ocean circulation may be derived by integrating the horizontal momentum equations over a large enough depth of the ocean to cover the entire *Ekman layer* (the near surface region of the ocean where vertical mixing of the momentum imparted by the wind is important). Consider horizontal momentum equations for a steady flow (no local time derivative) at small *Rossby number* (Coriolis force and, hence, rotation of Earth is important). Assumption of small Rossby number results in the nonlinear advection terms becoming negligible. Moreover, assume that both friction and Coriolis forces are important. The horizontal momentum equations with the aforementioned assumptions are

$$-\rho f v = -\frac{\partial p}{\partial x} + \frac{\partial \tau^x}{\partial x} \tag{16.1}$$

$$\rho f u = -\frac{\partial p}{\partial y} + \frac{\partial \tau^y}{\partial y} \tag{16.2}$$

Sverdrup integrated these equations [(16.1) and (16.2)] from the ocean surface to a depth at which the horizontal pressure gradient becomes zero (i.e., up to a depth level of no motion). Then, one obtains

$$-f \int_{-z_o}^{0} \rho v\, dz = - \int_{-z_o}^{0} \frac{\partial p}{\partial x}\, dz + [\tau^x]_{-z_o}^{0} \tag{16.3}$$

$$f \int_{-z_o}^{0} \rho u\, dz = - \int_{-z_o}^{0} \frac{\partial p}{\partial y}\, dz + [\tau^y]_{-z_o}^{0} \tag{16.4}$$

Equations (16.3) and (16.4) are written in terms of the vertically integrated zonal and meridional mass transports M^x and M_y, respectively and are

$$-fM^y = - \int_{-z_o}^{0} \frac{\partial p}{\partial x}\, dz + \tau^x_{\text{wind}}, \qquad M^y = \int_{-z_o}^{0} \rho v\, dz \tag{16.5}$$

and τ^x at $z = z_o$ is assumed zero (at level of no motion).

$$fM^x = - \int_{-z_o}^{0} \frac{\partial p}{\partial y}\, dz + \tau^y_{\text{wind}}, \qquad M^x = \int_{-z_o}^{0} \rho u\, dz \tag{16.6}$$

and τ^x and τ^y at $z = -z_o$ are assumed zero (at level of no motion). Taking $-\dfrac{\partial}{\partial y}$ of Equation (16.5) and adding to $\dfrac{\partial}{\partial x}$ of Equation (16.6), one obtains

$$\frac{\partial}{\partial y}[fM^y] + \frac{\partial}{\partial x}[fM^x] = -\frac{\partial \tau^x}{\partial y} + \frac{\partial \tau^y}{\partial x} \tag{16.7}$$

Equation (16.7) is rewritten as

$$\frac{\partial f}{\partial y}M^y + f\left[\frac{\partial M^x}{\partial x} + \frac{\partial M^y}{\partial y}\right] = \frac{\partial \tau^y}{\partial x} - \frac{\partial \tau^x}{\partial y} \tag{16.8}$$

As conservation of mass (continuity equation) is expressed as

$$\frac{\partial u}{\partial x} + \frac{\partial v}{\partial y} = 0 \tag{16.9}$$

multiplying Equation (16.9) by density of water ρ and integrating the resulting equation over depth from the ocean surface to the depth level of no motion, one obtains

$$\frac{\partial M^x}{\partial x} + \frac{\partial M^y}{\partial y} = 0 \tag{16.10}$$

Equation (16.10) enables one to define a mass transport stream function ψ as follows:

$$\frac{\partial \psi}{\partial x} = M^y \qquad \text{and} \qquad -\frac{\partial \psi}{\partial y} = M^x \tag{16.11}$$

Using Equation (16.10), Equation (16.8) can be rewritten as

$$\beta M^y = \frac{\partial \tau^y}{\partial x} - \frac{\partial \tau^x}{\partial y} = \vec{k} \cdot \nabla \times \vec{\tau}, \tag{16.12}$$

where β is the meridional variation of the Coriolis parameter and $\vec{\tau}$ is the surface wind stress vector. Equation (16.12) is the Sverdrup relation or Sverdrup model.

Assume a rectangular ocean basin with $0 \le x \le a, 0 \le y \le b$. With the knowledge of the prescribed surface wind stress vector $\vec{\tau}$, it is possible to calculate M_y from Equation (16.12) and the meridional current v from the definition of M_y. Knowledge of v will help in calculating the zonal current u from Equation (16.9) using

$$u(x,y) = -\int_0^a \frac{\partial v}{\partial y} \, dx \tag{16.13}$$

where it is assumed that $u(a,y) = 0$ along the eastern boundary. Relating the relative vorticity in the vertical direction ζ to the stream function ψ, one obtains

$$\frac{\partial^2 \psi}{\partial x^2} + \frac{\partial^2 \psi}{\partial y^2} = \frac{\partial M^y}{\partial x} - \frac{\partial M^x}{\partial y} = \zeta \tag{16.14}$$

Equation (16.12) and (16.14) together provide the equations that need to be solved from the knowledge of the applied surface wind stress and subject to the boundary conditions. To solve Equation (16.14), which is a Poisson equation, one requires appropriate boundary conditions. Assume a simple situation where the northern and southern boundaries are on the zero line of $\vec{k} \cdot \nabla \times \vec{\tau}$. From Equation (16.12), this corresponds to M^y being zero at both the northern and southern boundaries. Using Equation (16.11), one can apply $\psi = 0$ at both northern and southern boundaries. On the eastern boundary, $\psi = 0$ as $M^x = 0$. On the western boundary, one needs to approximate the boundary condition by solving the equation

$$\frac{\partial^2 \psi}{\partial x^2} = -\frac{\partial M^x}{\partial y}, \quad \psi(0,0) = 0, \quad \text{and} \quad \psi(0,b) = 0 \tag{16.15}$$

Equation (16.15) provides a tridiagonal system that can be solved by efficient tridiagonal solvers such as the Thomas algorithm. Discretizing the domain x and y by a uniform grid with $\Delta x = \Delta y = \Delta h$ and using j and k as indices in the x and y directions, we get

$$x_j = (j-1)\Delta h, \text{ where } j = 1, 2, \ldots, J \text{ and } J - 1 = a/\Delta h$$
$$y_k = (k-1)\Delta h, \text{ where } k = 1, 2, \ldots, K \text{ and } K - 1 = b/\Delta h$$

It is assumed that ψ is known at all k points for $j = 1$ and $j = J$ and ψ is known at all j points for $k = 1$ and $k = K$. At each grid point, one approximates $\psi(j\Delta h, k\Delta h) = Q_{j,k}$ and it is assumed that the curl of the mass transport $\zeta_{j,k}$ is known at each grid point. Approximating the second derivative of ψ with respect to x and y using the central finite difference, one obtains

$$\frac{\partial^2 \psi}{\partial x^2} = \frac{Q_{j+1,k} - 2Q_{j,k} + Q_{j-1,k}}{(\Delta h)^2} + O(\Delta h)^2 \tag{16.16}$$

$$\frac{\partial^2 \psi}{\partial y^2} = \frac{Q_{j,k+1} - 2Q_{j,k} + Q_{j,k-1}}{(\Delta h)^2} + O(\Delta h)^2 \tag{16.17}$$

Applying Equations (16.16) and (16.17) in Equation (16.14), one obtains the following set of linear equations:

$$Q_{j+1,k} + Q_{j-1,k} + Q_{j,k+1} + Q_{j,k-1} - 4Q_{j,k} = (\Delta h)^2 \zeta_{j,k} = L_{j,k} \tag{16.18}$$

for $j = 2, 3, \ldots, J-1$, $\quad k = 2, 3, \ldots, K-1$.

Sverdrup showed that the time-independent wind stress curl is directly related to the meridional mass transport in the ocean. In the real situation, the ocean basin may be divided into 100 grid levels in each of the two directions, taking the number of unknowns to the order 10^4. For example, if $J = 5$ and $K = 6$, the number of interior points would be $(J-2) \times (K-2) = 12$ and the unknown vector will have 12 elements. For this case, Equation (16.18) can be written in the matrix form as

$$
\begin{bmatrix}
-4 & 1 & 0 & 1 & 0 & 0 & 0 & 0 & 0 & 0 & 0 & 0 \\
1 & -4 & 1 & 0 & 1 & 0 & 0 & 0 & 0 & 0 & 0 & 0 \\
0 & 1 & -4 & 0 & 0 & 1 & 0 & 0 & 0 & 0 & 0 & 0 \\
 & & \vdots & & & \vdots & & & \vdots & & & \\
0 & 0 & 0 & 0 & 0 & 0 & 1 & 0 & 0 & -4 & 1 & 0 \\
0 & 0 & 0 & 0 & 0 & 0 & 0 & 1 & 0 & 1 & -4 & 1 \\
0 & 0 & 0 & 0 & 0 & 0 & 0 & 0 & 1 & 0 & 1 & -4
\end{bmatrix}
\begin{bmatrix}
Q_{2,2} \\
Q_{3,2} \\
Q_{4,2} \\
\vdots \\
Q_{2,5} \\
Q_{3,5} \\
Q_{4,5}
\end{bmatrix}
=
\begin{bmatrix}
L_{2,2} - Q_{1,2} - Q_{2,1} \\
L_{3,2} - Q_{3,1} \\
L_{4,2} - Q_{4,1} - Q_{5,2} \\
\vdots \\
L_{2,5} - Q_{1,5} - Q_{2,6} \\
L_{3,5} - Q_{3,6} \\
L_{4,5} - Q_{4,5} - Q_{5,5}
\end{bmatrix}
\tag{16.19}
$$

This set of equations is block tridiagonal.

$$
\begin{bmatrix}
B & I & \phi & \phi \\
I & B & I & \phi \\
\phi & I & B & I \\
\phi & \phi & I & B
\end{bmatrix}
\begin{bmatrix}
S_1 \\
S_2 \\
S_3 \\
S_4
\end{bmatrix}
=
\begin{bmatrix}
Z_1 \\
Z_2 \\
Z_3 \\
Z_4
\end{bmatrix}
\tag{16.20}
$$

where $S_i^T = [Q_{2,i}, Q_{3,i}, Q_{4,i}], \phi$ is a null matrix and

$$
B = \begin{bmatrix} -4 & 1 & 0 \\ 1 & -4 & 1 \\ 0 & 1 & -4 \end{bmatrix} \tag{16.21}
$$

16.2.1 Sverdrup model for ocean circulation having a zonal wind stress with meridional variation

It would be instructive to envisage the Sverdrup model for a special case of a wind stress having only the zonal component. As the observed winds over the tropical regions are easterlies whereas the observed winds over mid-latitudes are essentially westerlies in both hemispheres, it would be appropriate to assume that the observed wind stress over the tropical/mid-latitude oceans are essentially zonal. Considering that the aforementioned observed zonal stress varies with latitude, it would be appropriate to assume that the observed zonal wind stress is a function of meridional distance, i.e., y. The Sverdrup balance equation (16.12) is as follows:

$$
M^y = \frac{1}{\beta} \left(\frac{\partial \tau^y}{\partial x} - \frac{\partial \tau^x}{\partial y} \right) = \frac{1}{\beta} \vec{k} \cdot \nabla \times \vec{\tau} \tag{16.22}
$$

Assuming that the observed wind stress has only zonal components and no meridional components, the Sverdrup balance equation becomes

$$
M^y = -\frac{1}{\beta} \frac{\partial \tau^x}{\partial y} \tag{16.23}
$$

Assume an observed zonal wind stress that depends only on y as follows:

$$
\tau^x = -\tau \cos \left[\frac{\pi(y - y_{\text{trade}})}{L_y} \right] \tag{16.24}
$$

From Equations (16.11), the mass transport stream function ψ at the western boundary (ψ_{wb}) can be written as

$$
\psi_{wb} = \frac{1}{\beta} \int_{eb}^{wb} \vec{k} \cdot \nabla \times \vec{\tau} \, dx = \frac{\pi \tau L_x}{\beta L_y} \sin \left[\frac{\pi(y - y_{\text{trade}})}{L_y} \right] \tag{16.25}
$$

The value of the mass transport stream function ψ at the western boundary can be calculated at the western boundary using the following values in Equation (16.25): $\tau = 0.1 \, \text{N/m}^2$, $\beta = 1.98 \times 10^{-11} \, \text{m}^{-1}\text{s}^{-1}$, $L_y = 3{,}000 \, \text{km}$, $L_x = 8{,}000 \, \text{km}$.

16.3   Stommel Model for Ocean Circulation

Stommel in his seminal work in 1948 considers a rectangular ocean basin with ground plane having $(0 < x < a; 0 < y < b)$ and a time-independent height $D + h(x,y)$, where D is the mean average water height and $h(x,y)$ is the deviation of the water height from the mean average water height. Stommel further assumed that the water in the ocean is homogeneous $(\rho = \rho_o = constant)$, incompressible with no vertical movement. He further considered that the observed wind stress is essentially zonal, the latter varying only with the meridional distance y. Moreover, he assumed linear friction (Rayleigh friction) and Rayleigh friction coefficient to be constant. Assuming the Rossby number to be very small (inertial terms can be neglected as compared to the Coriolis force terms) and invoking the β plane approximation, Stommel's momentum and continuity equations are as follows:

$$-\rho_o f v = -\frac{\partial p}{\partial x} + \rho_o F_x \tag{16.26}$$

$$\rho_o f u = -\frac{\partial p}{\partial y} + \rho_o F_y \tag{16.27}$$

$$0 = -\frac{\partial p}{\partial z} - g\rho_0 \tag{16.28}$$

$$\frac{\partial u}{\partial x} + \frac{\partial v}{\partial y} = 0 \tag{16.29}$$

where F_x and F_y are related to wind stress and Rayleigh friction. Integrating Equation (16.28) with respect to z, one obtains

$$p(x,y,z,t) = p_o + \int_z^{D+h} -\rho_o g \, dz = p_o + \rho_o g [D + h(x,y,t) - z], \tag{16.30}$$

where the air pressure p_o is assumed to be constant and is independent of time and position. Substituting Equation (16.30) into Equations (16.26) and (16.27) and dividing by ρ_o, one obtains

$$-f v = -g\frac{\partial h}{\partial x} + F_x \tag{16.31}$$

$$f u = -g\frac{\partial h}{\partial y} + F_y \tag{16.32}$$

Integrating Equations (16.31) and (16.32) with respect to z from $z = 0$ to $z = D + h$, one obtains after assuming $D \sim D + h$

$$-fvD = -gD\frac{\partial h}{\partial x} + DF_x \tag{16.33}$$

$$fuD = -gD\frac{\partial h}{\partial y} + DF_y \tag{16.34}$$

It is assumed that both the zonal and the meridional components of velocity are independent of z. This follows from the assumption of homogeneous fluid. Stommel assumed that F_x is determined by the zonal wind stress and the x component of the Rayleigh friction force term, whereas F_y is determined by the y component of the Rayleigh friction force term as follows:

$$DF_x = -Ru - T\cos\frac{\pi y}{b} \quad \text{and} \quad DF_y = -Rv \tag{16.35}$$

Equations (16.33) and (16.34) become

$$-fvD = -gD\frac{\partial h}{\partial x} - \left(Ru + T\cos\frac{\pi y}{b}\right) \tag{16.36}$$

$$fuD = -gD\frac{\partial h}{\partial y} - Rv \tag{16.37}$$

Differentiating Equation (16.37) with respect to x and Equation (16.36) with respect to y and subtracting one from the other after invoking the β plane approximation, one obtains after division by D throughout

$$f\left(\frac{\partial u}{\partial x} + \frac{\partial v}{\partial y}\right) + \beta v = -\frac{R}{D}\left(\frac{\partial v}{\partial x} - \frac{\partial u}{\partial y}\right) - \frac{\pi T}{Db}\sin\frac{\pi y}{b} \tag{16.38}$$

Noting that the first term in the left-hand side vanishes from the continuity equation, and introducing a stream function ψ such that

$$v = \frac{\partial \psi}{\partial x}, \quad u = -\frac{\partial \psi}{\partial y}, \quad \text{and} \quad \frac{\partial v}{\partial x} - \frac{\partial u}{\partial y} = \nabla^2 \psi, \tag{16.39}$$

$$\frac{R}{D}\nabla^2\psi + \beta\frac{\partial \psi}{\partial x} = -\frac{\pi T}{Db}\sin\frac{\pi y}{b}. \tag{16.40}$$

Equation (16.40) is called the *Stommel equation*. An alternative form of the Stommel equation can be obtained by multiplying Equation (16.40) throughout by D/R:

$$\nabla^2\psi + \alpha\frac{\partial \psi}{\partial x} = -\gamma\sin\frac{\pi y}{b}, \tag{16.41}$$

where

$$\alpha = \frac{D\beta}{R} \quad \text{and} \quad \gamma = \frac{\pi T}{bR} \tag{16.42}$$

Stommel applied the following boundary conditions with ψ as zero at the boundary of the ocean basin, i.e.,

$$\psi(0,y,t) = \psi(a,y,t) = \psi(x,0,t) = \psi(x,b,t) = 0 \tag{16.43}$$

Using appropriate non-dimensional variables, it is possible to write Equation (16.41) for the one-dimensional case as follows:

$$\varepsilon \frac{\partial^2 \psi}{\partial x^2} + \frac{\partial \psi}{\partial x} = -1, \tag{16.44}$$

where ε is a measure of the non-dimensional friction parameter. Equation (16.44) is to be solved subject to the boundary conditions

$$\psi(x=0) = 0 \quad \text{and} \quad \psi(x=1) = 0 \tag{16.45}$$

Stommel's equation (16.44) is to be solved subject to the boundary condition (16.45). Equation (16.44) has an analytical solution given by

$$\psi = C\left(e^{-x/\varepsilon} - 1\right) - x \tag{16.46}$$

where

$$C = \frac{1}{e^{-1/\varepsilon} - 1}$$

One can find approximate numerical solutions to Equation (16.44) subject to the boundary condition, Equation (16.45), on a regular grid of $N+1$ points in the x direction, $x_i = (i-1)\Delta x$, where $\Delta x = 1/N$ with N as the number of grid cells in the x direction. Approximating the derivatives with respect to x using second-order finite differences, one obtains

$$\varepsilon \frac{\psi_{i-1} - 2\psi_i + \psi_{i+1}}{(\Delta x)^2} + \frac{\psi_{i+1} - \psi_{i-1}}{2\Delta x} = -1, \quad i = 2,3,\dots,N \tag{16.47}$$

$$\psi_i = 0, \quad i = 1, N+1 \tag{16.48}$$

Equations (16.47) and (16.48) can be written as a matrix equation

$$A\psi = b \tag{16.49}$$

where A is the sparse tridiagonal matrix, ψ is a vector of unknowns ψ_i and b is a known vector. The elements of the A matrix are given by

$$A_{1,1} = 1, \ A_{i,i-1} = \frac{\varepsilon}{(\Delta x)^2} - \frac{1}{2\Delta x}, \ A_{i,i} = -\frac{2\varepsilon}{(\Delta x)^2}, \ A_{i,i+1} = \frac{\varepsilon}{(\Delta x)^2} + \frac{1}{2\Delta x}, \ A_{N+1,N+1} = 1, i = 2.3...N$$

(16.50)

The sparse tridiagonal matrix A may be inverted easily using efficient algorithms such as Thomas algorithm.

For solving Stommel's equation, Equation (16.41), for the two-dimensional case subject to the boundary conditions, Equation (16.43), one can utilize a similar strategy as in the one-dimensional case, except that for solving Equation (16.41), one also requires the second derivative of ψ with respect to y to be approximated. Assigning the following values of the parameters

$D = 200\,\text{m}, \ a = 10^7\,\text{m}, \ b = 2\pi \times 10^6\,\text{m}, \ T = 0.1\,\text{m}^2/\text{s}^2, \ R = 0.02\,\text{m/s}, \ \beta = 2 \times 10^{-11}\,\text{m}^{-1}\text{s}^{-1}$ and using the central difference approximation for the second derivative of ψ with respect to y, one obtains

$$\frac{\psi_{i-1,j} - 2\psi_{i,j} + \psi_{i+1,j}}{(\Delta x)^2} + \frac{\psi_{i,j-1} - 2\psi_{i,j} + \psi_{i,j+1}}{(\Delta y)^2} + \alpha \left[\frac{\psi_{i+1,j} - \psi_{i-1,j}}{2\Delta x} \right] = -\gamma \sin \frac{\pi y_j}{b} \quad (16.51)$$

Equation (16.51) subject to boundary conditions, Equation (16.43), can be solved by matrix methods or by employing successive over relaxation methods.

16.3.1 Stommel model for ocean circulation having a zonal wind stress with meridional variation

Stommel's equation, Equation (16.40), can be rewritten as

$$\frac{R}{D}\nabla^2\psi + \beta\frac{\partial\psi}{\partial x} = \frac{\partial\tau_w^y}{\partial x} - \frac{\partial\tau_w^x}{\partial y} = -\frac{\pi T}{Db}\sin\frac{\pi y}{b}, \quad (16.52)$$

where it is assumed that the observed wind stress is zonal only, and the same is dependent on y only as given in Equation (16.35). Letting $K = R/D$, one obtains the following equation:

$$K\nabla^2\psi + \beta\frac{\partial\psi}{\partial x} = -\frac{\pi T}{Db}\sin\frac{\pi y}{b} \quad (16.53)$$

Equation (16.53) subject to the boundary conditions, Equation (16.43), has the following solution:

$$\psi = \frac{bT}{\pi K}\left\{1 - \left[\frac{\left(1 - e^{D_2 L}\right)e^{D_1 x} - \left(1 - e^{D_1 L}\right)e^{D_2 x}}{e^{D_1 L} - e^{D_2 L}}\right]\right\}\sin\frac{\pi y}{b}, \tag{16.54}$$

where

$$D_1 = -\frac{\beta}{2K} + \sqrt{\left(\frac{\beta}{2K}\right)^2 + \left(\frac{\pi}{b}\right)^2} \quad \text{and} \quad D_2 = -\frac{\beta}{2K} - \sqrt{\left(\frac{\beta}{2K}\right)^2 + \left(\frac{\pi}{b}\right)^2} \tag{16.55}$$

Typical values of the parameters are $\beta = 2 \times 10^{-11}\,\mathrm{m^{-1}s^{-1}}$, $K = 2 \times 10^{-6}\,\mathrm{s^{-1}}$, $a = 6,000\,\mathrm{km}$, $b = 3,000\,\mathrm{km}$.

16.4 Munk Model for Ocean Circulation

Section 16.3 brought out the important assumption of Stommel's model (over the Sverdrup model) in which the former assumed bottom friction to be the only counteracting friction force, thus giving rise to Rayleigh friction terms in the horizontal momentum equation while neglecting the viscous force term. The linear Rayleigh friction is not commonly employed in fluid dynamics situations. Hence, Munk improved over Stommel's model by modeling for the viscous force and neglecting the bottom friction term of Stommel. Hence, the main difference in the models of Stommel and Munk lies in their modeling of the counter-acting friction force. Munk's governing equations for the momentum and continuity equations are as follows:

$$-fv = -\frac{1}{\rho_o}\frac{\partial p}{\partial x} + A\left(\frac{\partial^2 u}{\partial x^2} + \frac{\partial^2 u}{\partial y^2}\right) + F_1 \tag{16.56}$$

$$fu = -\frac{1}{\rho_o}\frac{\partial p}{\partial y} + A\left(\frac{\partial^2 v}{\partial x^2} + \frac{\partial^2 v}{\partial y^2}\right) + F_2 \tag{16.57}$$

$$0 = -\frac{1}{\rho_o}\frac{\partial p}{\partial z} - g \tag{16.58}$$

$$\frac{\partial u}{\partial x} + \frac{\partial v}{\partial y} = 0 \tag{16.59}$$

where F_1 and F_2 are related to wind stress components and A is the eddy viscosity coefficient. Differentiating Equation (16.57) with respect to x and Equation (16.56) with respect to y and subtracting one from the other gives

$$f\left(\frac{\partial u}{\partial x} + \frac{\partial v}{\partial y}\right) + \beta v = A\left[\frac{\partial}{\partial x^2}\left(\frac{\partial v}{\partial x} - \frac{\partial u}{\partial y}\right) + \frac{\partial}{\partial y^2}\left(\frac{\partial v}{\partial x} - \frac{\partial u}{\partial y}\right)\right] + \left(\frac{\partial F_2}{\partial x} - \frac{\partial F_1}{\partial y}\right) \quad (16.60)$$

Noting that the first term in the left-hand side is zero (Equation (16.59)) and introducing the stream function ψ such that

$$v = \frac{\partial \psi}{\partial x}, \quad u = -\frac{\partial \psi}{\partial y}, \quad \text{and} \quad \frac{\partial v}{\partial x} - \frac{\partial u}{\partial y} = \nabla^2 \psi \quad (16.61)$$

The last two terms in RHS of Equation (16.60) refer to the vertical component of the curl of wind stress. Equation (16.60) then becomes

$$\beta\frac{\partial \psi}{\partial x} = A\left[\frac{\partial}{\partial x^2}\left(\nabla^2 \psi\right) + \frac{\partial}{\partial y^2}\left(\nabla^2 \psi\right)\right] + \left(\frac{\partial F_2}{\partial x} - \frac{\partial F_1}{\partial y}\right) = A\nabla^2\left(\nabla^2 \psi\right) + \left(\frac{\partial F_2}{\partial x} - \frac{\partial F_1}{\partial y}\right),$$
$$(16.62)$$

which becomes

$$\beta\frac{\partial \psi}{\partial x} = A\left(\nabla^4 \psi\right) + \left(\frac{\partial F_2}{\partial x} - \frac{\partial F_1}{\partial y}\right), \quad (16.63)$$

where ∇^4 is the *biharmonic operator*. Munk utilized the following:

$$\frac{\partial F_2}{\partial x} - \frac{\partial F_1}{\partial y} = \frac{\tau}{D\rho_o}, \quad (16.64)$$

where τ is the observed wind stress.

As the Munk equation, Equation (16.63), is of the fourth-order, boundary conditions are required on all the boundaries for ψ as well as the normal derivative of ψ.

16.4.1 Munk model for ocean circulation having a wind stress with both zonal and meridional variations

Let the forcing shear stress $\tau(x, y)$ be given by

$$\tau(x, y) = -[p\sin(\pi x) + 1 - p]\sin(\pi y), \quad 0 \leq p \leq 1 \quad (16.65)$$

Equation (16.65) reduces to the standard forcing form for $p = 0$ and for $p = 1$. For $p = 0$, the forcing form becomes $\tau(x, y) = -\sin(\pi y)$ and for $p = 1$, the forcing form becomes

$\tau(x,y) = -\sin(\pi x)\sin(\pi y)$.

Equation (16.63) after utilizing appropriate non-dimensional quantities with forcing given by Equation (16.65), gets modified to

$$\frac{\partial \psi}{\partial x} - \varepsilon \nabla^4 \psi = -[p\sin(\pi x) + 1 - p]\sin(\pi y), \tag{16.66}$$

where ε is a measure of the lateral diffusion and ψ is the stream function.

Consider the following solution:

$$\psi(x,y) = \sum_k \phi_k(x)\sin(k\pi y) \tag{16.67}$$

The stream function ψ in Equation (16.67) satisfies no mass flux as well as free-slip conditions at the zonal boundaries, that is, at $y = 0$ and $y = 1$, and, hence, one needs to focus attention only on the longitudinal structure of the model solution. One can invoke the following orthogonal property:

$$\int_0^1 \sin(l\pi y)\sin(k\pi y)\,dy = \frac{1}{2}\delta_{1k} \tag{16.68}$$

Substituting the form of the stream function ψ in Equation (16.67) in Equation (16.66), one obtains, using Equation (16.68), with a little algebra, the following fourth-order ordinary differential equation in ϕ:

$$\left(\frac{d^4}{dx^4} - 2\pi^2 \frac{d^2}{dx^2} - r\frac{d}{dx} + \pi^4\right)\phi(x) = r[p - 1 - p\sin(\pi x)], \tag{16.69}$$

where $\phi = \phi_1$ and $r = 1/\varepsilon$.

Equation (16.69) will be integrated using the no mass flux boundary condition derived as given by

$$\phi(0) = 0 \quad \text{and} \quad \phi(1) = 0 \tag{16.70}$$

together with additional boundary conditions

$$\pi p \left.\frac{d\phi}{dx}\right|_{x=1} + (1-p)\left.\frac{d^2\phi}{dx^2}\right|_{x=1} = 0 \tag{16.71}$$

$$\left.\frac{d^3\phi}{dx^3}\right|_{x=1} = 0 \tag{16.72}$$

Equation (16.69) can be solved subject to the boundary conditions [Equations (16.70) to (16.72)] by standard boundary value problem ordinary differential equation solvers.

16.5 Nonlinear Model for Ocean Circulation

The vertically averaged equations of motion (horizontal momentum equations and continuity equation in non-dimensional form) for an ocean of mean depth h in a β plane are as follows:

$$\frac{\partial u}{\partial t} + R\vec{V} \cdot \nabla u - fv = -\frac{\partial p}{\partial x} - \varepsilon u + \tau^x \tag{16.73}$$

$$\frac{\partial v}{\partial t} + R\vec{V} \cdot \nabla v + fu = -\frac{\partial p}{\partial y} - \varepsilon v + \tau^y \tag{16.74}$$

$$\frac{\partial u}{\partial x} + \frac{\partial v}{\partial y} = 0 \tag{16.75}$$

where $\vec{V}$ is the horizontal velocity vector with zonal component u and meridional component v, f is the Coriolis parameter, a is the radius of Earth, πL is the east–west and north–south extent of the ocean basin, ϕ_o is the tangent latitude for the β plane approximation, x and y are non-dimensionalized with scale L so that x and y lie in a range from 0 to π, p is non-dimensionalized pressure, t is non-dimensionalized time with respect to βL, where $\beta = (2\Omega \cos \phi_o)/a$, τ^x and τ^y are non-dimensionalized wind-stress with Ω being the angular velocity of the rotating earth. Rossby number R and friction parameter ε are both defined as

$$R = \frac{W}{D\beta^2 L^3} \quad \text{and} \quad \varepsilon = \frac{K}{\beta L}, \tag{16.76}$$

where D is the depth of the barotropic ocean, W is the amplitude of the wind-stress and K is the coefficient of bottom friction. Equation (16.75) provides for the introduction of a stream function ψ defined as

$$v = \frac{\partial \psi}{\partial x} \quad \text{and} \quad u = -\frac{\partial \psi}{\partial y} \tag{16.77}$$

Cross differentiating (differentiating Equation (16.74) with respect to x and Equation (16.73) with respect to y) and subtracting one from the other, one obtains the vertical component of the vorticity equation

$$\frac{\partial \zeta}{\partial t} + R\vec{V} \cdot \nabla \zeta + v = -\varepsilon \zeta + \nabla \times \tau, \tag{16.78}$$

where

$$\zeta = \frac{\partial v}{\partial x} - \frac{\partial u}{\partial y} = \nabla^2 \psi \tag{16.79}$$

One needs to solve Equation (16.78), which is written in terms of ψ subject to the following boundary conditions

$$\psi(0,y) = \psi(\pi,y) = \psi(x,0) = \psi(x,\pi) = 0 \tag{16.80}$$

The nonlinear term in Equation (16.78) can be written in the following more convenient form:

$$\vec{V}\cdot\nabla\zeta = \nabla\cdot(\vec{V}\zeta) \tag{16.81}$$

To solve Equation (16.78) subject to the boundary conditions, let the ocean basin ($0 \leq x \leq \pi$, $0 \leq y \leq \pi$) be divided into a uniform grid having grid sizes $\Delta x = \Delta y = \pi/(N-1)$, where the basin is divided into $N-1$ equal intervals in both the x and y directions. The quantity $\zeta_{i,j}$ refers to the value of ζ at the following grid point, $x_i = (i-1)\Delta x$ and $y_j = (j-1)\Delta y$.

Utilizing central finite differences for the first-order derivative in both time and space, one obtains the following difference equation:

$$\zeta_{i,j}^{n+1} = \zeta_{i,j}^{n-1} - \frac{R\Delta t}{2\Delta x\Delta y}J(\psi^n,\zeta^n) - \frac{\Delta t}{\Delta x}\left(\psi_{i+1,j}^n - \psi_{i-1,j}^n\right) = -\varepsilon\Delta t\left(\zeta_{i,j}^{n+1} + \zeta_{i,j}^{n-1}\right) + \Delta t(\nabla\times\tau)_{i,j}^n,$$

$$\tag{16.82}$$

where $J(\psi^n,\zeta^n)$ is the Jacobian term which is given as

$$\begin{aligned}J(\psi^n,\zeta^n) &= \left(\psi_{i+1,j+1}^n - \psi_{i-1,j+1}^n\right)\zeta_{i,j+1}^n - \left(\psi_{i+1,j-1}^n - \psi_{i-1,j-1}^n\right)\zeta_{i,j-1}^n \\ &\quad - \left(\psi_{i+1,j+1}^n - \psi_{i+1,j-1}^n\right)\zeta_{i+1,j}^n + \left(\psi_{i-1,j+1}^n - \psi_{i-1,j-1}^n\right)\zeta_{i-1,j}^n\end{aligned} \tag{16.83}$$

and

$$\zeta_{i,j}^n = \frac{1}{\Delta x\Delta y}\left(\psi_{i,j+1}^n + \psi_{i,j-1}^n + \psi_{i+1,j}^n + \psi_{i-1,j}^n - 4\psi_{i,j}^n\right) \tag{16.84}$$

The frictional term $\varepsilon\zeta$ term has been evaluated as an average of ζ at the previous time $n-1$ and the next time $n+1$. This manner of evaluation of the frictional term leads to a numerical procedure that is considered "stable" for the frictional term. The form of the Jacobian term ensures that it conserves both vorticity as well as kinetic energy. It is possible to integrate Equation (16.82) in a computer by employing a standard method of solution. However, it must be noted that at each time step, one requires to solve the Poisson equation, Equation (16.84), for ψ.

16.6   Vertical Coordinate for Ocean Models

16.6.1   Height (z) coordinate

The simplest and most obvious coordinate would be the height (z) coordinate with $z = 0$ corresponding to the surface and $z = -d$ corresponding to the bottom of the oceans. Over most of the oceans, the warm upper layers are mostly isolated from the colder deep layers. The transports across constant potential density surfaces are called the diapycnal transports and these diapycnal transports between layers are typically subtle. However, it is important for the numerical model to represent these transports accurately. For example, the thermohaline circulation involves warm water flowing to the far northern Atlantic, becoming colder and saltier, and then sinking. The dynamics of the aforementioned thermohaline circulation could be misrepresented in a numerical model if the model allows inaccurate or spurious diffusion of heat or salt between the upper and lower regions. It has been found that the height z coordinate can implicitly allow spurious, non-physical diapycnal transports as surfaces of constant z can intersect isopycnals, the latter giving rise to numerical diffusion which can then cause artificial transport between different water masses. The diffusion terms in transport equations are another potential source of spurious transport in height z models.

16.6.2   Isopycnic coordinate

A well-known vertical coordinate (other than z coordinate) that is commonly employed in ocean models is the "isopycnic" coordinate. The isopycnic coordinate refers to a vertical coordinate with potential density or some other quantity related to the potential density. The potential density of a fluid parcel at pressure P is defined as the density that the fluid parcel would have if it were adiabatically brought to a reference pressure P_o, usually ($P_o = 1$ bar or $100\,\text{kPa}$). While density changes with changing pressure in the oceans, potential density of a fluid parcel is conserved as the pressure experienced by the parcel changes (provided no mixing with other parcels or net heat flux occurs).

While employing isopycnic coordinates and, hence, isopycnic models, one seeks a quantity s that is (approximately) conserved along particle paths, i.e., one requires $ds/dt \approx 0$, where d/dt denotes the material derivative. This indicates that the quantity s is conserved following the motion. One also desires that the resultant surfaces of constant s to be approximate neutral surfaces. We assume that everywhere in the fluid $ds/dt = 0$. This assumption results in a surface of constant s to be a material surface. A material surface is one in which if a fluid parcel lies initially in such a surface, then with the passage of time, the fluid parcel will retain the same value of s and thus remain in the same material surface. The elevation of such a material surface can, however, vary with horizontal position and time; however, any two such material surfaces will always enclose the same mass of fluid. If the fluid domain is discretized with respect to s, for purposes of numerical solution, then the fluid is divided into physical layers that do not mix. Water masses with distinct physical properties (of temperature and salinity) are then distinguished automatically by the choice of coordinate system. Therefore, isopycnic models are also referred to as "layered" models. If s is chosen to be the reciprocal of potential density or a related quantity, then the aforementioned statements are approximately true. Due to these reasons, isopycnic coordinates are also known as *layered coordinates*.

Isopycnic coordinates also have its fair share of disadvantages, as it is observed that the potential density is not exactly a neutral variable, and under certain conditions potential density can also be non-monotonic with height, i.e., not monotonically changing as a function of z. Furthermore, if potential density is used as the vertical coordinate, then the form of the lateral pressure forcing in the momentum equations allows the possibility of numerical inaccuracy and, hence, eventual numerical instability. Isopycnals can intersect both the upper or lower boundaries of the fluid domain; this implies a loss of vertical resolution, especially at the higher latitudes. A loss of vertical resolution is especially prominent and a cause of concern in the mixed layer at the upper boundary of the ocean, as this layer is vertically homogeneous. When an isopycnal intersects the top (bottom) of the fluid domain, the layer above (below) the isopycnal approaches zero thickness; such vanishing of layers due to intersection of isopycnals can introduce algorithmic complications.

16.6.3   Sigma (σ) coordinate

The σ coordinate as the vertical coordinate is a terrain-following coordinate, which is chosen to vary linearly from $\sigma = 0$ at the upper boundary of the fluid to $\sigma = -1$ at the bottom boundary. The σ coordinate allows for an accurate representation of the topography of the ocean's bottom, and it facilitates the modeling of the bottom boundary layer. However, it appears that the σ coordinate surfaces can intersect isopycnals (constant potential density surfaces), which allows the possibility of spurious diapycnal transport, as in the case of the z-coordinate. A further limitation with the σ coordinate deals with the accurate computation of the lateral pressure gradient. When the pressure is written in σ coordinates, the lateral pressure gradient $\nabla_\sigma p = (\partial p/\partial x, \partial p/\partial y)$ represents differentiation along surfaces of constant σ. Conversion of the lateral pressure gradient along constant surface σ to a pressure gradient $\nabla_z p = (\partial p/\partial x, \partial p/\partial y)$ along constant z surface requires a correction term involving the hydrostatic balance and the slope of the σ surface. Over some regions, and in some situations, this correction can have a magnitude comparable to that of $\nabla_\sigma p$; this results in $\nabla_z p$ being the difference of two large quantities of opposite signs, and errors in each of the aforementioned two quantities can impact the value of $\nabla_z p$ significantly. Mostly limited area regional ocean models as well as coastal models utilize the σ coordinate.

16.6.4 Hybrid coordinate

Section 16.6.2 highlighted some of the limitations of employing isopycnic coordinates as vertical coordinates in ocean models. An attempt to overcome these limitations resulted in a new class of vertical coordinates called "hybrid coordinates." Hybrid coordinates duly combine features of all the three vertical coordinates that were earlier discussed in Section 16.6, (i) isopycnic coordinates, (ii) z coordinates, and (iii) σ coordinates. In the general framework that employs hybrid coordinates in ocean models, each coordinate layer is assigned a target density. If a given layer's target density lies within the range that is represented in the fluid at a given horizontal location and time, then the layer is assigned that density, and the layer at that location is isopycnic. However, assuming that a given layer's target density is not present in the fluid, then in a purely isopycnic framework having only isopycnic coordinates, the layer would reduce to zero thickness. Such situations arise naturally for a layer of relatively low density at a high latitude location. However, employing a hybrid coordinate, a layer such as the aforementioned is inflated in order to have positive thickness and thus improves the vertical resolution of the model.

16.7 Barotropic–Baroclinic Splitting

16.7.1 External and internal waves

External waves typically manifest at the interface between the air and the sea surface while internal waves can show up within the interior of the fluid medium. The external gravity wave manifests on the ocean surface when an object, say a stone is thrown over the ocean surface. This sets up the external gravity wave for which the restoring force is the difference in density between the water and air. Oceans are a stably stratified fluid system with density (increasing) varying continuously with depth. The aforementioned ocean system can be envisaged as a system of infinite layers of water, each of them having a differing water density. Hence, just as in the example of the external gravity wave manifesting at the air–sea surface interface, an infinite layer system of water with differing densities may support the manifestation of gravity waves (internal to the water medium) at each and every interface between any of the infinite layers of water. Such gravity waves that manifest internal to the medium are called internal gravity waves. The restoring force for such internal gravity waves is due to the variations of density within the fluid. This force is much weaker for the internal gravity wave, as the difference in densities within the fluid are smaller as compared to the difference in density between water and air. Hence, internal gravity waves move much more slowly than external gravity waves.

In the case of an external wave, all fluid layers thicken or thin by approximately the same proportion at a given horizontal location and time. For such an external wave, the behaviour of the free surface at the top of the fluid reveals the nature of the wave motion throughout the interior, and the horizontal velocity field is essentially independent of vertical position. Such motion is termed as "barotropic" or "external." However, an internal wave has undulations of surfaces of constant density within the fluid, and the free surface remains nearly level. Such motion is termed "baroclinic" or "internal."

It was indicated earlier that the external gravity waves are the fastest motions in a numerical model of ocean circulation. The external gravity waves travel much faster than any of the other motions that are present in the ocean system. Assume that one employs an explicit Eulerian finite difference scheme in the ocean numerical model. As the explicit scheme is conditionally stable, there is a restriction on the choice of the time step that can be employed. The choice of the time step is purely determined in terms of the fastest motions that are present, which happen to be the external gravity wave. It can be shown that the fastest motion associated with barotropic or external motions are essentially two-dimensional as the horizontal velocity field is essentially independent of vertical position for such motions. Real motions in the oceans have both the barotropic and the baroclinic components. This implies that the numerical solution of a complex three-dimensional ocean circulation system is severely constrained by a special set of two-dimensional motions, the latter corresponding to the barotropic or external component.

16.7.2   Barotropic–baroclinic subsystems

Due to the aforementioned reasons, it has therefore become common practice to split the dynamics of the ocean system into two subsystems; one a relatively simple two-dimensional system that represents the fast motions (barotropic or external) and the other a relatively complex three-dimensional system that represents the remaining (slow) processes (baroclinic or internal). The latter system is solved explicitly with a relatively long time step as the processes are inherently slower and the time step is not very much restrictive due to the slowness of the motion. The fast system (barotropic or external) is solved using an implicit finite difference scheme with the same time step as the slow system, the implicit scheme being unconditionally stable. Another option to solve the fast system is to solve it using an explicit finite difference scheme with shorter time steps. Such a splitting is traditionally referred to as a barotropic–baroclinic splitting. It turns out that the resulting algorithm is much more computationally efficient than solving the entire unsplit system explicitly with short steps.

16.8   Time Discretization

16.8.1   Leapfrog scheme

One of the most popular and widely applied time differencing scheme applied to both atmospheric and oceanic models is the leapfrog scheme. The leapfrog time differencing scheme is a central difference scheme which is also a three-level scheme. For a partial differential equation of the following form $(\partial f/\partial t) = F(f,t)$, the leapfrog finite difference form is defined as

$$f_{n+1} = f_{n-1} + 2\Delta t F(f_n, t_n)$$

The leapfrog scheme is a second-order accurate scheme in time; however, it has a limitation that it allows a non-physical computational mode consisting of saw-tooth oscillations in time t.

In the special case $F = 0$ of the aforementioned PDE, the leapfrog scheme becomes $f_{n+1} = f_{n-1}$. Hence, the leapfrog scheme for this case allows both constant solutions and saw-tooth solutions of the form $f_n = c(-1)^n$. One way of keeping the computational mode bounded is to apply an Asselin filter of the form

$$\overline{f}_n = \gamma f_{n+1} + (1 - 2\gamma) f_n + \gamma \overline{f}_{n-1},$$

where $\overline{f}_n$ is a smoother quantity and γ is a positive constant.

16.8.2   Two-time level finite difference scheme

The two-time level finite difference scheme is devoid of the computational mode and, hence, does not give rise to the saw-tooth oscillations. Due to these reasons, a two-time level finite difference scheme is often preferred in ocean models. The two-time level finite difference scheme when applied to barotropic ocean models is of the predictor–corrector type whereas the two-time level finite difference scheme when applied to baroclinic ocean models is typically staggered in time. For the staggered in time approach, the velocity components are defined at integer time steps, while the tracers, pressure, and density are usually defined at half-integer steps. The staggered in time approach enables second-order central differencing without allowing the computational mode to manifest.

An alternate two-time level finite difference scheme on a non-staggered grid for layered ocean models is also popular. In the aforementioned two-time level scheme, after an initial forward Euler time marching of the baroclinic velocity from time t_n to time t_{n+1}, all time discretization for the baroclinic equations involve centered differencing and unweighted averaging about the middle of the time interval $[t_n, t_{n+1}]$. The Coriolis terms are, however, implemented implicitly in this scheme.

16.9   Spatial Discretization and Horizontal Grids in Ocean Models

An obvious choice for the horizontal coordinate system for Planet Earth in global ocean models is the one based on latitude and longitude. However, this system of latitude and longitude admits a singularity at the North Pole in the ocean models (due to the merging of meridians near the north pole). A similar polar singularity over the south pole in ocean models is not a cause of concern as the South Pole is over land. The merging of the meridians near the North Pole would make it difficult to satisfy the *Courant–Friedrichs–Lewy condition*, if one were to employ explicit finite difference schemes.

To resolve the aforementioned polar singularity problem a coordinate system that does not place any singularities within the fluid domain can be employed. Such coordinate systems have been developed which provide for a class of orthogonal curvilinear grids with two coordinate poles that can be placed at arbitrary locations. In practice, the south coordinate pole can be located at the south geographical pole (which lies over the land anyway), whereas the latitude–longitude grid is deformed smoothly so that the north coordinate pole is also located on a land mass.

16.9.1   Spatial arrangement of the dependent variables in the horizontal

Similar to the atmospheric models where the staggered arrangement of the dependent variables is fairly common, the staggered arrangement of the dependent variables is also quite popular in ocean models. Among the five different types of Arakawa grids (Grid 'A', Grid 'B', Grid 'C', Grid 'D' and Grid 'E'), Grid 'A' is a non-staggered arrangement, whereas the remaining four grids (Grid 'B', Grid 'C', Grid 'D' and Grid 'E') are all examples of the staggered grids. As in the atmospheric models, Grid 'B' and Grid 'C' are the most widely used staggered grids in ocean models. The results of the linear analyzes of dispersion relations of numerical methods indicate that the 'C-grid' is more accurate than the 'B-grid' for relatively well-resolved waves, whereas the 'B-grid' is more accurate in the case of relatively coarser resolution. The resolution is defined in terms of the size of the grid spacing relative to the Rossby radius of deformation defined as c/f, where c is the speed of gravity waves and f is the Coriolis parameter. In short, the 'C-grid' is preferable for higher-resolution ocean models whereas the 'B-grid' is the preferred choice for coarser-resolution ocean simulations.

16.10 Various Approaches to Solving the Ocean Momentum Equations

One approach to solving the ocean momentum equations in the context of shallow water equations in a single layer is to use an advection scheme to implement the advective terms in the layer "u-component velocity equation" and in the analogous "v-component velocity equation." The Coriolis terms and the effects of pressure and viscosity are regarded as forcing terms that are implemented with standard differencing and/or averaging. In this formulation, the forcing is implemented with the following splitting in the forcings. One starts with the computed solution at time t_n, add half the forcing at time t_n (times Δt), apply an advection scheme to the result, and then add the other half of the forcing at time t_{n+1}(times Δt).

Another approach to solving the ocean momentum equations is to invoke Boussinesq approximation with the z coordinate by utilizing Arakawa "B-Grid." The space discretization so chosen must maintain energetic consistency in the model, i.e., the discrete nonlinear terms should have no effect on the total kinetic energy in the solution and that the exchanges between kinetic and potential energy are represented correctly. The aforementioned requirement of energy consistency constrains the fluxes at cell faces to be calculated in terms of centered second-order averages.

A third approach to solving the ocean momentum equations is to convert the dependent variables to velocity in the momentum equations and write the horizontal transport terms in terms of kinetic energy and vorticity using the "C-Grid." The kinetic energy and vorticity terms can then be approximated with centered finite differences.

Exercises 16 (Questions and answers)

1. Write down the steady equations of motion for the Ekman layer over the ocean surface along with the appropriate boundary conditions.

Answer: The steady equations of motion for the Ekman layer over the ocean surface where ocean waters are subject to a wind stress (τ^x, τ^y) and having interior flow velocity components $(\bar{u}, \bar{v})$ are given by

$$-f(v-\bar{v}) = \nu_E \frac{\partial^2 u}{\partial z^2}$$

$$f(u-\bar{u}) = \nu_E \frac{\partial^2 v}{\partial z^2}$$

where (u, v) are the flow components in the oceanic Ekman layer. The boundary conditions are as follows:

At $z = 0$ (ocean surface):

$$\rho_o \nu_E \frac{\partial u}{\partial z} = \tau^x \quad \text{and} \quad \rho_o \nu_E \frac{\partial v}{\partial z} = \tau^y$$

At $z \to \infty$, (interior flow) $u = \bar{u}$ and $v = \bar{v}$.

2. Write down the expression for the flow velocity components in the oceanic Ekman layer.

Answer: The expression for the flow velocity components in the oceanic Ekman layer subject to a wind stress (τ^x, τ^y) and having interior flow velocity components $(\bar{u}, \bar{v})$ are as follows:

$$u = \bar{u} + \frac{\sqrt{2}}{\rho_o f d} e^{z/d} [\tau^x \cos(z/d - \pi/4) - \tau^y \sin(z/d - \pi/4)]$$

$$v = \bar{v} + \frac{\sqrt{2}}{\rho_o f d} e^{z/d} [\tau^x \sin(z/d - \pi/4) + \tau^y \cos(z/d - \pi/4)]$$

where the departure of the flow fields from the interior flow is essentially determined by the wind stress components and where d is the Ekman layer depth, f is the Coriolis parameter and ρ_o is the density of water.

3. Show that the wind-driven horizontal transport in the surface oceanic Ekman layer is essentially determined by the applied surface wind stress components.

Answer: The wind-driven horizontal transport in the surface oceanic Ekman layer is defined as

$$U = \int_{-\infty}^{0} (u - \bar{u}) \, dz = \frac{1}{\rho_o f} \tau^y$$

$$V = \int_{-\infty}^{0} (v - \bar{v}) \, dz = -\frac{1}{\rho_o f} \tau^x$$

This expression shows that the wind-driven horizontal transport in the surface oceanic Ekman layer is oriented perpendicular to the wind stress always and to the right in the Northern Hemisphere and to the left in the Southern Hemisphere.

4. What are the alternate vertical coordinates (other than z coordinate) that can be utilized in ocean models?

    Answer: The "isopycnic" coordinate is a commonly employed vertical coordinate in ocean models and refers to a vertical coordinate with the potential density or some related quantity to the potential density. Whereas density changes with changing pressure in the oceans, potential density of a fluid parcel is conserved as the pressure experienced by the parcel changes (provided no mixing with other parcels or net heat flux occurs).

5. Is the ocean circulation affected by propagation of sound waves in the ocean? How are the sound waves filtered in ocean models?

    Answer: The ocean circulation is not affected by propagation of sound waves in the ocean. The sound waves can be filtered from the equations of motion using either the hydrostatic approximation or the Boussinesq approximation, the latter allowing the variation of density to manifest only in the buoyancy term.

6. Mention any limitations of the height z coordinate.

    Answer: "Diapycnal transport" refers to transport across such a constant potential density surface. Over most of the oceans, the warm upper layers are mostly isolated from the colder deep layers, and the diapycnal transports between layers are typically subtle. However, it is important for the numerical model to represent these transports accurately. For example, the dynamics of the thermohaline circulation could be misrepresented in a numerical model if the model allows inaccurate or spurious diffusion of heat or salt between the upper and lower regions. The height z coordinate can implicitly allow spurious, non-physical diapycnal transports as surfaces of constant z can intersect isopycnals, the latter giving rise to numerical diffusion can then cause artificial transport between different water masses. The diffusion terms in transport equations are other potential sources of spurious transport in height z models.

7. Mention the advantage of the isopycnic coordinate.

    Answer: In the case of isopycnic coordinates, the diapycnal transport is well represented. The transports of heat and salt are typically represented by advection–diffusion equations and such advection–diffusion equations do not contribute to spurious transport when isopycnic coordinates are employed.

8. Mention the advantages and limitations of the σ coordinate in ocean models.

    Answer: The σ coordinate allows for an accurate representation of the topography of the ocean's bottom and, hence, also facilitates the modeling of the bottom boundary layer.

However, the σ coordinate surfaces can intersect isopycnals, which allows the possibility of spurious diapycnal transport, as in the case of the z-coordinate. Another limitation of the σ coordinate deals with the accurate computation of the lateral pressure gradient in the σ coordinate, in which the lateral pressure gradient in the z coordinate will be the difference of two large quantities of opposite signs, and errors in these two quantities can impact the result drastically.

9. Mention the disadvantage of the hybrid coordinate.

   Answer: One disadvantage of the hybrid coordinates manifests during the seasonal cycle of the surface mixed layer. During winter, the surface mixed layer is relatively thick, owing to strong mixing caused by storms and also due to convective overturning. However, during the summer, the aforementioned forcing is much weaker, and consequently much of the fluid in this layer reverts to a stratified state. In a hybrid model, isopycnic layers are thus alternately destroyed and then recreated near the top of the fluid as part of the annual cycle. If the geometrically constrained, non-isopycnic layers are evenly distributed throughout a deep mixed layer during winter, they may need to be transported a long distance upward to reach their target densities in the spring or summer. These associated large transports of fluid between coordinate layers could contribute to numerical errors in representing such transports.

10. Why is the polar singularity problem not serious for the South Pole in ocean models?

    Answer: The polar singularity problem is not serious for the South Pole in ocean models as the South Pole is over land.

Appendix

Tridiagonal Matrix Algorithm

A system of simultaneous algebraic equations with nonzero coefficients only on the main diagonal, the lower diagonal, and the upper diagonal is called a *tridiagonal system of equations*. Consider a tridiagonal system of N equations with N unknowns, $u_1, u_2, u_3, \cdots u_N$ as follows:

$$
\begin{bmatrix}
d_1 & b_1 & & & & & \\
a_2 & d_2 & b_2 & & & & \\
& a_3 & d_3 & b_3 & & & \\
& & & \ddots & & & \\
& & & & \ddots & & \\
& & & & a_{N-1} & d_{N-1} & b_{N-1} \\
& & & & & a_N & d_N
\end{bmatrix}
\begin{bmatrix}
u_1 \\
u_2 \\
u_3 \\
\vdots \\
\vdots \\
u_{N-1} \\
u_N
\end{bmatrix}
=
\begin{bmatrix}
c_1 \\
c_2 \\
c_3 \\
\vdots \\
\vdots \\
c_{N-1} \\
c_N
\end{bmatrix}
\tag{A.1}
$$

A standard method for solving a system of linear, algebraic equations is Gaussian elimination. Thomas' algorithm, also called the tridiagonal matrix algorithm (TDMA) is essentially a shortened variant of the Gaussian elimination method to solve the tridiagonal system of equations.

The ith equation in the system may be written as

$$a_i u_{i-1} + d_i u_i + b_i u_{i+1} = c_i, \tag{A.2}$$

where $a_1 = 0$ and $b_N = 0$. Looking at the system of equations, we see that the ith unknown can be expressed in terms of $(i+1)^{th}$ unknowns. That is,

$$u_i = P_i u_{i+1} + Q_i \tag{A.3}$$

$$u_{i-1} = P_{i-1} u_i + Q_{i-1}, \tag{A.4}$$

where P_i and Q_i are constants. Note that if all the equations in the system are expressed in this fashion, the coefficient matrix of the system would transform to upper triangular matrix.

To determine the constants P_i and Q_i, we plug Equation (A.4) in (A.2) to yield

$$a_i P_{i-1} u_i + a_i Q_{i-1} + d_i u_i + b_i u_{i+1} = c_i$$

or

$$(d_i + a_i P_{i-1}) u_i + b_i u_{i+1} = c_i - a_i Q_{i-1}$$

or

$$u_i = \frac{-b_i}{d_i + a_i P_{i-1}} u_{i+1} + \frac{c_i - a_i Q_{i-1}}{d_i + a_i P_{i-1}} \tag{A.5}$$

Comparing Equations (A.3) and (A.5), we obtain

$$P_i = \frac{-b_i}{d_i + a_i P_{i-1}} \qquad Q_i = \frac{c_i - a_i Q_{i-1}}{d_i + a_i P_{i-1}} \tag{A.6}$$

These are the recurring relations for the constants P and Q. It shows that P_i can be calculated if P_{i-1} is known. To start the computation, we use the fact that $a_1 = 0$. Now, P_1 and Q_1 can be easily calculated because terms involving P_0 and Q_0 vanish. Therefore,

$$P_1 = \frac{-b_1}{d_1} \qquad Q_1 = \frac{c_1}{d_1} \tag{A.7}$$

Once the values of P_1 and Q_1 are known, we can use the recurring expressions for P_i and Q_i for all values of i.

To start the back substitution, we use the fact that $b_N = 0$. As a consequence, from Equation (A.6), we have $P_N = 0$. Therefore,

$$u_N = P_N u_{N+1} + Q_N = Q_N$$

Once the value of u_N is known, we use Equation (A.3) to obtain $u_{N-1}, u_{N-2}, \cdots u_1$.

Bibliography

Anderson, D., Tannehill, J. and Pletcher, R. (1984). *Computational Fluid Mechanics and Heat Transfer,* New York: McGraw-Hill Book Co.

Arakawa, A. (1966). Computational design for long-term numerical integration of the equations of fluid motion. *Journal of Computational Physics,* 1: 119–143.

——— (1988). Finite-difference methods in climate modeling. In *Physically-Based Modelling and Simulation of Climate and Climatic Change–Part I,* Springer, pp. 79–168.

Arakawa, A. and Lamb, V. R. (1977). Computational design of the basic dynamical processes of the UCLA general circulation model. *General Circulation Models of the Atmosphere,* 17 (Supplement C): 173–265.

Chan, J. and Williams, R. (1987). Analytical and numerical studies of the beta-effect in tropical cyclone motion, part I: Zero mean flow. *Journal of Atmospheric Sciences,* 44 (9): 1257–1265.

Charney, J. G. (1949). On a physical basis for numerical prediction of large-scale motions in the atmosphere. *Journal of Meteorology,* 6: 372–385.

Coiffier, J. (2011). *Fundamentals of Numerical Weather Prediction,* Cambridge: Cambridge University Press.

DeCaria, A. J. and Van Knowe, G. E. (2014). *A First Course in Atmospheric Numerical Modeling,* Sundog Publishing.

Donner, L. J., Schubert, W. H. and Somerville, R. (2018). *The Development of Atmospheric General Circulation Models: Complexity, Synthesis and Computation,* Cambridge: Cambridge University Press.

Döös, K. (2010). *Numerical Methods in Meteorology and Oceanography.* URL: http://doos.misu.su.se/pap/compnum.pdf.

Durran, D. R. (1999). *Numerical Methods for Wave Equations in Geophysical Fluid Dynamics,* Amsterdam: Springer Nature.

——— (2010). *Numerical Methods for Fluid Dynamics, With Applications to Geophysics,* Amsterdam: Springer Nature.

Ehrendorfer, M. (2012). *Spectral Numerical Weather Prediction Models*, Society for Industrial and Applied Mathematics.

Fornberg, B. (2009). *A Practical Guide to Pseudospectral Methods*, Cambridge: Cambridge University Press.

Ghil, M. and Robertson, A. W. (2000). Solving problems with GCMS: General circulation models and their role in the climate modeling hierarchy. In *International Geophysics* Vol. 70, pp. 285–325. DOI: 10.1016/S0074-6142(00)80058-3.

Haltiner, G. J. and Williams, R. T. (1980). *Numerical Prediction and Dynamic Meteorology*, New Jersey: J. Wiley and Sons.

Holton, J. (1992). *An Introduction to Dynamic Meteorology*, 3rd ed., San Diego: Academic Press.

Jacobson, M. Z. (2005). *Fundamentals of Atmospheric Modeling*, 2nd ed., Cambridge: Cambridge University Press.

John D. Anderson, J. (1995). *Computational Fluid Dynamics: The Basics with Applications*, New York: McGraw-Hill.

Kalnay, E. (2013). *Atmospheric Modeling, Data Assimilation and Predictability*, Cambridge: Cambridge University Press.

Kämpf, J. (2010). *Advanced Ocean Modelling: Using Open-source Software*, Heidelberg: Springer-Verlag.

Kowalik, Z. and Murty, T. S. (1993). *Numerical Modeling of Ocean Dynamics: Advanced Ocean Engineering*, Singapore: World Scientific Publishing Company.

Krishnamurti, T. N. (1995). Numerical weather prediction. *Annual Review of Fluid Mechanics*, 27: 195–225.

Krishnamurti, T. N., Bedi, H., Hardiker, V. and Watson-Ramaswamy, L. (2006). *An Introduction to Global Spectral Modeling*, ATSL Vol. 35, New York: Springer-Verlag.

Krishnamurti, T. N. and Bounoua, L. (1996). *An Introduction to Numerical Weather Prediction Techniques*, Boca Raton: CRC Press.

Leith, C. E. (1965). Numerical Simulation of the Earth's Atmosphere. *Methods Computational Physics*, 4: 1–28.

Machenhauer, B. (1991). *Spectral Methods, ECMWF Lecture notes*. URL: https://www.ecmwf.int/node/10901.

Machenhauer, B., Kaas, E. and Lauritzen, P. H. (2009). Finite-volume methods in meteorology. In *Computational Methods for the Atmosphere and the Oceans*, Amsterdam: Elsevier.

Manabe, S. (1985a). *Issues in Atmospheric and Oceanic Modeling, Part A: Climate Dynamics*, Advances in Geophysics, Vol. 28, Amsterdam: Science Direct.

——— (1985b). *Issues in Atmospheric and Oceanic Modeling, Part B: Weather Dynamics*, Advances in Geophysics, Vol. 28, Amsterdam: Science Direct.

Marras, S., Kelly, J. F., Moragues, M., Müller, A., Kopera, M. A., Vázquez, M., Giraldo, F. X., Houzeaux, G. and Jorba, O. (2016). A review of element-based Galerkin methods for numerical weather prediction: Finite elements, spectral elements, and discontinuous Galerkin. *Archives of Computational Methods in Engineering*, 23: 673–722.

McGregor, J., Leslie, L. and Gauntlett, D. (1978). The ANMRC limited-area model: Consolidated formulation and operational results. *Monthly Weather Review*, 106: 427–438.

Mesinger, F. and Arakawa, A. (1976). *Numerical Methods Used in Atmospheric Models*, Global Atmospheric Research Programme (GARP) WMO–ICSU Joint Organizing Committee.

Mullan, B. (2015). *Applied Principles in Atmospheric Models*, Callisto Reference.

——— (2019). *Atmospheric Modeling, Analysis and Applications*, Callisto Reference.

Müller, P. K. (2004). *Computer Modelling in Atmospheric and Oceanic Sciences*, Heidelberg: Springer-Verlag.

Murthy, J. Y. (2002). *Finite Volume Method*. URL: http://math.ubbcluj.ro/tgrosan/2018IntroCFDC11.pdf.

O'Brien, J. J. (1986). *Advanced Physical Oceanographic Numerical Modelling*, Amsterdam: Springer.

Phillips, N. A. (1959). An example of non-linear computational instability. In *The Atmosphere and the Sea in Motion*, Oxford: Oxford University Press, pp. 501–504.

Press, W. H. and Teukolsky, S. A. (1991). Multigrid methods for boundary value problems I. *Computers in Physics*, 5: 514–519.

Prusov, V. A. and Doroshenko, A. Y. (2017). *Computational Techniques for Modeling Atmospheric Processes*. DOI https://doi.org/10.1007/s10559-019-00115-w.

Randall, D. A. (2000). *General Circulation Model Development: Past, Present, and Future* Vol. 70, Cambridge, MA: Academic Press.

Rao, K. R., Kim, D. N. and Hwang, J.-J. (2010). *Fast Fourier Transform: Algorithms and Applications*, Amsterdam: Springer.

Richardson, L. (1922). *Weather Prediction by Numerical Process*, Cambridge: Cambridge University Press. Reprinted (1965), New York: by Dover, with a new introduction by Sydney Chapman.

Richtmyer, R. D. and Morton, K. W. (1967). *Difference Methods for Initial-Value Problems*, New York: Wiley Interscience Pub.

Riddaway, R. (2002). *Numerical Methods, ECMWF Lecture notes.* URL: https://www.ecmwf.int/node/16948.

Røed, L. P. (2019). *Atmospheres and Oceans on Computers: Fundamental Numerical Methods for Geophysical Fluid Dynamics*, Amsterdam: Springer.

Rosmund, T. E. and Faulkner, F. D. (1976). Direct solution of elliptic equations by block cyclic reduction and factorization, *Monthly Weather Review*, 104: 641–649.

Runborg, O. (2012). *Finite Volume Discretization of the Heat Equation.* URL https://www.csc.kth.se/utbildning/kth/kurser/DN2255/ndiff13/Lecture3.pdf.

Satoh, M. (2014). *Atmospheric Circulation Dynamics and General Circulation Models*, Heidelberg: Springer-Verlag.

Smith, R. K. and Ulrich, W. (2008). Lectures on Numerical Meteorology. URL: https://www.meteo.physik.uni-muenchen.de/roger/manuskripte/Numerical Meteorology.pdf.

Staniforth, A. and Côté, J. (1991). Semi-Lagrangian Integration Schemes for Atmospheric Models – A review. *Monthly Weather Review*, 119 (9): 2206–2223.

Steyn, D. G. (2015). *Introduction to Atmospheric Modelling*, Cambridge: Cambridge University Press.

Trenberth, K. E. (1992). *Climate System Modeling*, Cambridge: Cambridge University Press.

Vallis, G. K. (2017). *Atmospheric and Oceanic Fluid Dynamics, Fundamentals and Large-scale Circulations*, Cambridge: Cambridge University Press.

Warner, T. T. (2011). *Numerical Weather and Climate Prediction*, Cambridge: Cambridge University Press.

Washington, W. M. and Parkinson, C. L. (1986). *An Introduction to Three-dimensional Climate Modeling*, New York: University Science Books.

Index